BIOLOGICAL CONTROL
IN THE
WESTERN UNITED STATES

BIOLOGICAL CONTROL
IN THE
WESTERN UNITED STATES

ACCOMPLISHMENTS AND BENEFITS OF
REGIONAL RESEARCH PROJECT W-84,
1964–1989

TECHNICAL EDITORS

J. R. Nechols (*Executive Editor*)
L. A. Andres
J. W. Beardsley
R. D. Goeden
C. G. Jackson

UNIVERSITY OF CALIFORNIA

DIVISION OF AGRICULTURE AND NATURAL RESOURCES

1995

PUBLICATION 3361

ORDERING INFORMATION:

To order this publication, contact

Publications
Division of Agriculture and Natural Resources
University of California
6701 San Pablo Avenue
Oakland, California 94608-1239

Telephone (510) 642-2431
Within California call 1-800-994-8849
FAX (510) 643-5470

Publication 3361

Library of Congress Catalog Card Number 94-61791
ISBN 1-879906-25-2
ISBN 1-879906-21-X (pbk.)

ON THE COVER

Top: Adult ragwort flea beetle, *Longitarsus jacobaeae* (Coleoptera: Chrysomelidae). First imported from Europe and released in 1969 in California, the beetle has been chiefly responsible for the excellent sustained control of the poisonous weed tansy ragwort (*Senecio jacobaea*) in the northwestern United States. Adults feed on the foliage, the larvae on the roots. The impact of the latter is usually what kills the weed. *Photo by Charles E. Turner.*

Center: Female *Aphytis melinus* parasitizing a third-instar California red scale, *Aonidiella aurantii,* a pest of citrus in California and other semiarid regions of the world. This parasitoid is responsible for suppressing red scale to subeconomic levels in southern California. It also suppresses the scale in central California citrus when released augmentatively. *Photo by Max Badgley.*

Bottom: Larva of *Chrysoperla rufilabris,* one of the important green lacewing species that are commercially produced and distributed by insectaries in the Western Region. Green lacewings are widespread predators of lepidopteran eggs and larvae, aphids (shown here), and mites. Their usefulness as biological control agents has been increased by W-84 project research involving food sprays to enhance oviposition, effective storage and mass-production techniques, biotype selection, the release of genetically improved strains, and the systematics of adults and larvae. *Photo by Jack Kelly Clark.*

 Printed on Recycled Paper

CONTENTS

Part 3—WEED CASE HISTORIES

ASTERACEAE

Publisher's Note

To help readers retrieve information pertinent to their individual interests, several indices have been included. For added convenience, each case history in parts 2 and 3—on arthropod pests and weeds, respectively—begins with three standardized headings: the pest's common name; genus and species; and order and family.

PREFACE

In 1992 Western Regional Research Project W-84, "Biological Control in Pest Management Systems of Plants," was reorganized as Project W-185, officially ending the oldest regional research committee on biological control in the United States. This book summarizes and highlights the accomplishments and progress made by W-84 researchers, emphasizing the first 25 years of the project's 29-year history. Some of the case histories, however, have been updated to include significant findings since 1989.

As noted in the historical overview (chapter 1) and exemplified by the 3 general chapters and 79 case histories, W-84 was one of the largest, most productive, and most diverse regional research projects on biological control. This project attracted scientists from agricultural experiment stations and from USDA Agricultural Research Service (ARS) laboratories in almost all of the states within the Western Region, as well as two other states and two U.S. territories. The present committee has 26 voting members, but 35 to 40 scientists participate on a regular basis.

The productivity of this project is evidenced by the 237 publications—including 2 books, 30 book chapters, and over 180 peer-reviewed articles (RIC Regional Research Project Reviews)—produced by its members in 1988 and 1989. This averages out to about 7 publications per scientist per year.

The research conducted under the W-84 project has benefited from the broad interests and expertise of its cooperating scientists. This diversity is reflected in the large number of crop systems, pests, natural enemies, and research approaches reported in this volume. Natural enemies under investigation have included predators, parasitoids, herbivores, and, to a lesser extent, pathogens of over 90 arthropod and weed pests. Research in the field and laboratory has focused on all levels of organization—from agroecosystems to populations and from whole organisms to their cellular and molecular constituents.

Members past and present have gained national and international recognition for their contributions to biological control and integrated pest management as well as for advances made in all fields of biological science. Members' studies of pest and natural enemy nutrition, ecology, behavior, physiology, genetics, systematics, and evolution have provided a sound biological foundation on which to evaluate and implement biological control

and pest management programs. Moreover, many of these efforts have improved our fundamental knowledge of living systems. Clearly, the interaction between basic and applied research has had a catalytic effect on many of the successful biological control programs developed in the Western Region.

The W-84 project has been unique among regional projects on biological control in many respects. For example, it is extremely well represented, both scientifically and programmatically, in the biological control of arthropod and weed pests. In addition, because of its geographic location and history, a significant part of all research conducted under W-84 has been devoted to the management of introduced pests (that is, to "classical," or importation, biological control). These activities have been supported by the well-equipped quarantine and importation facilities that exist in many participating states, particularly California and Montana. W-84 research in classical biological control has extended from the tropics and subtropics to the north temperate zone. It has involved foreign cooperators in countries ranging from Canada to Australia and from the Philippines to Turkey. Cooperators at USDA-ARS laboratories overseas have provided frequent and invaluable assistance in the collection, screening, and shipment of natural enemies to the United States.

Well-acclaimed and highly successful research has also been conducted in the areas of natural enemy augmentation and conservation, for both indigenous and introduced pests. Finally, many investigations have focused on the evaluation and impact of biological control, as well as on its integration into more broadly based pest management programs.

Arguably, W-84 has made more contributions to the theory, practice, and success of biological control than has any other regional research project to date. It is our hope that this book will serve as a valuable long-term resource for all researchers and educators who are interested in biological control and in integrated pest management.

Executive Committee

J. R. Nechols (*Chair*)	C. G. Jackson
L. A. Andres	L. E. Ehler
J. W. Beardsley	J. P. McCaffrey
R. D. Goeden	J. M. Story

Contributors

M. T. ALINIAZEE, Department of Entomology, Oregon State University, Corvallis, Oregon

L. A. ANDRES, Biological Control of Weeds Lab, U.S. Department of Agriculture, Agricultural Research Service, Albany, California

K. Y. ARAKAWA, Department of Entomology, University of California, Riverside, California

B. A. BARRETT, Department of Entomology, University of Missouri, Columbia, Missouri

M. S. BARZMAN, Division of Biological Control, University of California, Albany, California

J. W. BEARDSLEY, Department of Entomology, University of Hawaii, Honolulu, Hawaii

T. S. BELLOWS, JR., Department of Entomology, University of California, Riverside, California

K. D. BIEVER, Yakima Agricultural Research Lab, U.S. Department of Agriculture, Agricultural Research Service, Yakima, Washington

G. W. BISHOP, Southwest Idaho Research & Extension Center, University of Idaho, Parma, Idaho

R. E. BROWN, Oregon Department of Agriculture, Salem, Oregon

J. F. BRUNNER, Tree Fruit Research Center, Washington State University, Wenatchee, Washington

R. M. BURKHART, Hawaii Department of Agriculture, Honolulu, Hawaii

L. E. CALTAGIRONE, Division of Biological Control, University of California, Albany, California

C. L. CAMPBELL, State of Hawaii, Department of Agriculture, Lihue, Hawaii

C. H. CHIU, Northern Marianas College Land Grant Programs, Saipan

E. M. COOMBS, Oregon Department of Agriculture, Salem, Oregon

K. M. DAANE, Division of Biological Control, University of California, Kearney Agricultural Center, Parlier, California

D. L. DAHLSTEN, Division of Biological Control, University of California, Albany, California

N. A. DAVIDSON, Department of Pesticide Regulation, Environmental Monitoring and Pest Management, Sacramento, California

D. W. DAVIS, Department of Biology, Utah State University, Logan, Utah

J. W. DEBOLT, Biological Control of Insects Lab, U.S. Department of Agriculture, Agricultural Research Service, Tucson, Arizona

S. H. DREISTADT, IPM Education and Publications, University of California, Davis, California

L. E. EHLER, Department of Entomology, University of California, Davis, California

J. J. ELLINGTON, Department of Entomology & Plant Pathology, New Mexico State University, Las Cruces, New Mexico

L. K. ETZEL, Division of Biological Control, University of California, Albany, California

D. L. FLAHERTY, University of California Cooperative Extension Service, Visalia, California

M. FONG, Division of Biological Control, University of California, Albany, California

R. D. GOEDEN, Department of Entomology, University of California, Riverside, California

D. GONZÁLEZ, Department of Entomology, University of California, Riverside, California

G. GORDH, Department of Entomology, University of Queensland, Queensland, Australia

K. S. HAGEN, Division of Biological Control, University of California, Albany, California

T. J. HENNEBERRY, Western Cotton Research Lab, U.S. Department of Agriculture, Agricultural Research Service, Phoenix, Arizona

M. P. HOFFMANN, Department of Entomology, Insectary Building, Cornell University, Ithaca, New York

M. A. HOY, Department of Entomology & Nematology, University of Florida, Gainesville, Florida

S. C. HOYT, Tree Fruit Experiment Station, Washington State University, Wenatchee, Washington

C. G. JACKSON, U.S. Department of Agriculture, Agricultural Research Service, Hilo, Hawaii

J. B. JOHNSON, Division of Entomology, University of Idaho, Moscow, Idaho

M. W. JOHNSON, Department of Entomology, University of Hawaii, Honolulu, Hawaii

V. P. JONES, Department of Entomology, University of Hawaii, Honolulu, Hawaii

W. A. JONES, Biological Control of Pests Research Lab, U.S. Department of Agriculture, Agricultural Research Service, Weslaco, Texas

C. E. KENNETT, Division of Biological Control, University of California, Albany, California

J. R. LEEPER, Product Delivery Systems Group, Dupont/Merck, Inc., Wilmington, Delaware

T. F. LEIGH, Department of Entomology, University of California, Davis, California

R. F. LUCK, Department of Entomology, University of California, Riverside, California

J. P. McCAFFREY, Department of Plant, Soil & Entomological Science, University of Idaho, Moscow, Idaho

P. B. McEVOY, Department of Entomology, Oregon State University, Corvallis, Oregon

J. A. McMURTRY, Department of Entomology, University of California, Riverside, California

G. P. MARKIN, U.S. Department of Agriculture, Forest Service, Hilo, Hawaii

D. W. MEALS, Division of Biological Control, University of California, Albany, California

R. H. MESSING, Department of Entomology, Kauai Research Center, University of Hawaii, Kapa'a, Hawaii

A. B. MOHAMMAD, Department of Entomology, Oregon State University, Corvallis, Oregon

D. M. NAFUS, College of Agriculture & Life Sciences, University of Guam, Mangilao, Guam

S. E. NARANJO, Western Cotton Research Lab, U.S. Department of Agriculture, Agricultural Research Service, Phoenix, Arizona

J. R. NECHOLS, Department of Entomology, Kansas State University, Manhattan, Kansas

R. M. NOWIERSKI, Department of Entomology, Montana State University, Bozeman, Montana

E. R. OATMAN, Department of Entomology, University of California, Riverside, California

G. S. PAULSON, Department of Biology, Washington State University, Pullman, Washington

R. W. PEMBERTON, Aquatic Weed Control Research Laboratory, U.S. Department of Agriculture, Agricultural Research Service, Ft. Lauderdale, Florida

K. S. PIKE, Irrigated Agriculture Research Extension, Washington State University, Prosser, Washington

G. L. PIPER, Department of Entomology, Washington State University, Pullman, Washington

N. E. REES, Rangeland Insect Lab, U.S. Department of Agriculture, Agricultural Research Service, Bozeman, Montana

N. J. REIMER, Department of Entomology, University of Hawaii, Honolulu, Hawaii

S. S. ROSENTHAL, Rangeland Insect Lab, Biological Control of Weeds Research Unit, U.S. Department of Agriculture, Agricultural Research Service, Bozeman, Montana

P. STARÝ, Institute of Entomology, Czechoslovak Academy of Sciences, České Budějovice, Czechoslovakia

J. M. STORY, Western Agricultural Research Center, Montana State University, Bozeman, Montana

T. S. SU, Department of Entomology, National Chung Hsing University, Taichung, Taiwan

R. L. TASSAN, Division of Biological Control, University of California, Albany, California

C. A. TAUBER, Department of Entomology, Cornell University, Ithaca, New York

M. J. TAUBER, Department of Entomology, Cornell University, Ithaca, New York

M. TOPHAM, Department of Entomology, University of Hawaii, Honolulu, Hawaii

C. E. TURNER, Biological Control of Weeds Lab, U.S. Department of Agriculture, Agricultural Research Service, Albany, California

G. K. UCHIDA, Department of Entomology, University of Hawaii, Honolulu, Hawaii

T. R. UNRUH, Yakima Agricultural Research Lab, U.S. Department of Agriculture, Agricultural Research Service, Yakima, Washington

S. L. WAGGY, Department of Entomology, University of Hawaii, Honolulu, Hawaii

T. F. WATSON, Department of Entomology, University of Arizona, Tucson, Arizona

P. H. WESTIGARD, Southern Oregon Experiment Station, Oregon State University, Medford, Oregon

L. T. WILSON, Department of Entomology, Texas A & M University, College Station, Texas

A. M. VARGO, American Samoa Community College, Pago Pago, American Samoa

E. R. YOSHIOKA, Hawaii Department of Agriculture, Hilo, Hawaii

Part 1

History and Impact of Regional Research Project W-84

1 / A Historical Overview of Regional Research Project W-84

J. A. McMurtry, L. A. Andres, T. S. Bellows, Jr., S. C. Hoyt, and K. S. Hagen

Biological control, or the suppression of pest organisms by their natural enemies, is the most acceptable long-range control tactic available for incorporation into pest management programs. This pest suppression method is based on the knowledge that in nature there is a balance between populations of pest arthropods or weeds and their natural enemies, and that in managed (for example, agricultural) ecosystems that balance may be disturbed, sometimes severely. When pests invade a new area that lacks their natural enemies, that balance is lost altogether. By manipulating environmental factors, we can restore balance and thereby produce more permanent, effective pest suppression. Our manipulation may involve one or more of the following ecological approaches: (1) the importation of exotic natural enemies (classical biological control); (2) the conservation of resident and introduced beneficial organisms; and (3) the mass production and periodic release of natural enemies.

Biological control of invertebrate pests and weeds is particularly desirable because the method is environmentally safe, energy self-sufficient, cost-effective, and often self-sustaining. Biological control projects have returned an estimated $30 for every $1 invested in research and application. The corresponding estimate associated with the employment of pesticides is only $3 to $5 for each $1 spent. Furthermore, benefits from the use of natural enemies in many cases accrue annually at no additional cost, whereas the deployment of chemicals represents a recurrent expense to the agricultural producer.

This book surveys research conducted under Regional Research Project W-84, "Biological Control in Pest Management Systems." The project was initiated in 1964 as part of an effort to coordinate biological control work by the various state agricultural experiment stations and the USDA Agricultural Research Service in the western United States. The project has fostered communication, cooperation, and a broader approach to the solution of pest problems using biological control tactics. It has also served as a forum for biological control workers and has provided a means of responding to important issues. The main objective has been to develop biological control procedures and incorporate them into pest management programs.

In the early years, Project W-84 included entomologists from agricultural experiment stations in Arizona, California, Colorado, Hawaii, Idaho, Oregon, and Washington. Presently, researchers in 12 of the 13 states along with 2 U.S. territories actively participate in this regional research project (fig. 1.1), now operating under the number W-185. Initially, the focus of the research was on parasites and predators to control insect and mite pests. The research emphasis has broadened to include a comparable effort in the biological control of weeds. The cooperation of personnel in various state agriculture departments has been important to the progress of the project, especially in phases involving introduced natural enemies. The project currently covers biological control activities in over 40 percent of the land mass of the United States.

When the W-84 project was reorganized as W-185 in 1992, it had four major research objectives: (1) to identify, introduce, disseminate, and establish natural enemies of pest arthropods and weeds of regional importance; (2) to evaluate the ecological and physiological bases for the pest and natural enemy interactions necessary to regulate pest populations; (3) to conserve and augment natural enemies; and (4) to evaluate the impact of biological control.

W-84 RESEARCH OBJECTIVES AND ACHIEVEMENTS

Objective A: The Identification and Introduction of Beneficial Organisms

Introducing and establishing beneficial organisms in areas where they do not naturally occur (often termed "classical" biological control) has resulted in many successes in biological control, ranging from partial to complete control. This activity has been a major focus of Project W-84 work. By fostering a regional approach, the project has facilitated interstate shipment and the

Figure 1.1. States and territories participating in Western Regional Project W-84.

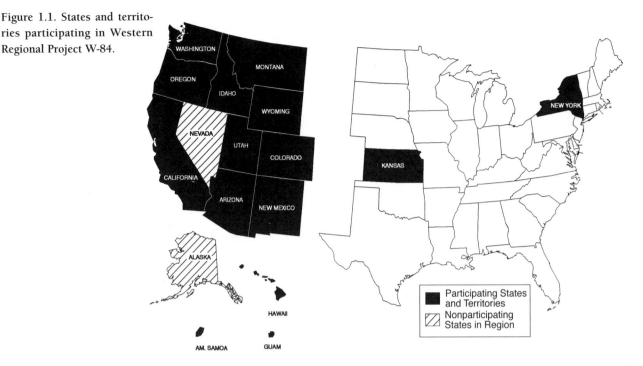

more rapid spread of beneficial species as well as a more coordinated evaluation of releases. Natural enemy cultures maintained by Project participants have also been important sources for beneficial organisms provided to laboratories outside the region and the country.

The introduction of exotic beneficial arthropods must comply with stringent federal regulations, and all such introductions must be confined to an approved quarantine facility. The University of California facilities at Berkeley and Riverside, the USDA facility at Albany, California, and, more recently, the Montana State University–USDA facility at Bozeman, Montana, have played key roles in meeting this major objective.

With respect to the biological control of arthropod pests, we have introduced, established, and done followup studies of natural enemies to control a large number of pests, including the alfalfa weevil (one of the major targets of the project), several aphids (for example, the pea aphid, blue alfalfa aphid, filbert aphid, and asparagus aphid), iceplant scale, various whiteflies, several psyllids, the elm leaf beetle, and spider mites.

In our work on the biological control of weeds, we have established phytophagous arthropods on 21 weed species in the western United States. Throughout the course of Project W-84, over two dozen weed species have been the targets of research, although in recent years our focus has narrowed to leafy spurge, diffuse and spotted knapweed, and a number of thistles. The success rate has been very high (90 to 99 percent) over extensive areas for some target species—for example, St. Johnswort, tansy ragwort, and puncturevine (in Hawaii). In other cases, substantial or partial control has been achieved in

selected areas of the weeds' ranges. Notable examples of partial success include the control of musk thistle, rush skeletonweed, and puncturevine (outside of Hawaii).

The discipline of taxonomy is of paramount importance to biological control, as an incorrect identification of either the natural enemy or the pest can result in the failure of a project. Several members of Project W-84 are authorities in the systematics of beneficial arthropods, including many of the major families of parasitic Hymenoptera and Diptera, the predatory insect families Coccinellidae and Chrysopidae, and the predaceous mite family Phytoseiidae. These are major research projects, and identification work on all these groups continues to be a part of the new W-185 project.

Objective B: Ecological and Physiological Studies

Progress in the effective use of biological control agents is dependent on a sound understanding of the parameters affecting natural enemy–pest interactions. Various types of laboratory and field studies on such aspects as biology, behavior, and population ecology have been emphasized by W-84, and this research has resulted in significant progress. Biological studies are instrumental in developing rearing techniques and interpreting field data, as well as in predicting natural enemy–pest interactions. Biological data are required on all exotic weed-, insect-, and mite-feeding arthropods being considered by federal and state regulatory agencies for release from quarantine.

Identifying and utilizing different natural enemy biotypes has been a major thrust of W-84 research programs. With both weed-feeding and entomophagous arthropods, especially those closely associated with their hosts, a biotype introduced from one host or region may be a successful biological control agent whereas one from another host or region is unsuccessful because of differences in such characteristics as host suitability, synchrony, or differential responses to physical factors. Artificial selection of improved biotypes or strains has proved feasible in developing pesticide-resistant strains of predators and parasites and establishing them in commercial orchards.

Other accomplishments, based on laboratory studies, have included expanding our knowledge of the nutrition and behavior of entomophagous arthropods. Elucidating the role of supplemental foods in the biologies of some of these arthropods has led to a better understanding of their population dynamics and to the possibility of manipulating these foods in the field (for instance, through the use of food sprays). Research showing the importance of certain chemical cues (kairomones) to parasites or predators in finding their hosts or prey has contributed greatly to an understanding of the interactions between natural enemies and their hosts.

Field studies on population dynamics of natural enemies in relation to pest population density and on other factors such as phenology, alternative or supplemental foods, weather, and pesticides have given considerable insight into the potential and limitations of natural enemies. Much effort has also gone into developing methods of measuring the population levels of pests and beneficials. A sampling program that gives statistically reliable estimates of population densities is a requirement in studies on natural enemy–pest interactions.

Objective C: Conservation and Augmentation

The conservation of natural enemies involves manipulating the environment, either by removing or reducing adverse factors or by adding requisites. Augmentation, on the other hand, is the direct manipulation of natural enemies, either by mass production and periodic colonization or by some type of planned genetic improvement. One example of W-84 research progress in conservation involved the use of food sprays to attract predators (for example, coccinellids or chrysopids) or weed control agents (for instance, seed-infesting flies) to particular fields or plants. Another example of progress

is a study that involved evaluating the benefits of cultural practices such as polycultures and cover crops to furnish alternative hosts or supplemental foods to maintain higher natural enemy populations. Another important method of conserving natural enemies has been to modify pest control practices, particularly by emphasizing selective pesticides and manipulating pesticide placement or rates. Examples of the successful augmentation of natural enemies include the mass releases of predatory mites for spider mite control on certain high-value crops, the use of *Trichogramma* wasps to control lepidopterous pests on tree crops, and the establishment and spread of laboratory-selected, pesticide-resistant predatory mites for spider mite control in tree crops.

Objective D: Impact

One of the most important—and difficult—challenges of biological control research is to estimate the mortality and degree of suppression (both long- and short-term) that a given natural enemy or natural enemy complex exerts on an arthropod or weed pest population. Project W-84 has made advances by using both experimental evaluation methods (for example, by chemically, biologically, or mechanically excluding natural enemies from some plots and making comparisons with plots in which the natural enemy is present) and analytical techniques (for instance, life tables). Prerequisites of such research include developing sampling techniques to adequately estimate the population densities of both the natural enemy and the pest arthropod or weed, and developing suitable analytical techniques.

The financial benefits of increased utilization of natural enemies have seldom been calculated, usually because of a lack of adequate documentation of cost-benefit and other data. An obvious long-term economic and environmental benefit is the reduction or even elimination of pesticide applications. A less obvious, but equally important, economic benefit has been the delay in the development of resistance in pests to pesticides through reduced selection pressure. To consider the whole economic picture, an agricultural economist must be involved. The W-84 project has accomplished this kind of interdisciplinary research in the evaluation of pesticide-resistant predatory mites in almond orchards.

2 / Integrated Pest Management: Contributions of Biological Control to Its Implementation

M. W. Johnson and L. T. Wilson

Using indigenous and exotic natural enemies in today's agriculture can provide benefits that pesticide control cannot (Wilson and Huffaker 1976). These advantages include (1) highly effective, inexpensive pest suppression, (2) continuous control at little cost following the initial introduction of exotic agents, (3) an absence of harmful effects on people and the environment, (4) the utility of natural enemies as "biotic insecticides" (as in the mass release of parasitoids or predators, or use of *Bacillus thuringiensis* Berliner), and (5) the inability of pests to develop significant resistance to biological control agents. When introduced agents are highly effective the results can be dramatic: the coccinellid predator *Rodolia cardinalis* (Mulsant) and cryptochetid parasitoid *Cryptochaetum iceryae* (Williston), introduced on southern California citrus over 100 years ago, provided a high level of control of cottony cushion scale, *Icerya purchasi* Maskell (Doutt 1964). Less effective natural enemies may provide substantial or partial biological control (DeBach 1964a), thereby reducing the need for pesticides.

Unfortunately, many arthropod pests targeted in classical biological control programs inhabit crops or agroecosystems that support other pest species for which biological control is poor or nonexistent. In addition, introducing and establishing effective natural enemies may not solve a pest problem if provisions are not taken to protect biological control agents from such agricultural practices as pesticide applications, weed control, and crop rotation. These same agricultural practices can take their toll on indigenous natural enemies. Growers, given the real or perceived possibility of economic loss and the absence of alternative management strategies (for example, resistant crop varieties or cultural controls), often opt for conventional pesticides. However, once pests develop resistance to a pesticide, growers usually increase the frequency and

dosage of applications. This often leads to such further complications as greater pesticide resistance, pest resurgences and upsets, environmental pollution, and human health hazards (Dover and Croft 1984). The concept of integrated pest management (IPM) originated in the late 1950s with the goal of more effectively controlling pest problems and reducing the undesirable side effects of pesticide use. As successful IPM systems were developed, the concept has gained acceptance. Scientists participating in Regional Project W-84 have contributed significantly to IPM's development and implementation; many of their projects are examples of effective systems that use natural enemies as an alternative to complete reliance on pesticides: for example, apple (Hoyt 1969), cherry (AliNiazee 1978), cotton (Sterling, El-Zik, and Wilson 1989), tomato (Oatman et al. 1983b), almond (Hoy 1985), and watermelon (Johnson et al. 1989).

The research of participating scientists has provided many solutions to the obstacles that limit the effective use of natural enemies in agricultural systems throughout the United States and Pacific Basin (Davis et al. 1979). This chapter reviews those research contributions of Regional Project W-84 scientists that feature biological control as a significant part of an IPM system.

Biological control can be improved in natural ecosystems and agroecosystems in three ways (DeBach 1964b). The first, known as classical biological control, involves introducing exotic natural enemies to control pests. The second, conservation, means modifying any environmental factors that are adverse to biological control. The third, called augmentation, focuses on improving the effectiveness of natural enemies. Given the numerous arthropod pests (more than 55) and crop systems (more than 10) investigated since the establishment of Regional Project W-84, it is beyond the scope of this chapter to discuss all pro-

We thank T. J. Henneberry, USDA-ARS, Phoenix, Arizona; W. A. Jones, USDA-ARS, Weslaco, Texas; K. S. Hagen and D. L. Dahlsten, University of California, Berkeley; M. A. Hoy, University of Florida, Gainesville; S. H. Dreistadt and L. E. Ehler, University of California, Davis; E. R. Oatman and E. F. Legner, University of California, Riverside; C. A. Tauber and M. J. Tauber, Cornell University, Ithaca; D. M. Nafus, University of Guam, Mangilao; J. W. Beardsley and V. P. Jones, University of Hawaii at Manoa; J. R. Nechols, Kansas State University; J. M. Story, Montana State University; S. S. Rosenthal, USDA-ARS, Bozeman, Montana; J. J. Ellington, New Mexico State University; M. T. AliNiazee, Oregon State University; and S. C. Hoyt and G. L. Piper, Washington State University, for their assistance and contributions. This is Journal Series No. 3583 of the Hawaii Institute of Tropical Agriculture and Human Resources, University of Hawaii.

ject contributions to IPM implementation. Therefore, we will focus on some of the most significant and innovative contributions covering a range of crop and natural systems. Contributions will be discussed with respect to the major objectives of the regional project.

THE INTRODUCTION, DISSEMINATION, AND ESTABLISHMENT OF NATURAL ENEMIES

When there are no effective indigenous natural enemies, IPM programs for introduced and indigenous pest species should include the use of exotic biological control agents. Although the concept of introducing beneficial organisms is simple, such factors as the logistical aspects of colonization and the environmental differences between the origin (or collection site) of the exotic species and the establishment site make the task difficult to accomplish. Analyzing why a given organism did not establish is therefore essential to the success of future introduction attempts.

A recent example of a highly successful classical biological control program achieved exclusively through the efforts of W-84 was the control of the filbert aphid *Myzocallis coryli* (Goetze) in Oregon (Messing and AliNiazee 1989) by using a French strain of the parasitoid *Trioxys pallidus* Haliday. The filbert aphid is of European origin and was accidentally introduced to the western United States nearly 100 years ago. It was a serious pest of hazelnuts and required two applications of organophosphate compounds for control. A large number of predators were recorded feeding on aphids in filbert orchards (AliNiazee 1980; Messing and AliNiazee 1985), but they did not provide reliable biological control (AliNiazee 1983). The filbert aphid was not a suitable host for California strains of *T. pallidus,* previously established to control the walnut aphid *Chromaphis juglandicola* (Kaltenbach). In 1984 an extensive search for host-specific natural enemies was conducted in western Europe, and individuals from three different *T. pallidus* populations were sent to Oregon for mass culture and colonization (Messing and AliNiazee 1989). A French population, released extensively during the growing seasons of 1984 to 1986, became established throughout Oregon's Willamette Valley and provided impressive control of the filbert aphid. Initial establishment took only 2 years. The parasitoid is now a major part of the filbert IPM program in Oregon and Washington.

Following the introduction of synthetic pesticides, mites in the family Tetranychidae became important pests in many crop systems, in part because effective natural enemies were destroyed by pesticide applications. Of particular significance to spider mite control in southern California was the establishment of the preda-

tory mite *Phytoseiulus persimilis* Athias-Henriot (McMurtry et al. 1978). Establishment was a result of experimental studies on the benefits of using mass releases of *P. persimilis* to control twospotted spider mite, *Tetranychus urticae* Koch. Insectary-produced *P. persimilis* individuals were released in commercial strawberry plantings in Ventura County, California, to reduce or eliminate pesticide applications for *T. urticae.* Two *P. persimilis* strains were originally introduced, a "Chilean" stock and an "Italian" stock. Between 1971 and 1977 releases of both stocks totaled 234,440 *P. persimilis* released in 13 fields. The predator was collected extensively in one of the original release sites during both the 1972–1973 and the 1973–1974 seasons. The Chilean stock probably established first. In the 1974 to 1975 season, collections outside release fields were made 4 kilometers from the original release sites. Subsequent population spread probably included individuals from both *P. persimilis* stocks; they dispersed from release sites onto surrounding crop and wild plant hosts. This predator became an important spider mite mortality factor in strawberry IPM programs in the Oxnard area of Ventura County.

Agromyzid leafminers in the genus *Liriomyza* have become significant pests in vegetable and ornamental crops throughout the world (Parrella 1987; Waterhouse and Norris 1987). After the accidental introduction of *Liriomyza trifolii* Burgess into the U.S. Territory of Guam, *Ganaspidium utilis* Beardsley was introduced from Hawaii and established on Guam. It became the dominant parasitoid on beans, where it parasitizes up to 78 percent of the *L. trifolii* larvae infesting the crop. Leafminer densities decreased dramatically and are no longer a problem in unsprayed bean plantings.

The imported cabbageworm *Artogeia* (=*Pieris*) *rapae* (L.), a common pest of cole crops, is often the target of preventive pesticide applications. Attempts to establish the exotic egg parasitoid *Trichogramma evanescens* Westwood (from Europe) and the larval parasitoid *Apanteles rubecula* Marshall (from British Columbia) on *A. rapae* infesting cabbage in southern California were unsuccessful (Oatman and Platner 1972). Because during each year of the study no colonized parasitoids were recovered after September, the project concluded that lack of diapause in southern California populations of *A. rapae* prevented these natural enemies from surviving to the next spring season. This study provided significant information on the specific biological parameters necessary for future importations of exotic biological control agents for control of *A. rapae* in southern California.

In 1985 *Nezara viridula* (L.), a pest of international importance, was found infesting tomato fields in northern California (Hoffmann, Wilson, and Zalom 1987). Populations of the parasitoid *Trissolcus basalis*

(Wollaston), collected from France, Italy, and Spain, were studied to quantify their basic life-table parameters to determine the likelihood of establishment (Awan et al. 1990). Subsequent field releases have resulted in the establishment of each geographical population and parasitism levels that in some cases reach 90 percent.

THE CONSERVATION OF NATURAL ENEMIES

Probably the greatest contributions to implementing IPM have been in the conservation of biological control agents. Arthropod pest species targeted for biological control are rarely the only species that cause significant economic losses in a given crop or crop complex. Even though complete biological control may be possible for the targeted pest species, chemical controls are often necessary to control one or more of the remaining pest species. Those pesticides used for other pests may, however, reduce the effectiveness and densities of natural enemies and cause pest population increases—or "pest upsets"—in the species formerly under biological control. Furthermore, when less than complete biological control is achieved, supplementary pesticide applications may be required for the pest targeted for biological control. Applications of broad-spectrum materials can also decimate populations of introduced or indigenous natural enemies. Frequently, these treatments cause resurgences of the target pest, especially if the pest has developed significant levels of pesticide resistance compared with the natural enemy. Herbicides applied against weeds may also affect phytophagous species introduced for weed control. Additionally, agricultural practices (rotation, sanitation, and so on) may decimate natural enemy populations. Many agroecosystems may lack specific prerequisites for effective biological control, such as the presence of supplemental food sources (for example, pollen, nectar, and honeydew) that some adult predators and parasitoids require for reproduction (Hagen and Bishop 1979). Although some work has been conducted on conservation of phytophagous arthropods attacking weed species (McCaffrey and Callihan 1988; Story, Boggs, and Good 1988), the greatest contributions have been made on conserving natural enemies of arthropod pests in agroecosystems.

The Conservation of Predators and Parasitoids

Approaches to conserving natural enemies within agroecosystems may be either direct or indirect. Direct approaches either modify adverse agricultural practices to alleviate their detrimental impact on natural enemies or manipulate natural enemy populations to increase their effectiveness. These approaches may be quite specific for a given agroecosystem or pest–natural enemy association. They include combining biological control with nondisruptive, nonchemical control methods, such as host plant resistance, cultural controls, and autocidal methods; using supplemental foods and behavioral chemicals to increase natural enemy effectiveness; and using selective chemical methods, such as physiological and ecological selectivity. Indirect approaches are methods that determine the necessity and timing of adverse practices. These generally apply to most crop production systems and may or may not reduce the impact of adverse practices on natural enemies. They include the use of sampling methods to estimate the abundance of pest species and the potential effect of associated natural enemies, as well as the use of economic thresholds (action thresholds, density treatment levels, and so on) to determine the need for management intervention. Because of the wide applicability of the indirect methods, we will discuss W-84 contributions in this area first.

Sampling techniques for pests and natural enemies. Without practical sampling techniques, it is impossible to utilize economic thresholds effectively or to estimate current and potential levels of biological control. When monitoring procedures are not robust, or when they require more time than a field scout or consultant can justify, the result is an excessive use of pesticides and poor timing of applications. In summarizing IPM practices on processing tomatoes, J. M. Antle and S. K. Park (1986) and J. I. Grieshop, F. G. Zalom, and G. Miyao (1986) found that if farmers used quantitative monitoring procedures in conjunction with realistic management thresholds, they could reduce pesticide use by 40 percent without affecting fruit quality or yield. Similar savings have been reported with cotton (Room 1979) and other crops (National Research Council 1989). Greater reductions in pesticide use and agricultural nonpoint source pollution have only been accomplished through restrictive and often disruptive externally imposed regulations.

W-84 researchers have made significant contributions in this area. They have concentrated on (1) determining the spatial patterns of pest and natural enemy species, (2) simplifying sampling programs for grower and consultant use, and (3) evaluating the effect of sampling programs (and use of economic thresholds) on the necessity for pesticide applications.

One of the first steps in any sampling program is to quantify the spatial pattern of pest species in their environment. Studies on cotton in Arizona show that the pink bollworm Pectinophora gossypiella (Saunders) prefers to oviposit on cotton bolls (when present) as compared with plant terminals and squares (Henneberry and Clayton 1982a). This information helped in developing simple but effective sampling techniques for pink bollworm (Hutchison et al. 1988). Studies in California show that the cotton bollworm *Heliothis zea* (Boddie) prefers to oviposit most of its eggs in the terminal of the plant. Upon hatching, the larvae move up the plant for the first few instars, then begin to move down the plant (Wilson, Gutierrez, and Leigh 1980). Synchronized with the distribution of prey, smaller predators such as *Orius tristicolor* (White) and *Geocoris* spp. are found higher on the plants than are larger predators such as *Chrysoperla* (=*Chrysopa*) *carnea* (Stephens) and *Nabis americoferus* Carayon, which are capable of attacking the latter larval instars (Wilson and Gutierrez 1980). This information was used to develop a sampling plan that provided cost-effective estimates of the treatment status of cotton bollworm in a fraction of the time required by conventional sampling procedures (Wilson, Gutierrez, and Leigh 1980).

In almonds, a presence-absence sampling technique was used to monitor *Tetranychus* spp. and predatory mites (Wilson et al. 1984); this method proved a cost-effective way to obtain the data needed to determine mean densities of pest and predatory mites, ratios of predator mites to spider mites, and accumulated spider mite days. This information was necessary for the implementation of mite management guidelines in almonds (Hoy 1985). Wilson and his co-workers have quantified the spatial pattern of a range of pests and natural enemy species, considered economic and ecological factors, and developed cost-effective sampling methods for almonds (Wilson et al. 1984; Zalom et al. 1984a,b), Brussels sprouts (Pickel et al. 1983; Wilson, Pickel et al. 1983), cassava (Braun et al. 1989), citrus (Zalom et al. 1985, 1986), cotton (Wilson and Gutierrez 1980; Wilson, Gutierrez, and Leigh 1980; Wilson, Leigh, and Maggi 1981; Wilson 1982, 1985; Wilson, Gutierrez, and Hogg 1982; Wilson and Room 1982, 1983; Wilson, González et al. 1983; Wilson, Leigh et al. 1983; Leigh, Maggi, and Wilson 1984; Wilson et al. 1984, 1987; Plant and Wilson 1985; Wilson, González, and Plant 1985; Pickett, Wilson, and González 1988; Wilson, Sterling et al. 1989; Hassan et al. 1990), and tomatoes (Wilson, Zalom et al. 1983; Zalom et al. 1983; Zalom, Wilson, and Smith 1983; Zalom, Wilson, and Hoffman 1986; Hoffmann et al. 1986).

In other W-84-related research, J. T. Trumble, E. R. Oatman, and V. Voth (1983a) determined that mixed populations of strawberry aphid, melon aphid, green peach aphid, and potato aphid could be estimated more efficiently by sampling the oldest trifoliate strawberry leaves as compared with whole-plant samples. M. W. Johnson et al. (1980) developed a simple technique for monitoring *Liriomyza* leafminers in tomatoes by placing styrofoam trays beneath the leaf canopy of the plants. They discovered a significant correlation between the second and third instar larvae in the tomato foliage versus the number of leafminer larvae pupating on the trays. This technique, which permits a more rapid assessment of foliar leafminer densities than is possible by inspecting tomato leaflets, simplified the monitoring process.

Many monitoring techniques such as bait and pheromone traps provide information on the mobile adult stages of pests. Studies on the phenology of adult apple maggot *Rhagoletis pomonella* Walsh (AliNiazee and Westcott 1987; Jones et al. 1989) and an evaluation of various traps (in combination with different lures) to monitor adult densities on apples and tart cherry (AliNiazee, Mohammad, and Booth 1987; Jones and Davis 1989) provided information that enables growers to time pesticide applications better. The best trap for apple maggot under Utah's conditions was a yellow panel trap baited with ammonium carbonate. Under Oregon's conditions, a Ladd trap—composed of a yellow panel with two red hemispheres and an apple volatile attractant—was best. Including apple volatiles substantially increased the performance of red sphere traps (AliNiazee, Mohammad, and Booth 1987). Attractant-baited traps were used successfully to sample the western cherry fruit fly *Rhagoletis indifferens* Curran (AliNiazee 1978). Using these traps to help time pesticide applications resulted in a 20 to 100 percent reduction in pesticide use in test orchards. This information will also increase precision when timing sprays and may reduce the severity of the effects of sprays on populations of the western orchard predatory mite, *Galendromus* (=*Metaseiulus* or *Typhlodromus*) *occidentalis* (Nesbitt), which attacks the twospotted spider mite often found on deciduous orchard crops.

Pheromone traps provide a simple technique to monitor adult populations. The rate of increase in numbers of adult pests trapped over a relatively short period can be used to predict injurious levels of their progeny. T. J. Henneberry and T. E. Clayton (1982b) developed a survey program using gossyplure-baited Delta pheromone traps to monitor pink bollworm populations in cotton more effectively. Average catches of male moths for 3 to 7 days between boll-sampling periods correlated strongly with oviposition on cotton bolls ($r = 0.86$), the percentages of infested bolls ($r = 0.82$), and the numbers of larvae per boll ($r = 0.82$). These researchers found the pheromone trap most useful for scheduling initial insecticide treatments for pink bollworm infestations within a cotton planting. R. A. Van Steenwyk et al. (1983), using Pherocon 1C pheromone traps, developed a program that growers can use to time the initiation of pesticide treatments for tomato pinworm, *Keiferia lycopersicella* (Walsingham). They found that pesticide applications can be delayed until 10 adult pinworms are caught per trap per night over a 1.5 week period. These findings help limit applications of broad-spectrum compounds such as methomyl that promote pest upsets of agromyzid leafminers (that is, *Liriomyza sativae* Blanchard and *L. trifolii*) by reducing effective parasitoids.

Most sampling techniques only provide density estimates for a few arthropod species at best. As part of W-84 project activities, J. Ellington et al. (1984) developed the "insectavac," a high-clearance, high-volume vacuuming platform to sample arthropod densities in cotton. By significantly reducing the hand labor usually required in sampling populations, the instrument enabled large sample sizes to be collected with relatively small increases in labor. It was particularly adept at sampling parasitoids, predators (for example, *Lygus* spp.), and first- to third-instar and adult *Heliothis* species.

Knowing an arthropod's spatial pattern can increase sample precision. J. A. Lynch and M. W. Johnson (1987) conducted field studies to evaluate the accuracy of random versus stratified sampling of watermelon foliage—with respect to leaf size and distance from the plant base—for *Liriomyza sativae* and *L. trifolii* larvae and associated parasitoids. Prior to full leaf-canopy establishment, there were significantly greater densities of *Liriomyza* larvae per leaf at the plant base than at the distal end of the vine. Stratifying according to leaf size significantly increased the precision of the sample mean estimates for both *Liriomyza* larvae and the eulophid parasitoid *Chrysonotomyia punctiventris* (Crawford). Stratifying based on the distance from the plant base also resulted in greater precision of *Liriomyza* density estimates prior to full leaf-canopy establishment.

The simpler the sampling methods, the more suitable they are for IPM consultants and growers. The development of presence-absence sampling programs was a significant step toward simplification. L. T. Wilson et al. (1984) and F. G. Zalom et al. (1984a,b) developed presence/absence sequential sampling techniques for *Tetranychus* spp. and *Galendromus* (=*Metaseiulus*) *occidentalis* in almonds in California. The sampling method for *G. occidentalis* was one of the few sampling programs that considered the presence of natural enemies in an IPM system. Different presence-absence decision tables were used depending on the presence of *G. occidentalis* in the orchard. V. P. Jones (1990) developed similar presence-absence sampling plans for twospotted spider mite and *G. occidentalis* on tart cherry in Utah. Growers can use these plans to more precisely time pesticide applications, thereby targeting predators when they are present.

S. H. Dreistadt and D. L. Dahlsten (1989) developed a presence-absence sampling program for eggs of the elm leaf beetle, *Xanthogaleruca luteola* (Müller) (Coleoptera: Chrysomelidae), on English and Siberian elm trees. By studying the maximum proportion of presence-absence samples infested with first-generation elm leaf beetle eggs, the scientists could predict cumulative damage to English elm, but not to Siberian elm. This sampling program could be very useful in determining the need for insecticidal treatments directed at elm leaf beetles (Dreistadt and Dahlsten 1989) and in pinpointing localities where biological control efforts should be maximized.

J. T. Trumble, E. R. Oatman, and V. Voth (1983a,b) developed sequential and presence-absence sampling plans that were effective in determining mean levels of mixed infestations of aphid species in strawberry plantings in southern California. Using a threshold of more than 30 percent infested trifoliate leaves would limit pesticide applications that might decimate *Phytoseiulus persimilis* released for twospotted spider mite control.

After developing economic thresholds and sampling methods, researchers still need to evaluate the methods' usefulness within a crop system. W. D. Hutchison et al. (1988) compared the efficacy of sampling eggs (with a presence-absence sampling plan) versus larvae in order to time pesticide applications for pink bollworm in cotton. Working in commercial cotton plantings, they found that using a threshold of 6 to 8 percent egg-infested bolls—as opposed to the conventional threshold of 5 percent larval-infested bolls—meant that fewer pesticide applications were needed. Sampling eggs resulted in a 35 percent reduction in insecticide costs over larval sampling, with no significant reduction in yield. This reduction in pesticide use increased the potential for conserving and augmenting biological control agents, and facilitated the implementation of more effective insecticide resistance management programs.

M. W. Johnson et al. (1989) implemented a monitoring program for the multispecies pest complex that occurs in commercial watermelon plantings in Hawaii. Their study revealed that growers could reduce insecticide applications directed at *Liriomyza* leafminers by over 90 percent, given the added control provided by the introduced parasitoids *Chrysonotomyia punctiventris* and *Ganaspidium utilis* (Johnson 1987). However, growers must attempt to conserve leafminer natural enemies if they use pesticide controls for *Thrips palmi* Karny, a recently introduced pest, and the melon aphid, which vectors nonpersistent cucurbit mosaic viruses. On more than 13 percent of the dates when a watermelon planting was sampled during a growing cycle, these two pests surpassed nominal density treatment levels (Johnson et al. 1989).

The development of economic thresholds. Given the basic premise that pesticides should not be applied unless necessary, it might seem odd that economic thresholds or their functional equivalents are not available for all crops. Using economic thresholds can limit the amount of pesticide applied to a crop and reduce its impact on natural enemies. Thresholds also define the levels of biological control necessary to maintain pest populations at subeconomic densities. However, determining economic thresholds is often logistically difficult and quite expensive. S. C. Hoyt, L. K. Tanigoshi, and R. W. Browne (1979) identified factors that hindered attempts to quantify yield responses of deciduous fruit crops to spider mite injury. These factors included the unreliability of pest populations occurring naturally in test plots and the difficulty of replicating specific pest population levels. Five experimental studies determined levels at which the feeding of the McDaniel spider mite, *Tetranychus mcdanieli* McGregor, reduced apple yields, and established economic injury levels (about a 12 percent yield loss) at 3,000 mite-days (approximately 120 mites per leaf).

In another study, the impact of twospotted spider mite was determined on strawberry crops in southern California. In winter plantings *T. urticae* densities of 25 mites per leaflet could be tolerated without significant yield loss (Oatman et al. 1981, 1982), whereas in summer plantings a density treatment level of 50 mites per leaflet (Wyman, Oatman, and Voth 1979) could be tolerated when the selective acaricide cyhexatin was used. This pesticide was recently dropped from California registration, and now the predatory mite *Phytoseiulus persimilis* Athias-Henriot has become more important in spider mite control (Oatman person. commun.). The density treatment levels are now valuable in providing a level of biological control necessary for acceptable strawberry production. Given the increased importance of *P. persimilis*, growers should, if possible, limit pesticides applied for other strawberry pests. However, growers often apply pesticides for aphids that infest strawberries, because commercial wholesalers reject fruit if individual aphids are present. In additional W-84 studies, J. T. Trumble, E. R. Oatman, and V. Voth (1983a) determined that mixed populations of strawberry aphid *Chaetosiphon fragaefolii* (Cockerell), melon aphid *Aphis gossypii* Glover, green peach aphid *Myzus persicae* (Sulzer), and potato aphid *Macrosiphum euphorbiae* (Thomas) contaminated harvested strawberry fruit when their densities surpassed 30 aphids per strawberry plant. The potential for aphid contamination was high when over 50 percent of whole-plant samples or more than 30 percent of leaf subsamples were infested. By using these thresholds, growers can limit the application of aphicides detrimental to *P. persimilis*.

Studies have provided guidelines for reducing pesticide applications on vegetable crops where preventive treatments are common. Common pests of broccoli grown in southern California include the cabbage looper *Trichoplusia ni* (Hübner), the diamondback moth *Plutella xylostella* (L.), and the imported cabbageworm *Artogeia* (=*Pieris*) *rapae* (L.). Yields were not reduced by combined lepidopterous pest densities of less than nine larvae per plant during the period from thinning to preheading of the broccoli (Wyman and Oatman 1977).

Limited pesticide applications on this crop would maximize the effects of beneficial species. In Guam, yield losses on beans due to *Liriomyza trifolii* infestations were nonlinear and best described by a quadratic model (Schreiner et al. 1986). When more than 40 miners per leaf were present, virtually all photosynthetic tissue was destroyed as miners matured, and no additional yield loss occurred as leafminer densities increased. Reduced pesticide applications to beans would enhance biological control by the introduced parasitoid *Ganaspidium utilis* and retard the development of pesticide resistance in the leafminer and other bean pests such as *Thrips palmi*.

Many factors, in combination with pest density, influence the amount of injury a plant can sustain from a given arthropod species before yield losses become economically significant. These factors include plant phenology and cultivar, seasonal conditions, water stress, and agronomic practices. Additional factors—such as the choice of registered pesticides available and the presence of natural enemies of primary and secondary pests—may affect the decision to apply pesticides to control a particular arthropod. Considering several of these factors, M. A. Hoy (1985) provided guidelines for determining the need for mite suppression and appropriate acaricide treatment in almonds. Depending on the month (May, June, July, or August), decisions to treat for spider mites were based primarily on the ratio of predatory mites to spider mites, spider mite density, and the number of accumulated mite-days. Acaricides (that is, propargite) were sometimes recommended at lower-than-label rates, to adjust the predator-to-prey ratios in favor of the natural enemies. Although they are complex, detailed guidelines such as Hoy developed are extremely valuable in integrating biological and chemical controls.

The integration of biological control and agricultural practices. Several factors must be considered when integrating biological control with cultural practices to reduce pest populations. Two major contributors to control decisions are the flora around the targeted crops and the impact of host plant resistance on natural enemies.

The importance of flora surrounding crops. Scientists participating in the W-84 project have shown that the biological control of a given species within a crop can be influenced by the adjacent crops and wild plants. J. R. Leeper (1974) and M. Topham and J. W. Beardsley (1976) demonstrated how important Euphorbiaceae weed species (*Euphorbia hirta* L., *E. glomerifera* [Millsp.], *E. geniculata* Ortega, and *E. heterophylla* L.) that surround sugarcane fields in Hawaii are to the biological control of the sugarcane weevil, *Rhabdoscelis obscurus* (Boisduval). These plants provide a nectar source and mating sites for adults of the tachinid parasitoid *Lixophaga sphenophori* (Villeneuve), which attacks the sugarcane weevil. Periodic herbicide applications eliminated feral nectar source plants from ditch banks and field margins, greatly reducing adult parasitoid populations in and around sugarcane fields. Recognizing the importance of *Euphorbia* spp. adjacent to cane plantings, growers modified their herbicidal applications to spare this weed group.

L. T. Wilson, C. H. Pickett, et al. (1989) and C. H. Pickett, L. T. Wilson, and D. L. Flaherty (1990), in a follow-up to earlier work by H. Kido et al. (1984), found that damage by the grape leafhopper *Erythroneura elegantula* Osborn could be reduced by a companion planting of French prune trees adjacent to the grape vineyards. A small mymarid parasitoid, *Anagrus epos* Girault, builds up in the prunes on the prune leafhopper *Dikrella californica* (Lawson). It then moves to the grapes and suppresses the grape leafhopper population before the population reaches economically damaging levels. By planting prune trees adjacent to and upwind of the vineyard, the need for grape leafhopper control can be all but eliminated.

In southern California, the ability of the introduced predatory mite *Phytoseiulus persimilis* to rapidly infest strawberry fields from adjacent weed hosts and lima bean fields contributed substantially to twospotted spider mite control on strawberry (McMurtry et al. 1978). D. P. Carroll and S. C. Hoyt (1986) showed that in Washington apple orchards the apple aphid *Aphis pomi* DeGeer was parasitized by parasitoids (*Lysiphlebus* sp. and *Praon* sp.) that develop on other aphid hosts. Their research showed that the parasitoids could not develop on apple aphid, but must develop on other aphid species within or outside the apple orchard. Therefore, interplanting apples with host plants that support alternative host aphids may increase the biological control of apple aphid.

Host plant resistance and biological control. Although biological control and host plant resistance have been referred to as the "cornerstones of pest management" (Huffaker 1985), studies that address specific interactions among the three trophic levels (resistant plants–pests–natural enemies) are rare. Studies sponsored by the W-84 project evaluated such interactions among insects associated with cotton, potato, and squash.

Nectariless cotton sustained less foliar injury by the cotton leafperforator *Bucculatrix thurberiella* Busck than did cotton with nectaries (Henneberry, Bariola, and Kittock 1977). In the same study, seasonal mean densities of *Lygus* spp. were also lower in nectariless cotton. However, general predators such as *Chrysopa* sp.,

Geocoris spp., and various coccinellids (important for pink bollworm control) were less prevalent in nectariless cotton than in nectaried cotton. T. J. Henneberry, L. A. Bariola, and D. L. Kittock (1977) suggested that in large plantings of nectariless cotton, the complex of interacting arthropod populations change drastically in favor of those species that feed predominantly on pollen and other plant materials rather than on nectar. I. K. Adjei-Maafo and L. T. Wilson (1983), however, later presented results suggesting that nectariless cotton, if planted in commercial-size areas, may not provide the degree of pest reduction anticipated from the results of small-scale research plots. When given a choice, most cotton arthropods, including *Heliothis* spp., prefer to feed and oviposit in nectaried cotton; however, commercial fields are so large that it is difficult for pests and beneficials to select the most suitable habitat.

J. J. Obrycki, M. J. Tauber, and W. M. Tingey (1983) and J. J. Obrycki and M. J. Tauber (1984, 1985) examined the interactions between natural enemies and potato plants bred for resistance to insect pests. Their investigations showed that the behavioral responses of natural enemies to resistant plants under laboratory and greenhouse conditions did not adequately predict the effects in the field. Under field conditions, the negative effects of the resistant plants on the beneficial insects were reduced, and natural enemies (both parasitoids and predators) caused substantial aphid mortality. Subsequent studies demonstrated that resistant potato plants had direct effects on the eulophid parasitoid *Edovum puttleri* Grissell, as well as indirect effects mediated through the parasitoid's host eggs of the Colorado potato beetle *Leptinotarsa decemlineata* (Say). In the field, certain resistance mechanisms (namely glandular trichomes) were compatible with the parasitoid, whereas others (other types of glandular trichomes) were not.

Studies on *Gryon pennsylvanicum* (Ashmead) and *Ooencyrtus* sp. 'light form', egg parasitoids of the squash bug *Anasa tristis* DeGeer, showed that squash bug–resistant plants (*Cucurbita moschata* cv. 'Waltham Butternut') had no impact on parasitoids reared in eggs produced from *A. tristis* reared on the resistant squash (Nechols, Tracy, and Vogt 1989). Specifically, no significant differences were found in parasitization rates, immature survivorship, developmental times, or sex ratios between parasitoids reared from eggs produced by squash bugs feeding on resistant and susceptible (*Cucurbita moschata* cv. 'Early Prolific Straightneck') plants.

These studies indicate that the interactions between natural enemies and pest-resistant plants are varied and complex. However, by elucidating the mechanisms underlying pest resistance and their effect on nontarget natural enemies, it is possible to integrate these two key pest management tactics.

The use of supplemental foods. Most predators require food as adults to produce eggs (Hagen and Bishop 1979). However, not all predaceous species require the same food items in the immature and adult stages; for example, several adult predators are not predaceous but rather feed on nectar, pollen, and honeydew. These adult foods may not be within fields harboring potential prey for predaceous larvae and nymphs. Additionally, some adult Hymenoptera and Diptera require food sources other than their hosts before becoming reproductively active. Thus, the food that the adults need may be absent from monoculture plantings, especially when fields are kept free of weeds and associated plant species (Hagen and Bishop 1979). Growers can improve these conditions by providing artificial foods to either retain, arrest, or stimulate natural enemies to oviposit (Hagen et al. 1971; Hagen and Hale 1974; Hagen 1976).

Scientists participating in Regional Project W-84 have made significant contributions to this area. K. S. Hagen et al. (1976) showed that, in alfalfa plantings where few aphids are present, species of coccinellids and chrysopids would reduce their dispersal/searching movements and aggregate in plantings treated with sucrose solution; in untreated alfalfa plantings they would disperse to other areas. In one day, these sucrose treatments increased the adult densities of *Chrysoperla* (=*Chrysopa*) *carnea* (Stephens) and *Hippodamia* spp. approximately 200- and 20-fold, respectively. Adding attractants to supplementary foods can also manipulate natural enemy populations in the field. Artificial honeydew can be made by combining Wheast—composed of the yeast *Kluyveromyces fragilis* plus a milk-whey substrate—and sugar (Hagen et al. 1976). The main attractant in the artificial honeydew, tryptophan, is highly effective in attracting *C. carnea* adults. Ben Saad and Bishop (1976) found that potato plants sprayed with molasses, honey, and tryptophan—either alone or combined with Wheast—attracted adults of several predator species including *C. carnea*, *Hippodamia* spp., and *Geocoris pallens* Stål.

In addition to attracting natural enemies and arresting their movements, supplemental foods can stimulate egg production and oviposition in the field. K. S. Hagen and R. L. Tassan (1970) found that *C. carnea* egg production could be increased significantly when adults fed on complex artificial diets containing proteins that simulated honeydews. These diets have been applied in field situations to stimulate *C. carnea* egg production (Hagen and Bishop 1979). The oviposition of *Hippodamia* spp. was enhanced by applying artificial honeydews to alfalfa plantings where low aphid densities supplemented the artificial diet (Hagen et al. 1971).

Despite its potential benefits, the use of supplemental food has not been been well investigated. Given the

need for greater conservation of natural enemies in many IPM systems, it appears that this is a promising avenue to explore.

The integration of biological control and pesticides. Conserving biological control agents within agroecosystems where pesticides are used requires a knowledge of pesticide impact on natural enemies, an identification of compounds that are nondisruptive (selective) to natural enemy effectiveness, and an analysis of the logistical and biological factors that influence the effect of pesticides on natural enemies.

The impact of pesticides. S. C. Hoyt, P. H. Westigard, and E. C. Burtis (1978) and M. T. AliNiazee and C. E. Cranham (1980) reported on the impact of pyrethroid sprays on field populations of the predator mites *Galendromus* (=*Metaseiulus*) *occidentalis* and *Typhlodromus pyri* Scheuten on apple and pear. Permethrin and fenvalerate were highly toxic to the predator in comparison with twospotted spider mite, McDaniel spider mite, and European red mite. Significant increases in populations of tetranychid species occurred after applications of permethrin. Using laboratory bioassays, M. A. Hoy and J. Conley (1987) determined the impact of numerous pesticides (for example, azinphosmethyl, benomyl, and carbaryl) commonly used in almonds and several unregistered compounds (such as dicofol and fenvalerate) on two strains of *G. occidentalis*. Their results indicated tremendous variation in the effects of various compounds on predator mite mortality. J. J. Obrycki, M. J. Tauber, and W. M. Tingey (1986) studied the pesticide impact of six compounds on the egg parasitoid *Edovum puttleri*, which attacks the Colorado potato beetle. They found that the mortality rate was affected by the developmental stage and also by the biotype. By taking into account the parasitoid's developmental stage and dosage response, the adverse effects of pesticide applications can be reduced. On tomato, M. W. Johnson, E. R. Oatman, and J. A. Wyman (1980a,b) showed the adverse effect of field applications of the broad-spectrum pesticides methomyl and chlorpyrifos on populations of *Diglyphus begini* (Ashmead) and *Chrysonotomyia punctiventris*, parasitoids of the leafminer *Liriomyza sativae*. Weekly pesticide applications resulted in significantly higher *L. sativae* densities, a result of the destruction of the parasitoids.

Different selection pressures among areas can cause natural enemy responses to pesticides to vary over their geographical range. M. A. Hoy and F. E. Cave (1988) surveyed five wild populations of *Trioxys pallidus* Haliday, a parasitoid of the walnut aphid *Chromaphis juglandicola* (Kaltenbach), to determine variability in responses to dosages of azinphosmethyl. There was an evident variation in responses, but tolerance to azinphosmethyl was slight. Colonies were laboratory selected for azinphosmethyl resistance, which resulted in a 7.5-fold increase in resistance. However, laboratory bioassays indicated that parasitoids would not survive field applications of azinphosmethyl. A field test showed otherwise, suggesting that studies to determine the impact of pesticides on natural enemies should be conducted under field conditions: laboratory tests may overestimate the impact of pesticides. In Hawaii, susceptibilities of the leafminer parasitoids *D. begini*, *C. punctiventris*, *G. utilis*, and *Halticoptera circulus* (Walker) were determined for the pyrethroids permethrin and fenvalerate (Mason and Johnson 1988). Compared with their leafminer hosts, *L. sativae* and *L. trifolii*, these parasitoids showed medium to high tolerances of the compounds tested. *D. begini* had significantly higher LC_{50}'s to both pyrethroids than did the leafminers. Later studies by R. J. Rathman et al. (1990) showed that Hawaiian populations of *D. begini* had significantly higher levels of resistance to fenvalerate (17-fold), permethrin (13-fold), oxamyl (20-fold), and methomyl (22-fold) than did a susceptible California strain. Resistant strains have potential in IPM programs in localities (for example, California and Arizona) where *D. begini* populations are frequently decimated by applications of broad-spectrum pesticides.

The identification of selective pesticides. Compounds that cause high mortality in pest species but not in natural enemy species play an important role in the integration of biological and chemical controls. Regional Project W-84 has done considerable work on identifying selective pesticides in fruit and vegetable IPM programs. S. C. Hoyt (1969) demonstrated that the chemical control of insects and the biological control of mites could be effectively integrated by utilizing selective pesticides that suppressed codling moth, *Cydia pomonella* (L.), but conserved the predatory mite *M. occidentalis*. Low spider mite population densities were needed to maintain *M. occidentalis* in orchards. Improper or ill-timed pesticide applications limited the growth of predator populations. On strawberry, G. G. Kennedy, E. R. Oatman, and V. Voth (1976) demonstrated the selective properties of tricyclohexyltin hydroxide and Pirimor for the control of twospotted mite and aphid pests, respectively. Although effective for controlling target species, these compounds had no measurable adverse effect on the predatory mite *Phytoseiulus persimilis*.

Using the biotic larvicide *Bacillus thuringiensis* Berliner var. *kurstaki* combined with mass releases of the egg parasitoid *Trichogramma pretiosum* Riley provided lepidopterous pest control on tomatoes without increasing populations of the leafminer *Liriomyza sativae*—

unlike methomyl treatments, which, by destroying parasitoids, do increase leafminer populations (Johnson, Oatman, and Wyman 1980a,b; Oatman, Platner et al. 1983). Using the ovicide chlordimeform provided high selectivity for lepidopterous eggs and had less impact on effective natural enemies of the leafminer than did methomyl. G. G. Kennedy, E. R. Oatman, and V. Voth (1976) demonstrated that a combination of Dipel (1.12 kilograms per hectare) and Pirimicarb (0.28 kilograms per hectare) applied to broccoli during the preheading period was as effective as methomyl at 0.5 kilograms per hectare in controlling the lepidopterous pests *Trichoplusia ni*, *Artogeia* (=*Pieris*) *rapae*, *Plutella xylostella*, and the aphids *Brevicoryne brassicae* (L.) and *Myzus persicae*, with less reduction in larval parasitism of *T. ni* and *P. xylostella* by *Copidosoma truncatellum* (Dalman) and *Diadegma insularis* (Cresson), respectively. Although control was not as effective as that provided by methomyl at 1.0 kilograms per hectare, there was no need to keep lepidopterous larval densities near zero given the relatively high economic threshold (about nine larvae per plant) during the preheading stage (Wyman and Oatman 1977).

Chemical alternatives to pesticides. Not all chemicals used to suppress arthropod populations function as pesticides, and some of them have great potential for integration with biological controls. In cotton, T. J. Henneberry et al. (1981) used the pink bollworm sex pheromone "gossyplure" to confuse male moths and disrupt mating, thereby reducing the need for conventional pesticide applications. The researchers confused the males by inundating cotton plantings with vinyl plastic multilayered dispensers (0.05-square-cm flakes), into which gossyplure was formulated. Dispensers were delivered by air at a mean rate of 1.6 flakes per square meter of ground area (equal to 46 grams of gossyplure per hectare). This technique resulted in 68 percent fewer pink bollworm larvae in cotton bolls than could be achieved by a conventional pesticide regime. Given the high specificity of pheromones and their nontoxic effects on natural enemies, mating disruption may be a useful technique in many agroecosystems where conservation of natural enemies is difficult because of frequent pesticide applications.

The introduction of new pesticides into IPM programs. When introducing new pesticides into established, successful IPM systems, one of the least considered—but very important—requirements is a management policy. Incorporating a new compound into an IPM program can be devastating if all economic and ecological ramifications (such as pest upsets) are not considered. B. A. Croft and S. C. Hoyt (1978) described three possible courses of actions when introducing new pesticides into effective IPM systems. The most useful and least disruptive method was for growers to continue with an effective IPM program until pesticide resistance problems appeared in currently used materials, or until the program became economically noncompetitive. Once the program shows signs of ineffectiveness, new pesticides should be carefully incorporated into the system. The two researchers suggested that cost/benefit comparisons be made of programs utilizing new materials versus current materials. These analyses should consider not only direct costs but also the costs of indirect features such as pest upsets and pesticide resistance. J. C. Headley and M. A. Hoy (1987) demonstrated the value of a cost/benefit analysis in their study examining the profitability of using a pesticide-resistant strain of *G. occidentalis* to control *Tetranychus* spp. in combination with lower-than-label acaricide rates in almonds. Their study showed that growers who used the IPM program would save $60 to $110 per hectare, as compared with growers who only used acaricides at label rates.

Given the decline in the development of new pesticide compounds, B. A. Croft and S. C. Hoyt (1978) recommended conserving new materials that become available. Growers can conserve materials by implementing a strategy of resistance management; for example, the effectiveness of environmentally desirable pesticides can be prolonged by using them in a rotation program that limits the selective pressures directed against each chemical. Every IPM program requiring the use of pesticides should consider resistance management. B. A. Croft, S. C. Hoyt, and P. H. Westigard (1987) called for better implementation of existing IPM programs and continued inputs into dynamic, ever-changing IPM systems. These actions would reduce the negative effect of new pesticides on effective IPM systems.

The Conservation of Beneficial Herbivores

Herbicides can affect arthropods as well as the plant species targeted for suppression. Eliminating introduced phytophagous biological controls can result in a resurgence of a given weed species. Thus, research is necessary to improve integration of phytophagous agents with herbicide use. On spotted knapweed, *Centaurea maculosa* Lamarck, J. M. Story, K. W. Boggs, and W. R. Good (1988) determined the optimal timing of the herbicide 2,4-D for compatibility with two seed head flies, *Urophora affinis* (Frauenfeld) and *U. quadrifasciata* (Meigen). They found that herbicide applications made during the knapweed rosette stage in the spring season (May) had no effect on the adult emergence of either fly species. However, applications coinciding with the flower-bud and flowering stages signifi-

cantly reduced the June emergence of *U. affinis*. J. P. McCaffrey and R. H. Callihan (1988) also found that spring applications of the herbicides picloram and 2,4-D and a combination of the two compounds were compatible with the *U. affinis* and *U. quadrifasciata*. Flies were able to develop within knapweed seedheads produced during the spring by plants that had survived fall applications of picloram.

THE AUGMENTATION OF NATURAL ENEMIES

Introduced and indigenous natural enemies are not always effective, even when conservation practices are implemented. Several factors may affect the natural enemy's ability to keep pest levels below economic thresholds. These factors include a low dispersal ability, poor synchrony with pest populations in time and space, the containment of pest populations within closed systems (such as a greenhouse), and disruptive pesticide use. Sometimes these problems can be solved by periodic releases of mass-reared natural enemies. Such releases are termed either "inoculative" or "inundative" (DeBach and Hagen 1964). In inoculative releases, pest control is dependent upon the progeny of the released agent for more than one generation following the release. Pest suppression in inundative releases, in contrast, is achieved mainly by the released agents, not their progeny. Inundatively released natural enemies are thought of as "biotic insecticides"; compared with inoculative releases, the control they provide control is usually short-term. In agroecosystems where pesticide use cannot be stopped or significantly modified, attempts may be made to develop resistant strains of natural enemies through laboratory selection. Although the concept is old, this latter approach is relatively new. Selected strains may be either established in areas or used in mass releases. W-84 scientists have been in the forefront of both of these research areas.

Augmentative Releases of Natural Enemies

E. R. Oatman et al. (1977) conducted studies on releasing the predator mites *Phytoseiulus persimilis, Amblyseius californicus* (McGregor), and *Galendromus* (=*Metaseiulus*) *occidentalis* for twospotted spider mite control on strawberry in southern California. Release rates were 90,720 predators per hectare for each species. *P. persimilis* was found to be more efficient than *A. californicus* and *G. occidentalis*, probably due to its high levels of mobility, prey consumption, and reproductive potential. Although strawberry fruit yields were significantly greater in all release plots as compared with an untreated check, *P. persimilis* release plots exhibited significantly greater yields than the other two release treatments. Strawberry yields recorded in the *P. persimilis* treatment were within the range reported in commercial plantings during that growing season.

On tomatoes in California, E. R. Oatman and G. R. Platner (1971) conducted studies on inundative releases of *Trichogramma pretiosum* for control of *Heliothis zea*, *Trichoplusia ni* (Hübner), and the tobacco hornworm *Manduca sexta* (Johannson). In their 1969 experiment, parasitization of *H. zea*, *T. ni*, and *M. sexta* eggs was about five-, seven-, and tenfold higher, respectively, in release plots as compared with control plots, with 2.1 and 7.2 percent fruit damage in release and control plots, respectively. In later studies on fresh-market tomatoes, inundative releases of *T. pretiosum* controlled the same pest species (Oatman and Platner 1978). In another experiment, augmentative releases of *T. pretiosum* were combined with applications of *Bacillus thuringiensis* Berliner var. *kurstaki* on fresh-market tomatoes. Control of lepidopterous fruit injury was not significantly different from that achieved with weekly applications of methomyl (Oatman, Wyman et al. 1983).

Augmentative releases can be effective in urban and natural habitats as well as in agroecosystems. A major elm pest, the elm leaf beetle *Xanthogaleruca luteola* (Müller), is the third most important urban forest pest in the western United States and fifth in importance nationwide (Kielbaso and Kennedy 1983). *Oomyzus gallerucae* (Fonscolombe) is a eulophid egg parasitoid introduced to control the elm leaf beetle. D. L. Dahlsten et al. (1990) reported that an early spring (May) release of 1,000 *O. gallerucae* individuals in 1986 resulted in 95 percent parasitism of elm leaf beetle by August of that year. Evidence indicated that the release also helped reduce beetle damage to Siberian elm in the release area. Additionally, the inoculative release contributed to reductions in the overwintering elm leaf beetle population: only one *X. luteola* egg cluster was found at the release site in the spring of 1987.

Identifying effective natural enemies that can be economically mass reared for augmentative purposes is also important. J. L. Tracy and J. R. Nechols (1987, 1988) analyzed the biological attributes of *Ooencyrtus anasae* (Ashmead) and *Ooencyrtus* sp., encyrtid egg parasitoids of the squash bug *Anasa tristis* DeGeer, to determine the best species for potential use in augmentative releases. Based on immature development, sex ratio, a faster rate of oocyte production at low temperatures, and a higher innate capacity of increase at high temperatures, *O. anasae* was suggested as the better species to use in augmentative release programs in Kansas. Further analysis, which included two strains of *Ooencyrtus* sp., *O. anasae*, and the scelionid *Gryon pennsylvanicum* (Ashmead), indicated that *G. pennsylvanicum* might be a better candidate for an augmentation program, based on its intrinsic rate of increase, reproductive rate, parasitization rate,

broad range of host stages accepted for parasitism, and survival at temperatures commonly recorded when squash bug is present in the field (Nechols, Tracy, and Vogt 1989).

The Use of Laboratory-Selected Strains

Much of this work has been conducted on natural enemies attacking pests of orchard crops. M. A. Hoy et al. (1984) reported on the effectiveness of mass rearing and releasing a laboratory-selected pesticide-resistant strain of *G. occidentalis* in commercial almond orchards. Applying selective acaricides at lower-than-label rates helped adjust spider mite–predator ratios in favor of the predator. M. A. Hoy, P. H. Westigard, and S. C. Hoyt (1983) released and evaluated a laboratory-selected, pyrethroid- and organophosphate-resistant strain of *G. occidentalis* in Oregon and Washington pear orchards. Predators were established in both locations, but spider mite control was low in Washington because of high permethrin rates. This research demonstrated that laboratory-selected strains can be established in the wild without significantly reducing selected polygenic traits. Releases of a permethrin-OP-resistant strain enabled growers to use pyrethroids against codling moth without promoting the spider mite problems commonly associated with pyrethroid use. This study documented the potential value of developing pesticide-resistant natural enemy strains in advance of the use of new pesticides in specific cropping systems.

THE EVALUATION OF NATURAL ENEMIES

Knowing how natural enemies regulate pest populations within integrated pest management systems helps researchers determine (1) the value and shortcomings of existing natural enemies, (2) the need for introducing new natural enemies into a system, and (3) the need to manipulate the environment (conservation) or natural enemy (augmentation) to make the resident species more effective (DeBach, Huffaker, and MacPhee 1976). The effectiveness of indigenous natural enemies should be appraised before the introduction of exotic agents, in order to assess the need for additional biological control or to identify conditions that preclude effective biological control (Zwölfer, Ghani, and Rao 1976). Natural enemies are introduced to control pest species in urban and natural environments as well as in agroecosystems, and valuable insights can be obtained by examining biological control in urban and natural ecosystems. The following pages will describe in detail some W-84 project contributions to the evaluation of biological control agents of arthropods.

Biological Control in Agroecosystems

D. P. Carroll and S. C. Hoyt (1984) conducted studies on the natural enemies of the apple aphid *Aphis pomi* DeGeer in apple orchards. First they used exclusion cages to determine the natural enemy impact. Two parasitoids, *Lysiphlebus* sp. and *Praon* sp., were observed to oviposit on the aphid, but neither one was able to complete development to the adult stage. These agents provided early-season control of fundatrix aphid colonies. The study provided a better understanding of the role natural enemies play in controlling aphid densities throughout the year. The study also described various habitats with respect to their richness of aphid predators. The number of aphids during the spring season depended on parasitization or predation (*Coccinella transversoguttata* Fald.) in early spring, and on the predation of fall populations the previous year. In summer poor apple aphid control occurred, because predators were not synchronized with aphid populations. Because spring aphid colonies were too small to stimulate predator oviposition, the first seasonal generation of predators developed on different aphid hosts, in habitats other than apple, producing a "predator gap" in apple from early June to mid-July. Apple aphid densities in apple can reach extremely high numbers before predators arrive in late summer. The degree of biological control of apple aphid therefore largely depended on the habitats and prey species surrounding the orchards. By manipulating nearby plant species and orchard ground cover, improved biological control was achieved.

R. H. Messing and M. T. AliNiazee (1985) reported a total of 55 aphidophagous predators in filbert orchards and suggested that collectively they were adequate to provide effective biological control of filbert aphids under undisturbed conditions. They studied the feeding response and control impact of one coccinellid, *Adalia bipunctata* (L.), and three mirid species—*Deraeocoris brevis* (Uhler), *Heterotoma meriopterum* (Scopoli), and *Compsidolon salicellum* (Herrich-Schaeffer) (Messing and AliNiazee 1986). Their data showed that these predators consumed from 5 to 65 aphids daily, with the coccinellid consuming more than the mirids.

It is often extremely difficult to differentiate morphologically between biotypes or races of introduced parasitoids. For example, different biotypes of the parasitoid *Trioxys pallidus* attack the filbert aphid and the walnut aphid in the Pacific Northwest. R. H. Messing and M. T. AliNiazee (1988) found that host-preference characteristics play an important role in the biotype development of *T. pallidus*. They suggested that host morphology and allelochemicals may be important components of the race-formation process in this parasitoid species. Differences in isozyme banding patterns have been used with some success to differentiate biotypes and races often considered to be morphologically indistinguishable (for example, see Unruh et al. 1989). Unfortunately, the number of parasitoids that must be collected or reared for an isozyme analysis can be extremely high and, as a result, time-consuming and expensive. C. H. Pickett et al. (1987, 1989) were able to make use of ecological theory involving parasitoid attack behavior in developing a biological assay for distinguishing geographical races of *Anagrus epos*, a parasitoid of *Erythroneura* spp. on grapes. They found that the geographic race from the San Joaquin Valley of California attacks the grape leafhopper *Erythroneura elegantula* about nine times more frequently than it does the variegated leafhopper, *E. variabilis* Beamer. In contrast, the geographic race from the Coachella Valley attacks the variegated leafhopper about three times more frequently than it does the grape leafhopper. By analyzing races from central and southern California, and from Arizona and Mexico, C. H. Pickett et al. (1987, 1989) were able to detect consistent differences comparing each race. This same technique was then used to follow populations after field release to evaluate which races were able to establish.

E. R. Oatman, J. A. Wyman et al. (1983) evaluated natural enemies attacking *Heliothis zea*, *Trichoplusia ni*, *Spodoptera exigua* and *Manduca sexta* in southern California tomato plantings. They found that parasitization rates of *H. zea* larvae ranged from 72.5 to 89.5 percent and egg parasitization ranged from 51.5 to 55.7 percent. *M. sexta* was poorly parasitized with egg and larval parasitization, ranging from 41.9 to 49.5 percent and 0 to 14.1 percent, respectively. Larval parasitization rates for *T. ni* and *S. exigua* fell between those for *M. sexta* and *H. zea*. The ichneumonid *Hyposoter exiguae* Viereck was the most common parasitoid reared from larvae, and only *T. pretiosum* was reared from eggs. Efforts to conserve or augment these natural enemies should be a prime objective of tomato IPM programs in southern California. Additional studies on fresh-market tomatoes examined parasitization levels of the tomato pinworm *Keiferia lycopersicella* (Oatman, Wyman, and Platner. 1979). Seasonal parasitization levels were low (1.6 to 36.8 percent); *Sympiesis stigmatipennis* Girault was the parasitoid most commonly reared from collected larvae. These studies indicated that additional control measures (for example, pesticides and the pheromone confusion technique) were necessary to prevent economic losses.

In Hawaii, studies on *Liriomyza* leafminers showed significantly different numbers of parasitoid species attacking leafminers infesting onions compared with adjacent plantings of beans (Johnson and Mau 1986; Herr 1987). The varying abundance of leafminer parasitoids among host plants infested by leafminers may result from differences in host habitat preferences (Johnson and Hara 1987).

T. J. Henneberry and T. E. Clayton (1982a) quantified levels of pink bollworm egg predation in Arizona cotton during a field season. Eggs oviposited on an artificial substrate in the laboratory were placed in the field for a three-day period on each of eight test dates ranging from July to September. Egg consumption by general predators (for example, *Chrysopa*, *Nabis*, *Geocoris*, and *Orius* spp.) ranged from 35 to 95 percent, and was greater than 55 percent during seven of the eight test periods.

Another study assessed biological control of the potato tuberworm *Phthorimaea operculella* (Zeller) in southern California, based on larvae collected from foliage (Oatman and Platner 1974). Thirteen parasitoid species were reared, of which only the braconid *Orgilus lepidus* Musesbeck was introduced (from Argentina, in 1963 and 1964). Seasonal parasitization levels ranged from 29.4 to 77.4 percent. Despite relatively high levels of parasitization in some cases, exposed potato tubers were commonly infested by the tuberworm. These results indicated that biological control was of only partial value, and that preventive cultural measures were also necessary for control.

The Evaluation of Biological Control in Urban and Natural Systems

Two coccinellid predators, *Diomus pumilio* Weise and *Harmonia conformis* (Boisduval), were introduced to control infestations of the acacia psyllid, *Psylla uncatoides* (Ferris & Klyver), on ornamental and shade trees and on native *Acacia* species in California and Hawaii, respectively (Leeper and Beardsley 1976). Evaluation studies on the island of Hawaii showed that *D. pumilio* did not establish, but significant reductions in acacia psyllid populations on *Acacia* spp. followed the introduction of *H. conformis*. Densities of psyllid eggs and small nymphs dropped by more than two-thirds, and serious twig terminal dieback was eliminated on surveyed trees.

In California, only *D. pumilio* was established, but a distinct reduction in psyllid numbers was observed. Biological control by this predator has saved the California Department of Transportation more than $55,000 per year in reduced pesticide applications to acacias lining freeways (Pinnock et al. 1978). These studies point out the importance of multiple-species introductions to achieve biological control in areas of widely diverse habitat.

The introduced parasitoids *Oomyzus brevistigma* (Gahan) and *Erynniopsis antennata* (Rondani) were examined with respect to their impact on the elm leaf beetle infesting elm trees in northern California urban forests. Studies indicated that the eulophid pupal parasitoid *O. brevistigma* was patchily distributed in northern California, with mean and maximum apparent parasitism rates of 1.2 and 22 percent, respectively (Dreistadt and Dahlsten 1990). This natural enemy was considered to be of little benefit in suppressing elm leaf beetle populations in northern and southern California (Dahlsten et al. 1990). The tachinid larval and larval-adult parasitoid *E. antennata* was found in 11 of 12 northern California survey sites during 1986 and 1987. At four of them it exhibited a maximum apparent parasitism of elm leaf beetle greater than 40 percent. According to R. F. Luck and G. T. Scriven (1976), *E. antennata* was significant in reducing elm leaf beetle densities in southern California, but its effectiveness was limited by a secondary parasitoid, *Baryscapus erynniae* (Domenichini), and by a lack of synchrony with its host's first generation. Further studies are necessary to completely assess the impact of *E. antennata* on elm leaf beetle in northern California.

Studies were also conducted to understand the effectiveness of parasitoids attacking gall midges developing within terminal galls on coyote brush, *Baccharis pilularis* DC. L. E. Ehler (1987) suggested that quantifying the patch-exploitation efficiency of natural enemies with regard to their host's habitats might provide evidence of natural enemy effectiveness. Natural enemies that exploit host patches with the highest densities may be more effective than biological control agents whose patch exploitation is not correlated with host density.

This may prove a valuable prediction tool for both classical biological control introductions and augmentation programs. Other studies of this system indicate that species of parasitoids that are ineffective in their native homes may nonetheless be of value in classical biological control and inundative release programs if their ineffectiveness is due largely to interspecific interactions rather than to low reproductive capacity (Ehler 1986). Given this premise, it may be possible to restructure natural enemy guilds to alter their overall effectiveness in controlling an individual organism (Ehler 1985).

CONCLUSIONS

Scientists participating in Regional Project W-84 have made numerous contributions to the conceptual development and implementation of IPM systems that feature biological control as a major component. They have made significant advances in natural enemy conservation by identifying agronomic factors that impede biological control and developing practical strategies to circumvent those problems. The scientists have also developed cost-effective sampling programs for a range of crops, which contributes significantly to reducing the use of pesticides. To a lesser degree, they have made advances in developing economic thresholds that take into account environmental and biotic factors. Given the loss of previously effective compounds to eventual pest resistance, the decline in available pesticide controls due to governmental regulations (in turn contributing to the lack of development of new materials), and the national campaign to reduce chemical inputs in agriculture (namely, low-input sustainable agriculture), the demand for effective IPM programs will grow. Biological control will become the foundation for many of these programs, due in part to the contributions of Regional Project W-84.

REFERENCES CITED

Adjei-Maafo, I. K., and L. T. Wilson. 1983. Factors affecting the relative abundance of arthropods on nectaried and nectariless cotton. *Environ. Entomol.* 12:349–52.

AliNiazee, M. T. 1978. The western cherry fruit fly. 3. Developing a management program by utilizing attractant traps as monitoring devices. *Can. Entomol.* 110:1133–39.

———. 1980. Filbert insect and mite pests. Oregon State Agricultural Experiment Station Bulletin 643.

———. 1983. Pest status of filbert (hazelnut) insects: A ten-year study. *Can. Entomol.* 115:1155–62.

AliNiazee, M. T., and C. E. Cranham. 1980. Effect of four synthetic pyrethroids on the predatory mite *Typhlodromus pyri* and its prey *Panonychus ulmi* on apple in southeastern England. *Environ. Entomol.* 9:436–39.

AliNiazee, M. T., and R. L. Westcott. 1987. Flight period and seasonal development of the apple maggot, *Rhagoletis pomonella,* in Oregon. *Ann. Entomol. Soc. Am.* 80:823–28.

AliNiazee, M. T., A. B. Mohammad, and S. R. Booth. 1987. Apple maggot response to traps in an unsprayed orchard in Oregon. *J. Econ. Entomol.* 80:1143–48.

Antle, J. M., and S. K. Park. 1986. The economics of IPM in processing tomatoes. *Calif. Agric.* 40(3–4):31–32.

Awan, M. S., L. T. Wilson, and M. P. Hoffmann. 1990. Comparative biology of three geographic populations of *Trissolcus basalis* (Wollaston) (Scelionidae: Hymenoptera). *Environ. Entomol.* 19:387–92.

Ben Saad, A. A., and G. W. Bishop. 1976. Effect of artificial honeydews on insect communities in potato fields. *Environ. Entomol.* 5:453–57.

Braun, A. R., J. M. Guerrero, A. C. Bellotti, and L. T. Wilson. 1989. Within-plant distribution of *Mononychellus tanajao* (Bondar) (Acari: Tetranychidae) on cassava: Effect of clone and predation on aggregation. *Bull. Entomol. Res.* 79:235–49.

Carroll, D. P., and S. C. Hoyt. 1984. Natural enemies and their effects on apple aphid, *Aphis pomi* DeGeer (Homoptera: Aphididae), colonies on young apple trees in central Washington. *Environ. Entomol.* 13:469–81.

———. 1986. Hosts and habitats of parasitoids (Hymenoptera: Aphidiidae) implicated in biological control of apple aphid (Homoptera: Aphidiidae). *Environ. Entomol.* 15:1171–78.

Croft, B. A., and S. C. Hoyt. 1978. Considerations for the use of pyrethroid insecticides for deciduous fruit pest control in the U.S.A. *Environ. Entomol.* 7:627–30.

Croft, B. A., S. C. Hoyt, and P. H. Westigard. 1987. Spider mite management on pome fruits, revisited: Organotin and acaricide resistance management. *J. Econ. Entomol.* 80:304–11.

Dahlsten, D. L., S. H. Dreistadt, J. R. Geiger, S. M. Tait, D. L. Rowney, G. Y. Yokota, and W. A. Copper. 1990. Elm leaf beetle biological control and management in northern California. Final Report to California Department of Forestry and Fire Protection.

Davis, D. W., S. C. Hoyt, J. A. McMurtry, and M. T. AliNiazee. 1979. *Biological control and insect pest management.* Oakland: University of California Division of Agricultural Sciences, Publication 4096.

DeBach, P. 1964a. Successes, trends, and future possibilities. In *Biological control of insect pests and weeds,* ed. P. DeBach, 673–713. London: Chapman & Hall.

———. 1964b. The scope of biological control. In *Biological control of insect pests and weeds,* ed. P. DeBach, 3–20. London: Chapman & Hall.

DeBach, P., and K. S. Hagen. 1964. Manipulation of entomophagous species. In *Biological control of insect pests and weeds,* ed. P. DeBach, 429–58. London: Chapman & Hall.

DeBach, P., C. B. Huffaker, and A. W. MacPhee. 1976. Evaluation of the impact of natural enemies. In *Theory and practice of biological control,* eds. C. B. Huffaker and P. S. Messenger, 255–85. New York: Academic Press.

Doutt, R. L. 1964. The historical development of biological control. In *Biological control of insect pests and weeds,* ed. P. DeBach, 21–42. London: Chapman & Hall.

Dover, M., and B. Croft. 1984. *Getting tough: Public policy and the management of pesticide resistance.* Study 1, World Resources Institute.

Dreistadt, S. H., and D. L. Dahlsten. 1989. Density-damage relationship and presence-absence sampling of the elm leaf beetle, *Xanthogaleruca luteola* (Coleoptera: Chrysomelidae), in northern California. *Environ. Entomol.* 18:849–53.

———. 1990. Distribution and abundance of *Erynniopsis antennata* (Diptera: Tachinidae) and *Tetrastichus brevistigma* (Hymenoptera: Eulophidae), two intoduced elm leaf beetle parasitoids in northern California. *Entomophaga* 35:527–36.

Ehler, L. E. 1985. Species-dependent mortality in a parasite guild and its relevance to biological control. *Environ. Entomol.* 14:1–6.

———. 1986. Distribution of progeny in two ineffective parasites of a gall midge (Diptera: Cecidomyiidae). *Environ. Entomol.* 15:1268–71.

———. 1987. Patch-exploitation efficiency in a torymid parasite (Hymenoptera: Torymidae) of a gall midge (Diptera: Cecidomyiidae). *Environ. Entomol.* 16:198–201.

Ellington, J., K. Kiser, M. Cardenas, J. Duttle, and Y. López. 1984. The insectavac: A high-clearance, high-volume arthropod vacuuming platform for agricultural ecosystems. *Environ. Entomol.* 13:259–65.

Grieshop, J. I., F. G. Zalom, and G. Miyao. 1986. Exploratory study on the adoption of the IPM tomato worm monitoring program by tomato growers in Yolo County: Descriptive statistics. IPM Implementation Group, Davis: University of California Division of Agriculture and Natural Resources.

Hagen, K. S. 1976. Role of nutrition in insect management. *Tall Timbers Conf. Ecol. Animal Contr. by Habitat Management* 6:221–61.

Hagen, K. S., and G. W. Bishop. 1979. Use of supplemental foods and behavioral chemicals to increase the effectiveness of natural enemies. In *Biological control and insect pest management,* eds. D. W. Davis, S. C. Hoyt, J. A. McMurtry, and M. T. AliNiazee, 49–60. Oakland: University of California Division of Agricultural Sciences, Publication 4096.

Hagen, K. S., and R. Hale. 1974. Increasing natural enemies through use of supplementary feeding and non-target prey. In *Proceedings of the Summer Institute for the Biological Control of Plant Insects and Diseases,* eds. F. C. Maxwell and F. A. Harris, 170–81. Jackson: University of Missouri Press.

Hagen, K. S, and R. L. Tassan. 1970. The influence of Food Wheast and related *Saccharomyces fragilis* yeast product on the fecundity of *Chrysopa carnea. Can. Entomol.* 102:808–11.

Hagen, K. S., E. F. Sawall, Jr., and R. L. Tassan. 1970. The use of food sprays to increase effectiveness of entomophagous insects. *Tall Timbers Conf. Ecol. Animal Contr. by Habitat Management* 2:59–81.

Hagen, K. S., P. Greany, E. F. Sawall, Jr., and R. L. Tassan. 1976. Tryptophan in artificial honeydews as a source of attractant for adult *Chrysopa carnea. Environ. Entomol.* 5:458–68.

Hassan, S. T. S., L. T. Wilson, and P. R. B. Blood. 1990. *Heliothis armigera* (Hübner) and *H. punctigera* (Wallengren) (Lepidoptera: Noctuidae) oviposition on okra-leaf and smooth-leaf cotton. *Environ. Entomol.* 19:710–16.

Headley, J. C., and M. A. Hoy. 1987. Benefit/cost analysis of an integrated mite management program for almonds. *J. Econ. Entomol.* 80:555–59.

Henneberry, T. J., and T. E. Clayton. 1982a. Pink bollworm: Seasonal oviposition, egg predation, and square and boll infestations in relation to cotton plant development. *Environ. Entomol.* 11:663–66.

———. 1982b. Pink bollworm of cotton (*Pectinophora gossypiella* [Saunders]): Male moth catches in gossyplure-baited traps and relationships to oviposition, boll infestation, and moth emergence. *Crop Protect.* 1:497–504.

Henneberry, T. J., L. A. Bariola, and D. L. Kittock. 1977. Nectariless cotton: Effect on cotton leafperforator and other cotton insects in Arizona. *J. Econ. Entomol.* 70:797–99.

Henneberry, T. J., J. M. Gillespie, L. A. Bariola, H. M. Flint, P. D. Lingren, and A. F. Kydonieus. 1981. Gossyplure in laminated plastic formulations for mating disruption and pink bollworm control. *J. Econ. Entomol.* 74:376–81.

Herr, J. C. 1987. Influence of intercropping on the biological control of *Liriomyza* leafminers in green onions. Master's thesis, University of Hawaii at Manoa, Honolulu.

Hoffmann, M. P., L. T. Wilson, and F. G. Zalom. 1987. Control of stink bugs in tomatoes. *Calif. Agric.* 41(5–6):4–6.

Hoffmann, M. P., L. T. Wilson, F. G. Zalom, and L. McDonough. 1986. Lures and traps for monitoring tomato fruitworm. *Calif. Agric.* 40(9–10):17–18.

Hoy, M. A. 1985. Almonds (California). In *Spider mites: Their biology, natural enemies and control,* Vol. 1, eds. B. W. Helle and M. W. Sabelis, 299–310. Amsterdam: Elsevier Science Publications.

Hoy, M. A., and F. E. Cave. 1988. Guthion-resistant strain of walnut aphid parasite. *Calif. Agric.* 42(4):4–5.

Hoy, M. A., and J. Conley. 1987. Toxicity of pesticides to western predatory mite. *Calif. Agric.* 41(7–8):12–14.

Hoy, M. A., P. H. Westigard, and S. C. Hoyt. 1983. Release and evaluation of a laboratory-selected, pyrethroid-resistant strain of the predaceous mite *Metaseiulus occidentalis* (Acari: Phytoseiidae) in southern Oregon pear orchards and a Washington apple orchard. *J. Econ. Entomol.* 76:383–88.

Hoy, M. A., W. W. Barnett, L. C. Hendricks, D. Castro, D. Cahn, and W. J. Bentley. 1984. Managing spider mites in almonds with pesticide-resistant predators. *Calif. Agric.* 38(7–8):18–20.

Hoyt, S. C. 1969. Integrated chemical control of insects and biological control of mites on apple in Washington. *J. Econ. Entomol.* 62:74–86.

Hoyt, S. C., L. K. Tanigoshi, and R. W. Browne. 1979. Economic injury level in relation to mites on apple. *Recent Adv. Acarology* 1:3–12.

Hoyt, S. C., P. H. Westigard, and E. C. Burtis. 1978. Effects of two synthetic pyrethroids on the codling moth, pear psylla, and various mite species in northeast apple and pear orchards. *J. Econ. Entomol.* 71:431–34.

Huffaker, C. B. 1985. Biological control in integrated pest management: An entomological perspective. In *Biological control in agricultural IPM systems,* eds. M. A. Hoy and D. C. Herzog, 67-88. New York: Academic Press.

Hutchison, W. D., C. A. Beasley, T. J. Henneberry, and J. M. Martin. 1988. Sampling pink bollworm (Lepidoptera: Gelechiidae) eggs: Potential for improved timing and reduced use of insecticides. *J. Econ. Entomol.* 81:673–78.

Johnson, M. W. 1987. Parasitization of *Liriomyza* spp. (Diptera: Agromyzidae) infesting commercial watermelon plantings in Hawaii. *J. Econ. Entomol.* 80:56–61.

Johnson, M. W., and A. H. Hara. 1987. Influence of host crop on parasitoids (Hymenoptera) of *Liriomyza* spp. (Diptera: Agromyzidae). *Environ. Entomol.* 16:339–44.

Johnson, M. W., and R. F. L. Mau. 1986. Effects of intercropping beans and onions on populations of *Liriomyza* spp. and associated parasitic Hymenoptera. *Proc. Hawaii. Entomol. Soc.* 27:95–103.

Johnson, M. W., E. R. Oatman, and J. A. Wyman. 1980a. Effects of insecticides on populations of the vegetable leafminer and associated parasites on summer pole tomatoes. *J. Econ. Entomol.* 73:61–66.

———. 1980b. Effects of insecticides on populations of the vegetable leafminer and associated parasites on fall pole tomatoes. *J. Econ. Entomol.* 73:67–71.

Johnson, M. W., R. F. L. Mau, A. P. Martinez, and S. Fukuda. 1989. Foliar pests of watermelon in Hawaii. *Trop. Pest Manage.* 35:90–96.

Johnson, M. W., E. R. Oatman, J. A. Wyman, and R. A. Van Steenwyk. 1980. A technique for monitoring *Liriomyza sativae* in fresh market tomatoes. *J. Econ. Entomol.* 73:552–55.

Jones, V. P. 1990. Sampling and dispersion of the twospotted spider mite (Acari: Tetranychidae) and the western orchard predatory mite (Acari: Phytoseiidae) on tart cherry. *J. Econ. Entomol.* 83:1376–80.

Jones, V. P., and D. W. Davis. 1989. Evaluation of traps for apple maggot (Diptera: Tephritidae) populations associated with cherry and hawthorn in Utah. *Environ. Entomol.* 18:521–25.

Jones, V. P., D. W. Davis, S. L. Smith, and D. B. Allred. 1989. Phenology of apple maggot (Diptera: Tephritidae) associated with cherry and hawthorn in Utah. *J. Econ. Entomol.* 82:788–92.

Kennedy, G. G., and E. R. Oatman. 1976. *Bacillus thuringiensis* and Pirimicarb: Selective insecticides for use in pest management on broccoli. *J. Econ. Entomol.* 69:767–72.

Kennedy, G. G., E. R. Oatman, and V. Voth. 1976. Suitability of Plictran and Pirimor for use in a pest management program on strawberries in southern California. *J. Econ. Entomol.* 69:269–72.

Kido, H., D. L. Flaherty, D. F. Bosch, and K. A. Valero. 1984. French prune trees as overwintering sites for the grape leafhopper egg parasite. *Am. J. Enol. Vitic.* 35:156–60.

Kielbaso, J. J., and M. K. Kennedy. 1983. Urban forestry and entomology: A current appraisal. In *Urban entomology: Interdisciplinary perspectives,* eds. G. W. Frankie and C. S. Koehler, 423–40. New York: Praeger.

Leeper, J. R. 1974. Adult feeding behavior of *Lixophaga sphenophori,* a tachinid parasite of the New Guinea sugarcane weevil. *Proc. Hawaii. Entomol. Soc.* 21:403–12.

Leeper, J. R., and J. W. Beardsley, Jr. 1976. The biological control of *Psylla uncatoides* (Ferris & Klyver) (Homoptera: Psyllidae) on Hawaii. *Proc. Hawaii. Entomol. Soc.* 22:307–21.

Leigh, T. F., V. L. Maggi, and L. T. Wilson. 1984. Development and use of a machine for recovery of arthropods from plant leaves. *J. Econ. Entomol.* 77: 271–76.

Luck, R. F., and G. T. Scriven. 1976. The elm leaf beetle, *Pyrrhalta luteloa,* in southern California: Its pattern of increase and its control by introduced parasites. *Environ. Entomol.* 5:409–16.

Lynch, J. A., and M. W. Johnson. 1987. Stratified sampling of *Liriomyza* spp. and associated hymenopterous parasites on watermelon. *J. Econ. Entomol.* 80:1254–61.

Mason, G. A., and M. W. Johnson. 1988. Tolerance to permethrin and fenvalerate in hymenopterous parasitoids associated with *Liriomyza* spp. (Diptera: Agromyzidae). *J. Econ. Entomol.* 81:123–26.

McCaffrey, J. P., and R. H. Callihan. 1988. Compatibility of picloram and 2,4-D with *Urophora affinis* and *U. quadrifasciata* (Diptera: Tephritidae) for spotted knapweed control. *Environ. Entomol.* 17:785–88.

McMurtry, J. A., E. R. Oatman, P. A. Phillips, and C. W. Wood. 1978. Establishment of *Phytoseiulus persimilis* (Acari: Phytoseiidae) in southern California. *Entomophaga* 23:175–79.

Messing, R. H., and M. T. AliNiazee. 1985. Natural enemies of filbert aphids. *J. Entomol. Soc. Brit. Columbia* 82:14–18.

———. 1986. Impact of predaceous insects on the filbert aphid, *Myzocallis coryli,* in western Oregon. *Environ. Entomol.* 15:1037–41.

———. 1988. Hybridization and host suitability of *Trioxys pallidus. Ann. Entomol. Soc. Am.* 81:6–9.

———. 1989. Introduction and establishment of *Trioxys pallidus* in Orgeon, U.S.A., for control of filbert aphid *Myzocallis coryli* (Homoptera: Aphididae). *Entomophaga* 34:153–63.

National Research Council. 1989. *Alternative agriculture.* Washington, D.C.: National Academy Press.

Nechols, J. R., J. L. Tracy, and E. A. Vogt. 1989. Comparative ecological studies of indigenous egg parasitoids (Hymenoptera: Scelionidae; Encyrtidae) of the squash bug, *Anasa tristis* (Hemiptera: Coreidae). *J. Kansas Entomol. Soc.* 62:177–88.

Oatman, E. R., and G. R. Platner. 1971. Biological control of the tomato fruitworm, cabbage looper, and hornworms on processing tomatoes in southern California, using mass releases of *Trichogramma pretiosum. J. Econ. Entomol.* 64:501–6.

———. 1972. Colonization of *Trichogramma evanescens* and *Apanteles rubecula* on the imported cabbageworm on cabbage in southern California. *Environ. Entomol.* 1:348–51.

———. 1974. Parasitization of the potato tuberworm in southern California. *Environ. Entomol.* 3:262–64.

———. 1978. Effect of mass releases of *Trichogramma pretiosum* against lepidopterous pests on processing tomatoes in southern California, with notes on host egg population trends. *J. Econ. Entomol.* 71:896–900.

Oatman, E. R., J. A. Wyman, and G. R. Platner. 1979. Seasonal occurrence and parasitization of the tomato pinworm on fresh market tomatoes in southern California. *Environ. Entomol.* 8:661–64.

Oatman, E. R., J. A. McMurtry, F. E. Gilstrap, and V. Voth. 1977. Effect of releases of *Amblyseius californicus, Phytoseiulus persimilis,* and *Typhlodromus occidentalis* on the twospotted spider mite on strawberry in southern California. *J. Econ. Entomol.* 70:45–47.

Oatman, E. R., J. A. Wyman, H. W. Browning, and V. Voth. 1981. Effects of releases and varying infestation levels of the twospotted spider mite on strawberry yield in southern California. *J. Econ. Entomol.* 74:112–15.

Oatman, E. R., J. A. Wyman, R. A. Van Steenwyk, and M. W. Johnson. 1983. Integrated control of the tomato fruitworm (Lepidoptera: Noctuidae) and other lepidopterous pests on fresh market tomatoes in southern California. *J. Econ. Entomol.* 76:1363–69.

Oatman, E. R., F. V. Sances, L. F. LaPre, N. C. Toscano, and V. Voth. 1982. Effects of different infestation levels of the twospotted spider mite on strawberry yield in winter plantings in southern California. *J. Econ. Entomol.* 75:94–96.

Oatman, E. R., G. R. Platner, J. A. Wyman, R. A. Van Steenwyk, M. W. Johnson, and H. W. Browning. 1983. Parasitization of lepidopterous pests on fresh market tomatoes in southern California. *J. Econ. Entomol.* 76:452–55.

Obrycki, J. J., and M. J. Tauber. 1984. Natural enemy activity on glandular pubescent potato plants in the greenhouse: An unreliable predictor of effects in the field. *Environ. Entomol.* 13:679–83.

———. 1985. Seasonal occurrence and abundance of aphid predators and parasitoids on pubescent potato plants. *Can. Entomol.* 117:1231–37.

Obrycki, J. J., M. J. Tauber, and W. M. Tingey. 1983. Predator and parasitoid interaction with aphid-resistant potatoes to reduce aphid densities: A two-year field study. *J. Econ. Entomol.* 76:456–62.

———. 1986. Comparative toxicity of pesticides to *Edovum puttleri* (Hymenoptera: Eulophidae), an egg parasitoid of the Colorado potato beetle (Coleoptera: Chrysomelidae). *J. Econ. Entomol.* 79:948–51.

Parrella, M. P. 1987. Biology of *Liriomyza. Annu. Rev. Entomol.* 32:201–24.

Pickel, C., R. C. Mount, F. G. Zalom, and L. T. Wilson. 1983. Monitoring aphids on Brussels sprouts. *Calif. Agric.* 37(5–6):24–25.

Pickett, C. H., L. T. Wilson, and D. L. Flaherty. 1990. Role of refuges in crop protection, with reference to plantings of French prune trees in a grape agroecosystem. In *Monitoring and integrated management of arthropod pests of small fruit crops,* eds. N. J. Bostanian, L. T. Wilson, and T. J. Dennehy. Dorset, U.K.: Intercept Press.

Pickett, C. H., L. T. Wilson, and D. González. 1988. Population dynamics and within-plant distribution of the western flower thrips (Thysanoptera: Thripidae), an early-season predator of spider mites infesting cotton. *Environ. Entomol.* 17:551–59.

Pickett, C. H., L. T. Wilson, D. L. Flaherty, and D. González. 1989. Measuring the host preference of parasites: An aid in evaluating biotypes of *Anagrus epos* (Hym: Mymaridae). *Entomophaga* 34(4):551–58.

Pickett, C. H., L. T. Wilson, D. González, and D. L. Flaherty. 1987. Biological control of variegated grape leafhopper. *Calif. Agric.* 41(7–8):14–16.

Pinnock, D. E., K. S. Hagen, D. V. Cassidy, R. J. Brand, J. E. Milstead, and R. L. Tassan. 1978. Integrated pest management in highway landscapes. *Calif. Agric.* 32(2):33–34.

Plant, R. E., and L. T. Wilson. 1985. A Bayesian method for sequential sampling and forecasting in agricultural pest management. *Biometrics* 41:203–14.

Rathman, R. J., M. W. Johnson, J. A. Rosenheim, and B. E. Tabashnik. 1990. Carbamate and pyrethroid resistance in the leafminer parasitoid *Diglyphus begini* (Hymenoptera: Eulophidae). *J. Econ. Entomol.* 83:2153–58.

Room, P. M. 1979. A prototype 'on-line' system for management of cotton pests in the Namoi Valley, New South Wales. *Protect. Ecol.* 1:245–61.

Schreiner, I., D. Nafus, and Claron Bjork. 1986. Control of *Liriomyza trifolii* (Burgess) (Diptera: Agromyzidae) on yard-long (*Vigna unguiculata*) and pole beans (*Phaseolus ulgaris*) on Guam: Effect on yield loss and parasite numbers. *Trop. Pest Manage.* 32:333–37.

Sterling, W. L., K. M. El-Zik, and L. T. Wilson. 1989. Biological control of pest populations. In *Integrated pest management in cotton production,* eds. R. E. Frisbie, K. M. El-Zik, and L. T. Wilson, Chap. 7, 155–90. New York: Wiley.

Story, J. M., K. W. Boggs, and W. R. Good. 1988. Optimal timing of 2,4-D applications for compatibility with *Urophora affinis* and *U. quadrifasciata* (Diptera: Tephritidae) for control of spotted knapweed. *Environ. Entomol.* 17:911–14.

Topham, M., and J. W. Beardsley. 1975. Influence of nectar source plants on the New Guinea sugarcane weevil parasite, *Lixophaga sphenophori* (Villeneuve). *Proc. Hawaii. Entomol. Soc.* 22:145–54.

Tracy, J. L., and J. R. Nechols. 1987. Comparisons between the squash bug egg parasitoids *Ooencyrtus anasae* and *O.* sp. (Hymenoptera: Encyrtidae): Development, survival, and sex ratio in relation to temperature. *Environ. Entomol.* 16:1324–29.

———. 1988. Comparison of thermal responses, reproductive biologies, and population growth potentials of the squash bug egg parasitoids *Ooencyrtus anasae* and *O.* sp. (Hymenoptera: Encyrtidae). *Environ. Entomol.* 17:636–43.

Trumble, J. T., E. R. Oatman, and V. Voth. 1983a. Development and estimation of aphid populations infesting annual winter plantings of strawberries in California. *J. Econ. Entomol.* 76:496–501.

———. 1983b. Thresholds and sampling for aphids in strawberries. *Calif. Agric.* 37(11–12):20–21.

Unruh, T., W. White, D. González, and J. Woolley. 1989. Genetic relationships among 17 *Aphidius* populations including six species. *Ann. Entomol. Soc. Am.* 82:754–768.

Van Steenwyk, R. A., E. R. Oatman, N. C. Toscano, and J. A. Wyman. 1983. Pheromone traps to time tomato pinworm control. *Calif. Agric.* 37(7–8):22–24.

Waterhouse, D. F., and K. R. Norris. 1987. *Biological control: Pacific prospects.* Melbourne: Inkata Press.

Wilson, F., and C. B. Huffaker. 1976. The philosophy, scope, and importance of biological control. In *Theory and practice of biological control,* eds. C. B. Huffaker and P. S. Messenger, 3–15. New York: Academic Press.

Wilson, L. T. 1982. Development of an optimal monitoring program in cotton: Emphasis on spider mites and *Heliothis* spp. *Entomophaga* 27:45–50.

———. 1985. Estimating the abundance and impact of arthropod natural enemies in IPM systems. In *Biological control in agricultural IPM systems,* eds. M. A. Hoy and D. C. Herzog, 303–22. New York: Academic Press.

Wilson, L. T., and A. P. Gutierrez. 1980. Within-plant distribution of predators on cotton: Comments on sampling and predator efficiencies. *Hilgardia* 48:3–11.

Wilson, L. T., and P. M. Room. 1982. The relative efficiency and reliability of three methods for sampling arthropods in Australian cotton fields. *J. Aust. Entomol. Soc.* 21:175–81.

———. 1983. Clumping patterns of fruit and arthropods in cotton, with implications for binomial sampling. *Environ. Entomol.* 12:50–54.

Wilson, L. T., D. González, and R. E. Plant. 1985. Predicting sampling frequency and economic status of spider mites on cotton. In *Proc. 1985 Beltwide Cotton Prod. Res. Conf., New Orleans, La.,* eds. T. Cotton Nelson and J. M. Brown, 168–70. Memphis, Tenn.: National Cotton Council of America.

Wilson, L. T., A. P. Gutierrez, and D. B. Hogg. 1982. Within-plant distribution of cabbage looper, *Trichoplusia ni* (Hübner) on cotton: Development of a sampling plan for eggs. *Environ. Entomol.* 11:251–54.

Wilson, L. T., A. P. Gutierrez, and T. F. Leigh. 1980. Within-plant distribution of the immatures of *Heliothis zea* (Boddie) on cotton. *Hilgardia* 48:12–23.

Wilson, L. T., T. F. Leigh, and V. Maggi. 1981. Presence-absence sampling of spider mite densities on cotton. *Calif. Agric.* 35(7–8):10.

Wilson, L. T., M. A. Hoy, F. G. Zalom, and J. M. Smilanick. 1984. Sampling mites in almonds: I. The within-tree distribution and clumping pattern of mites with comments on predator-prey interactions. *Hilgardia* 52:1–13.

Wilson, L. T., T. F. Leigh, D. González, and C. Foristiere. 1983. Distribution of *Lygus hesperus* (Knight) (Miridae: Hemiptera) on cotton. *J. Econ. Entomol.* 77:1313–19.

Wilson, L. T., C. H. Pickett, D. L. Flaherty, and T. A. Bates. 1989. French prune trees: Refuge for grape leafhopper parasite. *Calif. Agric.* 43(2):7–8.

Wilson, L. T., C. H. Pickett, T. F. Leigh, and J. R. Carey. 1987. Spider mite (Acari: Tetranychidae) infestation foci: Cotton yield reduction. *Environ. Entomol.* 16:614–17.

Wilson, L. T., D. González, T. F. Leigh, V. Maggi, C. Foristiere, and P. Goodell. 1983. Within-plant distribution of spider mites (Acari: Tetranychidae) on cotton: A developing implementable monitoring program. *Environ. Entomol.* 12:128–34.

Wilson, L. T., F. G. Zalom, R. Smith, and M. P. Hoffman. 1983. Monitoring for fruit damage in processing tomatoes: Use of a dynamic sequential sampling plan. *Environ. Entomol.* 12:835–39.

Wilson, L. T., W. L. Sterling, D. R. Rummel, R. E. Frans, and J. E. DeVay. 1989. Quantitative sampling principles in cotton IPM. In *Integrated pest management in cotton production,* eds. R. E. Frisbie, K. M. El-Zik, and L. T. Wilson, chap. 5, 85–120. New York: Wiley.

Wilson, L. T., C. Pickel, R. C. Mount, and F. G. Zalom. 1983. Presence-absence sequential sampling for cabbage aphid and green peach aphid (Homoptera: Aphididae) on Brussels sprouts. *J. Econ. Entomol.* 76:476–79.

Wyman, J. A., and E. R. Oatman. 1977. Yield responses in broccoli plantings with *Bacillus thuringiensis* at various lepidopterous larval density treatment levels. *J. Econ. Entomol.* 70:821–24.

Wyman, J. A., E. R. Oatman, and V. Voth. 1979. Effects of varying twospotted spider mite infestation levels on strawberry yield. *J. Econ. Entomol.* 72:747–53.

Zalom, F. G., L. T. Wilson, and M. P. Hoffmann. 1986. Impact of feeding by tomato fruitworm, *Heliothis zea* (Boddie) (Lepidoptera: Noctuidae), and beet armyworm, *Spodoptera exigua* (Hübner) (Lepidoptera: Noctuidae), on processing tomato fruit quality. *J. Econ. Entomol.* 79:822–26.

Zalom, F. G., L. T. Wilson, and R. Smith. 1983. Oviposition patterns by several lepidopterous pests on processing tomatoes in California. *Environ. Entomol.* 12:1133–37.

Zalom, F. G., M. A. Hoy, L. T. Wilson, and W. W. Barnett. 1984a. Sampling mites in almonds. II. Presence-absence sequential sampling for *Tetranychus* mite species. *Hilgardia* 52:14–24.

———. 1984b. Sampling *Tetranychus* spider mites in almonds. *Calif. Agric.* 38(5–6):17–19.

Zalom, F. G., L. T. Wilson, M. P. Hoffmann, W. H. Lange, and C. V. Weakley. 1983. Monitoring lepidopterous pest damage to processing tomatoes. *Calif. Agric.* 37(3–4):25–26.

Zalom, F. G., C. E. Kennett, N. V. O'Connell, D. L. Flaherty, J. G. Morse, and L. T. Wilson. 1985. Distribution of *Panonychus citri* (McGregor) and *Euseius tularensis* Congdon on Central California orange trees with implications for binomial sampling. *Agric. Ecosyst. & Environ.* 14:119–29.

Zalom, F. G., L. T. Wilson, C. E. Kennett, N. V. O'Connell, D. L. Flaherty, and J. G. Morse. 1986. Presence-absence sampling of citrus red mite. *Calif. Agric.* 40(3–4):15–16.

Zwölfer, H., M. A. Ghani, and V. P. Rao. 1976. Foreign exploration and importation of natural enemies. In *Theory and practice of biological control,* eds. C. B. Huffaker and P. S. Messenger, 189–207. New York: Academic Press.

3 / THE CONTRIBUTIONS OF BIOLOGICAL CONTROL TO POPULATION AND EVOLUTIONARY ECOLOGY

R. F. LUCK, M. J. TAUBER, AND C. A. TAUBER

Biological control in general and the research conducted by W-84 members in particular has contributed substantially to population and evolutionary ecology. W-84 research has centered on empirical tests of theory. Its largest contributions to ecological theory concern population dynamics and predator-prey interactions; in evolutionary ecology W-84 research has focused on the genetic variation and evolutionary adaptation of natural enemy populations. In addition, W-84 members have contributed greatly to the systematics and comparative biology of a number of phytophagous and entomophagous taxa that are important to biological control.

In the first part of this chapter, "Population Ecology," we review the historical development of population dynamics and predator-prey theory as it relates to biological control, highlighting two W-84 projects that have tested aspects of this theory. We also review the contributions that several W-84 projects have made to the understanding of interspecific competition, phenology, insect-plant interactions, and tritrophic interactions. The second part of the chapter, "Evolutionary Ecology," discusses the W-84 project's contributions to evolutionary theory, especially as that theory relates to the evolution of life history and the diversification of biological control agents. Finally, a third section reviews the systematic work—crucial to the development of biological control—that has derived from the W-84 project.

Figure 3.1. The classical view of biological control, in which an introduced arthropod or weed occurs at pestiferous densities until one or more natural enemies are introduced. The natural enemies reduce and then maintain pest densities at levels that do not cause economic loss.

POPULATION ECOLOGY

Population Dynamics and Predator-Prey Interactions

Historical overview. Biological control uses arthropod natural enemies and pathogens to suppress and regulate pests (both arthropods and weeds) of agriculture and forestry. Initially, biological control was only concerned with introducing exotic natural enemies to suppress pests that were usually of foreign origin—an approach frequently referred to as classical biological control. The early successes of this technique spawned a theoretical school of population dynamics based on the general premise that biotic agents—principally parasitoids and predators—regulate phytophagous arthropod populations, and that phytophagous arthropods in turn regulate plant populations, often in conjunction with the effects of plant competition. According to these paradigms, population regulation is the result of a stable equilibrium between the natural enemy and its prey, or host. Classical biological control sought to re-establish this equilibrium by introducing the exotic pest's natural enemies from the pest's area of endemicity (see fig. 3.1).

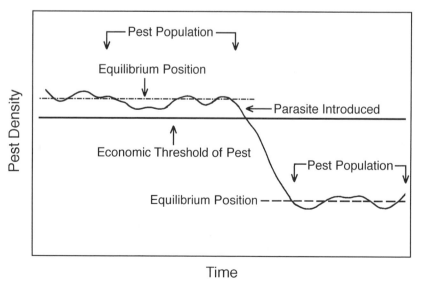

Among those who have espoused this view are L. D. Howard and W. F. Fisk (1911); A. J. Nicholson (1933); H. S. Smith (1935); A. J. Nicholson and V. A. Bailey (1935); M. E. Solomon (1949); C. B. Huffaker and P. S. Messenger (1964); C. G. Varley, G. R. Gradwell, and M. P. Hassell (1973); M. P. Hassell (1978); and D. R. Strong, J. H. Lawton, and R. Southwood (1984).

Predictably, an alternative school developed, the adherents of which viewed natural enemies or competition as largely unimportant, arguing instead that weather was the principal determinant of population density. Not surprisingly, this point of view was frequently referred to as the abiotic school of thought. Among those who articulated these ideas are W. R. Thompson (1939), H. G. Andrewartha and L. C. Birch (1954), P. J. den Boer (1978), and J. P. Dempster (1983).

Classical biological control projects act as experimental tests of these hypotheses or paradigms. The large number of successful projects clearly attest to the general validity of the biotic school's paradigm. About 290 exotic introductions have suppressed arthropod pest populations permanently, another 800 partly (in a portion of their geographic range) or intermittently (during some parts of the season, in some years, or in some crops) (Greathead 1986). However, because biological control is an applied field, economic criteria are the arbiter of success. Thus, though a number of introductions reduced the density of the pest, they did not bring it to levels that were economically acceptable. Nevertheless, many of these "unsuccessful" classical biological control projects still confirmed the biotic school's view of population regulation.

W-84 project contributions. It is now generally believed that classical biological control can permanently suppress pest populations. Because natural enemies obtained from areas where a native pest is not abundant—in effect, areas where the predator-prey interaction was indigenous—are often successful at controlling a pest, indigenous natural enemies were also recognized as important regulating agents. This led to the idea that natural enemies form the foundation of integrated pest management (IPM) and should be conserved. Two other observations supported this view. First, in several cases native predators and parasitoids either prevented agents introduced for weed control from becoming established, or limited their effectiveness (Goeden and Louda 1976; Müller, Nuessly, and Goeden 1990), which suggested that resident natural enemies suppressed populations of introduced phytophagous arthropods. Second, the widespread use of broad-spectrum pesticides frequently caused increased densities of previously innocuous pests (Lord 1947; Huffaker and Kennett 1966; Hoyt 1969; Ehler, Eveleens, and van den Bosch 1973; Eveleens, van den Bosch, and Ehler 1973; Luck and Dahlsten 1975;

Luck, van den Bosch, and Garcia 1977). IPM uses various tactics to conserve indigenous natural enemies and enhances their effectiveness with cultural practices and selective tactics.

However, not all introduced or resident populations of natural enemies can lower host or prey populations to levels that are economically acceptable. Natural enemy populations may be adversely affected by climate and weather (DeBach, Fisher, and Landi 1955), or they may be poorly synchronized with their prey/host or sporadically distributed. Such observations have led to the development of augmentative biological control. In this approach, insectary-reared natural enemies are released to supplement indigenous populations, to initiate populations each season, to re-establish populations eliminated by pesticides or other causes, or to initiate populations where they are absent (DeBach and Hagen 1964; Hoy 1982; Hoy, Westigard, and Hoyt 1983; Hoy et al. 1990; Moreno and Luck 1992). Augmentation can also include the release of selected and mass-reared insecticide-resistant natural enemies in crops where there are no alternatives to broad-spectrum sprays for controlling a key pest—for instance, the navel orangeworm in almonds and the codling moth in apples and pears (Hoy 1982; Hoy, Westigard, and Hoyt 1983; Hoy et al. 1990).

The success of biological control in general and of classical biological control in particular has confirmed the notion that parasites and predators help regulate arthropod densities. As a result, biological control became grist for theoretical population ecologists interested in predator-prey and population dynamics theory—for example, A. J. Nicholson (1933); A. J. Nicholson and V. A. Bailey (1935); H. S. Smith (1935); C. G. Varley (1947); C. S. Holling (1959); M. P. Hassell and C. G. Varley (1969); C. G. Varley, G. R. Gradwell, and M. P. Hassell (1973); M. P. Hassell (1978); J. R. Beddington, C. A. Free, and J. H. Lawton (1978); J. K. Waage and M. P. Hassell (1982); W. W. Murdoch, J. Chesson, and P. L. Chesson (1985); and W. W. Murdoch (1990). From their theoretical studies and from empirical observations, population ecologists evolved the paradigm that arthropod populations are maintained at a stable equilibrium largely through a density-dependent interaction. A review of six classical biological projects that had sufficient data for analysis supported this view. The review showed that pest densities dropped more than 98 percent after biological control agents were introduced (Beddington, Free, and Lawton 1978). Models of the predator-prey interactions in these projects suggested that regulation at a stable equilibrium was the result of a natural enemy's tendency to aggregate at certain prey or host patches, but not necessarily at the denser patches. The tendency to aggregate at some patches and not at others provides the host or prey populations with a refuge. Theoretically, such refugia should stabilize populations. However, a re-analysis of those same six projects plus an examination of three additional ones suggested that only one population evinced a stable equilibrium (California red scale, *Aonidiella aurantii* [Mask.]). The other eight were equivocal and could be explained just as easily by an alternative process, such as local—that is, patch—extinctions (Murdoch, Chesson, and Chesson 1985).

A couple of important questions emerge from this: What are the mechanisms that regulate arthropod and weed populations? Is stability or the notion of an equilibrium density the best way to characterize these interactions? For example, various processes or mechanisms have been suggested as stabilizing agents of population densities: (1) the parasitoid's tendency to aggregate at denser host patches (in effect, density dependence) or to aggregate independent of host density, (2) the existence of refuges for the host population, (3) a decrease in the percentage of hosts parasitized with increasing host density (inverse density dependence) within a generation, (4) asynchrony between the parasitoid and host population, and (5) sex-ratio variation in the parasitoid population via local parental control (local mate competition or sib-mating) or with size-dependent sex allocation (larger hosts are allocated female offspring, small hosts male offspring). These mechanisms are important to biological control because they address such practical questions as which criteria should be used to choose an effective natural enemy for introduction or mass production, or how to identify a resident natural enemy that would be of value in an IPM program.

An effective natural enemy is usually depicted as (1) having a high searching capacity—namely, the ability to find hosts when they are scarce; (2) being reasonably host specific; (3) having a high potential rate of increase—namely, a short generation time and high fecundity; (4) being able to occupy all of the host's or prey's habitat (Doutt and DeBach 1964); (5) being synchronous with the host or prey population; (6) engaging in intraspecific competition among members of the parasitoid population, a phenomenon often referred to as mutual interference; and (7) aggregating on host patches (Waage and Hassell 1982). However, these criteria are too ambiguous to be of practical value.

W-84 project case studies. To determine whether these attributes truly characterize effective natural enemies, we need long-term studies and experimental manipulations that focus on the population dynamics of biological control agents and their hosts or prey. Two such studies have been done within the W-84 project. One involved the biological control of a weed, tansy ragwort, by the cinnabar moth and a flea beetle. The other involved the biological control of an insect pest of citrus, the California red scale, by three parasitoids and a predaceous ladybird beetle. In part, these projects were used as model systems because the biology of the organisms involved was well studied, the pests had been biologically controlled, and the interactions between the organisms provided tractable systems for testing hypotheses.

Biological control of a weed, tansy ragwort. The control of tansy ragwort, *Senecio jacobaea* L., by two introduced insects—*Tyria jacobaeae* L., the cinnabar moth, and *Longitarsus jacobaeae* (Waterhouse), a flea beetle—in the Pacific Northwest is a classic example of successful biological weed control. Field experiments sought to answer three general questions underlying the theory and practice of biological weed control: Do effective natural enemies impose a low, stable equilibrium at a local spatial scale, or is it more appropriate to describe this phenomenon as local pest extinction? Should one rely on a single "best" natural enemy species or on the cumulative effects of multiple species? How does the effect of natural enemies compare with that of plant competition on the dynamics of weed populations?

The biological control of weeds seeks both to introduce natural enemies that reduce and regulate a weed population and to replace the weed with more desirable vegetation. In conducting a series of annual observations at a single site, P. B. McEvoy (McEvoy 1985; McEvoy, Cox, and Coombs 1991) showed that following the introduction of natural enemies, ragwort declined by 99 percent and that dense stands of the weed were replaced by a plant community composed predominantly of perennial grasses (see fig. 3.2). A survey of 42 sites (6 sites for 10 years or more, 16 sites for between 5 and 10 years, and 20 sites for less than 5 years) that were generally representative of the conditions and habitats where ragwort control is desired, showed substantial declines—including zero flowering plants during at least one survey—in 31 of the sites. Three sites showed a 67 percent drop in ragwort density. The remaining eight sites had only been surveyed for 2 or 3 years, a period too short to determine the effectiveness of the natural enemy (McEvoy, Cox, and Coombs 1991).

To discover which mechanisms caused the decline in the ragwort population, researchers conducted a perturbation experiment (McEvoy et al. 1990). They created incipient ragwort outbreaks and then measured the individual and combined effects that the natural enemies and plant competition had on controlling them. The study consisted of 96 square plots, 0.25 square meters in size, that were grouped into four blocks and assigned one of two disturbance treatments (a fall 1986 or spring 1987 disturbance), three plant competition levels (the vegetation other than ragwort was either removed, clipped, or left unaltered), two cinnabar moth

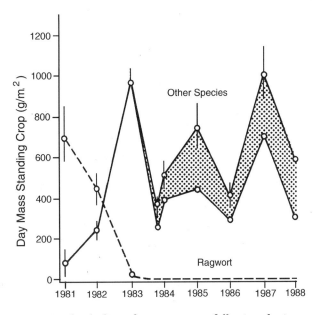

Figure 3.2. The decline of tansy ragwort following the introduction of the cinnabar moth, *Tyria jacobaeae* **L., and a flea beetle,** *Longitarsus jacobaeae* **(Waterhouse), and ragwort's replacement with other plant species, mainly grasses. Other species are divided into two bands representing forbs (the shaded area) and grasses (the unshaded area below it) (McEvoy, Cox, and Coombs 1991).**

levels (either exposed to, or protected from, the moths), and two flea beetle levels (either exposed to, or protected from, the beetles). Each treatment combination was replicated four times. Then ragwort density, biomass, and reproduction were compared.

Plant competition, the flea beetle, and the cinnabar moth each had a negative effect on ragwort performance (McEvoy et al. 1990, unpub. data). The separate effects of the beetle and plant competition were relatively strong, whereas that of the cinnabar moth alone was relatively weak. The timing of the outbreak—that is, in spring or fall—had no effect. If the ragwort populations were protected from competition and herbivory, they persisted indefinitely, whereas those that were not protected were eliminated, except for the seeds that remained in the soil. As the number of factors increased—namely, where there were combinations of competition, flea beetles, and cinnabar moth—the ragwort population decreased more rapidly. Together the flea beetle and the cinnabar moth reduced survivorship and seed production more than either agent did alone (McEvoy et al. 1990). Younger plants suffered the most mortality (McEvoy, Cox, and Coombs 1991).

Ragwort was controlled because the herbivores extirpate most of the plants. The population appears to persist as a seed bank that buffers the effect of this herbivory. In Oregon, seed density at soil depths below 8

centimeters did not appear to decline over the course of a 5-year study (McEvoy, Cox, and Coombs 1991), and seed density was similar to that found at the same depth in English pastures, part of ragwort's native range (Chippindale and Milton 1934). Disturbance renewed the space required for plant establishment—the primary limiting resource in the ragwort system—and set the stage for colonization and occupancy. Thus, ragwort density was dependent on a balance between disturbance, colonization, plant competition, and insect herbivory (McEvoy unpub. data).

The outcome of the Oregon studies on tansy ragwort suggested that local extinction of plants from herbivory and competition, and an invulnerable age class (the seed bank as a refugium) determined the small-scale population dynamics that, when viewed on a larger scale, provided a stable pest density. When sporadic and temporary soil disturbances caused increases in ragwort density, natural enemies and plant competition returned the ragwort density to its original level. The natural enemies' localized response to increased plant density appeared to be the mechanism by which biological control is achieved. The studies in Oregon also suggested that several species of natural enemies can complement each other by attacking different stages of the pest—thereby eliminating some invulnerable age classes—at different times in the pest's life cycle, thereby eliminating temporal refugia. Thus, complementary action can be important to effective biological control. The same association has been responsible for the control of tansy ragwort in northern California (Pemberton and Turner 1990).

Biological control of an arthropod pest, California red scale. California red scale, introduced into California in the early 1870s, has been the subject of a biological control effort that has lasted over a hundred years. Starting before the turn of the century, over 52 species or biotypes of natural enemies have been introduced (Luck 1986). Biological control has been achieved progressively, beginning early on with a small section of coastal California around Santa Barbara (DeBach and Sisojevic 1960). Biological control in southern California improved when *Aphytis lingnanensis* Compere was introduced from China in 1947, but it was not successful until 1956 or 1957, when *A. melinus* DeBach was imported from northwestern India and southwestern Pakistan (Rosen and DeBach 1979). Exclusion experiments have confirmed that in southern California (excluding the deserts) red scale is under excellent biological control (DeBach, Rosen, and Kennett 1971; DeBach, Huffaker, and MacPhee 1976). These populations have manifested stable equilibria (Murdoch, Chesson, and Chesson 1985). However, California red scale does remain a problem in San Joaquin Valley citrus (Flaherty, Pehrson, and Kennett 1973). Because biological control was achieved after its introduction, *A. melinus* was considered the agent responsible for suppressing the scale populations. However, field experiments failed to detect spatial density dependence or host aggregation by *A. melinus* on denser branches within trees or on whole trees (see fig. 3.3; see also Smith and Maelzer 1986; Reeve and Murdoch 1986).

Another factor thought to stabilize populations is the presence of refugia (Murdoch and Oaten 1975). However, field experiments ruled this out as a likely explanation (see fig. 3.4). *A. melinus* spends most of its time searching the peripheral 0.5 meters of the tree (Gregory 1985); hence, scales that occur in this part of the tree—the leaves and the fruit—are heavily parasitized, whereas those on twigs and branches suffer less parasitization (Yu, Luck, and Murdoch 1990). It was hypothesized that the scale populations on the twigs and scaffolding branches supplied the crawlers that colonized the fruits and leaves, and that this is what re-established the scale population on these substrates each spring after they laregely had been annihilated by *A. melinus* in the fall. Researchers predicted that trees without such refugia would manifest more variable and perhaps higher densities of scales in their peripheries than would those with refugia. In one experiment all scales were removed from the interior of a random set of five trees. As checks, scales were allowed to remain on five other trees. The results differed from those expected, in that trees without refugia—where there were no scales on the branches—had lower densities of scale on the leaves and fruits than those with a refuge (fig. 3.4). More importantly, the scale densities on these trees showed a seasonal change of two orders of magnitude, with the low densities occurring during the winter, high densities during the summer. Both sets of trees—those with and without refugia—showed similar patterns and ranges in their seasonal changes in scale densities (fig. 3.4).

A third possible explanation for red scale stability was that changes in scale density affect the sex ratio of *A. melinus*; that is, an increase in scale density leads to the production of more female wasps, whereas a decrease in density leads to more male wasps. The mechanism proposed for this sex ratio shift was local mate competition (Hassell, Waage, and May 1983). However, preliminary experiments ruled out this type of simple shift (Reeve and Murdoch 1986).

The potential explanation for stability that has emerged as a hypothesis is a combination of host size–dependent sex allocation by *A. melinus* (females are allocated to large hosts); a preference by *A. melinus* for large scales (in effect, a preference for virgin third-instar females occurring on the leaves and fruits);

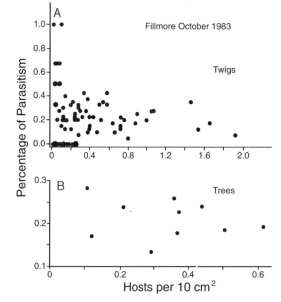

Figure 3.3. The relationship between the density of the California red scale, *Aonidiella aurantii* (Mask.), (A) on branches within trees and (B) on trees, and the percentage of scales that are parasitized by *Aphytis melinus* DeBach, the scale's putative biological control agent. These data indicate that a density-dependent relationship does not exist.

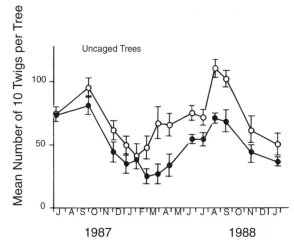

Figure 3.4. The density of the California red scale, *Aonidiella aurantii* (Mask.), on five trees from which the scale was physically removed from the interior (closed circles), and the density on five trees in which the scale was allowed to remain in the interior (open circles). This experiment sought to test the hypothesis that a refuge, by supplying colonists to the periphery of the tree, would stabilize the scale population and decrease its equilibrium density in the periphery, compared with trees without such a refuge. (Scales in the tree's interior are not heavily parasitized by *Aphytis melinus*; those in the tree's periphery are.) These data show that the prediction was not realized. Both populations were equally variable, and the population without the refuge was less dense (Luck and Murdoch unpub. data).

increased parasitization of red scale in the autumn, when the scale's development is slowed; and reduced recruitment of young scales in late autumn, due to cool conditions. When *A. melinus* populations are low relative to those of red scale but red scale is reasonably abundant, the parasitoid prefers large hosts (virgin female red scale) on the leaves and fruit. About two-thirds of these hosts are allocated female progeny by the wasp, and the remainder are allocated male progeny. Because an *A. melinus* generation occurs every 15 to 20 days in the late summer to early autumn, and because initially most of the progeny are female, the parasitoid population increases rapidly. The scale's slower development in autumn compared with summer (due to autumn's cool nights) increases the period during which it is at risk to parasitism. As the larger scales on the leaves and fruits are exploited, only smaller scales (second-instar females and males) remain, which are less preferred by *A. melinus*. Smaller scales parasitized by the wasp are allocated mostly male progeny (at a ratio of approximately 90 percent males).

Intraspecific competition among *A. melinus* also occurs in the form of superparasitism. Previously parasitized hosts are allocated additional eggs, which kills many of the eggs laid earlier. Thus, the slower scale development in autumn coupled with changes in the sex of the *Aphytis* progeny, plus the probing and host feeding on small scales (which kills them) and the reduced recruitment of scale progeny all together cause a rapid decline in the scale population, followed by a rapid decline in the female wasp population. Less efficient parasitism occurs in the summer because of the scale's rapid development, much of which occurs at night when *A. melinus* is inactive. These factors may explain the population increase during the summer. Thus, the interaction between the scale and *A. melinus* is not nearly so simple as was initially thought.

It is also clear that other natural enemy species supplement the biological control achieved with *A. melinus*. In coastal California *Encarsia perniciosi* (Tower) parasitizes a substantial percentage of the scale on scaffolding branches and twigs, and a ladybird beetle, *Rhyzobius lophanthae* (Blaisdell), feeds on dense scale patches in both the tree's exterior and interior. In the inland valleys of California, the parasitoid *Comperiella bifasciata* (Howard) supplements the parasitism of *A. melinus*. Since biological control generally was not achieved in southern California until *A. melinus* was established, it is clear that, by themselves, these other species of natural enemies are incapable of suppressing red scale. But it also is uncertain whether *A. melinus* alone can fully suppress red scale populations. Unfortunately, no differential exclusion method has been devised that would allow researchers to determine the role of each natural enemy. We expect, however, that they are all important in the scale's suppression, and that their effect has been complementary: as each species was introduced and established, the scale was controlled in more areas.

Interspecific Competition

The biological control of the California red scale also involved a classic case of competitive displacement and a case of resource partitioning among the exotic parasitoids. *Aphytis lingnanensis*, the second of three *Aphytis* species introduced into southern California for scale control, was displaced by *A. melinus*, the third species introduced (DeBach and Sunby 1963; DeBach 1966; Rosen and DeBach 1979). On the basis of laboratory experiments and field observations, P. DeBach concluded that *A. melinus* displaced *A. lingnanensis* because it was a better searcher, not because it exploited the scale population more efficiently when the required scale stages were scarce. This posed a dilemma, because biological control theory viewed control as the result of an interaction in which the natural enemy drove down the density of the host population to levels that limited the density of the natural enemies (in this case, two *Aphytis* populations). P. DeBach concluded that resource limitation did not occur. So how had biological control been achieved?

Subsequent research provided an explanation that was more consistent with the traditional view of biological control (Luck, Podoler, and Kfir 1982; Luck and Podoler 1985; Yu 1986). *A. melinus* displaced *A. lingnanensis* in southern California because it produced daughters on slightly smaller scale insects than did *A. lingnanensis*. When an *Aphytis* parasitizes a host, it halts the host's growth. Consequently, when the California red scale population becomes scarce in the autumn, *A. melinus* can produce more daughters than *A. lingnanensis* because it can use smaller hosts. In doing so it prevents the scales from attaining the size required by *A. lingnanensis* for the production of its daughters, placing a crucial resource in short supply. Thus, it was the difference in the host size on which a species produces female progeny that determined displacement, not the host size they generally parasitized. Although one species of parasitoid was displaced by a second, biological control improved. Introductions of congeners that are relatively specific to the pest thus enhance biological control.

Resource partitioning also played a role in the interaction between *Encarsia perniciosi* and *A. melinus* (Yu, Luck, and Murdoch 1990). *E. perniciosi* prefers to parasitize late first and early second instars of red scale (if such scales are scarce, it reluctantly parasitizes late second or virgin third instars), whereas *A. melinus* prefers virgin third-instar female scales (parasitizing late second-instar males and females when necessary). Moreover, *E. perniciosi* mostly uses scales inhabiting the scaffolding branches and twigs, whereas *A. melinus* rarely uses those scales, probably because they are too small to support the production of female progeny. *E. perniciosi* is an internal parasitoid that emerges from a scale at the same time that an unparasitized scale of the same cohort undergoes a second molt. In contrast, *A. melinus* is an external parasitoid that prefers later stages of scale. By parasitizing the younger scale on the scaffolding branches, *E. perniciosi* largely avoids competition with *A. melinus*; in fact, its parasitization complements that by *A. melinus* of the scale on the leaves and fruit and appears to result in better overall scale suppression.

Why *E. perniciosi* is restricted to the coastal regions and inland coastal valleys of southern California is not understood. Also, the presence of *E. perniciosi* does not prevent *A. melinus* from parasitizing scale previously parasitized by *E. perniciosi*. Although multiple parasitism harms *E. perniciosi*—when *A. melinus* parasitizes a young scale containing an *E. perniciosi* egg or larva, it kills that egg or larva and usurps the scale insect for the development of its own progeny—*A. melinus* does not appear to be affected. Again, as with the tansy ragwort example from Oregon, it appears that the presence of additional natural enemies, or multiple introductions of primary, relatively specific parasitoid species, enhances the biological control of a pest.

Phenology

Phenology—the seasonal occurrence and activity of organisms, whether plants, pests, or natural enemies—is a vital aspect of IPM and biological control. Incorporating phenology into the planning and implementation of biological control can increase the efficiency and effectiveness of many procedures, including (1) timing foreign exploration to coincide with the seasonal cycles of the natural enemies in their native home; (2) choosing biotypes whose life cycles are well synchronized to those of their host and that are well adapted to seasonal changes in the new habitat; (3) averting diapause during quarantine and rearing or mass-rearing; (4) releasing and manipulating natural enemies so that they are well synchronized with the pest in the field; and (5) understanding the population dynamics of pests and natural enemies in their seasonal environments.

Projects under W-84 sponsorship have focused on these issues and have contributed significantly to the theory and study of insect seasonality. By synthesizing and critically evaluating a vast and scattered body of modern literature on the subject, major reviews have made phenologi-

cal concepts and research findings readily available to researchers in applied and theoretical ecology (Tauber, Tauber, and Masaki 1984, 1986). Also, the projects helped incorporate seasonality into general life-history studies of both parasitoids and predators (for example, Tauber and Tauber 1981, 1982, 1989; Tauber et al. 1983).

Some of the W-84-related studies helped clarify long-standing misconceptions about the role of low temperatures during insect dormancy (Tauber and Tauber 1976; Tauber et al. 1982) as well as about the physiological changes that occur during dormancy (see fig. 3.5). Additional studies have elucidated the geographical variability and the genetic basis for a variety of environmental responses that regulate insect voltinism and dormancy (for example, Nechols, Tauber, and Tauber 1987; Tauber, Tauber, and Nechols 1987). Finally, other studies have quantified the ecophysiological and genetic mechanisms that underlie seasonal cycles in predators (Obrycki and Tauber 1981, 1982; Obrycki et al. 1983), in parasitoids (Obrycki and Tauber 1979; Nechols and Kikuchi 1985; Tracy and Nechols 1988; Nechols, Tracy, and Vogt 1989), and in pests (Nechols 1987, 1988; Tauber, Tauber, Gollands, et al. 1988; Tauber, Tauber, Obrycki, et al. 1988a,b; Tauber et al. 1989; Biever and Chauvin 1990). These findings are essential to developing predictive phenological models for pests and their natural enemies, as well as for timing biological control and other IPM procedures.

Insect-Plant Interactions

The interaction between phytophagous arthropods and their natural enemies is often mediated by the plant. Because plants provide the site of the interaction as well as the chemicals—such as plant-based synomones—that mediate the interactions, they can influence the success of biological control (Hagen, Dadd, and Reese 1984; Hagen 1986, 1987). In several case studies, members of the W-84 committee have demonstrated the importance of tritrophic interactions (also see chap. 2). For example, K. S. Hagen and coworkers demonstrated that volatile kairomones associated with honeydew and synomones emanating from growing plants help attract *Chrysoperla* (=*Chrysopa*) *carnea* (Stephens) females to their oviposition site (Hagen 1986). Kairomones and synomones also help the phytoseiid predator *Galendromus* (=*Metaseiulus*) *occidentalis* (Nesbitt) find its prey (Hoy and Smilanick 1981).

Another W-84 study illustrated the complexities of tritrophic interactions. In this study, pest-resistant potato plants exerted both direct (via entrapment by glandular trichomes) and indirect effects on *Edovum puttleri* Grissell, a parasitoid of Colorado potato beetle eggs (Ruberson et al. 1989). When *E. puttleri* were reared on eggs from Colorado potato beetles that had

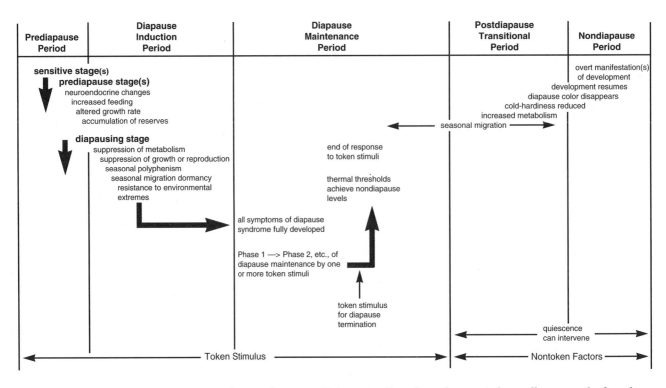

Figure 3.5. Diagrammatic representation of insect dormancy. Token stimuli, such as photoperiod, usually govern the first three periods. During the last two periods, the immediate conditions of the environment (not token stimuli) regulate growth and development (Tauber, Tauber, and Masaki 1986).

fed on resistant potato plants, the longevity of female parasitoids was reduced, as was the number of Colorado potato beetle eggs that were killed by the adult parasitoids.

Plants can also provide shelters or other resources that increase the effectiveness of natural enemies. For example, R. W. Pemberton and C. W. Turner (1989), two W-84 project participants, recently reported finding a number of plants with domatia—small invaginations and hair tufts at vein junctions on the undersides of leaves in many woody dicots. Of 32 plant species investigated, 87 percent had domatia that were occupied by beneficial species of mites (primarily species of Phytoseiidae and Tydeidae, but also of Bdellidae, Cheyletidae, and Stigmaeidae). These authors hypothesized that a facultative mutualism exists between plants with domatia and beneficial mites. The leaf domatia serve as shelters and nurseries for mites, which prey on phytophagous arthropods and pathogens colonizing the plants. This mutualism may be important in some agricultural crop plants such as grapes or walnuts.

Another W-84 project reported plant-pest–natural enemy interactions that had significant economic consequences (Johnson and Hara 1987). Various polyphagous species of Liriomyza leafminers (Diptera: Agromyzidae) have become important pests on vegetable and ornamental crops. It appears that the para-sitoids associated with many of these Liriomyza species may be extremely selective about the host plant or its habitat. Of the 40 parasitoid species reported to attack the four leafminers, one—Diglyphus begini (Ashmead) (Hymenoptera: Eulophidae)—was dominant in more than 60 percent of the 12 vegetable crops. Two other species, Halticoptera circulus (Walker) (Hymenoptera: Pteromalidae) and Chrysonotomyia punctiventris (Crawford) (Hymenoptera: Eulophidae), were common in 20 to 30 percent of the crops, while six other species were found in other vegetable crops. Although a plant–parasitoid-species association exists, the basis for this association is not understood. Nevertheless, in order to use these parasitoid species as augmentative biological control agents, it is crucial to know their plant preferences. Relying on a single species of para-sitoid for the control of leafminers would result in only partial success, because one species can be effective only on some vegetables.

All of these studies point to two crucial issues: (1) in order to integrate biological control and host-plant resistance as complementary pest management tactics, we need to understand the interactions of natural enemies with plants and with herbivorous pests; and (2) such a tritrophic perspective provides an ecologically and evolutionarily sound framework within which selection for plant resistance should be conducted.

The Ecological and Evolutionary History of Natural Enemy Guilds

How natural enemy species interact over ecological and evolutionary time is relevant to both community ecology and biological control. The parasites and predators of an insect pest can often be classified into guilds—groups of species that exploit a host in a similar way (compare with Root 1967). W-84 researchers have contributed substantially to the development of the guild concept, which has proven useful in analyzing interspecific interactions among a host's natural enemies.

L. E. Ehler (1992) recognized three types of predator and parasite guilds: natural, restructured, and anthropogenic (or synthetic). Natural guilds consist of species that share a long evolutionary history with both the host species and each other. Such guilds can be restructured by adding one or more exotic species, such as occurs in the classical biological control of a native pest. Anthropogenic guilds are the result of human activity, such as occurs in the classical biological control of an exotic species (for example, the group of introduced parasites associated with California red scale), or in a multiple-species release of commercially raised natural enemies. In restructured guilds, and certain anthropogenic ones, not all members have a common evolutionary history. This can have important implications for community ecology and biological control, as the following example illustrates.

In chapter 39 L. E. Ehler summarizes the introduction of *Encarsia aurantii* (Howard) for control of the obscure scale, *Melanaspis obscura* (Comstock). In this case, the evolutionary history of the attendant parasite guild played a critical role in the decision to import this particular species. In its native home, the obscure scale is exploited by a complex of natural enemies, including a diverse parasite guild. However, this guild was apparently restructured during the last hundred years with the fortuitous introduction of *E. aurantii*, a species believed to be native to Asia. This species not only invaded the parasite guild but became one of its dominant members, and it appears to have had a major impact on the host in the latter's native home. As noted in chapter 39, *E. aurantii* is an opportunistic species, and the introduction strategy was to release this opportunist free of its competitors.

These investigations raise two fundamental issues in community ecology. First, the successful invasion by *E. aurantii* clearly suggests that native parasite guilds are not necessarily at equilibrium in respect to species richness (Ehler person. commun.). This has obvious implications for classical biological control of native pests, especially those that already have a diverse parasite guild. Second, we must ask ourselves whether we can expect restructured parasite guilds (or restructured enemy communities) to conform to current ecological theory, which was developed primarily from the study of natural, coevolved systems. This critical question clearly needs investigation.

EVOLUTIONARY ECOLOGY

Biological control provides an excellent opportunity for viewing the interactions of natural enemies with their hosts and crops from the perspectives of both genetics and evolution. Such a broad approach can contribute enormously to the theory and practice of biological control as well as to the testing of population genetics theory in the field.

Virtually all the major procedures used in biological control—for example, exploration, quarantine, rearing, and release—can influence the genetic makeup and thus quality of natural enemies. An impediment to bringing genetic principles and biological control closer together has been the dearth of information on the genetic variation in natural enemies and the selective pressures that mold the interactions between natural enemies and their hosts (Diehl and Bush 1984; Hoy 1985; Tauber, Hoy, and Herzog 1985; Tauber, Tauber, and Masaki 1986, chap. 10). Researchers in the W-84 project have sought to address this issue. Their studies cover a range of topics concerning natural enemies, including (1) identifying and characterizing variation in the traits that determine survival and reproductive success, (2) selecting and modifying genetically based traits, (3) elucidating the environmental and genetic bases for variation, and (4) quantifying the genetic structure of geographically and ecologically diverse populations in the field. The following sections highlight the findings of those studies; to learn their application to specific management practices in the field, refer to the chapters cited therein.

Genetic Variability in Natural Enemies

Historically, biological control practitioners have been sensitive to differences within species of natural enemies, and some have emphasized the importance of genetic variability—biotypes—to the success of biological control projects (for example, Flanders 1931, 1950; van den Bosch and Messenger 1973; Caltagirone 1985). Moreover, they have used a variety of approaches to identify and characterize the variability in natural enemies. In addition to the analysis of ecological traits that we will discuss next, W-84 researchers have used modern molecular methods in this regard. For example, Tom Unruh (1995) summarized the use of molecular techniques in classical biological control, and his research addressed questions concerning the structure of genetic variation within and among species of natural enemies.

In one study, molecular variation among *Aphidius* species and populations was assessed to clarify systematic relationships in that difficult group and to identify which of several introduced biotypes of *A. ervi* Haliday became established in California following a decade of releases to control pea aphids (Unruh et al. 1986, 1989). Molecular markers were also used to measure the loss of genetic variation in natural enemy populations during laboratory culture and colonization in the field. Unruh et al. (1983) showed that genetic variation was lost at four times the rate expected in cultured *Aphidius* populations, and that the reason for the loss was probably high variance among individuals in terms of reproductive success.

Another study that used molecular methods re-evaluated the early unsuccessful attempts in California to establish colonies of *Rhinocyllus conicus* (Froelich)—a European weevil that feeds on at least 11 species of thistles (Hawkes et al. 1972). Only after the discovery and appropriate use of specific biotypes did *R. conicus* become established on several species of thistles (Goeden and Ricker 1977, 1978; Goeden 1978; Zwölfer and Preiss 1983; Goeden, Ricker, and Hawkins 1985). Characterization of the variation in isozyme patterns among the several introduced European biotypes allowed the researchers to identify the biotypes that switched to native North American thistles; this switch was found to be consistent with the host range of that biotype in Europe (Unruh and Goeden 1987).

Other W-84 researchers used experimental, rather than molecular, approaches to the evaluation of genetic variation in natural enemies. Building on well-known examples—such as the differential abilities of French and Iranian biotypes of *Trioxys pallidus* Haliday to survive and suppress walnut aphids in California (van den Bosch et al. 1979; Flint 1980), and the differential encapsulation of *Bathyplectis curculionis* (Thomson) by the alfalfa weevil and the Egyptian alfalfa weevil (Salt and van den Bosch 1967)—studies from the W-84 project have provided direct evidence that recognizing and using genetic variation appropriately are important to biological control. A summary follows.

Parasitoids. Comparative studies of geographically and ecologically diverse populations have revealed genetic variation in the ecological traits of more than 50 species from more than 14 families of hymenopteran parasitoids (Ruberson, Tauber, and Tauber 1989; Unruh and Messing 1993). This variation is of major importance to biological control, and, significantly, over a quarter of the reports originate with W-84 members. Among recent examples, Mexican and Colombian populations of *Edovum puttleri*, a eulophid parasitoid of the Colorado potato beetle, were shown to differ in a variety of responses to biotic and abiotic factors, such as survivor-

ship under low temperature, acceptance of old eggs as hosts and the sex ratio of progeny in old hosts, net reproductive rate and intrinsic rate of increase, adult longevity, and susceptibility to pesticides (Obrycki, Tauber, and Tingey 1986; Obrycki et al. 1987; Ruberson, Tauber, and Tauber 1987, 1988, 1989). As a result of these studies, researchers have been able to make specific recommendations for using *E. puttleri* biotypes in the integrated pest management of the Colorado potato beetle (see chap. 45).

On a smaller geographic scale—that is, within a single county—comparative studies of an aphelinid demonstrated that changes in habitat resulted in evolutionary change. Insecticide resistance in 13 populations of *Aphytis melinus*, a biological control agent of the California red scale, correlated with the history of chemical pesticide use, both within groves and countywide (Rosenheim and Hoy 1986). Selection for pesticide resistance in another parasitoid, *Trioxys pallidus* Haliday, resulted in the parasitoid's ability to survive field applications of Guthion. The parasitoid also exhibited cross resistance to other pesticides, a trait attributable to more than one gene (Hoy and Cave 1988, 1991). These and related studies have provided the basis for the continued use of these parasitoids to control walnut aphids in IPM programs involving chemical pesticides (see chap. 33 and Hoy, Cave, and Caprio 1991).

Not all parasitoid variation can be attributed to genetic causes. In fact, W-84 studies have implicated extranuclear factors in the expression of traits—such as the sex of offspring and the mode of sexual reproduction—that we usually consider to be transmitted via nuclear genetic material. For example, although genetic factors may be involved in the expression of thelytokous or arrhenotokous reproduction by *Muscidifurax raptor* Girault and Sanders, extranuclear microorganisms or the chemicals they produce clearly have a role in the process (Legner 1987a,b). The reproduction of another species of *Muscidifurax*, *M. raptorellus* Kogan and Legner, is also influenced by extranuclear factors, but the effects are very different from those in *M. raptor*. Females with the genotype for solitary oviposition (one egg per host) express gregarious oviposition (more than one egg per host) after mating with males of the gregarious genotype. Conversely, females with the genotype for gregarious oviposition reduce the magnitude of their gregarious oviposition after mating with males of the solitary genotype (Legner 1987c,d,e, 1988, 1989a,b,c, 1990, 1991a,b). Other studies of parthenogenetic (thelytokous) *Trichogramma* biotypes have indicated that parthenogenesis is induced in most forms by a microorganism in the cytoplasm (Stouthamer, Luck, and Hamilton 1990). Eliminating the microorganism by exposing the immature parasitoids to elevated temperatures over several generations, or by feeding female parasitoids particular antibiotics, causes parthenogenic lines to revert to normal arrhenotokous reproduction—that is, male and female offspring, with females arising only from fertilized eggs.

In the field, thelytokous and arrhenotokous populations coexist. The parthenogens may mate with males from the arrhenotokous populations and incorporate the male's genome as fertilized diploid (female) eggs. The parthenogens also produce unfertilized diploid female eggs. The F_1 fertilized eggs from the parthenogens are heterozygous for the paternal and maternal genome, but the F_2 offspring are homozygous for either the maternal or paternal genome (Luck, Stouthammer, and Nunney 1993; Stouthamer and Kazmer unpub. data). The taxonomic and genetic implications of this form of parthenogenesis are discussed elsewhere (Stouthamer, Luck, and Hamilton 1990; Luck, Stouthammer, and Nunney 1993, respectively). The evolutionary significance of these studies is great, for they allow us to investigate how and when selection acts on traits that lie at the core of parasitoid-host interactions.

W-84 studies have documented the ability of parasitic Hymenoptera to adjust the sex ratio of their progeny to environmental conditions, either in the form of host quality or in the number of females contributing progeny to a host patch (Nunney and Luck 1988; Luck, Stouthammer, and Nunney 1992; Nadel and Luck 1992). When parasitoids allocate progeny to hosts that occur in patches, the progeny are expected to have a strongly female-biased sex ratio if those progeny mate only among themselves before mated females emigrate to a new patch. (Males remain in the natal patch.) However, the more females that allocate offspring to a patch, the less female-biased is the sex ratio of their offspring (for a detailed explanation see Luck, Nunney, and Stouthammer in press). Also, when host size varies and female parasitoid offspring lose more reproductive potential than do male parasitoids by being small, female offspring should be allocated to large hosts and males to small hosts. This is generally the case with ectoparasitoids and a few endoparasitoids that use hosts that are in a nongrowing stage—in effect, when they are in the egg or pupal stage (Luck, Stouthammer, and Nunney 1992). Issues of sex allocation and sex determination are important to biological control because they determine the proportion of female offspring, thus affecting the vigor of mass-reared colonies; this, in turn, can affect the genetic variability of the released population (Luck, Stouthammer, and Nunney 1993; Luck, Nunney, and Stouthammer in press).

Predators. Although researchers have done fewer detailed studies on predaceous insect species than on parasitic ones, investigations have found genetic variation in suites of traits that contribute strongly to fitness—for example, seasonal cycles, developmental requirements, host specificity, and pesticide susceptibility. Under the auspices of the W-84 project, geographic populations of lacewings in the genera *Chrysopa* and *Chrysoperla* (see the following discussion) were shown to differ in various developmental and reproductive traits that are critical to predator-prey interactions. *Chrysopa oculata* Say exhibited some developmental and seasonal traits that vary with geographic origin, as well as other developmental and seasonal traits that are relatively constant across a wide range of North American populations, from southern Canada to northern Mexico (Nechols, Tauber, and Tauber 1987; Tauber, Tauber, and Nechols 1987). Other investigations found that genetically controlled traits characterize prey associations and prey specificity in *Chrysopa quadripunctata* Burmeister and its relatives, which are widespread and important predators in orchards (Tauber and Tauber 1987a; Milbrath, Tauber, and Tauber 1993). Such studies provide the kind of information needed to choose the appropriate geographic populations for specific climatic and agricultural conditions. They also identify traits that can limit the ability of a species to colonize or expand its geographic range, as well as traits that can be improved by artificial selection.

Studies with the predaceous mite *Galendromus* (=*Metaseiulus*) *occidentalis* (Nesbitt) demonstrated that predators can evolve physiological resistance to as many pesticides as can most major arthropod pests (Hoy 1987). However, predators in the field exhibit a delay in developing resistance; this delay underscores how important various ecological and behavioral traits are in the evolution of these species (see chap. 8). Basic studies on the mode of reproduction (parahaploidy), dispersal, diapause, and searching and reproductive behavior seek to overcome these limits so that resistant strains can be maintained in the field (see chap. 8 for a summary of research on spider mite biological control in almonds).

The Genetic Structure of Natural Enemy Populations

Some of the most perplexing problems in evolutionary genetics concern how genetic variation is partitioned among populations and maintained within populations. These fundamental issues have broad implications for biological control: how genetic variation is expressed and selected in field populations affects the outcome of the foreign exploration, quarantine, release, and colonization of natural enemies. Thus, it is essential to develop sound procedures for characterizing and dealing with genetic variability.

Under the auspices of the W-84 project, researchers at Cornell University have conducted a comprehensive characterization of genetic variation in an economically important natural enemy species. Their investigation focused on green lacewings in the *Chrysoperla carnea* (Stephens) species complex—a group of predators that is commercially mass-reared and released in a variety of agricultural and horticultural situations. The results indicate that these predators exhibit enormous variability throughout their broad North American distribution, which ranges from Alaska to Mexico and the East Coast to the West Coast. The findings also illustrate how a comprehensive understanding of genetic variation can improve the efficiency and effectiveness of natural enemies. An assessment of contributions made over the last 15 years reveals five categories of investigation, detailed in the following pages.

1: Variation in life-history traits, habitat associations, and seasonal cycles within the group. Some *C. carnea* populations are multivoltine, others univoltine; still others contain both univoltine and multivoltine individuals (Tauber and Tauber 1982, 1986a). Some populations occur in fields and row crops, others in deciduous fruit and ornamental trees, and still others in conifers (Tauber and Tauber 1993). *C. carnea* also harbors genetic variation for susceptibility to a range of pesticides (Grafton-Cardwell and Hoy 1985). All of these traits affect the role of the predators as biological control agents; they determine both where the populations occur as well as when they reproduce and feed on prey.

2: Patterns of geographic variation. In general, West Coast populations show much more variability than do those in the midwestern or eastern United States (Tauber and Tauber 1986a). This variability is expressed in the full range of traits that influence the seasonal occurrence and population dynamics of the predator, and it includes genetic polymorphisms—genetic variation among individuals within a single population—as well as qualitative and quantitative differences among geographic populations in their responses to photoperiod, temperature, and food (see fig. 3.6).

After achieving an understanding of this genetic variability, researchers identified specific biotypes—that is, geographic populations—that are likely to be effective natural enemies in specific agronomic situations (Tauber and Tauber 1993). For example, one biotype in the species complex is commercially mass-reared and distributed for agricultural and urban use. Although this biotype can be very effective in row and field crops, several genetically controlled behavioral and phenological traits make it unsuitable for general use in orchards or parklands. However, other biotypes may be suitable for such situations.

3: Ecophysiological and genetic bases for variation. The seasonal cycles of reproduction and activity—that is, voltinism—in this group of predators are determined by specific responses to photoperiod and food; these responses vary greatly within and among populations (Tauber and Tauber 1982, 1986a,b, 1987b). The photoperiodic component of the life cycle is determined by simple Mendelian inheritance, whereas the responses to prey are under polygenic control. A crucial point is that the genes that control the photoperiodic component of the life cycle also control whether or not the responses to prey are expressed. This information is critical to understanding how genetic variability is subject to natural selection. Moreover, it allows us to predict how the seasonal cycles of the various biotypes would be expressed under new environmental conditions—in effect, in a release area away from the collection site. Such information is essential in matching biotypes to particular climates.

4: Intrapopulation variation. The researchers conducted extensive work on the genetic relationships between various key traits that control the life histories and seasonal cycles of *C. carnea* (Tauber and Tauber 1986b, 1987b). This includes a comprehensive analysis of intrapopulation variation in the various phenological and life-history traits of nine geographical populations. A significant finding is that the variation among geographical populations is based on homologous genes throughout North America. The results provide a strong genetic basis for assessing the evolutionary adaptation of natural enemies to constantly changing environments (Tauber and Tauber 1992). For example, they demonstrate that the seasonal cycles of western populations are not prevented from evolving because of a lack of genetic variation. Rather, the characteristic seasonal cycles are maintained by natural selection acting on variable but correlated suites of genetically controlled traits.

From the studies it is clear that altering the habitat—for example, through agricultural practices—can lead to evolutionary changes in the seasonal cycle and life history of the predators, but that those changes are limited by the range of variation and covariation among traits. The studies also indicated that some, but not all, populations carry genetic variation for desired traits; therefore, for mass-rearing or for introduction to countries outside of North America, the source—the geographic locality and habitat—of *C. carnea* stock is critical.

5: Speciation. The research done on the *C. carnea* system has strong implications for theoretical aspects of speciation, especially for the mechanisms that may be involved. Early comparative and genetic studies led to the controversial proposal that sympatric speciation might account for the existence of two seasonally distinct and reproductively isolated populations in the northeastern United States (Tauber and Tauber 1977a,b). It was proposed that the initial step in specia-

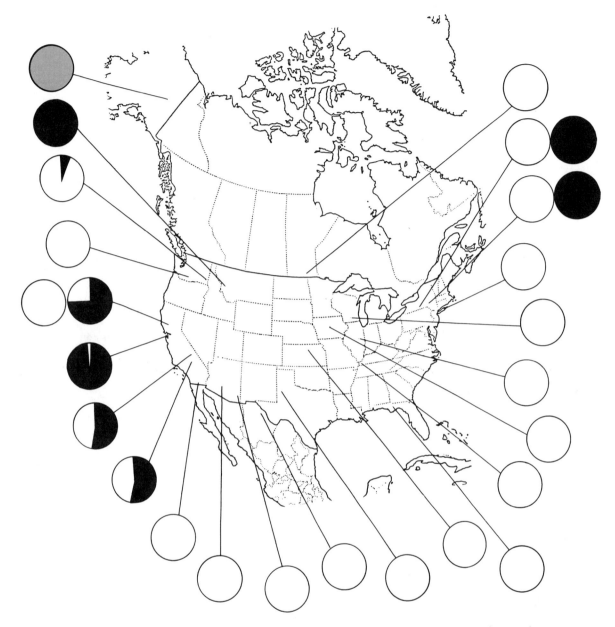

Figure 3.6. Variation among North American populations of the *Chrysoperla carnea* species complex. Dark sections represent the proportion of univoltine individuals; light sections represent the proportion with a multivoltine life cycle. The Alaskan population has a unique photoperiodic requirement that ensures univoltine reproduction (Tauber and Tauber 1986a). Reprinted with permission of the *Canadian Journal of Zoology*.

tion involved habitat diversification and that this was followed by the evolution of seasonal asynchrony. Subsequent studies demonstrated that individual populations—such as those in western North America—harbor the type of genetic variation that enables these steps to occur (Tauber and Tauber 1986a,b, 1987b, 1992). When the researchers did a comprehensive review of speciation models in relation to the biology of many groups of insects, they found that this mode of speciation may not be unusual (Tauber and Tauber 1989).

Summary. W-84 studies have shown that genetic variation can present some immediate problems for biological control. However, the research also demonstrates that this variability can have long-term advantages for biological control that far outweigh the disadvantages. In order to realize these advantages, we need a much fuller understanding of genetic variation in natural enemies. We also need to transfer relevant principles from population genetics into the theory and practice of biological control.

SYSTEMATICS AND THE COMPARATIVE BIOLOGY OF NATURAL ENEMIES AND THEIR HOSTS

Theoretical and applied biologists generally agree that a strong systematics foundation is indispensable to biological control efforts. First, systematic studies provide precise names of natural enemy species as well as keys and manuals for identifying species. The unique names allow scientists to communicate their findings as well as provide access to information on geographic distributions, host associations, and seasonal cycles. The value of proper identification of natural enemies and their hosts—and, conversely, the problems that arise when systematics is neglected—is well described in the literature (see, for example, Delucchi, Rosen, and Schlinger 1976; DeBach and Rosen 1991).

Second, systematics is the primary means by which we can discern the evolutionary relationships among organisms. Such information is essential in discovering new and potentially valuable biological control agents and in choosing appropriate species and biotypes for biological control.

Several researchers within the W-84 group have contributed significantly to developing systematic materials for parasitic, predaceous, and herbivorous natural enemies, as well as pests.

Parasitoids

The most numerous and most frequently used natural enemies belong to the parasitic Hymenoptera. Of an estimated 500,000 species worldwide, only about 10 percent have been described (DeBach and Rosen 1991). W-84-related studies are improving this ratio. G. Gordh's work has emphasized taxonomic descriptions, revisions, catalogs, and biological studies of a variety of parasitic Hymenoptera, with emphasis on the Encyrtidae and Bethylidae (Beardsley and Gordh 1988; Gordh 1985a,b, 1986, 1987, 1988a,b, 1990; Gordh and Medved 1986; Gordh and Moczär 1990; Gordh and Orth 1987; Gordh and Wills 1989; Hawkins and Gordh 1986; Tachikawa and Gordh 1987; Walker and Gordh 1989). J. W. Beardsley, also in connection with the W-84 project, has dealt with the taxonomy of parasitoid wasps used in biological control projects in Hawaii (Beardsley 1969, 1976, 1985, 1986a, 1988, 1990; Beardsley and Tsuda 1991; Beardsley and Uchida 1991).

Predators

The neuropteran family Chrysopidae is used in a variety of biological control situations, and several species are mass-reared and released to control aphids (see chaps. 27–33) and other arthropod pests. However, the systematics of the group needs revision; the larval stages, especially, are poorly known and difficult to identify. Comparative morphological studies have led to the development of descriptions, keys, and manuals for identifying larval stages in three genera: *Chrysoperla*, *Yumachrysa* (=*Suarius*), and *Anomalochrysa* (Tauber 1974, 1975; Tauber, Johnson, and Tauber 1992). In addition, manuals for identifying larvae of the various neuropteran families have been published (Tauber 1991a,b). Studies have also assessed, on both worldwide and national levels, the status of systematic research on the Neuroptera, Raphidioptera, and Megaloptera; they have also established priorities for future funding and systematic research on these groups (Tauber and Adams 1989). Comparative biological investigations of species in two chrysopid genera, *Chrysopa* and *Chrysoperla*, provide information that is valuable to the biosystematics of the groups, as well as to biological control. One study examined species-specific differences in the relative humidity levels that *C. carnea* and *C. rufilabris* require for reproduction (Tauber and Tauber 1983). These studies have also led to specific recommendations for using the two species in biological control (Tauber and Tauber 1993).

Predaceous mites of the family Phytoseiidae are important predators of phytophagous mites and thrips (see chaps. 5–13). Taxonomic studies of phytoseiids from various parts of the world have emphasized the faunas of areas being explored for biological control agents, and they have assessed how these species relate to the known world fauna (for example, McMurtry and de Moraes 1989). One problem we sometimes need to address is whether geographic populations that show slight but consistent morphological differences are in fact different species. This problem has been especially evident for the genus *Eusieus*. Laboratory cross-breeding studies to determine the degree of reproductive compatibility have often helped with taxonomic decisions (for example, Congdon and McMurtry 1985; McMurtry, Badii, and Congdon 1985; McMurtry and Badii 1989). Scientists are currently preparing a key to the genera of the world.

Herbivores of Weedy Plants

Gall-forming and flower-infesting flies in the family Tephritidae have been imported from Europe and Central America to help control several important asteraceous weeds in western North America, including several states in the W-84 region (see chaps. 62–64, 70, 72). R. D. Goeden and associates in Riverside, California, are conducting a detailed biosystematic study of one of the important genera in this group, *Procecidochares*. Comprehensive studies of the taxonomy and comparative morphology of immature stages, biology, mating behavior, host-plant relations, and natural enemies of little-known native species of nonfrugivorous

Tephritidae have helped in characterizing this important family (Goeden 1985, 1987a,b, 1988, 1989, 1990; Goeden and Blanc 1986; Goeden and Headrick 1990, 1991; Headrick and Goeden 1990a,b, 1991).

Pests

In many instances the systematics of pests is so poorly known that it hinders attempts at biological control. Systematists in the W-84 project recognize the importance of this problem. For example, J. W. Beardsley has published taxonomic papers on homopterous pests (Beardsley 1965, 1966, 1970, 1975, 1979, 1986b).

REFERENCES CITED

Andrewartha, H. G., and L. C. Birch. 1954. *The distribution and abundance of animals.* Chicago: University of Chicago Press.

Beardsley, J. W. 1965. Notes on the pineapple mealybug complex with descriptions of two new species (Homoptera: Pseudococcidae). *Proc. Hawaii. Entomol. Soc.* 29(1):55–68.

———. 1966. The Coccoidea of Micronesia (Homoptera). *Insects of Micronesia* 6(7):377–562.

———. 1969. The *Anagyrina* of the Hawaiian Islands (Hymenoptera: Encyrtidae) with descriptions of two new species. *Proc. Hawaii. Entomol. Soc.* 20(2):287–310.

———. 1970. *Aspidiotus destructor* Signoret, an armored scale pest new to the Hawaiian Islands. *Proc. Hawaii. Entomol. Soc.* 20(3):505–8.

———. 1975. Homoptera: Coccoidea. *Insects of Micronesia* Suppl. 6(9):1–6.

———. 1976. A synopsis of the Encyrtidae of the Hawaiian Islands with keys to genera and species (Hymenoptera: Chalcidoidea). *Proc. Hawaii. Entomol. Soc.* 22(2):181–228.

———. 1979. The current status of the names of Hawaiian aphids. *Proc. Hawaii. Entomol. Soc.* 23(1):45–50.

———. 1985. Notes on Hawaiian Alloxystidae and Cynipidae (Hymenoptera: Cynipoidea). *Proc. Hawaii. Entomol. Soc.* 25:49–52.

———. 1986a. Taxonomic notes on the genus *Gapaspidium* Weld (Hymenoptera: Cynipoidea: Eucoilidae). *Proc. Hawaii. Entomol. Soc.* 26:35–39.

———. 1986b. Taxonomic notes on *Pseudococcus elisae* Borchsenius, a mealybug new to the Hawaiian fauna. *Proc. Hawaii. Entomol. Soc.* 26:31–34.

———. 1988. Eucoilid parasites of agromyzid leafminers in Hawaii (Hymenoptera: Cynipoidea). *Proc. Hawaii. Entomol. Soc.* 28:33–47.

———. 1989. Hawaiian Eucoilidae (Hymenoptera: Cynipoidea), key to genera and taxonomic notes on apparently nonendemic species. *Proc. Hawaii. Entomol. Soc.* 29:165–93.

Beardsley, J. W., and G. Gordh. 1988. A new *Parechthrodyinus* Girault, 1916 attacking *Xylococculus* Morrison, 1927 in California, with a discussion of the host relationship (Hymenoptera: Encyrtidae; Homoptera: Margarodidae). *Proc.*

Hawaii. Entomol. Soc. 28:161–68.

Beardsley, J. W., and D. M. Tsuda. 1991. *Marietia pulchella* (Howard) (Hymenoptera: Aphelinidae), a primary parasite of *Conchaspis angraeci* Cockerell (Homoptera: Conchaspididae). *Proc. Hawaii. Entomol. Soc.* 30:151–53.

Beardsley, J. W., and G. K. Uchida. 1991. Parasites associated with the leucaena psyllid, *Heteropsylla cubana* Crawford, in Hawaii. *Proc. Hawaii. Entomol. Soc.* 30:155–56.

Beddington, J. R., C. A. Free, and J. H. Lawton. 1978. Characteristics of successful natural enemies in models of biological control of insect pests. *Nature* 273:513–19.

Biever, K. D., and R. L. Chauvin. 1990. Prolonged dormancy in a Pacific northwest population of the Colorado potato beetle, *Leptinotarsa decemlineata* (Say) (Coleoptera: Chrysomelidae). *Can. Entomol.* 122:175–77.

Caltagirone, L. E. 1985. Identifying and discriminating among biotypes of parasites and predators. In *Biological control in agricultural IPM systems,* eds. M. A. Hoy and D. C. Herzog, 189–200. Orlando, Fla.: Academic Press.

Chippindale, H. G., and W. E. J. Milton. 1934. On the viable seeds present in the soil beneath pastures. *J. Ecol.* 22:508–31.

Congdon, B. D., and J. A. McMurtry. 1985. Biosystematics of *Euseius* on California citrus and avocado with the description of a new species (Acari: Phytoseiidae). *Internat. J. Acarol.* 11:23–30.

DeBach, P. 1966. The competitive displacement and coexistence principles. *Annu. Rev. Entomol.* 11:183–212.

DeBach, P., and K. S. Hagen. 1964. Manipulation of entomophagous species. In *Biological control of insect pests and weeds,* ed. P. DeBach, 429–58. London: Chapman & Hall.

DeBach, P., and D. Rosen. 1991. *Biological control by natural enemies,* 2nd ed. London: Cambridge University Press.

DeBach, P., and P. Sisojevic. 1960. Some effects of temperature and competition on the distribution and relative abundance of *Aphytis lingnanensis* and *A. chrysomphali* (Hymenoptera: Aphelinidae). *Ecology* 41:153–60.

DeBach, P., and R. A. Sundby. 1963. Competitive displacement between ecological homologues. *Hilgardia* 34:105–66.

DeBach, P., T. W. Fisher, and J. Landi. 1955. Some effects of meteorological factors on all stages of *Aphytis lingnanensis*, a parasite of the California red scale. *Ecology* 36:743–53.

DeBach, P., C. B. Huffaker, and A. W. MacPhe. 1976. Evaluation of the impact of natural enemies. In *Theory and practice of biological control,* eds. C. B. Huffaker and P. S. Messenger, 255–85. New York: Academic Press.

DeBach, P., D. Rosen, and C. E. Kennett. 1971. Biological control of coccids by introduced natural enemies. In *Biological control,* ed. C. B. Huffaker, 165–94. New York: Plenum Press.

Delucchi, V. L., D. Rosen, and E. I. Schlinger. 1976. Relationship of systematics to biological control. In *Theory and practice of biological control,* eds. C. B. Huffaker and P. S. Messenger, 81–91. New York: Academic Press.

Dempster, J. P. 1983. The natural control of populations of butterflies and moths. *Biol. Rev.* 58:461–81.

den Boer, P. J. 1978. Spreading of risk and stabilization of animal numbers. *Acta Biotheor.* 18:165–94.

Diehl, S. R., and G. L. Bush. 1984. An evolutionary and applied perspective of insect biotypes. *Annu. Rev. Entomol.* 29:471–504.

Doutt, R. L., and P. DeBach. 1964. Some biological control concepts and questions. In *Biological control of insect pests and weeds,* ed. P. DeBach, 118–67. London: Chapman & Hall.

Ehler, L. E. 1992. Guild analysis in biological control. *Environ. Entomol.* 21:26–40.

Ehler, L. E., K. G. Eveleens, and R. van den Bosch. 1973. An evaluation of some natural enemies of cabbage looper on cotton in California. *Environ. Entomol.* 2:1009–15.

Eveleens, K. G., R. van den Bosch, and L. E. Ehler. 1973. Secondary outbreak induction of beet armyworm by experimental insecticide applications in cotton in California. *Environ. Entomol.* 2:497–503.

Flaherty, D. L., J. E. Pehrson, and E. E. Kennett. 1973. Citrus pest management studies in Tulare County. *Calif. Agric.* 27(11):3–7.

Flanders, S. E. 1931. Temperature relationships of *Trichogramma minutum* as a basis for racial segregation. *Hilgardia* 44:1–25.

———. 1950. Races of apomictic parasitic Hymenoptera introduced into California. *J. Econ. Entomol.* 43:719–20.

Flint, M. L. 1980. Climatic ecotypes in *Trioxys complanatus,* a parasite of the spotted alfalfa aphid. *Environ. Entomol.* 9:501–7.

Goeden, R. D. 1978. Initial analyses of *Rhinocyllus conicus* (Froelich) (Coleoptera: Curculionidae) as an introduced natural enemy of milk thistle (*Silybum marianum* [L.] Gaertner) and Italian thistle (*Carduus pycnocephalus* L.) in southern California. USDA Publication 771-106/02.

———. 1985. Host-plant relations of *Trupanea* spp. (Diptera: Tephritidae) in southern California. *Proc. Entomol. Soc. Wash.* 87:564–71.

———. 1987a. Host-plant relations of native *Urophora* spp. (Diptera: Tephritidae) in southern California. *Proc. Entomol. Soc. Wash.* 89:269–74.

———. 1987b. Life history of *Trupanea conjuncta* (Adams) on *Trixis californica* Kellogg in southern California (Diptera: Tephritidae). *Pan-Pacific Entomol.* 63:284–91.

———. 1988. Gall formation by the capitulum-infesting fruit fly, *Tephritis stigmatica* (Diptera: Tephritidae). *Proc. Entomol. Soc. Wash.* 90:37–43.

———. 1989. Host-plants of *Neaspilota* in California (Diptera: Tephritidae). *Proc. Entomol. Soc. Wash.* 91:164–68.

———. 1990. Life history of *Eutreta diana* (Osten Sacken) on *Artemisia tridentata* Nuttall in southern California (Diptera: Tephritidae). *Pan-Pacific Entomol.* 66:24–32.

Goeden, R. D., and F. L. Blanc. 1986. New synonymy, host, and California records in the genera *Dioxyna* and *Paroxyna* (Diptera: Tephritidae). *Pan-Pacific Entomol.* 62:88–90.

Goeden, R. D., and D. Headrick. 1990. Notes on the biology and immature stages of *Stenopa affinis* Quisenberry (Diptera: Tephritidae). *Proc. Entomol. Soc. Wash.* 92:641–48.

———. 1991. Life history and descriptions of immature stages of *Tephritis baccharis* (Coquillett) on *Baccharis salicifolia* (Ruiz & Pavon) Persoon in southern California (Diptera: Tephritidae). *Pan-Pacific Entomol.* 67:86–98.

Goeden, R. D., and S. M. Louda. 1976. Biotic interference with insects imported for weed control. *Annu. Rev. Entomol.* 21:325–42.

Goeden, R. D., and D. W. Ricker. 1977. Establishment of *Rhinocyllus conicus* on milk thistle in southern California. *Weed. Sci.* 25:288–92.

———. 1978. Establishment of *Rhinocyllus conicus* (Coleoptera: Curculionidae) on Italian thistle in southern California. *Environ. Entomol.* 7:787–89.

Goeden, R. D., D. W. Ricker, and B. A. Hawkins. 1985. Ethological and genetic differences among three biotypes of *Rhinocyllus conicus* (Coleoptera: Curculionidae) introduced into North America for the biological control of asteraceous thistles. In *Proceedings of the 6th International Symposium of Biological Control of Weeds, 19-25 August 1984, Vancouver, British Columbia,* ed. E. S. Delfosse, 181–89. Ottawa: Agriculture Canada.

Gordh, G. 1985a. A new species of *Paratetracnemoidea* Girault, 1915, found in North America, with a discussion of generic placement (Hymenoptera: Encyrtidae). *Fla. Entomol.* 8:587–94.

———. 1985b. *Uropoda* sp. phoretic on *Elater leconti* Horn. *Pan-Pacific Entomol.* 61(2):154.

———. 1986. A new species of *Goniozus* Foerster, 1851 from southern Africa parasitizing sugar cane borer, *Eldana saccharina* Walker, and taxonomic notes on species of the genus in Africa (Hymenoptera: Bethylidae; Lepidoptera: Pyralidae). *J. Entomol. Soc. So. Africa* 49:257–65.

———. 1987. A taxonomic study of Nearctic *Meromyzobia* Ashmead, 1900 (Hymenoptera: Encyrtidae). *Pan-Pacific Entomol.* 63(1):16–36.

———. 1988a. A new species of *Goniozus* from India and taxonomic notes on related species (Hymenoptera: Bethylidae). *Indian J. Entomol.* 48(4):361–65.

———. 1988b. A new species of *Goniozus* Foerster from India (Hymenoptera: Bethylidae) used in biological control of *Diaphania indica* (Saunders) (Lepidoptera: Pyralidae). *Pan-Pacific Entomol.* 64(2):173–82.

———. 1990. *Apenesia evansi* sp.n. (Hymenoptera: Bethylidae) from Australia with comments on phoretic copulation in bethylids. *J. Aust. Entomol. Soc.* 29:167–70.

Gordh, G., and R. E. Medved. 1986. Biological notes on *Goniozus pakmanus* Gordh (Hymenoptera: Bethylidae), a parasite of pink bollworm, *Pectinophora gossypiella* (Saunders) (Lepidoptera: Gelechiidae). *J. Kansas Entomol. Soc.* 59(4):723–34.

Gordh, G., and L. Moczàr. 1990. A catalog of the world Bethylidae. *Mem. Am. Entomol. Inst.* 46:1–364.

Gordh, G., and R. W. Orth. 1987. Insects and mites: Techniques for collection and preservation. *Proc. Entomol. Soc. Wash.* 84:

842–47.

Gordh, G., and L. Wills. 1989. Anatomical notes on *Uropoda* sp., a phoretic mite infesting dung-inhabiting beetles in southern California (Acari: Uropodidae; Coleoptera: Tenebrionidae, Histeridae). *Pan-Pacific Entomol.* 65(4):410–13.

Grafton-Cardwell, E. E., and M. A. Hoy. 1985. Intraspecific variability in response to pesticides in the common green lacewing, *Chrysoperla carnea* (Stephens) (Neuroptera: Chrysopidae). *Hilgardia* 53(6):32.

Greathead, D. J. 1986. Parasitoids in classical biological control. In *Insect Parasitoids*, eds. J. K. Waage and D. J. Greathead, 290–318. New York: Academic Press.

Gregory, W. A. 1985. In-flight response of citrus-inhabiting Aphelinidae (Hymenoptera) to trap colors and plant structures. Ph.D. diss., University of California, Riverside.

Hagen, K. S. 1986. Ecosystem analysis: Plant cultivars (HPR), entomophagous species and food supplements. In *Interactions of plant resistance and parasitoids and predators of insects*, eds. D. J. Boethel and R. D. Eikenbary, 151–97. Chichester, U.K.: Ellis Horwood.

———. 1987. Nutritional ecology of terrestrial insect predators. In *Nutritional ecology of insects, mites, and spiders*, eds. F. Slansky, Jr. and J. G. Rodriguez, 533–77. New York: Wiley.

Hagen, K. S., R. H. Dadd, and J. Reese. 1984. The food of insects. In *Ecological entomology*, eds. C. B. Huffaker and R. L. Rabb, 79–112. New York: Wiley.

Hassell, M. P. 1978. *The Dynamics of Arthropod-Prey Systems.* Princeton, N.J.: Princeton University Press.

Hassell, M. P., and C. G. Varley. 1969. New inductive population model for insect parasites and its bearing on biological control. *Nature* 233:1133–37.

Hassell, M. P., J. K. Waage, and R. M. May. 1983. Variable parasitoid sex ratios and their effect on host-parasitoid dynamics. *J. Anim. Ecol.* 52:889–904.

Hawkes, R. B., L. A. Andres, and T. H. Dunn. 1972. Seed weevil released to control milk thistle. *Calif. Agric.* 26(12):14.

Hawkins, B. A., and G. Gordh. 1986. Bibliography of the world literature of the Bethylidae (Hymenoptera: Bethyloidea). *Insecta Mundi* 1(4):261–83.

Headrick, D., and R. D. Goeden. 1990a. Description of the immature stages of *Paracantha gentilis* (Diptera: Tephritidae). *Ann. Entomol. Soc. Am.* 83:776–85.

———. 1990b. Life history of *Paracantha gentilis* Hering (Diptera: Tephritidae). *Ann. Entomol. Soc. Am.* 83:776–85.

———. 1991. Life history of *Trupanea californica* Malloch (Diptera: Tephritidae) on *Gnaphalium* spp. in southern California. *Proc. Entomol. Soc. Wash.* 91:559–70.

Holling, C. S. 1959. The components of predation as revealed by a study of small mammal predation of the European pine sawfly. *Can. Entomol.* 91:293–320.

Howard, L. O., and W. F. Fisk. 1911. The importation into the United States of the parasites of the gypsy-moth and the brown-tail moth. *Bull. Bur. Ent. USDA.* 91:1–312.

Hoy, M. A. 1982. Aerial dispersal and field efficacy of a genetical-ly improved strain of the spider mite predator *Metaseiulus occidentalis. Entomol. Exp. Appl.* 32:205–12.

———. 1985. Improving establishment of arthropod natural enemies. In *Biological control in agricultural IPM systems*, eds. M. A. Hoy and D. C. Herzog, 151–66. New York: Academic Press.

———. 1987. Developing insecticide resistance in insect and mite predators and opportunities for gene transfer. In *Biotechnology in agricultural chemistry*, eds. H. M. LeBaron, R. O. Mumma, R. C. Honneycutt, and J. H. Duesing, chap. 10, 125–38. Am. Chemical Soc. Symp. Series 334.

Hoy, M. A., and F. E. Cave. 1988. Guthion-resistant strain of walnut aphid parasite. *Calif. Agric.* 40(4):4–5.

———. 1991. Genetic improvement of a parasitoid: Response by *Trioxys pallidus* to laboratory selection with azinphosmethyl. *BioControl Sci. Tech.* 1:31–41.

Hoy, M. A., and S. A. Smilanick. 1981. Non-random prey location by the phytoseiid predator, *Metaseiulus occidentalis*: Differential response to several spider mite species. *Entomol. Exp. Appl.* 29:241–53.

Hoy, M. A., F. E. Cave, and M. A. Caprio. 1991. Guthion-resistant parasite ready for implementation in walnuts. *Calif. Agric.* 45(5):29–31.

Hoy, M. A., P. H. Westigard, and S. C. Hoyt. 1983. Release and evaluation of a laboratory-selected pyrethroid-resistant strain of the predaceous mite *Metaseiulus occidentalis* (Acari: Phytoseiidae) in a southern Oregon pear orchard and a Washington apple orchard. *J. Econ. Entomol.* 76:383–88.

Hoy, M. A., F. E. Cave, R. H. Beede, J. Grant, W. H. Krueger, W. J. Olson, K. M. Spollen, W. W. Barnett, and L. C. Hendricks. 1990. Release, dispersal, and recovery of a laboratory-selected strain of the walnut aphid parasite, *Trioxys pallidus* (Hymenoptera: Aphidiidae) resistant to azinphosmethyl. *J. Econ. Entomol.* 83:89–96.

Hoyt, S. C. 1969. Integrated control of insects and biological control of mites on apple in Washington. *J. Econ. Entomol.* 62:74–86.

Huffaker, C. B., and C. E. Kennett. 1966. Studies of two parasites of olive scale, *Parlatoria oleae* (Colvee) IV. Biological control of *P. oleae* through the compensatory action of two introduced parasites. *Hilgardia* 37:283–335.

Huffaker, C. B., and P. S. Messenger. 1964. The concept and significance of natural control. In *Biological control of insect pests and weeds*, ed. P. DeBach, 74–117. London: Chapman & Hall.

Johnson, M. W., and A. H. Hara. 1987. Influence of host crop on parasitoids (Hymenoptera) of *Liriomyza* spp. (Diptera: Agromyzidae). *Environ. Entomol.* 16:339–44.

Legner, E. F. 1987a. Transfer of thelytoky to arrhenotokous *Muscidifurax raptor* Girault & Sanders (Hymenoptera: Pteromalidae). *Can. Entomol.* 119:265–71.

———. 1987b. Pattern of thelytoky acquisition in *Muscidifurax raptor* Girault & Sanders (Hymenoptera: Pteromalidae). *Bull. Soc. Vect. Ecol.* 12(2):1–11.

———. 1987c. Inheritance of gregarious and solitary oviposition in *Muscidifurax raptorellus* Kogan & Legner (Hymenoptera: Pteromalidae). *Can. Entomol.* 119:791–808.

———. 1987d. Some quantitative aspects of inheritance in breeding synanthropic fly parasitoids. *Bull. Soc. Vect. Ecol.* 12(2):528–33.

———. 1987e. Further insights into extranuclear influences on behavior elicited by males in the genus *Muscidifurax* (Hymenoptera: Pteromalidae). *Proc. Calif. Mosq. and Vect. Contr. Assoc., Inc.* 55:127–30.

———. 1988. *Muscidifurax raptorellus* (Hymenoptera: Pteromalidae) females exhibit postmating oviposition behavior typical of the male genome. *Ann. Entomol. Soc. Am.* 81:522–27.

———. 1989a. Wary genes and accretive inheritance in Hymenoptera. *Ann. Entomol. Soc. Am.* 82:245–49.

———. 1989b. Paternal influences in males of *Muscidifurax raptorellus* (Hymenoptera: Pteromalidae). *Entomophaga* 34:307–20.

———. 1989c. Phenotypic expressions of polygenes in *Muscidifurax raptorellus* (Hymenoptera: Pteromalidae), a synanthropic fly parasitoid. *Entomophaga* 34:523–30.

———. 1990. Estimations of gene number, heritability, and dominance in quantitative inheritance, with special reference to Hymenoptera. *Proc. Calif. Mosq. and Vect. Contr. Assoc., Inc.* 57:95–105.

———. 1991a. Estimations of number of active loci, dominance, and heritability in polygenic inheritance of gregarious behavior in *Muscidifurax raptorellus* (Hymenoptera: Pteromalidae). *Entomophaga* 36:1–18.

———. 1991b. Recombinant males in the parasitic wasp *Muscidifurax raptorellus* (Hymenoptera: Pteromalidae). *Entomophaga* 36:173–81.

Lord, F. T. 1947. The influence of spray programs on the fauna of apple orchards in Nova Scotia: II. Oystershell scale, *Lepidosaphes ulmi* (L.). *Can. Entomol.* 81:202–30.

Luck, R. F. 1986. Biological control as an alternative pest control tactic: The case of California red scale on California citrus. In *National Research Council Committee on Application of Ecological Theory to Environmental Problems (CAETEP): Concepts and case studies,* 165–86. Washington, D.C.: National Academy of Science Press.

Luck, R. F., and D. L. Dahlsten. 1975. Natural decline of a pine needle scale (*Chionaspis pinifoliae* [Fitch]) outbreak at South Lake Tahoe, California, following cessation of adult mosquito control with malathion. *Ecology* 56:893–904.

Luck, R. F., and H. Podoler. 1985. Competitive exclusion of *Aphytis lingnanensis* by *A. melinus*: Potential role of host size. *Ecology* 66:904–13.

Luck, R. F., L. Nunney, and R. Stouthamer. In press. Factors affecting sex ratio and quality in the culturing of parasitic Hymenoptera: An evolutionary perspective. In *Principles and practices in biological control,* eds. T. W. Fisher, T. S. Bellows, G. Gordh, D. L. Dahlsten, L. E. Caltagirone, and C. B. Huffaker. Berkeley: University of California Press.

Luck, R. F., H. Podoler, and R. Kfir. 1982. Host selection and egg allocation behavior by *Aphytis melinus* and *A. lingnanensis*: Comparison of two facultatively gregarious parasitoids. *Ecol. Entomol.* 7:397–408.

Luck, R. F., R. Stouthamer, and L. P. Nunney. 1993. Sex determination and sex ratio patterns in parasitic Hymenoptera. In *Evolution and diversity of sex ratio in Haplodiploid insects and mites,* eds. D. L. Wrensch and M. A. Ebbert, 442–76. London: Chapman & Hall.

Luck, R. F., R. van den Bosch, and R. Garcia. 1977. Chemical insect control—A troubled pest management strategy. *BioScience* 27:606–11.

McEvoy, P. B. 1985. Depression in ragwort (*Senecio jacobaea*) abundance following introduction of *Tyria jacobaeae* and *Longitarsus jacobaeae* on the central coast of Oregon. In *Proceedings of the 7th International Symposium of Biological Control of Weeds, 19-25 August 1984, Vancouver, British Columbia,* ed. E. S. Delfosse, 57–64. Ottawa: Agriculture Canada.

———. 1989. Application of the theoretical basis for IPM. In *Proc. Nat. IPM Symposium/Workshop, 25–28 April 1989, Las Vegas, Nevada,* 6–12.

McEvoy, P. B., C. Cox, and E. Coombs. 1991. Successful biological control of ragwort, *Senecio jacobaea,* by introduced insects in Oregon. *Ecol. Appl.* 1:430–42.

McEvoy, P. B., C. Cox, R. R. James, and N. T. Rudd. 1990. Ecological mechanisms underlying successful biological weed control: Field experiments with ragwort, *Senecio jacobaea.* In *Proeedings of the. 7th International Symposium of Biological Control of Weeds, 6-11 March 1988, Rome, Italy,* ed. E. S. Delfosse, 53–64. Rome: Ministero dell'Agricoltura e delle Foreste/Melbourne: CSIRO.

McMurtry, J. A., and M. H. Badii. 1989. Reproductive compatibility in widely separated populations of three species of phytoseiid mites. *Pan-Pacific Entomol.* 65:397–402.

McMurtry, J. A., and G. J. de Moraes. 1989. Some phytoseiid mites from Peru with descriptions of four new species. *Internat. J. Acarol.* 15:179–88.

McMurtry, J. A., M. H. Badii, and B. D. Congdon. 1985. Studies on a *Euseius*-species complex on avocado in Mexico and Central America, with a description of a new species (Acari: Phytoseiidae). *Acarologia* 26:107–16.

Milbrath, L. R., M. J. Tauber, and C. A. Tauber. 1993. Prey specificity in *Chrysopa*: An interspecific comparison of larval feeding and defensive behavior. *Ecology* 74:1384–93.

Moreno, D. S., and R. F. Luck. 1992. Augmentative releases of *Aphytis melinus* (Hymenoptera: Aphelinidae) to suppress California red scale (Homoptera: Diaspididae) in southern California lemon orchards. *J. Econ. Entomol.* 85:1112–19.

Müller, H., G. S. Nuessly, and R. D. Goeden. 1990. Natural enemies and host-plant asynchrony contributing to the failure of the introduced moth, *Coleophora parthenica* Meyrick (Lepidoptera: Coleophoridae), to control Russian thistle. *Agric. Ecosys. and Environ.* 32:133–42.

Murdoch, W. W. 1990. The relevance of pest-enemy models to biological control. In *Critical issues in biological control,* eds. M. Mackauer, L. E. Ehler, and J. Roland, 1–24. Andover, U.K.: Intercept Press.

Murdoch, W. W., and A. Oaten. 1975. Predation and population stability. *Adv. Ecol. Res.* 9:1–131.

Murdoch, W. W., J. Chesson, and P. L. Chesson. 1985. Biological control in theory and practice. *Am. Nat.* 125:344–66.

Nadel, H., and R. F. Luck. 1992. Dispersal and mating structure of a parasitoid with a female-biased sex ratio: Implications for theory. *Evol. Ecol.* 6:270–78.

Nechols, J. R. 1987. Voltinism, seasonal reproduction, and diapause in the squash bug (Heteroptera: Coreidae) in Kansas. *Environ. Entomol.* 16:269–73.

———. 1988. Photoperiodic responses of the squash bug (Heteroptera: Coreidae): Diapause induction and maintenance. *Environ. Entomol.* 17:427–31.

Nechols, J. R., and R. S. Kikuchi. 1985. Host selection of the spherical mealybug (Homoptera: Pseudococcidae) by *Anagyrus indicus* (Hymenoptera: Encyrtidae): Influence of host stage on parasitoid oviposition, development, sex ratio, and survival. *Environ. Entomol.* 14:32–37.

Nechols, J. R., M. J. Tauber, and C. A. Tauber. 1987. Geographical variability in ecophysiological traits controlling dormancy in *Chrysopa oculata* (Neuroptera: Chrysopidae). *J. Insect Physiol.* 33:627–33.

Nechols, J. R., J. L. Tracy, and E. A. Vogt. 1989. Comparative ecological studies of indigenous egg parasitoids (Hymenoptera: Scelionidae; Encyrtidae) of the squash bug, *Anasa tristis* (Hemiptera: Coreidae). *J. Kansas Entomol. Soc.* 62:177–88.

Nicholson, A. J. 1933. The balance of animal populations. *J. Anim. Ecol.* 2:132–78.

Nicholson, A. J., and V. A. Bailey. 1935. An outline of the dynamics of animal populations. *Aust. J. Zool.* 2:9–65.

Nunney, L., and R. F. Luck. 1988. Factors influencing the optimal sex ratio in a structured population. *Theor. Popul. Biol.* 31:1–31.

Obrycki, J. J., and M. J. Tauber. 1979. Seasonal synchrony of the parasite *Perilitus coccinellae* and its host *Coleomegilla maculata.* *Environ. Entomol.* 8:400–5.

———. 1981. Phenology of three coccinellid species: Thermal requirements for development. *Ann. Entomol. Soc. Am.* 74:31–36.

———. 1982. Thermal requirements for development of *Hippodamia convergens* (Coleoptera: Coccinellidae). *Ann. Ent. Soc. Am.* 75:678–83.

Obrycki, J. J., M. J. Tauber, and W. M. Tingey. 1986. Comparative toxicity of pesticides to *Edovum puttleri* (Hymenoptera: Eulophidae), an egg parasitoid of the Colorado potato beetle (Coleoptera: Chrysomelidae). *J. Econ. Entomol.* 79:948–51.

Obrycki, J. J., M. J. Tauber, C. A. Tauber, and B. Gollands. 1983. Environmental control of the seasonal life cycle of *Adalia*

bipunctata. Environ. Entomol. 12:416–21.

———. 1987. Developmental responses of the Mexican biotype of *Edovum puttleri* (Hymenoptera: Eulophidae) to temperature and photoperiod. *Environ. Entomol.* 16:1319–23.

Pemberton, R. W., and C. W. Turner. 1989. Occurrence of predatory and fungivorous mites in leaf domatia. *Am. J. Bot.* 76:105–12.

———. 1990. Biological control of *Senecio jacobaea* in northern California, an enduring success. *Entomophaga* 35:71–77.

Reeve, J. D., and W. W. Murdoch. 1986. Biological control by the parasitoid *Aphytis melinus* and population stability of the California red scale. *J. Anim. Ecol.* 55:1069–82.

Root, R. B. 1967. The niche exploitation pattern of the blue-gray gnat-catcher. *Ecological Monographs* 37:317–50.

Rosen, D., and P. DeBach. 1979. *Species of Aphytis of the world.* The Hague: Junk.

Rosenheim, J. A., and M. A. Hoy. 1986. Intraspecific variation in levels of pesticide resistance in field populations of a parasitoid, *Aphytis melinus* (Hymenoptera: Aphelinidae): The role of past selection pressures. *J. Econ. Entomol.* 79:1161–73.

Ruberson, J. R., M. J. Tauber, and C. A. Tauber. 1987. Biotypes of *Edovum puttleri* (Hymenoptera: Eulophidae): Responses to developing eggs of the Colorado potato beetle (Coleoptera: Chrysomelidae). *Ann. Entomol. Soc. Am.* 80:451–55.

———. 1988. Reproductive biology of two biotypes of *Edovum puttleri,* an egg parasitoid of the Colorado potato beetle. *Entomol. Exp. Appl.* 46:211–19.

———. 1989. Intraspecific variability in hymenopteran parasitoids: Comparative studies of two biotypes of the egg parasitoid *Edovum puttleri* (Hymenoptera: Eulophidae). *J. Kansas Entomol. Soc.* 62:189–202.

Ruberson, J. R., M. J. Tauber, C. A. Tauber, and W. M. Tingey. 1989. Interactions at three trophic levels: *Edovum puttleri* Grissell (Hymenoptera: Eulophidae), the Colorado potato beetle, and insect-resistant potatoes. *Can. Entomol.* 121:841–51.

Salt, G., and R. van den Bosch. 1967. The defense reactions of three species of *Hypera* (Coleoptera: Curculionidae) to an ichneumon wasp. *J. Invert. Pathol.* 9:164–77.

Smith, A. D. M., and E. A. Maelzer. 1986. Aggregation of parasitoids and density-independence of the wasp *Aphytis melinus* and its host, the red scale, *Aonidiella aurantii. Ecol. Entomol.* 11:425–34.

Smith, H. S. 1935. The role of biotic factors in the determination of population densities. *J. Econ. Entomol.* 28:873–98.

Solomon, M. E. 1949. The natural control of animal populations. *J. Anim. Ecol.* 18:1–35.

Stouthamer, R., R. F. Luck, and W. D. Hamilton. 1990. Antibiotics cause parthenogenetic *Trichogramma* (Hymenoptera: Trichogrammatidae) to revert to sex. *Proc. Nat. Acad. Sci.* 87:2424–27.

Strong, D. R., J. H. Lawton, and R. Southwood. 1984. *Insects on plants: Community pattern and mechanisms.* Cambridge, Mass.: Harvard University Press.

Tachikawa, T., and G. Gordh. 1987. A new genus and species of

Encyrtidae (Hymenoptera: Chalcidoidea) parasitic on *Idiococcus* (Homoptera: Pseudococcidae) in Japan. *Shikoku Ent. Soc.* 18:305–9.

Tauber, C. A. 1974. Systematics of North American chrysopid larvae: *Chrysopa carnea* group (Neuroptera). *Can. Entomol.* 106:1133–53.

———. 1975. Larval characteristics and taxonomic position of the lacewing genus *Suarius. Ann. Entomol. Soc. Am.* 68:695–700.

———. 1991a. Raphidiodea. In *An introduction to immature insects of North America,* ed. F. W. Stehr, vol. 2, chap. 32, 123–25. Dubuque, Ia.: Kendall/Hunt.

———. 1991b. Neuroptera. In *An introduction to immature insects of North America,* ed. F. W. Stehr, vol. 2, chap. 33, 126–43. Dubuque, Ia.: Kendall/Hunt.

Tauber, C. A., and P. A. Adams. 1989. Systematics of Neuropteroidea: Present status and future needs. In *Systematics of the North American insects and arachnids: Status and needs,* eds. M. Kosztarab and C. W. Schaefer, 151–64. Virginia Agr. Expt. Sta. Info. Ser. 90-1. Blacksburg, Virginia: Virginia Polytechnic Institute and State University.

———. 1992. Phenotypic plasticity in *Chrysoperla:* Genetic variation in the sensory mechanism and in correlated reproductive traits. *Evolution* 46:1754–73.

Tauber, C. A., and M. J. Tauber. 1977a. Sympatric speciation based on allelic changes at three loci: Evidence from natural populations in two habitats. *Science* 197:1298–99.

———. 1977b. A genetic model for sympatric speciation through habitat diversification and seasonal isolation. *Nature* 268:702–5.

———. 1981. Insect seasonal cycles: Genetics and evolution. *Annu. Rev. Ecol. Syst.* 12:281–308.

———. 1982. Evolution of seasonal adaptations and life history traits in *Chrysopa:* Response to diverse selective pressures. In *Evolution and genetics of life histories,* eds. H. Dingle and J. P. Hegmann, 51–72. New York: Springer Verlag.

———. 1986a. Ecophysiological responses in life-history evolution: Evidence for their importance in a geographically widespread insect-species complex. *Can. J. Zool.* 64:875–84.

———. 1986b. Genetic variation in all-or-none life-history traits of the lacewing *Chrysoperla carnea. Can. J. Zool.* 64:1542–44.

———. 1987a. Food specificity in predacious insects: A comparative ecophysiological and genetic study. *Evol. Ecol.* 1:175–86.

———. 1987b. Inheritance of seasonal cycles in *Chrysoperla. Genetical Research, Cambridge* 49:215–23.

———. 1989. Sympatric speciation in insects: Perception and perspective. In *Speciation and its consequences,* eds. D. Otte and J. A. Endler, 302–44. Sunderland, Mass.: Sinauer.

Tauber, C. A., J. B. Johnson, and M. J. Tauber. 1992. Larval and developmental characteristics of the endemic Hawaiian lacewing, *Anomalochrysa frater* (Neuroptera: Chrysopidae). *Ann. Entomol. Soc. Am.* 85:200–6.

Tauber, C. A., M. J. Tauber, and J. R. Nechols. 1987. Thermal requirements for development in *Chrysopa oculata:* A geo-

graphically stable trait. *Ecology* 68:1479–87.

Tauber, C. A., M. J. Tauber, B. Gollands, R. J. Wright, and J. J. Obrycki. 1988. Preimaginal development and reproductive responses to temperature in two populations of the Colorado potato beetle (Coleoptera: Chrysomelidae). *Ann. Entomol. Soc. Am.* 81:755–73.

Tauber, M. J., and C. A. Tauber. 1976. Insect seasonality: Diapause maintenance, termination, and postdiapause development. *Annu. Rev. Entomol.* 21:81–107.

———. 1983. Life history traits of *Chrysopa carnea* and *Chrysopa rufilabris* (Neuroptera: Chrysopidae): Influence of humidity. *Ann. Entomol. Soc. Am.* 76:282–85.

———. 1993. Adaptations to temporal variation in habitats: Categorizing, predicting, and influencing their evolution in agroecosystems. In *Evolution of insect pests: Patterns of variation,* eds. K. C. Kim and B. A. McPheron, 103–27. New York: Wiley.

Tauber, M. J., M. A. Hoy, and D. C. Herzog. 1985. Biological control in agricultural IPM systems: A brief overview of the current status and future prospects. In *Biological control in agricultural IPM systems,* eds. M. A. Hoy and D. C. Herzog, 3–9. Orlando, Fla.: Academic Press.

Tauber, M. J., C. A. Tauber, and S. Masaki. 1984. Adaptations to hazardous seasonal conditions: Dormancy, migration, and polyphenism. In *Ecological entomology,* eds. C. B. Huffaker and R. L. Rabb, 149–83. New York: Wiley-Interscience.

———. 1986. *Seasonal adaptations of insects.* Oxford: Oxford University Press.

Tauber, M. J., C. A. Tauber, J. R. Nechols, and R. G. Helgesen. 1982. A new role for temperature in insect dormancy: Cold maintains diapause in temperate zone Diptera. *Science* 218:690–91.

Tauber, M. J., C. A. Tauber, J. R. Nechols, and J. J. Obrycki. 1983. Seasonal activity of parasitoids: Control by external, internal, and genetic factors. In *Diapause and life cycle strategies in insects,* eds. V. K. Brown and I. Hodek, 87–108. The Hague: Junk.

Tauber, M. J., C. A. Tauber, J. J. Obrycki, B. Gollands, and R. J. Wright. 1988a. Voltinism and the induction of aestival diapause in the Colorado potato beetle, *Leptinotarsa decemlineata* (Coleoptera: Chrysomelidae) *Ann. Entomol. Soc. Am.* 81:748–54.

———. 1988b. Geographical variation in responses to photoperiod and temperature by *Leptinotarsa decemlineata* (Coleoptera: Chrysomelidae) during and after dormancy. *Ann. Entomol. Soc. Am.* 81:764–73.

Tauber, M. J., C. A. Tauber, J. R. Ruberson, A. J. Tauber, and L. P. Abrahamson. 1989. Dormancy in *Lymantria dispar* (L.) (Lepidoptera: Lymantriidae): An analysis of photoperiodic and thermal responses. *Ann. Entomol. Soc. Am.* 83:494–503.

Thompson, W. R. 1939. Biological control and the theories of the interactions of populations. *Parasitology* 31:299–388.

Tracy, J. L., and J. R. Nechols. 1988. Comparison of thermal responses, reproductive biologies, and population growth potentials of the squash bug egg parasitoids *Ooencyrtus anasae* and *O.* sp. (Hymenoptera: Encyrtidae). *Environ. Entomol.* 17:636–43.

Unruh, T. R. In press. Molecular methods in classical biological control. In T. W. Fisher (ed.), *Principles and application of biological control*, chap. 13. Berkeley: University of California Press.

Unruh, T. R., and R. D. Goeden. 1987. Electrophoresis helps to identify which race of the introduced weevil, *Rhinocyllus conicus* (Coleoptera: Curculionidae), has transferred to two native southern California thistles. *Environ. Entomol.* 16:979–83.

Unruh, T. R., and R. H. Messing. 1993. Intraspecific biodiversity in Hymenoptera: Implications for conservation and biological control. In *Hymenoptera and biodiversity*, eds. J. LaSalle and I. D. Gauld, chap. 2, 27–52. Wallingford, U.K.: CAB International.

Unruh, T. R., W. White, D. González, and R. F. Luck. 1986. Electrophoretic studies of parasitic *Hymenoptera* and implications for biological control. *Misc. Publ. Entomol. Soc. Am.* 61:150–63.

Unruh, T. R., W. White, D. González, and J. B. Woolley. 1989. Genetic relationships among seventeen *Aphidius* populations (Hymenoptera: Aphidiidae) including six species. *Ann. Entomol. Soc. Am.* 82:754–68.

Unruh, T. R., W. White, D. González, G. Gordh, and R. F. Luck. 1983. Heterozygosity and effective size in laboratory populations of *Aphidius ervi* (Hymenoptera: Aphidiidae). *Entomophaga* 28:245–58.

van den Bosch, R., and P. S. Messenger. 1973. *Biological control.* London: Intext.

van den Bosch, R., R. Hom, P. Matteson, B. D. Frazer, P. S. Messenger, and C. S. Davis. 1979. Biological control of the walnut aphid in California: Impact of the parasite *Trioxys pallidus. Hilgardia* 47:1–13.

Varley, C. G. 1947. The natural control of population balance in the knapweed gall-fly (*Urophora jaceana*). *J. Anim. Ecol.* 16:139–87.

Varley, C. G., G. R. Gradwell, and M. P. Hassell. 1973. *Insect population ecology.* Berkeley: University of California Press.

Waage, J. K., and M. P. Hassell. 1982. Parasitoids as biological control agents—A fundamental approach. *Parasitology* 84:241–68.

Walker, G., and G. Gordh. 1989. The occurrence of apical labial sensilla in the Aleyrodidae and evidence for a contact chemosensory function. *Entomol. Exp. Appl.* 51:215–24.

Yu, D. S. 1986. The interactions between California red scale, *Aonidiella aurantii* (Maskell), and its parasitoids in citrus groves of inland southern California. Ph.D. diss., University of California, Riverside.

Yu, D. S., R. F. Luck, and W. W. Murdoch. 1990. Competition, resource partitioning and coexistence of an endoparasitoid, *Encarsia perniciosi,* and an ectoparasitoid, *Aphytis melinus,* of the California red scale. *Ecol. Entomol.* 15:469–80.

Zwölfer, H., and M. Preiss. 1983. Host selection and oviposition behavior in west-European ecotypes of *Rhinocyllus conicus* Froel. (Coleoptera: Curculionidae). *Zeit. angew. Entomol.* 95:113–22.

4 / THE ECONOMIC, ENVIRONMENTAL, AND SOCIOPOLITICAL IMPACT OF BIOLOGICAL CONTROL

M. T. ALINIAZEE

Biological control can be defined as the deliberate use of natural enemies—predators, parasites (parasitoids), and pathogens—to suppress and maintain populations of pest species at levels that do not cause economic injury. This control is achieved by augmenting natural enemies (which includes inundative releases as a curative measure), by conserving natural enemies, or by employing the "classical" approach, which includes selecting and importing natural enemies from foreign lands and then mass-rearing, releasing, and establishing them. These deliberate efforts are complemented by natural control—that is, by the suppressive actions of naturally occurring enemies and other environmental factors.

Over the past half-century the remarkable impact of these endeavors has been documented in detail (DeBach 1964; Huffaker and Messenger 1976; Ridgway and Vinson 1977; Waterhouse and Norris 1987; Fisher et al. in press). For example, F. Simmonds (1967) has estimated that seven biological control programs in the Commonwealth countries have returned an average of 1,000 percent annually over expenditures. Since P. DeBach's (1964) analysis of the benefits of biological control, there have been a number of other highly successful programs in California as well as in other parts of the world. K. H. Reichelderfer (1981) discussed a number of factors that have a direct impact—either positive or negative—on the economic feasibility of biological control (see table 4.1). Other factors, including public concern about the effects of pesticides, may also be counted among the economic benefits of biological control.

This chapter discusses the impact of the biological control work conducted under the auspices of the W-84 project during the past 25 years—an impact that has indeed been enormous. For example, the 1964 cost-benefit analysis presented by P. DeBach suggested that five successful biological control projects in California had provided a net savings of nearly $110 million between 1923 and 1959. Although DeBach did not include the cost of privately implemented programs in the expenditures, neither did he include any savings accruing from social and environmental benefits. In any case, the savings are impressive.

It is difficult to estimate the full impact of biological control efforts, because the indirect social and environmental benefits are seldom measured or even documented. For example, most growers—who are the direct beneficiaries of biological control and other forms of crop protection—are primarily concerned with the immediate economic effect of biological control (namely, the costs incurred and monetary profits derived on a short-term basis). The general public, on the other hand, is often interested in the longer-term social and environmental impacts and is less affected by short-term crop losses unless they are significant enough to cause market prices to rise.

Table 4.1. The relative impact of some specific factors on the economic feasibility of biological control.

Factors	Impact on economic feasibility
Pest severity	inverse
Pest spectrum in the field	inverse
Pest density consistency	positive
Technical feasibility of biological control agents	positive
Crop price and yield	positive
Risk of biological control	inverse
Costs of nonbiological control methods	positive
All costs of biological control methods	inverse
Net benefit of biological vs. alternative controls	positive
Existence of community and regional organization	positive

My sincere thanks to Drs. Stanley Hoyt, Marjorie Hoy, and Dennis Isaacson for their contribution to and review of this manuscript. I am also indebted to Dr. James Nechols for his patience, encouragement, and editorial assistance during the manuscript's preparation.

In the discussion that follows, I analyze some of the economic, environmental, social, and political impacts of biological control based on four case examples, all conducted in the Western Region. They include (1) a classical—that is, importation—program of biological control of an insect pest, (2) a natural enemy conservation and augmentation program, (3) an inoculative release program based on the genetic manipulation of a predator, and (4) a weed control program that used insects as natural enemies. This is not meant to be an exhaustive treatment of the subject, because several other important examples would also illustrate the impact of biological control in the Western Region. However, because of time and space constraints and because the other work was not conducted under the auspices of the W-84 project, I have not included those other studies here.

THE BIOLOGICAL CONTROL OF THE FILBERT APHID IN OREGON

The filbert aphid, *Myzocallis coryli*, is a key pest of nearly 35,000 acres of cultivated hazelnuts (also called filberts) in North America. A native of Europe, this monoecious aphid was accidentally introduced into the United States nearly one hundred years ago. It is now well established throughout the hazelnut-growing areas of the United States and Canada.

The History of the Program

After a detailed analysis of the pest status of different insects on hazelnuts in Oregon, M. T. AliNiazee (1983a) concluded that aphid control costs were counterproductive because they resulted in the devastation of natural enemies. Research showed that the aphid population increased only where chemicals were used extensively to control other pests. As a result of its extensive exposure to pesticides, the filbert aphid developed a resistance to many commonly used organophosphates, carbamates, and synthetic pyrethroids (AliNiazee 1983b; Katundu and AliNiazee 1990). Introducing newer insecticides has provided only a temporary solution: a more permanent answer lies in biological control.

Investigating the natural enemies of the filbert aphid, R. H. Messing and M. T. AliNiazee (1985) found that although some generalist predators were highly effective, they were disrupted by the use of chemicals and constrained by weather conditions. The researchers found no host-specific parasitoid (Messing and AliNiazee 1986). M. T. AliNiazee conducted a preliminary search for natural enemies in southern Italy during 1983 and found *Trioxys pallidus* in commercial hazelnut orchards (AliNiazee 1991). European entomologists

had reported the presence of *Trioxys* species on *M. coryli* earlier (Starý 1978). During 1984 R. H. Messing conducted an extensive European search for natural enemies (Messing and AliNiazee 1989). He found *T. pallidus* in good numbers and identified it as the key mortality factor of the filbert aphid in Europe. A project to import *T. pallidus* to Oregon was initiated in 1984, and early colonization and releases were made during 1984 and 1985.

Preliminary surveys indicated that the parasitoid became established within a year of its release (Messing and AliNiazee 1989). Further studies have shown that the parasitoid is highly effective: in areas where it was released and became established, aphid populations have dropped dramatically (AliNiazee 1991). Currently, *T. pallidus* is an extremely important part of the hazelnut agroecosystem in Oregon and Washington. For a general review of this filbert aphid program, also see chapter 28.

The Economic and Environmental Impact

The economic impact of classical biological control programs can be dramatic since, in many cases, the initial costs are low to moderate, and the benefits may be substantial and continue indefinitely. In the case of the filbert aphid program, the initial costs were extremely low and the benefits have been extremely high.

Reduced pesticide costs. Since the filbert aphid is one of the major pests of hazelnuts (AliNiazee 1980), growers have traditionally used different organophosphate insecticides to control it. A survey conducted in Oregon in the early 1980s showed that growers sprayed an average of 1.5 times per year to control this pest (AliNiazee et al. unpub. data). Using a conservative projected cost of $25 per acre per spray application and an estimate of about 80 percent of commercial orchards treated, every year Oregon hazelnut growers spent the following sum on chemical control:

$$24{,}000 \text{ acres} \times 1.5 \text{ applications} \times \$25 \text{ per acre} = \$900{,}000$$

In 1991—nearly 6 years after the release of *T. pallidus*—about 25 percent of Oregon's commercial growers had the parasitoid in their orchards and were reporting effective biological control (AliNiazee 1991). This represents a savings of nearly $225,000 a year in insecticide costs alone.

Reduced application costs. The cost of labor and equipment for spraying fruit tree crops averages about $20 per acre. Using this figure, the 1991 savings in application costs is an estimated $180,000. Since the range of the parasitoid is expanding rapidly, the savings should rise substantially during the next few years.

Reduced crop damage. Although many have justified the use of insecticides against the filbert aphid as a curative measure, studies have shown that the sprays actually have negative effects. For example, because they destroy natural enemies, and because the pests develop resistance to them, the sprays have increased the densities of at least three other pests in the filbert system as well as, over time, the filbert aphid populations themselves (AliNiazee 1983a,b). Studies have shown that orchards treated on a regular basis have the highest aphid densities. Consequently, chemical treatments have increased crop damage and yield losses. Biological control based on *Trioxys pallidus*, on the other hand, has reduced crop damage. Thus, in the long term, biological control results not only in less aphid damage but also in fewer losses to other pests such as leafrollers and scale insects, because their natural enemies are conserved.

Reduced pesticide resistance. Although this factor is difficult to measure, it is reasonable to assume that using fewer pesticides will delay the onset of resistance in both target and nontarget pests. If the pests are already resistant, reduced use might even reverse the resistance. Both of these phenomena are expected to have dramatic short- and long-term economic benefits for hazelnut production. Thus, *T. pallidus*–based control of the filbert aphid will undoubtedly influence the development of resistance in a number of hazelnut pests.

Reduced environmental contamination and residues on fruit. Perhaps the least appreciated aspect of biological control is that it reduces environmental contamination. The filbert aphid program has decreased toxic pesticide use by nearly 36,000 pounds a year. Another 5,000 pounds of pesticides used to control other pests such as leafrollers have also been cut. As the parasitoid spreads to new areas of Oregon's Willamette Valley, these cuts will increase dramatically. The result is much lower pesticide residues on treated nuts.

The promulgation of sustainable agriculture and organic farming. In recent years in North America, increasing emphasis has been placed on developing and adopting low-input, sustainable agriculture (LISA). The biological control of the filbert aphid should be a cornerstone of any LISA program in Oregon hazelnut production. In fact, some large-acreage farmers who are considering the LISA approach depend on *Trioxys pallidus* as their primary source of biological control. Many organic farmers also depend on this parasitoid—along with cultural practices—as their major pest-control component.

The biological control program against the filbert aphid has provided many economic benefits, and a cost-benefit analysis is impressive. The initial cost of about $8,900 (which included a graduate student's round-trip airfare, other transportation costs, general expenses, and salary, aside from the shipment costs) has already resulted in large savings for the hazelnut growers. Even if we include the ancillary costs (the salary of the primary investigator and the establishment costs of the first two years), no more than $30,000 has been invested in the program. Approximately $300,000 to $400,000 are being saved each year in spray costs alone. The total benefits so far are estimated at over $1 million and could reach nearly $10 million by the year 2000.

The Social and Political Impact

Although it is hard to measure the sociopolitical impact of biological control programs, recent publicity about toxic residues on food products has enhanced both the public's and growers' appreciation of biological control. As a result, a number of politicians have shown interest in alternative pest controls, including IPM and biological control. Legislators in the Oregon State legislative session of 1990 to 1992 talked extensively about IPM and biological control in their agricultural subcommittee and emphasized this approach for controlling other pests in Oregon. Furthermore, politicians from the agricultural counties of the state now speak about biological control as desirable and emphasize their support for such programs. Consequently there has been some increase in funding for research on alternative technologies and a stricter enforcement of restrictions on pesticide use. However, the publicity and enthusiasm surrounding biological control can lead to an oversimplification of the technology and result in unreasonable expectations.

THE BIOLOGICAL CONTROL OF SPIDER MITES ON APPLES IN WASHINGTON

Another important biological control program conducted exclusively under the auspices of W-84 dealt with spider mite management in apple orchards in Washington. The program, directed by S. C. Hoyt of Washington State University at Wenatchee, sought to conserve beneficial predatory mites—particularly *Galendromus* (=*Typhlodromus* or *Metaseiulus*) *occidentalis*—through a very selective use of pesticides. To understand the profound impact of this IPM program on the Washington apple industry, it is important to understand the circumstances that existed before the program was established.

The History of the Program

In 1964, the state of Washington had approximately 90,000 acres of apple orchards. At that time long-term storage in controlled atmospheres was still in its infancy, and most of the fruit was sold in domestic markets. Growers controlled mites entirely with chemicals, spraying between two and seven—for an average of about three—times a year. Even with this extensive use of acaricides, control was often inadequate because of the mites' high reproductive rates and fairly widespread pesticide resistance (Hoyt 1969; see also chap. 7).

In the majority of orchards, the McDaniel spider mite, *Tetranychus mcdanieli*, was the most severe mite pest. Despite the extensive use of acaricides, high pest populations usually developed late in the season. The foliage in many orchards showed a brown discoloration, which translated into poor fruit color and reduced fruit size. Many apples did not mature properly, had a short storage life, and contained live mites in the calyx, making them difficult to market. The relief afforded by the development of new acaricides was only temporary. Furthermore, the high residue of acaricides on the fruit was a serious problem. Growing apples under these conditions was not a winning proposition, and some growers gave voice to their sense of impending disaster.

By 1989, an integrated pest management program had been widely adopted. A survey of that year's crop indicated that only 6.7 percent of the acreage had been treated with an acaricide, and that the average for those growers was just one application per year (J. F. Brunner person. commun.). Mite damage resulting in poor fruit color, small fruit, and shortened storage life seldom occurred. McDaniel spider mites were rarely found in the calyx of apples. As a result of this program and of other favorable market conditions, the apple industry had expanded to over 160,000 acres. Almost 20 percent of the crop—12,000,000 boxes—was shipped abroad, and long-term storage in controlled atmospheres (by then a well-established practice) allowed year-round marketing.

The Economic and Environmental Impact

Although it is difficult to evaluate the true costs and value of the apple mite management program, the following is a reasonably accurate assessment of the major benefits and costs incurred.

Reduced pesticide costs. Using the estimated treated acreage averages for 1964 and 1989 given earlier and a conservative figure for acaricide costs of $30 per acre (adjusted for inflation), the approximate 1964 pesticide costs would have been

$$90,000 \text{ acres} \times 3 \text{ applications} \times \$30 \text{ per acre} = \$8,100,000$$

By comparison, the approximate 1989 costs would have been

$$6.7\% \times 160,000 \text{ acres} = 10,720 \text{ treated acres} \times \$30 \text{ per acre} = \$321,600$$

These calculations indicate a savings of more than $7,778,000 per year.

Reduced application costs. A conservative estimate for pesticide application is $20 per acre. Comparing 1989 with 1964, approximately $5,185,000 was saved in application costs. However, since applications are frequently made for more than one pest at a time, it is more likely that about 50 percent of that value, or $2,590,000, was actually saved.

Reduced environmental contamination and fruit residues. Using the preceding figures and assuming an average of 2 pounds of active ingredient (AI) per acre, the total acaricide used in 1964 would have been

$$90,000 \text{ acres} \times 3 \text{ applications} \times 2 \text{ lb AI per acre} = 540,000 \text{ lb AI}$$

The comparable figure for 1989 would be

$$6.7\% \times 160,000 \text{ acres} = 10,720 \text{ acres} \times 2 \text{ lb AI per acre} = 21,440 \text{ lb AI}$$

These calculations indicate that 518,560 fewer pounds were used—a 96 percent decrease. An additional benefit was that the majority of the 1989 fruit was free of acaricide residue. Although we have no basis for measuring the value of this change, in today's climate of environmental and food safety concerns, these types of pesticide reductions are very important.

Reduced crop damage. It is hard to measure the reduction in crop damage resulting from the biological control of mites as opposed to their chemical control, because of variation in damage from year to year and orchard to orchard and because other stresses on trees can cause effects similar to those caused by mites. Additionally, other market factors influence the value of fruit each year. However, the downgrading of fruit, small fruit size, and reduced storage life caused by mite damage are now seldom encountered, and there is little question that the biological control of mites has increased revenues from fruit sales by several million dollars a year. In years with a high potential for mite population development, the benefit from reduced crop damage is likely to be even more significant than the monetary benefits that have already been described. For example, if just 10 percent of the 1989 crop had been downgraded by one grade due to mite damage, revenue losses would have been $7 to $14 million. In a bad mite year, a higher percentage of the crop would be downgraded, and there would be additional losses due to smaller fruit size and reduced storage life.

Increased fruit exports. Before the use of the IPM program, it was common to find McDaniel spider mites in the calyx of apples, a problem that seldom occurs now. Because many countries do not allow fruit infested with pests to be imported, the change to biological control has helped expand Washington's apple exports. The state now exports approximately 12,000,000 bushels per year. Although it is impossible to determine the exact monetary value of the mite IPM program to fruit exports, it certainly has been substantial. For example, without the export markets, current production would exceed domestic use, resulting in either a smaller industry or lower prices.

The Washington apple industry achieved its present size (for example, 1989 sales of $436,000,000) because of several developments, including the controlled atmosphere storage for year-round marketing, improved horticultural practices that increased yield and fruit quality, and the development of export markets. The biological control program was essential for each of these developments.

Public agencies have borne the costs of the state program's research, development, implementation, and maintenance. Table 4.2 shows estimates of those costs, using current salary, benefits, and operations figures.

Table 4.2. Costs associated with Washington's IPM program for the apple industry.

Research and development	$630,000
Implementation	$180,000
Maintenance	$125,000
Total	$935,000

Overall, costs to the Washington apple industry for integrated mite management have been relatively minor and, for the most part, of short duration. Included in these costs was the initial change from a chemically based program to one of biological control—a change requiring the gradual buildup of predator populations. Because acaricides could not be used during that period, an estimated 30 percent of the orchards sustained considerable mite damage in the first year; approximately 5 percent of the orchards required 2 years to establish biological control. However, the extent of the damage was usually no more than that observed when acaricides were used. In the remaining 65 percent of the orchards, the change was made with minimal mite damage.

A second cost stemmed from the sequential sampling of mite and predator populations that was recommended during the first 2 years of the program. This sampling and counting cost about $6 per acre per season. Due to a shortage of time and of trained personnel, only about 10 percent of the apple acreage was sampled, resulting in a cost of some $54,000 per year. As the apple rust mite and the predatory mite *Galendromus occidentalis* became established, biological control became stable and sampling was generally abandoned.

A third cost resulted from the elimination from commercial recommendations of several broad-spectrum pesticides that are toxic to predaceous mites. Fortunately, it has been possible to maintain effective control of all arthropod and disease pests of apples with selective chemicals and biological control methods. Therefore, this changeover has remained a potential rather than a real cost to apple producers. As reregistration curtails the use of certain chemicals, or as chemicals become ineffective when the pests—including aphids, white apple leafhoppers, western tentiform leafminers, and codling moths—develop resistance, pesticide options are declining.

Taking into account the program's short-term transition costs coupled with the ability to substitute selective chemicals for broad-spectrum pesticides of nonmite pests, during its 26 years the program has been implemented at a negligible cost to growers and consumers. In fact, the benefits in any one year have far exceeded the entire cost of the program. For these reasons the use of biological control as well as of other alternative tactics, such as mating disruption with pheromones, will probably continue to be essential for the control of mites.

THE BIOLOGICAL CONTROL OF SPIDER MITES IN CALIFORNIA ALMONDS

The release of large numbers of effective natural enemies—known as inoculative or inundative augmentation—on a regular basis to control arthropod pests is not a new idea (see DeBach 1964; Ridgway and Vinson 1977). However, using the augmentative approach successfully requires the ability to mass-produce and release high-quality—that is, potentially effective—natural enemies at reasonable costs. M. A. Hoy and coworkers, under the sponsorship of W-84, have conducted research that has laid the foundation for the successful control of spider mites on almonds in California. Their program is based on inoculative releases of genetically improved natural enemies (see also chap. 8).

The History of the Program

Several species of mites are important pests of almonds. These include the Pacific spider mite, *Tetranychus pacificus* McGregor; the twospotted spider mite, *Tetranychus urticae* Koch; the strawberry mite, *Tetranychus turkestani* Ugarov & Nikolski; the brown mite, *Bryobia rubrioculus* (Scheuten); the citrus red mite, *Panonychus citri* (McGregor); and the European red mite, *Panonychus ulmi* (Koch) (Hoy et al. 1982). These pests are kept in check by a number of beneficial species, the most important of which is a phytoseiid mite, *Galendromus* (=*Typhlodromus* or *Metaseiulus*) *occidentalis* (Nesbitt).

Under field conditions in different parts of North America, *G. occidentalis* has developed a resistance to a number of organophosphorus pesticides (AliNiazee et al. 1974; Hoy et al. 1979; Hoy and Knop 1979; Croft and AliNiazee 1983; Hoy and Conley 1987). It has also been possible to rear this predatory mite in large numbers under laboratory conditions, genetically selecting it for resistance (Hoy and Knop 1981; Roush and Hoy 1981a; Hoy 1985). Researchers were able to develop a population resistant to organophosphates, carbaryl, and sulfur (Roush and Hoy 1981a,b; Hoy et al. 1982; Hoy and Standow 1982; Hoy, Castro, and Cahn 1982). When they released this population in California almond orchards, it became established, overwintered, survived carbaryl and organophosphate applications, and controlled spider mites (Hoy 1982; Hoy et al. 1982; Hoy 1991). Unlike in the Washington apple program, however, selective acaricides and monitoring were necessary even after the predators became established, because almond trees are much more sensitive to feeding damage than are apples. This made it necessary to develop an integrated program to minimize the use of pesticides for the control of insect pests as well as of spider mites.

The laboratory-selected strain of *G. occidentalis* may represent the first case in which a genetically manipulated natural enemy proved effective in the field. The spread and efficacy of these laboratory-reared, pesticide-resistant predators were monitored (Roush and Hoy 1981a; Hoy 1982) and found to be superior to native populations. Researchers noticed no loss in vigor in the carbaryl-OP-sulfur-resistant strain as a result of laboratory rearing (Hoy et al. 1982). Approximately 67 percent of California almond acreage is managed by an integrated mite management (IMM) program, which includes the use of genetically manipulated predators.

The Economic and Environmental Impact

Reduced pesticide costs. J. C. Headley and M. A. Hoy (1987) based their economic analysis of the IMM program on three assumptions about adoption rates. First, they assumed that 75 percent of the nearly 400,000 acres of California almonds would require pesticide treatment for mite control, and that 25 percent of that area (about 75,000 acres) would be involved in the program in the first year. Second, they assumed that in the first year growers would need to release genetically selected *G. occidentalis* in 20 percent of those 75,000 acres. Third, they speculated that by the second year 50 percent of the spider mite–infested area (200,000 acres)

would be managed according to IMM guidelines, and that 20 percent of the area involved in this program would require releases of predators. During the third year, 75 percent of the acreage (300,000 acres) would be managed under IMM guidelines, and 20 percent would require predator releases. A large part of the acreage would rely on naturally occurring (that is, unselected) *G. occidentalis*, along with established populations of genetically selected, pesticide-resistant *G. occidentalis*. The authors theorized that the genetically manipulated predators would persist in almond orchards for years if they were not sprayed with pesticides to which they were susceptible.

Before having access to the integrated mite control program, almond growers applied an average of 1.5 miticide treatments per year (Headley and Hoy 1985). Assuming a cost of about $30 per acre for the chemicals, and assuming that the program was implemented in 67 percent of the 400,000 acres, the reduction in pesticide costs alone would be calculated as follows:

$$268,000 \text{ acres} \times 1.5 \text{ applications} \times \$30 \text{ per acre} = \$12,060,000$$

Similarly, assuming spray application costs of $20 per acre would realize the following savings:

$$268,000 \text{ acres} \times 1.5 \text{ applications} \times \$20 \text{ per acre} = \$8,040,000$$

The combined savings, then, would be $20,100,000 per year. Even after subtracting the costs of releasing and monitoring the predatory mites—which was estimated at $20 and $10 per acre, respectively—the savings would be phenomenal.

Reduced environmental contamination and fruit residues. Another result of this successful biological control program is the reduction of total pesticide usage and residues on treated almonds. For example, using the program on just 50 percent of the acreage would result in a reduction of 300,000 to 600,000 pounds of toxicant per year. A 90 percent adoption rate would reduce toxicant use by 540,000 to 1,080,000 pounds per year. An additional benefit would be the likely reduction in the amount of pesticides used for secondary pests, since their natural enemies would be conserved. The overall reduction in residues on treated products would also be substantial. Based on the recent estimate of nearly 67 percent of acreage being dedicated to the program, its benefits are already very impressive.

Reduced crop damage. Even where acaricides are used, untimely sprays and poor application techniques can cause crops substantial damage. In some cases, the damage may not be discovered early enough to fix the problem. A program based on *G. occidentalis*, in contrast, can provide more timely and effective control, because the predators respond to increasing pest densities and keep them in check. Environmental and cultural factors that affect pests also affect predators—a fact that in most cases results in an excellent balance between predator and pest densities. In general, an integrated mite management program using *G. occidentalis* can provide good control with less damage than can a one-sided chemical control program.

Reduced pesticide resistance. IPM researchers and practitioners alike are very concerned with pesticide resistance and its management. Whereas the onset of resistance in pests can be disastrous, in a beneficial species it is actually advantageous. Shifting from an insecticide-based program to a biological control program has two major benefits in this context. First, it may delay or, in some cases, reverse resistance in the target pests by reducing the selective pressure from pesticides. Second, it may reduce the exposure of secondary pests to pesticides, thus suppressing the development of other resistant strains. Although it is difficult to estimate the impact of reduced pesticide use in almond culture, it may delay the onset of resistance in pests such as scale insects and the navel orangeworm. If so, this would be a very important secondary benefit.

The promulgation of sustainable and organic farming methods. Recently, a number of reports and public forums have focused on insecticide residues on food products. Both the public and the private sector are now placing increasing emphasis on sustainable and organic farming, and the demand for and high prices of organic produce suggest that the market is booming. With an effective biological control program against spider mites, California almond growers can now begin to think about types of production that involve minimal pesticides.

The Social and Political Impact

The social impact of this program is beginning to emerge in California. Growers who adopt the program can reduce the cost of production, eliminate some pesticide sprays, and enjoy a higher income. Consumers, too, welcome the news of reduced pesticide use. However, a small proportion of the pesticide industry might experience a decline in income.

THE BIOLOGICAL CONTROL OF TANSY RAGWORT

Tansy ragwort, *Senecio jacobaea* L., is an important weed pest found along the Pacific Coast in British Columbia,

Washington, Oregon, and northern California (Frick and Holloway 1964). Infestations of ragwort date back to the early 1900s (Harris et al. 1971; Isaacson 1978a), when the pest was accidentally introduced from Europe. Since then the weed has expanded its geographic distribution rapidly, increased in density, and invaded pastureland throughout western North America, where it has displaced desirable vegetation. In some areas damage has been severe, resulting in cattle losses and prompting ranchers to invest in costly preventive measures.

The History of the Program

Chemical, mechanical, and cultural controls have been used individually and collectively against tansy ragwort, with mixed results. Chemical control, though effective, is often not economical (Isaacson 1973). Mechanical control—namely, physically removing the weed—is only partly successful, and can be applied only in areas with isolated and minor infestations. Mowing has been suggested, but its results are variable. Other cultural control programs—planting competitive, aggressive grasses and employing other soil and range management strategies—are suitable under certain conditions but are not widely used. In contrast, the biological control of tansy ragwort has been highly successful in both northern California (Pemberton and Turner 1990) and Oregon (McEvoy, Cox, and Coombs 1991). Many W-84 members have been involved in research on this weed and have been instrumental in the program's success (see chap. 71).

The biological control program began in 1951 with the importation and release of two European natural enemies: the cinnabar moth, *Tyria jacobaeae* L., and a seed head fly, *Pegohylemyia seneciella* Meade (Frick and Holloway 1964). The cinnabar moth rapidly became established in many areas and at some sites caused nearly 100 percent defoliation within a short period (Hawkes and Johnson 1978). Later a flea beetle, *Longitarsus jacobaeae* (Waterhouse), was imported from Italy; it was released in the United States during 1968, 1969, and 1970 (Frick and Johnson 1973). By 1970 it had become established, and further colonization at other sites was initiated. Early results with this natural enemy have been very impressive.

The success of the biological control of tansy ragwort is mainly due to the complementary action of the cinnabar moth and the flea beetle (Hawkes and Johnson 1978; McEvoy 1985; Piper 1985). The economic impact has been enormous, in terms of both reduced weed infestation and decline in cattle mortality. Most of the data presented here are from Oregon, where the program began in 1960 and is still proving highly successful.

The Economic and Environmental Impact

Reduced pest infestations. Between August 1972 and January 1973, an estimated 3 to 9 million acres were infested by tansy ragwort (Isaacson 1978b). Later, in 1976, an aerial survey indicated that in western Oregon over 3 million acres of pastureland were infested (Isaacson 1978a). However, after the natural enemies became established, infestation rates dropped sharply; in many areas, tansy ragwort has become a much less serious problem. Oregon's Department of Agriculture has estimated that, as of 1988, in some areas tansy ragwort has declined by over 70 percent (Brown 1990). The decline of the weed has resulted in more suitable pastures for livestock.

In a detailed survey of the sites in western Oregon, R. E. Brown (1990) and P. B. McEvoy, C. C. Cox, and E. M. Coombs (1991) reported that weed densities at different study sites declined from an average of 9.9 flowering plants per 50 samples of 0.25 square meters to fewer than 0.5 flowering plants. Similarly, the density of viable seed buried in the soil declined from 40,000 seeds per square meter in 1981 to 4,800 in 1988 (McEvoy, Cox, and Coombs 1991). This remarkable reduction in the density of tansy ragwort is attributed entirely to the introduced natural enemies.

Reduced program costs and reduced use of pesticides. The costs and benefits of the tansy ragwort biological control program have not been fully worked out. However, a cursory examination suggests that this project has saved western cattle ranchers millions of dollars. For example, in the 1970s losses to the western Oregon livestock industry exceeded $4 million a year (Brown 1990). This estimate does not include indirect losses due to animal weight reduction and health care costs, nor to associated traditional weed control costs, including the use of herbicides. Thus, total monetary loss during periods of heavy infestation could have approached $6 to $8 million annually. Similarly, a fairly conservative estimate of a 60 percent reduction in tansy ragwort infestation through biological control (Brown 1990) suggests a savings of $2.4 to $4.8 million annually to the Oregon livestock industry.

Another result of this very successful program has been a substantial reduction in pesticide use, thus improving environmental quality. For example, in Oregon's Curry County, R. E. Brown (1990) reports that the chemical control of tansy ragwort declined rapidly between 1983 and 1985. During 1983 a total of 573 gallons of 2,4-D was used (at a cost of $4,584), compared with 132 gallons in 1985 (at a cost of $1,058). Other counties have experienced similar cost reductions.

SUMMARY

W-84 research has had a substantial economic, environmental, and sociopolitical impact on the western region. A careful review of the records shows that this research has played a major role in developing biological control concepts and promulgating integrated pest management. The research has also had a direct impact on the development and funding of many regional and national initiatives, such as the USDA/NSF-sponsored Huffaker project in the early 1970s. It is likely that we will never be able to measure all the direct and indirect benefits of the work that W-84 scientists have done. However, the small sampling of projects reviewed in table 4.3 shows the vast economic savings and the immeasurable social and environmental benefits that are the result of relatively minor investments in biological control.

Table 4.3. The economic benefits of four biological control programs in the Western Region.

Agricultural commodity	Target pest	Biological control approach	Natural enemies	Success rate	Potential cost:benefit
Filberts	*Myzocallis coryli* (filbert aphid)	classical	*Trioxys pallidus*	very high	1:10,000 or more
Apples	spider mites*	conservation/ augmentation	*Galendromus* (=*Metaseiulus* or *Typhlodromus*) *occidentalis*	high	1:1,000 or more
Almonds	spider mites*	augmentation	*G. occidentalis*	high	1:1,000 or more
Pasture/ rangeland	*Senecio jacobaea* (tansy ragwort)	classical	*Tyria jacobaeae* *Longitarsus jacobaeae*	very high	1:10,000 or more

*See text for list of species.

REFERENCES CITED

AliNiazee, M. T. 1980. *Filbert insect and mite pests.* Oregon State Agricultural Experiment Station Bulletin 643.

————. 1983a. Carbaryl resistance in the filbert aphid (Homoptera: Aphididae). *J. Econ. Entomol.* 76:1002–4.

————. 1983b. Pest status of filbert (hazelnut) insects: A ten-year study. *Can. Entomol.* 115:1155–62.

————. 1985. Pests of hazelnuts in North America. A review of their bionomics and ecology. In *Proc. Internat. Congress of Hazelnuts, Avellino, Italy, September 1983,* 463–76.

————. 1991. Biological control of the filbert aphid, *Myzocallis coryli,* in hazelnut orchards. *Proc. Oregon, Washington, and British Columbia Nut Growers Soc.* 76:46–53.

AliNiazee, M. T., E. M. Stafford, and E. H. Kido. 1974. Management of grape pests in central California vineyards: II. Toxicity of some commonly used chemicals to *Tetranychus pacificus* and its predator *Metaseiulus occidentalis. J. Econ. Entomol.* 67:543–48.

Brown, R. E. 1990. Biological control of tansy ragwort (*Senecio jacobaea* L.) in western Oregon, U.S.A., 1975–1987. In *Proceedings of the 7th International Symposium of Biological Control of Weeds, 6–11 March 1988, Rome, Italy,* ed. E. S. Delfosse, 299–306. Rome: Ministero dell'Agricoltura e delle Foreste/Melbourne: CSIRO.

Carlson, G. A. 1988. Economics of biological control of pests. *Am. J. Alter. Agric.* 3:110–16.

Croft, B. A., and M. T. AliNiazee. 1983. Differential resistance to insecticides in *Typhlodromus arboreus* and associated phytoseiid mites of apples in Willamette Valley, Oregon. *Environ. Entomol.* 12:1420–24.

DeBach, P. 1964. *Biological control of insect pests and weeds.* New York: Reinhold.

Fisher, T. W., T. S. Bellows, G. Gordh, D. L. Dahlsten, L. E. Caltagirone, and C. B. Huffaker. In press. *Principles and applications of biological control.* Berkeley: University of California Press.

Frick, K. E., and J. R. Holloway. 1964. Establishment of the cinnabar moth, *Tyria jacobaeae,* on tansy ragwort in the western United States. *J. Econ. Entomol.* 57:52–54.

Harris, P. A., T. S. Wilkinson, M. E. Neary, and L. S. Thompson. 1971. *Senecio jacobaea* L. tansy ragwort (Compositae) 1959–1968. In *Biological control programmes against insects and weeds in Canada,* 97–104. Commonwealth Institute of Biological Control Technical Communication 4.

Hawkes, R. B., and G. R. Johnson. 1978. *Longitarsus jacobaeae* aids moth in the biological control of tansy ragwort. In *Proceedings of the 4th International Symposium of Biological Control of Weeds, 30 August–2 September 1976, Gainesville, Florida,* ed. T. E. Freeman, 193–96. Gainesville: University of Florida Institute of Food and Agricultural Sciences.

Headley, J. C., and M. A. Hoy. 1985. The economics of integrated mite management in California almonds. *Calif. Agric.* 40(1–2):28–30.

————. 1987. Benefit/cost analysis of an integrated mite management program for almonds. *J. Econ. Entomol.* 80:555–59.

Hoy, M. A. 1982. Aerial dispersal and field efficacy of a genetically improved strain of the spider mite predator *Metaseiulus occidentalis,* Acarina: Phytoseiidae. *Entomol. Exp. Appl.* 32:205–12.

————. 1985. Recent advances in genetics and genetic improvement of the Phytoseiidae. *Annu. Rev. Entomol.* 30:345–70.

————. 1991. Genetic improvement of Phytoseiids: In theory and practice. In *Modern acarology,* eds. F. Dusbabek and V. Bukva, vol. 1, 175–84. The Hague: Academia Prague and SPB Academic Publishing.

Hoy, M. A., and J. Conley. 1987. Toxicity of pesticides to western predatory mite. *Calif. Agric.* 41(7–8):12–14.

Hoy, M. A., and N. F. Knop. 1979. Studies on pesticide resistance in the phytoseiid *Metaseiulus occidentalis* in California. In *Recent advances in acarology,* ed. J. G. Rodriguez, vol. 1, 89–94. New York: Academic Press.

————. 1981. Selection for and genetic analysis of permethrin resistance in *Metaseiulus occidentalis:* Genetic improvement of a biological control agent. *Entomol. Exp. Appl.* 30:10–18.

Hoy, M. A., and K. A. Standow. 1982. Inheritance of resistance to sulfur in the spider mite predator *Metaseiulus occidentalis. Entomol. Exp. Appl.* 31:316–23.

Hoy, M. A., D. Castro, and D. Cahn. 1982. Two methods for large scale production of pesticide-resistant strains of the spider mite predator *Metaseiulus occidentalis* (Nesbitt) (Acarina: Phytoseiidae) *Zeit. angew. Entomol.* 94:1–9.

Hoy, M. A., D. Flaherty, W. Peacock, and D. Culver. 1979. Vineyard and laboratory evaluations of methomyl, dimethoate, and permethrin for a grape pest management program in the San Joaquin Valley of California. *J. Econ. Entomol.* 72:250–55.

Hoy, M. A., W. M. Barnett, W. O. Reil, D. Castro, D. Cahn, L. Hendricks, R. Coviello, and W. J. Bentley. 1982. Large scale releases of pesticide-resistant spider mite predators. *Calif. Agric.* 36(1–2):8–10.

Hoyt, S. C. 1969. Integrated chemical control of insects and biological control of mites on apple in Washington. *J. Econ. Entomol.* 62:74–86.

————. 1976. Pesticide resistance in phytoseiids. In *Studies in biological control,* ed. V. L. Delucchi, 210–12. London: Cambridge University Press.

Huffaker, C. B., and P. S. Messenger. 1976. *Theory and practice of biological control.* New York: Academic Press.

Isaacson, D. L. 1973. A life table for the cinnabar moth, *Tyria jacobaeae,* in Oregon. *Entomophaga* 18(3):291–303.

————. 1978a. *Inventory of the distribution and abundance of tansy ragwort.* Pacific Northwest Regional Commission Land Resource Inventory Demonstration Project. Salem: Oregon Department of Agriculture.

————. 1978b. The role of biological control agents in integrated control of tansy ragwort. In *Proceedings of the 4th International Symposium of Biological Control of Weeds, 30 August–2 September 1976, Gainesville, Florida,* ed. T. E.

Freeman, 189–92. Gainesville: University of Florida Institute of Food and Agricultural Sciences.

Katundu, J., and M. T. AliNiazee. 1990. Variable resistance of *Myzocallis coryli* to insecticides in the Willamette Valley, Oregon. *J. Econ. Entomol.* 83:41–47.

McEvoy, P. B. 1985. Depression in ragwort (*Senecio jacobaea*) abundance following introduction of *Tyria jacobaeae* and *Longitarsus jacobaeae* on the central coast of Oregon. In *Proceedings of the 6th International Symposium of Biological Control of Weeds, 19–25 August 1984, Vancouver, British Columbia,* ed. E. S. Delfosse, 57–64. Ottawa: Agriculture Canada.

McEvoy, P. B., C. C. Cox, and E. M. Coombs. 1991. Successful biological control of ragwort, *Senecio jacobaea* L., by introduced insects in Oregon. *Ecol. Appl.* 1:430–42.

Messing, R. H., and M. T. AliNiazee. 1985. Natural enemies of *Myzocallis coryli* (Homoptera: Aphididae) in Oregon hazelnut orchards. *J. Entomol. Soc. Brit. Columbia* 82:14–18.

———. 1986. Impact of predaceous insects on filbert aphid, *Myzocallis coryli* (Homoptera: Aphididae). *Environ. Entomol.* 15:1037–41.

———. 1989. Introduction and establishment of *Trioxys pallidus* (Hym.: Aphidiidae) in Oregon, U.S.A. for control of filbert aphid *Myzocallis coryli* (Hom.: Aphididae). *Entomophaga* 34:153–63.

Pemberton, R. W., and C. E. Turner. 1990. Biological control of *Senecio jacobaea* L. in northern California, an enduring success. *Entomophaga* 35:71–77.

Piper, G. L. 1985. Biological control of weeds in Washington: Status report. In *Proceedings of the 6th International Symposium of Biological Control of Weeds, 19–25 August 1984, Vancouver, British Columbia,* ed. E. S. Delfosse, 817–26. Ottawa: Agriculture Canada.

Reichelderfer, K. H. 1981. Economic feasibility of biological control of crop pests. In *Biological control in crop protection,* ed. G. C. Papavizas, 403–17. BARC Symposium 5. Totowa, N.J.: Allanheld, Osmun.

Ridgway, R. L., and S. B. Vinson. 1977. *Biological control by augmentation of natural enemies.* New York: Plenum Press.

Roush, R. T., and M. A. Hoy. 1981a. Genetic improvement of *Metaseiulus occidentalis:* Selection with methomyl, dimethoate, and carbaryl, and genetic analysis of carbaryl resistance. *J. Econ. Entomol.* 74:138–41.

———. 1981b. Laboratory, glasshouse, and field studies of artificially selected carbaryl resistance in *Metaseiulus occidentalis. J. Econ. Entomol.* 74:142–47.

Simmonds, F. J. 1967. The economics of biological control. *J. Royal Soc. Arts* 880–98.

Starý, P. 1978. Parasitoid spectrum of the arboricolous callaphidid aphids in Europe. *Acta Entomol. Bohemoslovaca* 75:164–77.

Waterhouse, D. F., and K. K. Norris. 1987. *Biological control: Pacific prospects.* Melbourne: Inkata Press.

PART 2

ARTHROPOD CASE HISTORIES

5 / AVOCADO BROWN MITE

J. A. McMURTRY

avocado brown mite
Oligonychus punicae (Hirst)
Acari: Prostigmata: Tetranychidae

INTRODUCTION

The avocado brown mite, common in Mexico and Guatemala, was first discovered in California in the 1920s. The species confines its feeding mainly to the upper surfaces of the avocado leaf, beginning along the midrib, and in depressions in the leaf surface. However, during heavy infestations (about 80 to 100 adult females or 200 to 300 total postembryonic stages per leaf), the mites also colonize the lower leaf surfaces and fruit. Infestations can cause partial defoliation, especially on the Hass variety of avocado (McMurtry and Johnson 1966; McMurtry, Johnson, and Scriven 1969; McMurtry 1985). However, populations usually do not reach such high levels; chemical control is therefore rarely used for this pest.

RESULTS

Objective A: The Identification and Introduction of Beneficial Organisms

Researchers have explored various areas of Mexico and Central America for phytoseiid mite predators closely associated with the *Oligonychus* species. However, none of the six phytoseiids—including three relatively specialized ones in the *Galendromus* (=*Typhlodromus*) *occidentalis* group, *G. helveolus*, *G. porresi*, and *G. annectens*—that were introduced from that region (McMurtry 1989) became established. These species may have been prevented from becoming established by the same kind of interspecific competition found in the augmentative release experiments (see objective C).

Objective B: Ecological and Physiological Studies

The coccinellid *Stethorus picipes* Casey is the major factor responsible for suppressing *O. punicae* populations at medium to high densities. This coccinellid can usually control the mite before severe bronzing occurs—that is,

if it matches the increases of *O. punicae* when the pest is still at relatively low or medium densities (about 10 to 20 adult females per leaf). If *S. picipes* shows a lag and doesn't increase quickly enough, the result is usually a heavy infestation and severe bronzing of foliage (McMurtry and Johnson 1966; McMurtry 1985). The other major predator, the phytoseiid *Euseius hibisci* (Chant), does not usually respond soon enough to overtake an increasing *O. punicae* population, possibly because it lacks the tendency to congregate and oviposit on infested leaves, or because it cannot gain access to prey on the upper leaf surface, which is frequently protected by silken webbing (McMurtry and Johnson 1966).

Objective C: The Conservation and Augmentation of Beneficial Organisms

Experimental releases of *Stethorus picipes* at the rate of 400 adults per tree kept *O. punicae* populations below damaging levels (McMurtry et al. 1969). However, because producing *Stethorus* is so costly, this procedure is not economically feasible.

Because phytoseiid mites can be reared much more cheaply than *Stethorus* species—they require considerably fewer mites and some can be reared on pollen—researchers investigated augmentative releases of those predators. They released phytoseiid species thought to be more effective than the resident species into orchards, or parts of orchards, that historically had heavier-than-average infestations. In experiments conducted during three seasons, nine candidate species were released over a 4-week period at rates of 1,200 per tree. In no instance did the releases significantly affect the densities of the avocado brown mite populations or of the phytoseiid populations (McMurtry, Johnson, and Badii 1984). Nevertheless, these experiments demonstrated an interesting principle. Of the predators released, three of four specific mite predators and one of five of the more general predators showed marked increases and made up more than 20 percent of the phytoseiid population. However, this increase apparently retarded the increase of the native *E. hibisci* population: the combined densities of *E. hibisci* and the released species were never significantly higher than that of *E. hibisci* alone on the nonrelease trees. These results suggested interspecific competition (McMurtry, Johnson, and Badii 1984).

RECOMMENDATIONS

These studies have shown that various indigenous predators are usually able to control avocado brown mite on avocado plantings in California. However, further studies are needed to examine why natural control sometimes does not take effect until the mite densities have exceeded 200 pests per leaf. The determination of developmental thresholds and degree-day calculations for the mite and for its major predators should help us to answer that question.

REFERENCES CITED

McMurtry, J. A. 1985. Avocado. In *Spider mites, their biology, natural enemies, and control*, eds. W. Helle and M. W. Sabelis, 327–32. New York: Elsevier.

———. 1989. Utilizing natural enemies to control pest mites on citrus and avocado in California, U.S.A. In *Progress in acarology*, vol. 2, eds. C. P. Channabasavanna and C. A. Viraktamath, 325–36. Proc. 7th Internat. Congr. Acarol. New Delhi: Oxford & IBH.

McMurtry, J. A., and H. G. Johnson. 1966. An ecological study of the spider mite, *Oligonychus punicae* (Hirst), and its natural enemies. *Hilgardia* 37:363–402.

McMurtry, J. A., H. G. Johnson, and M. H. Badii. 1984. Experiments to determine effects of predator releases on populations of *Oligonychus punicae* on avocado in California. *Entomophaga* 29:11–19.

McMurtry, J. A., H. G. Johnson, and G. T. Scriven. 1969. Experiments to determine effects of mass releases of *Stethorus picipes* on the level of infestation of the avocado brown mite. *J. Econ. Entomol.* 62:1216–21.

6 / CITRUS RED MITE

J. A. McMURTRY

> **citrus red mite**
> *Panonychus citri* (McGregor)
> Acari: Prostigmata: Tetranychidae

INTRODUCTION

Biology and Pest Status

For many years the citrus red mite has been considered a major pest of California citrus. The mite feeds on leaves, fruit, and sometimes green twigs, causing a bronzing or silvering of the plant surface. Severe infestations can result in partial defoliation, especially during periods of hot, dry wind in the late summer and fall.

Historical Notes

The most common native predators of this mite are *Stethorus picipes* Casey, a coccinellid beetle, and *Euseius tularensis* Congdon, a phytoseiid mite that uses (and may prefer) other foods, especially pollen and thrips, in addition to spider mites (Congdon and McMurtry 1988). *Euseius hibisci* (Chant), a closely related species with similar feeding habits, occurs in the coastal orchards of southern California (Congdon and McMurtry 1985). Researchers, hypothesizing a vacant niche that could be exploited by a more specialized phytoseiid predator of tetranychid mites, conducted a program of introducing and colonizing exotic species.

RESULTS

Objective A: The Identification and Introduction of Beneficial Organisms

Over 20 species of Phytoseiidae have been introduced and released in California (McMurtry 1989). However, only 3 or 4—located after an exploration of 13 countries—have been tetranychid-specific phytoseiids found on citrus. The most highly specialized species, *Phytoseiulus persimilis* Athias-Henriot, did become established in California, but it was recovered in only a few coastal citrus orchards with colonies of *Tetranychus urticae* Koch that had dispersed from adjacent strawberry fields (McMurtry et al. 1978).

Only two other introduced species seem to have become established on citrus in California: *Typhlodromus rickeri* Chant and *Euseius stipulatus* (Athias-Henriot). *T. rickeri* was first released in 1962. After it initially became established (McMurtry 1969), it was not detected again until the early 1980s, when it was recovered in a few orchards near the coast, several kilometers from any release sites (McMurtry 1982). Although further studies are needed, this species is probably not abundant enough to have a significant impact on mite populations.

Euseius stipulatus was introduced from the Mediterranean area (McMurtry 1977) in 1972. Less than a year later, researchers observed that it was displacing the native *E. hibisci* on release trees and adjacent trees in a lemon orchard in San Diego County (McMurtry 1978). Presently, it is established in many coastal citrus orchards, and it appears that this species will eventually replace *E. hibisci* and *E. tularensis* in all the coastal citrus orchards in southern California.

Why has *E. stipulatus* displaced the native *E. hibisci*? The biology of the two species is similar: both develop and reproduce rapidly on pollen and both have somewhat lower reproductive rates on citrus thrips. However, *E. stipulatus* has a higher intrinsic rate of increase on the citrus red mite and can use a wider range of the mite's stages (Friese 1985).

Although these differences are important, it is unlikely that they are the only factors involved in this apparent "competitive displacement." Even in the virtual absence of tetranychid mites, displacement occurred. Because various studies have shown that the plant itself may play an important role in the dynamics of *Euseius* species (Porres, McMurtry, and March 1975; Congdon 1985), it is possible that *E. stipulatus* is better adapted to citrus than is *E. hibisci*. This may be due to its probable long-term association with citrus in the Mediterranean area, which could, conceivably, date back one to two millennia. Conversely, *E. hibisci*, a native of California and Mexico, may be better adapted to the avocado ecosystem and thus have a competitive advantage on avocado, where *E. stipulatus* has not become established. (McMurtry 1989).

Objective D: Impact

In the last 20 years, the pest status of the citrus red mite in California has declined, probably for two reasons. First, the biological control of California red scale on citrus in southern California has dramatically reduced the use of broad-spectrum pesticides, which has allowed an increase in predator populations and a lower average level of the citrus red mite. Second, past economic thresholds for this pest proved far too low. Better understanding of the thresholds has reduced the use of acaracides and allowed predator populations and the citrus red mite to reach equilibrium.

RECOMMENDATIONS

The best control strategy is to preserve natural enemies by limiting acaracide use to emergency situations, and then to employ only materials that are not highly toxic to mite predators, such as narrow-range oil at low volume. Future studies should focus on manipulating or conserving natural enemies.

REFERENCES CITED

Congdon, B. D. 1985. Distribution and ecology of California *Euseius* species. Ph.D. diss., University of California, Riverside.

Congdon, B. D., and J. A. McMurtry. 1985. Biosystematics of *Euseius* on California citrus and avocado with the description of a new species (Acari: Phytoseiidae). *Int. J. Acarol.* 11:23–30.

———. 1988. Prey selectivity in *Euseius tularensis* Congdon (Acari: Phytoseiidae). *Entomophaga* 33:281–87.

Friese, D. E. 1985. Factors influencing competition between *Euseius hibisci* and *Euseius stipulatus* (Acarina: Phytoseiidae). Ph.D. diss., University of California, Riverside.

McMurtry, J. A. 1969. Biological control of citrus red mite in California. In *Proc. 1st Internat. Citrus Symp.*, vol. 2, 855–62.

———. 1977. Some predaceous mites (Phytoseiidae) on citrus in the Mediterranean region. *Entomophaga* 22:19–30.

———. 1978. Biological control of citrus mites. In *Proc. Internat. Soc. Citriculture* 1977, 456–59.

———. 1982. The use of the phytoseiids for biological control: Progress and future prospects. In *Recent advances in knowledge of the Phytoseiidae,* ed. M. A. Hoy, 23–48. Oakland: University of California Division of Agricultural Sciences, Publication 3284.

———. 1989. Utilizing natural enemies to control pest mites on citrus and avocado in California, U.S.A. In *Progress in acarology,* vol. 2, eds. C. P. Channabasavanna and C. A. Viraktamath, 325–36. Proc. 7th Internat. Congr. Acarol. New Delhi: Oxford & IBH.

McMurtry, J. A., E. R. Oatman, P. A. Phillips, and C. W. Wood. 1978. Establishment of *Phytoseiulus persimilis* (Acari: Phytoseiidae) in southern California. *Entomophaga* 23:175–79.

Porres, M. A., J. A. McMurtry, and R. B. March. 1975. Investigations of leaf sap feeding by three species of phytoseiid mites by labelling with radioactive phosphoric acid ($H_3{}^{32}PO_4$). *Ann. Entomol. Soc. Am.* 68:871–72.

7 / MITE COMPLEX ON APPLES

S. C. HOYT

mite complex *Tetranychus* spp. and *Panonychus ulmi* (Koch) Acari: Prostigmata: Tetranychidae	mite complex *Aculus schlechtendali* (Nalepa) Acari: Prostigmata: Eriophyidae

INTRODUCTION

Biology and Pest Status

The phytophagous mite complex on apple in Washington includes the apple rust mite, *Aculus schlechtendali* (Nalepa) (Eriophyidae), and three species of Tetranychidae: the McDaniel spider mite, *Tetranychus mcdanieli* McGregor; the twospotted spider mite, *Tetranychus urticae* Koch; and the European red mite, *Panonychus ulmi* (Koch). Although the four species feed primarily on foliage, aspects of their biologies differ, including their preferences for foliage type, methods of overwintering, and temporal distributions.

McDaniel spider mite. This species overwinters as orange-colored adult females, primarily in the duff at the tree base or under bark scales on the trunk. The females emerge between late March and early May, feeding on the first green tissue of opening buds on the trunk and on the main leaders of the tree. They produce profuse webbing that, if the mite population is high, may completely enclose the blossoms. As the summer generations are produced, the mites usually inhabit the spur leaves in the center of the tree, gradually spreading to spurs on the periphery and eventually invading the older leaves on terminal shoots. Early infestations are more common on the lower surface of leaves, but as populations increase they can spread to the upper surface. Populations increase gradually through summer, peaking in late August. Where natural or chemical control is lacking, peak populations can exceed 300 mites per leaf.

Twospotted spider mite. Despite the many similarities between the habits of McDaniel spider mites and twospotted spider mites, there are some important differences. Although the two species overwinter in similar areas and emerge at about the same time, the twospotted spider mite commonly feeds and reproduces on several species of weed in the cover crop. Later in the season, when the cover crops mature or are mowed, the mites invade the trees. As a result, twospotted spider mite populations tend to be lower and, because they are dependent on the presence of appropriate weeds, more sporadic in occurrence.

European red mite. European red mites overwinter as eggs deposited on spurs, limbs, and the bases of shoots. The eggs hatch between mid-April and mid-May, and the first generation inhabits the primary spur leaves. Later generations spread to all spur leaves but may also inhabit leaves on terminal shoots. This species develops more slowly and lays fewer eggs than does the McDaniel spider mite. Also, unless disturbed by pesticides, its populations peak and decline earlier in the season— usually in late July. As a result, its numbers seldom reach the levels reported for the McDaniel spider mite.

All three species cause their damage primarily by destroying leaf function. This can reduce fruit growth (Hoyt, Tanigoshi, and Browne 1979) and can also affect fruit maturation. By reducing a tree's vigor and storage reserves, the species can also reduce the following year's bloom and the fruit set. Additionally, overwintering forms may be present in the calyx of harvested fruit, which creates a quarantine problem, particularly in export markets.

Apple rust mite. The apple rust mite, an eriophyid, is much smaller and has different feeding habits than the three tetranychids already discussed. Consequently, the tolerance level for apple rust mite populations is much higher. Rust mites overwinter as adults under buds and in crevices. The mites emerge in late March and early April, feeding primarily on the developing green tissue. On Golden Delicious, they feed on the young fruit just prior to and during bloom, causing russeting at the calyx end. The population development is bimodal, peaking first in late June to early July, then declining to very low levels, and finally reaching a second but lower peak in late August or early September. Summer populations are highest on fairly young terminal foliage, but relatively high populations may also occur on spur foliage.

Because all the species inhabit leaves and occupy similar niches, there is interspecific as well as intraspecific competition among them. I have described some of those interactions elsewhere (Hoyt 1969a).

Historical Notes

Mites had been a problem in Washington apple production since the early 1900s. In the 1940s, with the advent of synthetic organic pesticides, the use of specific acaricides greatly improved mite control. However, because mites have numerous generations per year and a high reproductive potential, and because of the selective pressure of the chemicals, the McDaniel spider mite and the European red mite developed resistance to acaricides (Hoyt 1976a). At first, new chemicals were developed rapidly enough to overcome the problem, but by the mid-1960s the McDaniel spider mite was resistant to virtually all registered acaricides. Growers applied expensive control programs, but control was poor and many orchards were extensively damaged.

Foreseeing this problem, researchers had already begun to study alternative controls in 1961. Predators, which were present late in the season only, had little effect on mite populations. Our initial studies determined that certain pesticides were preventing the predators from appearing earlier in the season. In 1962 and 1963 we found that when no sprays were used, the predators—primarily *Galendromus* (=*Metaseiulus* or *Typhlodromus*) *occidentalis* (Nesbitt)—could control the mites at acceptable levels (Hoyt 1965, 1969b).

With the beginning of Western Regional Project W-84 in 1964 and the added funding that it provided, our research was greatly expanded. We were then able to determine which chemicals and application rates could control pests without killing the predaceous mites. By the end of 1964, we had cataloged the toxicity of all essential chemicals. In 1965 we tested complete insect and disease control programs to evaluate the biological control of mites on apple. The programs proved generally successful (Hoyt 1969b), and in that same year commercial application of the programs began.

Originally we had not intended to release the programs to growers until we had more research experience, but a severe frost had left one part of the Yakima Valley with almost no crop. Since the growers had heard of the program and were anxious to save money, they asked if we could help. The local extension agent and I provided advice that the growers carefully followed. This major test proved highly successful, and by 1967 a large percentage of the state's apple orchards were integrating biological control of mites with chemical control of other pests. The costs of pest control dropped dramatically, and losses from mite damage declined to near zero (Hoyt and Caltagirone 1971). The McDaniel spider mite, which was the predominant apple pest from the late 1950s through the mid-1960s, all but disappeared from most orchards. The apple rust mite became the predominant mite species, serving as a food source that maintained populations of predators. At times the European red mite was troublesome, but populations were generally low and continued to decline over time.

Not all growers chose to, or could, use this program. This was particularly true after cyhexatin was registered for mite control in the 1970s. This compound was selective enough that, in spite of its use, a fortuitous integrated program developed. However, as the mites developed a resistance to cyhexatin in the early 1980s, growers used higher rates and more frequent applications, which killed the predators (Croft, Hoyt, and Westigard 1987). Mite problems again became common in those orchards. Since 1985, however, predators have been re-established and a high percentage of the acreage is now using biological control of mites.

Since 1964 researchers have conducted a number of studies to improve or protect the program, integrating the biological control of mites with the chemical control of insects. Those studies are summarized here.

RESULTS

Objective A: The Identification and Introduction of Beneficial Organisms

Although we have not introduced exotic natural enemies into Washington for this project, we have sent *G. occidentalis* from Washington to other states and to foreign countries. In 1987 live specimens were sent to the People's Republic of China. After being reared and released in China, those predators became established and are controlling *Tetranychus viennensis* Zacher on apple in Shaanxi Province (Deng Xiong person. commun.). *G. occidentalis* was also sent to Sweden, Australia, and Colorado, but we have had no report of the results. Live predators were sent to California for genetic improvement (Hoy, Westigard, and Hoyt 1983) and for comparative studies with other strains of *G. occidentalis* (Croft 1970, 1971; Croft and Barnes 1971; Croft and McMurtry 1972). The Washington strain was selected for the California studies because of its resistance to or tolerance of pesticides.

Objective B: Ecological and Physiological Studies

To understand the competitive or predaceous interactions among mite species on apple, researchers conducted detailed studies of the mites' life histories, distributions, population dynamics, and feeding behavior (Hoyt 1969a,b, 1970; Tanigoshi, Browne, and Hoyt

1975; Tanigoshi et al. 1975a,b, 1976; Tanigoshi, Hoyt, and Croft 1983). These studies formed the basis for a model of temperature-dependent rate phenomena in arthropods (Logan et al. 1976). They also led to a better understanding of the positive and negative interactions between predators and prey on apples, and to improved sampling methods.

Objective C: The Conservation and Augmentation of Beneficial Organisms

The studies on pesticide selectivity had two main purposes: to conserve the predator *G. occidentalis*, and to shift the balance between predator and prey in the predator's favor. Initial studies identified which pesticides, rates, or application techniques were selective enough to favor biological control (Hoyt 1969b), and to which pesticides predators had developed resistance (Hoyt 1976b). Later studies found that pyrethroids had a negative effect on predators and thus on biological control (Hoyt, Westigard, and Burts 1978; Croft and Hoyt 1978), and that organotin acaricides harmed the interaction between predators and prey (Croft, Hoyt, and Westigard 1987). M. A. Hoy, P. H. Westigard, and S. C. Hoyt (1983) reported on the laboratory selection and improvement of pyrethroid resistance in a strain of *G. occidentalis*, but so far the strain is not resistant enough to be of practical use.

Studies currently under way to protect the integrated control of mites center on the chemicals used to control the western tentiform leafminer, *Phyllonorycter elmaella* Doganlar & Mutuura, a pest problem that developed in the 1980s. Of the registered chemicals only pyrethroids, methomyl, and oxamyl are effective, and the first two are highly destructive to integrated mite-control programs. Oxamyl is moderately toxic to predators, and researchers are studying which methods of use are compatible with biological control.

To protect mite predators where a large complex of insects must be controlled, we need a constant study of insect-control options.

Objective D: Impact

The commercial use of predatory mites has virtually eliminated the McDaniel spider mite as a pest in integrated-control orchards and has extensively suppressed populations of the European red mite. The reduction in pesticide use and application costs has been documented for one large orchard (Hoyt and Caltagirone 1971), but quite a few similar examples could also be documented. In many orchards, growers have not used any specific acaricides in over 20 years. The reduction in crop damage has been harder to quantify because of the variation in damage from year to year and orchard to orchard as well as of the similarity between the indirect effect of mite damage on fruit and damage caused by other factors. There is little doubt, however, that this program has increased revenues from fruit sales by several million dollars per year, in addition to saving Washington growers approximately $10,000,000 per year in pesticide and application costs.

An additional benefit of the biological control of mites has been the delay in the development of resistance to acaricides. Although resistance episodes were common in Washington in the 1950s and 1960s (Hoyt 1976a), only one case has been documented since 1970 (Croft, Hoyt, and Westigard 1987). Although other factors may have played a role, researchers believe the lower selection pressure due to the reduced acaricide applications was very important.

RECOMMENDATIONS

The future goals of this project are to maintain and enhance the biological control of mites on apples in Washington. Many factors could influence mite control in the future, including the further regulation and reduction in pesticide options for insect control, major changes in orchard systems, and large-scale plantings of new apple varieties. Changes in insect-control procedures can have either a positive or negative effect on mite populations. Since we are unsure what effects legislation and regulation, as well as pesticide resistance, will have on the present programs of insect control, we know only that change will occur and that we must respond to it.

REFERENCES CITED

Croft, B. A. 1970. Comparative studies on four strains of *Typhlodromus occidentalis* (Acarina: Phytoseiidae): I. Hybridization and reproductive isolation studies. *Ann. Entomol. Soc. Am.* 63:1558–63.

————. 1971. Comparative studies on four strains of *Typhlodromus occidentalis* (Acarina: Phytoseiidae): V. Photoperiodic induction of diapause. *Ann. Entomol. Soc. Am.* 64:962–64.

Croft, B. A., and M. M. Barnes. 1971. Comparative studies on four strains of *Typhlodromus occidentalis*: III. Evaluations of releases of insecticide-resistant strains into an apple orchard ecosystem. *J. Econ. Entomol.* 64:845–50.

Croft, B. A., and S. C. Hoyt. 1978. Considerations for the use of pyrethroid insecticides for deciduous fruit pest control in the U.S.A. *Environ. Entomol.* 7:627–30.

Croft, B. A., and J. A. McMurtry. 1972. Comparative studies on four strains of *Typhlodromus occidentalis* Nesbitt (Acarina: Phytoseiidae): IV. Life history studies. *Acarologia* 13:461–70.

Croft, B. A., S. C. Hoyt, and P. H. Westigard. 1987. Spider mite management on pome fruits revisited: Organotin and acaricide resistance management. *J. Econ. Entomol.* 80:304–11.

Hoy, M. A., P. H. Westigard, and S. C. Hoyt. 1983. Release and evaluation of a laboratory-selected pyrethroid-resistant strain of the predaceous mite *Metaseiulus occidentalis* in southern Oregon pear orchards and a Washington apple orchard. *J. Econ. Entomol.* 76:383–88.

Hoyt, S. C. 1965. A possible new approach to mite control on apples. *Wash. State Hort. Assoc. Proc.* 61:127–28.

———. 1969a. Population studies of five mite species on apple in Washington. In *Proc. 2nd Internat. Congr. Acarology, Sutton Bonington, England, 1967,* 117–33.

———. 1969b. Integrated chemical control of insects and biological control of mites on apple in Washington. *J. Econ. Entomol.* 62:74–86.

———. 1970. The effect of short feeding periods by *Metaseiulus occidentalis* (Nesbitt) on fecundity and mortality of *Tetranychus mcdanieli* McGregor. *Ann. Entomol. Soc. Am.* 63:1382–84.

———. 1976a. Specific acaricides and carbaryl. In *History of fruit growing and handling in the United States of America and Canada,* eds. D. V. Fisher and W. H. Upshall, 275–79. Kelowna, B.C., Canada: Regatta City Press.

———. 1976b. Pesticide resistance in phytoseiids. In *Studies in biological control*, ed. V. L. Delucchi, 210–12. Cambridge: Cambridge University Press.

Hoyt, S. C., and L. E. Caltagirone. 1971. The developing programs of integrated control of pests of apples in Washington and peaches in California. In *Biological control,* ed. C. B. Huffaker, 395–421. New York: Plenum Press.

Hoyt, S. C., L. K. Tanigoshi, and R. W. Browne. 1979. Economic injury level studies in relation to mites on apple. In *Recent advances in acarology,* ed. J. G. Rodriguez, vol. 1, 3–12. New York: Academic Press.

Hoyt, S. C., P. H. Westigard, and E. C. Burts. 1978. Effects of two synthetic pyrethroids on codling moth, pear psylla, and various mite species in the northwest apple and pear orchards. *J. Econ. Entomol.* 71:431–34.

Logan, J. A., D. J. Wollkind, S. C. Hoyt, and L. K. Tanigoshi. 1976. An analytic model for description of temperature-dependent rate phenomena in arthropods. *Environ. Entomol.* 5:1133–40.

Tanigoshi, L. K., R. W. Browne, and S. C. Hoyt. 1975. A study of the dispersion pattern and foliage injury by *Tetranychus mcdanieli* (Acarina: Tetranychidae) in simple apple ecosystems. *Can. Entomol.* 107:439–46.

Tanigoshi, L. K., S. C. Hoyt, and B. A. Croft. 1983. Basic biology and management components for mite pests and their natural enemies. In *Integrated management of insect pests of pome and stone fruits*, eds. B. A. Croft and S. C. Hoyt, 153–202. New York: Wiley.

Tanigoshi, L. K., R. W. Browne, S. C. Hoyt, and R. F. Lagier. 1976. Empirical analysis of variable temperature regimes on life-stage development and population growth of *Tetranychus mcdanieli* (Acarina: Tetranychidae). *Ann. Entomol. Soc. Am.* 69:712–16.

Tanigoshi, L. K., S. C. Hoyt, R. W. Browne, and J. A. Logan. 1975a. Influence of temperature on population increase of *Tetranychus mcdanieli* (Acarina: Tetranychidae). *Ann. Entomol. Soc. Am.* 68:972–78.

———. 1975b. Influence of temperature on population increase of *Metaseiulus occidentalis* (Acarina: Phytoseiidae). *Ann. Entomol. Soc. Am.* 68:979–86.

8 / SPIDER MITES ON ALMONDS

M. A. HOY

> **spider mites**
> *Tetranychus* spp.
> Acari: Prostigmata: Tetranychidae

INTRODUCTION

Biology and Pest Status

Many growers consider spider mites to be primary, not secondary, pests of almonds. The approximately 400,000 acres of almonds in California's Central Valley are infested by six species of tetranychids: the European red mite, *Panonychus ulmi* (Koch); the citrus red mite, *Panonychus citri* (McGregor); the brown almond mite, *Bryobia rubrioculus* (Scheuten); the twospotted spider mite, *Tetranychus urticae* Koch; the Pacific spider mite, *Tetranychus pacificus* McGregor; and the strawberry mite, *Tetranychus turkestani* Ugarov & Nikolski. Because it is difficult to distinguish between *T. urticae, T. pacificus* and *T. turkestani*, in the field they are treated as a species group.

Various researchers have estimated the impact of *Tetranychus* species on both the growth and yield of almond trees (Barnes and Andrews 1978; Andrews and LaPre 1979; Welter et al. 1984). To prevent tree defoliation, many growers will apply an acaricide such as propargite, either when they see webbing or stippling from spider mite feeding or simply on a preventive calendar basis. However, this is not only costly, but it also may be disruptive to the biological control of *Tetranychus* mites.

Historical Notes

Several different natural enemy species have been identified in California almond orchards (Hoy, Ross, and Rough 1978; Hoy et al. 1979), but the most important is the phytoseiid mite *Galendromus* (=*Metaseiulus* or *Typhlodromus*) *occidentalis* (Nesbitt), particularly in pesticide-treated orchards. In our work we concentrated on evaluating a genetically manipulated strain of *G. occidentalis*, in order to determine whether it could become an effective predator of spider mites in almond orchards.

RESULTS

Objective A: The Identification and Introduction of Beneficial Organisms

The Australian lady beetle, *Stethorus histrio* Chazeau (=*S. nigripes* Kapur), was released in California almond orchards as a potential biological control agent for *Panonychus ulmi* (Hoy and Smith 1982). We have no evidence that the beetle established itself.

Objective B: Ecological and Physiological Studies

To conduct mode of inheritance tests, we studied the genetic system (parahaploidy) of this predator (Hoy 1979; Nelson-Rees, Hoy, and Roush 1980) and discovered two rickettsial-like organisms in *G. occidentalis*, one of which is pathogenic and can be detrimental to mass production (Hess and Hoy 1982). We demonstrated for the first time that this predator has an aerial dispersal mechanism, which can explain dispersal within the orchard (Hoy 1982; Hoy, Van de Baan, et al. 1984; Hoy, Groot, and Van de Baan 1985). We also investigated fundamental mating behavior to determine whether mating isolation can be enhanced through selection for nonmating (Hoy and Cave 1985a, 1988) or by identifying thelytokous strains (Hoy and Cave 1986). Reproductive isolation could allow the release and maintenance of genetically improved strains with traits that are quantitatively determined where there is no geographic isolation from the native strains. We also assessed activity levels of strains of *G. occidentalis* using a computerized quantification system (Mueller-Beilschmidt and Hoy 1987), which provided a new method for assaying the quality of genetically manipulated phytoseiids as well as parasites and insect predators. We analyzed the prey-seeking behavior of *G. occidentalis* and demonstrated different activity on residues deposited by *Tetranychus* spider mite species or by *Panonychus* or *Bryobia* species (Hoy and Smilanick 1981).

We also found that traits other than pesticide resistance can be genetically manipulated. Through laboratory selection and crosses, we developed a nondiapausing (and pesticide-resistant) strain of *G. occidentalis* (Hoy 1984a, 1985a). This genetically improved strain performed well in small plot trials in the greenhouse and should be evaluated for efficacy in large-scale trials in commercial rosehouses in California (Field and Hoy 1986).

Recently, the selection of *G. occidentalis* with abamectin has demonstrated another point. Selection for resistance to carbaryl or permethrin has sometimes been attributed to sampling field populations of this species that had been "preselected" with these or related products—namely, selection with permethrin might have been successful because the predators had been previously exposed to DDT. However, abamectin is a new acaricide-insecticide-nematacide that affects the GABA system of arthropods, and there is no reason to believe that any previously used pesticides could have preselected this predator for resistance to abamectin. Nevertheless, we were able to select an abamectin-resistant strain (Hoy and Ouyang 1989). Currently, *G. occidentalis* is resistant to organophosphates (OPs), carbamates, sulfur, pyrethroids, and abamectin; there are rumors that it has become resistant to chlorinated hydrocarbons (DDT), although this has never been documented. The number of resistances to different classes of pesticides that has been demonstrated in *G. occidentalis* rivals the resistances found in the most notorious pest arthropods. This suggests that in predators physiological mechanisms are not always limiting factors in the development of resistance; various ecological, biological, and behavioral attributes may delay the development of resistance in arthropod natural enemies in the field. It is possible that these limits can be more readily overcome under laboratory conditions (Hoy 1987).

Objective C: Conservation and Augmentation

Galendromus occidentalis developed resistance to OPs and sulfur in the field and to carbaryl and permethrin via laboratory selections (Hoy and Knop 1979, 1981; Roush and Hoy 1980, 1981a,b; Hoy 1985a). Through laboratory crosses and selections, a carbaryl-OP-sulfur-resistant (COS) strain was developed. The strain was released in almond orchards in California and survived relevant pesticide applications; it controlled spider mites, overwintered, and spread within the orchards (Roush and Hoy 1981b; Hoy 1982; Hoy, Barnett, et al. 1982; Hoy, Barnett, et al 1984; Hoy, Van de Baan, et al. 1984; Hoy, Groot, and Van de Baan 1985). Thus, the laboratory-selected biotype was judged efficacious, and research began on how to use it.

Large-scale tests with the COS strain were conducted for three years in commercial almond orchards, confirming that the strain was fit and effective. To ensure that it could be used in an integrated mite management program, we developed two mass-rearing techniques (Hoy, Castro, and Cahn 1982) and two sampling methods (Wilson et al. 1984; Zalom, Hoy, et al. 1984; Zalom, Wilson, et al. 1984). We also developed methods to use low rates of selective acaricides and determined which pesticides were selective for this predator (Roush and Hoy 1978; Grafton-Cardwell and Hoy 1983; Hoy 1984a, 1985b; Hoy and Conley 1987; Zalom et al. 1987). Commercial releases in almonds began in 1984 and have continued in subsequent years. The project consisted of Phase I (developmental) and Phase II (implementation) (Hoy 1989); we estimated costs and benefits for both phases (Headley and Hoy 1987).

After 1984, we conducted maintenance research (Phase III), which was critical if the integrated mite management program (IMMP) were to be maintained; we did not estimate costs and benefits for this phase. We also developed a protocol for testing the impact of microbial pathogens on *G. occidentalis* (Hoy 1986); tested three new acaricides (clofentezine, hexythiazox, beta-exotoxin of *Bacillus thuringiensis*) for their selectivity to *G. occidentalis* (Hoy and Cave 1985b; Hoy and Ouyang 1986, 1987); and selected a strain of *G. occidentalis* resistant to abamectin (Hoy and Ouyang 1989). We needed to learn how to use these new acaricides after the Pacific spider mite acquired resistance to cyhexatin, fenbutatin-oxide, and propargite in the field. Such maintenance research will probably be required periodically to maintain the IMMP in almonds.

Objective D: Impact

A cost-benefit analysis of Phases I and II concluded that the genetic improvement project with this predator was not only unique, but extraordinarily cost effective (Headley and Hoy 1985, 1987). Project development costs were calculated fairly completely, because much of the work was conducted in my laboratory. We used different assumptions for the adoption of the IMMP for 425,000 acres of California almonds, in which the pesticide-resistant predator plays a key role. If only 25 percent of the acreage used the program, the benefits would be $13,920,000 per year. Currently, about 65 percent of almond growers have adopted the program, and savings are conservatively estimated to be $21,256,000 per year. These benefits represent an annual return of 500 to 600 percent per year on the initial research investment, even allowing for a discount of 12 percent for future benefits and compounding development costs at 12 percent, primarily by saving on acaricide costs. These benefits are to the grower and have no impact on yield. We did not calculate potential benefits to the environment from reduced pesticide applications. Economic analyses of biological control projects are still rare; this economic analysis adds to our knowledge of the benefits of a program based on biological control of mites. It is also a milestone in that it documents the benefits of a genetically improved natural enemy and provides a model for documenting the benefits of other genetic improvement projects.

RECOMMENDATIONS

1. Because pests, registered pesticides, and crop management programs change, the integrated mite management program in almonds will require intermittent research to remain useful to almond growers.

2. The carbaryl-OP-sulfur-resistant and nondiapausing strain of *G. occidentalis* performed well in small plot trials in greenhouse roses. This strain should be evaluated in commercial plots to determine whether it can be used effectively in commercial rosehouses.

3. The successful development and implementation of pesticide-resistant strains of *G. occidentalis* in almond orchards indicates that genetic manipulation of arthropod natural enemies can reduce pesticide applications. Genetic manipulation of other arthropod natural enemies may be a useful method to increase the use of biological control in pest management programs.

4. Genetic improvement programs should not be initiated unless (a) naturally occurring biotypes with the desired attribute(s) do not exist; (b) genetic variability for the target trait is present; and (c) an implementation method can be perceived in advance.

5. Genetic improvement should become an occasional tactic in biological control, utilized to solve specific problems.

REFERENCES CITED

Andrews, K. L., and L. F. LaPre. 1979. Effects of Pacific spider mite on physiological processes of almond foliage. *J. Econ. Entomol.* 72:651–54.

Barnes, M. M., and K. L. Andrews. 1978. Effects of spider mites on almond tree growth and productivity. *J. Econ. Entomol.* 71:555–58.

Field, R. P., and M. A. Hoy. 1986. Evaluation of genetically improved strains of *Metaseiulus occidentalis* (Nesbitt) (Acarina: Phytoseiidae) for integrated control of spider mites on roses in greenhouses. *Hilgardia* 54(2):1–32.

Grafton-Cardwell, E. E., and M. A. Hoy. 1983. Comparative toxicity of avermectin B1 to the predator *Metaseiulus occidentalis* (Nesbitt) and the spider mites *Tetranychus urticae* Koch and *Panonychus ulmi* (Koch). *J. Econ. Entomol.* 76(6):1216–20.

Headley, J. C., and M. A. Hoy. 1985. The economics of integrated mite management in California almonds. *Calif. Agric.* 40(1–2):28–30.

———. 1987. Benefit/cost analysis of an integrated mite management program for almonds. *J. Econ. Entomol.* 80:555–59.

Hess, R. T., and M. A. Hoy. 1982. Microorganisms associated with the spider mite predator *Metaseiulus* (*Typhlodromus*) *occidentalis* (Nesbitt): Electron microscope observations. *J. Invert. Pathol.* 40:98–106.

Hoy, M. A. 1979. Parahaploidy of the "arrhenotokous" predator, *Metaseiulus occidentalis* (Acarina: Phytoseiidae), demonstrated by X–irradiation of males. *Entomol. Exp. Appl.* 26:97–104.

———. 1982. Aerial dispersal and field efficacy of a genetically-improved strain of the spider mite predator *Metaseiulus occidentalis* (Acarina: Phytoseiidae). *Entomol. Exp. Appl.* 32:205–12.

———. 1984a. Genetic improvement of a biological control agent: Multiple pesticide resistances and nondiapause in *Metaseiulus occidentalis* (Nesbitt). In *Acarology VI*, eds. D. A. Griffiths and C. E. Bowman, Vol. 2, 673–79. Chichester, U.K.: Ellis Horwood.

———. 1984b. *Managing mites in almonds—An integrated approach*. IPM Publication 1. Davis: University of California.

———. 1985a. Recent advances in genetics and genetic improvement of the Phytoseiidae. *Annu. Rev. Entomol.* 30:345–70.

———. 1985b. Almonds: Integrated mite management for California almond orchards. In *Spider mites, world crop pests*, eds. M. Helle and M. Sabelis, 299–310. Amsterdam: Elsevier.

———. 1986. *Interim protocol for testing the effects of microbial pathogens on predatory mites*. Washington, D.C.: Environmental Protection Agency.

———. 1987. Developing insecticide resistance in insect and mite predators and opportunities for gene transfer. In *Biotechnology in agricultural chemistry*, eds. H. M. LeBaron, R. O. Mumma, R. C. Honneycutt, and J. H. Duesing, 125–38. American Chemical Society Symposium Series 334.

———. 1989. Integrating biological control into agricultural IPM systems: Reordering priorities. In *Proc. Nat. IPM Symp., April 1989*, 41–57. Las Vegas, Nevada.

Hoy, M. A., and F. E. Cave. 1985a. Mating behavior in four strains of *Metaseiulus occidentalis* (Acari: Phytoseiidae). *Ann. Entomol. Soc. Am.* 78:588–93.

———. 1985b. Laboratory evaluation of avermectin as a selective acaricide for use with *Metaseiulus occidentalis* (Nesbitt) (Acarina: Phytoseiidae). *Exp. Appl. Acarol.* 1:139–52.

———. 1986. Screening for thelytoky in the parahaploid phytoseiid *Metaseiulus occidentalis*. *Exp. Appl. Acarol.* 2:273–76.

———. 1988. Premating and postmating isolation among populations of *Metaseiulus occidentalis* (Nesbitt)(Acarina: Phytoseiidae). *Hilgardia* 56(6):1–20.

Hoy, M. A., and J. Conley. 1987. Toxicity of pesticides to western predatory mite. *Calif. Agric.* 41(7–8):12–14.

———. 1989. Propargite resistance in Pacific spider mite (Acari: Tetranychidae): Stability and mode of inheritance. *J. Econ. Entomol.* 82:11–16.

Hoy, M. A., and N. F. Knop. 1979. Pesticide resistance in the phytoseiid *Metaseiulus occidentalis* (Nesbitt) in California: A progress report. In *Recent advances in acarology VI.* 1:89–94.

———. 1981. Selection for and genetic analysis of permethrin resistance in *Metaseiulus occidentalis*: Genetic improvement of a biological control agent. *Entomol. Exp. Appl.* 30:10–18.

Hoy, M. A., and Y. L. Ouyang. 1986. Selectivity of the acaricides clofentezine and hexythiazox to the phytoseiid predator *Metaseiulus occidentalis* (Nesbitt) (Acari: Phytoseiidae). *J. Econ. Entomol.* 79:1377–80.

———. 1987. Toxicity of beta-exotoxin of *Bacillus thuringiensis* to *Tetranychus pacificus* McGregor and *Metaseiulus occidentalis* (Nesbitt) (Acari: Tetranychidae and Phytoseiidae). *J. Econ. Entomol.* 80:507–11.

———. 1989. Selection of the western predatory mite, *Metaseiulus occidentalis* (Acari: Phytoseiidae), for resistance to abamectin. *J. Econ. Entomol.* 82:35–40.

Hoy, M. A., and J. M. Smilanick. 1981. Non-random prey location by the phytoseiid predator *Metaseiulus occidentalis*: Differential responses to several spider mite species. *Entomol. Exp. Appl.* 29:241–53.

Hoy, M. A., and K. A. Smith. 1982. Evaluation of *Stethorus nigripes* (Coleoptera: Coccinellidae) for biological control of spider mites in California almond orchards. *Entomophaga* 27(3):301–10.

Hoy, M. A., D. Castro, and D. Cahn. 1982. Two methods for large scale production of pesticide-resistant strains of the spider mite predator *Metaseiulus occidentalis* (Nesbitt) (Acarina: Phytoseiidae). *Zeit. angew. Entomol.* 94:1–9.

Hoy, M. A., J. Conley, and W. Robinson. 1988. Cyhexatin and fenbutatin-oxide resistance in Pacific spider mite (Acari: Tetranychidae): Stability and mode of inheritance. *J. Econ. Entomol.* 81:57–64.

Hoy, M. A., J. J. R. Groot, and H. E. van de Baan. 1985. Influence of aerial dispersal on persistence and spread of pesticide-resistant *Metaseiulus occidentalis* in California almond orchards. *Entomol. Exp. Appl.* 37:17–31.

Hoy, M. A., N. W. Ross, and D. Rough. 1978. Impact of NOW insecticides on mites in northern California almond orchards in 1977. *Calif. Agric.* 32(5):11–12.

Hoy, M. A., H. E. van de Baan, J. J. R. Groot, and R. P. Field. 1984. Aerial movements of mites in almonds: Implications for pest management. *Calif. Agric.* 38(9):21–23.

Hoy, M. A., R. T. Roush, K. B. Smith, and L. W. Barclay. 1979. Spider mites and predators in San Joaquin Valley almond orchards. *Calif. Agric.* 33(10):11–13.

Hoy, M. A., W. W. Barnett, L. C. Hendricks, D. Castro, D. Cahn, and W. J. Bentley. 1984. Managing spider mites in almonds with pesticide-resistant predators. *Calif. Agric.* 38(7–8):18–20.

Hoy, M. A., W. W. Barnett, W. O. Reil, D. Castro, D. Cahn, L. Hendricks, R. Coviello, and W. J. Bentley. 1982. Large scale releases of pesticide-resistant spider mite predators. *Calif. Agric.* 36(1–2):8–10.

Mueller-Beilschmidt, D., and M. A. Hoy. 1987. Activity levels of genetically manipulated and wild strains of *Metaseiulus occidentalis* (Nesbitt) (Acarina: Phytoseiidae) compared as a method to assay quality. *Hilgardia* 55(6):1–23.

Nelson-Rees, W. A., M. A. Hoy, and R. T. Roush. 1980. Heterochromatinization, chromatin elimination, and haploidization in the parahaploid mite *Metaseiulus occidentalis* (Nesbitt) (Acarina: Phytoseiidae). *Chromosoma* 77:263–76.

Roush, R. T., and M. A. Hoy. 1978. Relative toxicity of permethrin to a predator, *Metaseiulus occidentalis*, and its prey, *Tetranychus urticae*. *Environ. Entomol.* 7:287–88.

———. 1980. Selection improves Sevin resistance in spider mite predator. *Calif. Agric.* 34(1):11–14.

———. 1981a. Genetic improvement of *Metaseiulus occidentalis*: Selection with methomyl, dimethoate, and carbaryl and genetic analysis of carbaryl resistance. *J. Econ. Entomol.* 74:138–41.

———. 1981b. Laboratory, glasshouse, and field studies of artificially selected carbaryl resistance in *Metaseiulus occidentalis*. *J. Econ. Entomol.* 74:142–47.

Welter, S. C., M. M. Barnes, I. P. Ting, and Y. J. Hayashi. 1984. Impact of various levels of late spider mite (Acari: Tetranychidae) feeding damage on almond growth and yield. *Environ. Entomol.* 13:52–55.

Wilson, L. T., M. A. Hoy, F. G. Zalom, and J. M. Smilanick. 1984. Sampling mites in almonds: I. Within-tree distribution and clumping patterns of mites with comments on predator-prey interactions. *Hilgardia* 52(7):1–13.

Zalom, F. G., L. T. Wilson, M. A. Hoy, W. W. Barnett, and J. M. Smilanick. 1984. Sampling *Tetranychus* spider mites in almonds. *Calif. Agric.* 38(5–6):17–19.

Zalom, F. G., M. A. Hoy, L. T. Wilson, and W. W. Barnett. 1984. Sampling mites in almonds: II. Presence-absence sequential sampling for *Tetranychus* mite species. *Hilgardia* 52(7):14–24.

Zalom, F. G., R. A. van Steenwyk, W. J. Bentley, R. Coviello, R. E. Rice, W. W. Barnett, M. A. Hoy, and M. M. Barnes. 1987. *Almond pest management guidelines*. IPM IMPACT Pest Management Database, IPM Manual Group. Davis: University of California.

9 / Spider mites on Cotton and Grapes

T. F. LEIGH, L. T. WILSON, D. L. FLAHERTY, AND D. GONZÁLEZ

spider mites
Tetranychus spp. and Eotetranychus willamettei (McGregor)
Acari: Prostigmata: Tetranychidae

INTRODUCTION

Biology and Pest Status

Spider mites are pests of a wide range of crops worldwide, though their outbreaks are more common in arid regions or elsewhere during periods of low rainfall. In the United States, this pest group is prevalent in the drier regions of Nebraska, Kansas, Oklahoma, and Texas, as well as in the Pacific Coast states. Well-watered crops are less likely to develop severe spider mite infestations. Outbreaks are also associated with pest-predator imbalances, which are frequently created by insecticides. Thus, after several synthetic organic insecticides were introduced for crop protection, outbreaks became commonplace.

The most common spider mite species on field crops in California's San Joaquin Valley are the Pacific spider mite, *Tetranychus pacificus* McGregor, the strawberry mite, *Tetranychus turkestani* Ugarov & Nikolski, and the twospotted spider mite, *Tetranychus urticae* Koch. The Willamette mite, *Eotetranychus willamettei* (McGregor)—like the Pacific spider mite—is common on grapes. All three *Tetranychus* species reproduce throughout the year on actively growing crops and weeds, averaging 12 to 16 generations per year. On field crops, the strawberry mite is most common in late spring and fall, but it can be highly destructive throughout the growing season. Twospotted and Pacific mites are most common during the hotter midsummer period.

Historical Notes

To control spider mites, growers have used a variety of synthetic organic insecticides, including dicofol and propargite as well as the organophosphate oxydemeton-methyl. Recently, abamectin has been heavily used. Most spider mite populations have developed resistance to the organophosphates as well as to dicofol and propargite. Many researchers and farm advisors have recommended that growers use selective acaricides and refrain from unnecessary pesticide use to avoid disrupting natural biological control.

Several species of thrips—particularly the western flower thrips, *Frankliniella occidentalis* (Pergande)—are the most frequent spring predators of spider mites in field crops (Trichilo and Leigh 1986). During summer, bigeyed bugs (*Geocoris* spp.) and the minute pirate bug, *Orius tristicolor* (White), are the principal insect predators (González et al. 1982) and may be assisted by coccinellids and cecidomyids. In vineyards, two predaceous mites in the family Phytoseiidae—the western predatory mite, *Galendromus* (=*Metaseiulus*) *occidentalis* (Nesbitt), and *Amblyseius californicus* (McGregor)—are common. Occasionally, these predators also occur in cotton fields. Both predators are used commercially in vineyards and are being evaluated for use on cotton and corn.

RESULTS

Objective B: Ecological and Physiological Studies

Researchers evaluated the spatial distribution of spider mites and their associated predators within individual cotton and grape plants as well as within fields or vineyards. They then used those data to develop commercial monitoring programs based on presence-absence sampling (Wilson 1982; González and Wilson 1982; Wilson et al. 1983; Wilson, Flaherty, et al. 1987; Wilson, Pickett, et al. 1987; Pickett et al. 1988; Wilson et al. 1991; Flaherty et al. 1992). They also developed a formula for estimating absolute spider mite density from presence-absence data.

A study of the biological interactions between spider mites and the western predatory mite evaluated the short- and long-term consequences of predation and competition in the face of predator suppression and augmentation (Hanna and Wilson 1991). The Pacific mite, the economically more important pest species, was negatively affected by the presence of the Willamette mite, which puts it under greater predation pressure. In the absence of drought stress or pesticide-induced disruption, predation and competition shift species dominance in favor of the Willamette mite.

Objective C: The Conservation and Augmentation of Beneficial Organisms

Following the methods of V. M. Stern (1969), strips of alfalfa were interplanted in cotton to enhance naturally occurring biological control. The alfalfa was inoculated with spider mites and, later, with the western predatory mite. The goal was to determine whether the western predatory mite would disperse into the cotton and control infestations. However, although the predaceous bigeyed and minute pirate bugs distributed themselves uniformly across the cotton field, the western predatory mite dispersed less than 20 rows into the cotton (Corbett, Leigh, and Wilson 1991).

After conducting inoculative releases of the western predatory mite and *A. californicus* into cotton, we have established a clear relationship between the release rate and the rate of decline of the tetranychid mite infestation. The two predators proved equally efficacious—however, the earliest releases provided the greatest control and required the least number of phytoseiids.

RECOMMENDATIONS

1. Evaluate information developed on grape pests as a basis for sampling procedures and decision making, as well as to define those strategies.

2. Generate foundation information for vineyard and field crops that will enable us to estimate the impact of naturally occurring biological control. Data of particular importance are prey consumption rates, searching ability, and the interaction of biotic, environmental, and edaphic factors on the predator-prey system—information that will permit us to quantify biological control.

3. Develop a better understanding of the interactions between introduced and naturally occurring predators or parasites.

4. Reassess economic thresholds when pest management is based on natural and supplemental biological control rather than on chemical control.

5. Develop cultures of parasites and predators that are resistant to—or tolerant of—the insecticides necessary to control key pests.

REFERENCES CITED

Corbett, A., T. F. Leigh, and L. T. Wilson, 1991. Evaluation of alfalfa interplants as a source of *Metaseiulus occidentalis* (Acari: Phytoseiidae) for management of spider mites in cotton. *Biological Control* 1:188–96.

Flaherty, D. L., L. T. Wilson, S. Welter, and R. Hanna. 1992. Spider mites. In *Grape pest management*, ed. D. L. Flaherty. Oakland: University of California Division of Agriculture and Natural Resources, Publication 3343.

González, D., and L. T. Wilson. 1982. A food-web approach to economic thresholds: A sequence of pest/predaceous arthropods on California cotton. *Entomophaga* 27:31–43.

González, D., B. R. Patterson, T. F. Leigh, and L. T. Wilson. 1982. Mites: A primary food source for two predators in San Joaquin Valley cotton. *Calif. Agric.* 36(3–4):18–19.

Hanna, R., and L. T. Wilson. 1991. Prey preference by *Metaseiulus occidentalis* (Acari: Phytoseiidae) and the role of prey aggregation. *Biological Control* 1:51–58.

Pickett, C. H., L. T. Wilson, and D. González. 1988. Population dynamics and within-plant distribution of the western flower thrips (Thysanoptera: Thripidae), an early-season predator of mites infesting cotton. *Environ. Entomol.* 17:551–59.

Stern, V. M. 1969. Interplanting alfalfa in cotton to control lygus bugs and other insect pests. In *Proc. Tall Timbers Conf. on Ecological Animal Control by Habitat Management*, No. 1, ed. R. Komarek, 55–69.

Trichilo, P. J., and T. F. Leigh. 1986. Predation on spider mite eggs by the western flower thrips, *Frankliniella occidentalis* (Thysanoptera: Thripidae), an opportunist in a cotton agroecosystem. *Environ. Entomol.* 15:821–25.

Wilson, L. T. 1982. Development of an optimal monitoring program in cotton: Emphasis on spider mites and Heliothis spp. *Entomophaga* 27:45–50.

Wilson, L. T., C. H. Pickett, T. F. Leigh, and J. R. Carey. 1987. Spider mite (Acari: Tetranychidae) infestation foci: Cotton yield reduction. *Environ. Entomol.* 16:614–17.

Wilson, L. T., D. L. Flaherty, R. Hanna, G. Leavitt, and W. Peacock. 1987. Presence-absence sampling in grape systems. In *Proc. Table Grape Seminars, 1–4.* Tulare County: University of California Cooperative Extension Service.

Wilson, L. T., P. J. Trichilo, D. L. Flaherty, R. Hanna, and A. Corbett. 1991. Natural enemy–spider mite interactions: Comments on implications for population assessment. *Modern Acarology* 1:167–73.

Wilson, L. T., D. González, T. F. Leigh, V. Maggi, C. Foristiere, and P. Goodell. 1983. Within-plant distribution of spider mites (Acari: Tetranychidae) on cotton: A developing implementable monitoring program. *Environ. Entomol.* 12:128–34.

10 / Twospotted Spider Mite on Strawberry

J. A. McMurtry and E. R. Oatman

> **twospotted spider mite**
> *Tetranychus urticae* Koch
> Acari: Prostigmata: Tetranychidae

INTRODUCTION

Biology and Pest Status

Strawberries are a major cash crop in California; most of the acreage is in the coastal plains or valleys of the central and southern parts of the state. Of the more than 40 pests listed for the crop, twospotted spider mite, *Tetranychus urticae* Koch, is generally considered the most serious.

Historical Notes

The twospotted spider mite's resistance to most acaricides, combined with residue problems from frequent sprayings, have made biological control of this pest attractive if not essential. W-84 research has demonstrated the benefits and the feasibility of controlling the mite with releases of *Phytoseiulus persimilis* Athias-Henriot.

RESULTS

Objective A: The Identification and Introduction of Beneficial Organisms

After initial studies of *P. persimilis* in experimental plots proved encouraging, researchers began releases of the commercially produced phytoseiids in Ventura County; their goal was to reduce or eliminate acaricide sprays for the twospotted spider mite. Between 1971 and 1977 about 235,000 predators were released in 13 different fields (P. Phillips person. commun.). In 1972 *P. persimilis* was found in large numbers in a strawberry field about 4 kilometers from the nearest release site. After surveys were extended in 1976 to include other crops, *P. persimilis* was found in several lima bean fields, some as far as 20 kilometers from the nearest release fields. Researchers observed that, in some cases, this predator had virtually eliminated spider mites. *P. persimilis* was also collected from eight different weed species infested with the mite. The predator's high mobility (Oatman

1965, 1970) and its ability to colonize a variety of infested host plants (sometimes under adverse conditions, such as on foliage with heavy road-dust deposits and around fields receiving frequent pesticide applications) probably contributed to its establishment and dispersal (McMurtry et al. 1978). Although weeds and other crops are probably important reservoirs for *P. persimilis* movement into strawberry fields, augmentative releases are usually necessary to control the mites.

Objective C: The Conservation and Augmentation of Beneficial Organisms

In successful trials in small field plots (Oatman 1965), *P. persimilis* suppressed twospotted spider mite effectively at release rates of 800,000 per hectare—about 13 predators per plant (Oatman and McMurtry 1966). Experiments with lower release rates showed that under standard cultural practices the predator could suppress the mites at rates of 5 or 10 per plant (Oatman et al. 1976). However, because purchasing *P. persimilis* at these numbers is economically marginal or prohibitive, researchers investigated the potential of two native predator species, *Amblyseius* (=*Neoseiulus*) *californicus* (McGregor) and *Galendromus* (=*Typhlodromus* or *Metaseiulus*) *occidentalis* (Nesbitt). These two predatory mites have a greater ability to persist at low prey densities, and are potentially more economical to rear than *P. persimilis*. However, even though these predators (especially *A. californicus*) gave encouraging results, *P. persimilis* still exercised the best control, because of its superior dispersal powers in strawberry fields and its ability to respond more quickly to increases in prey density (Oatman et al. 1977a,b).

Researchers have observed that, with extensive monitoring and early detection of mite infestations, *P. persimilis* can control the mites at release rates considerably lower than five predators per plant. Still, a combination of predator releases and acaricide applications on an as-needed basis may be required (G. T. Scriven person. commun.).

RECOMMENDATIONS

1. We need more information on the minimum effective release rates of *P. persimilis* and on the best release strategies—such as spot-releases where mites are detected versus releases throughout a field.

2. The effect of one versus two or three predator species on twospotted spider mite populations should be studied.

3. Other potential predators—including *Phytoseiulus fragariae* Denmark & Schicha and *Phytoseiulus longipes* Evans—should be evaluated under a variety of field conditions.

REFERENCES CITED

McMurtry, J. A., E. R. Oatman, P. A. Phillips, and C. W. Wood. 1978. Establishment of *Phytoseiulus persimilis* (Acari: Phytoseiidae) in southern California. *Entomophaga* 23:175–79.

Oatman, E. R. 1965. Predaceous mite control of twospotted spider mite on strawberry. *Calif. Agric.* 19(2):6–7.

———. 1970. Integration of *Phytoseiulus persimilis* with native predators for control of the twospotted spider mite on rhubarb. *J. Econ. Entomol.* 63:1177–80.

Oatman, E. R., and J. A. McMurtry. 1966. Biological control of the two-spotted spider mite on strawberry in southern California. *J. Econ. Entomol.* 59:433–39.

Oatman, E. R., F. E. Gilstrap, and V. Voth. 1976. Effect of different release rates of *Phytoseiulus persimilis* (Acarina: Phytoseiidae) on the twospotted spider mite on strawberry in southern California. *Entomophaga* 21:269–73.

Oatman, E. R., J. A. McMurtry, F. E. Gilstrap, and V. Voth. 1977a. Effect of releases of *Amblyseius californicus*, *Phytoseiulus persimilis*, and *Typhlodromus occidentalis* on the twospotted spider mite on strawberry in southern California. *J. Econ. Entomol.* 70:45–47.

———. 1977b. Effect of releases of *Amblyseius californicus* on the twospotted spider mite on strawberry in southern California. *J. Econ. Entomol.* 70:638–40.

11 / YELLOW SPIDER MITE

M. T. ALINIAZEE AND P. H. WESTIGARD

> **yellow spider mite**
> *Eotetranychus carpini borealis* (Ewing)
> Acari: Prostigmata: Tetranychidae

INTRODUCTION

Biology and Pest Status

The yellow spider mite, *Eotetranychus carpini borealis* (Ewing), is a major pest of fruit trees in the western areas of Oregon and Washington (AliNiazee 1979a) and parts of southern Oregon (Westigard and Berry 1970). It is particularly harmful on apples and pears. In most years it causes damage to apples in Oregon's Willamette Valley that equals or exceeds that of the European red mite, *Panonychus ulmi* (Koch), and the twospotted spider mite, *Tetranychus urticae* Koch. In chemically treated orchards, population levels of up to 200 to 300 mites per leaf are not uncommon (AliNiazee 1979a). Such high mite densities can defoliate trees and seriously damage fruit quality. Relatively cool spring and mild early summer temperatures apparently favor this mite species; populations increase dramatically during the early season and decline in late August and September.

Yellow spider mites overwinter as mature females. Their bright yellow color distinguishes them from the lemon-orangish and larger twospotted spider mite females. On apple and pear trees, the females overwinter under loose bark and in the cracks and crevices of small limbs—approximately the same locations where twospotted spider mites hibernate. During late spring and early summer, the female mites move to opening buds and begin to lay eggs. Depending on the temperature, eggs hatch in 2 to 3 weeks, and the nymphs begin feeding on the expanding foliage. As temperatures increase, the generation time is reduced to about 2 weeks; during the summer, generations overlap. P. H. Westigard and D. W. Berry (1970) have reported that yellow mite females produce an average of 36 eggs over a maximum life span of 27 days. The researchers also reported a mid-July peak of about 12.5 mites per leaf, followed by a slow population decline.

Historical Notes

Most of the biological control work on the yellow mite has involved conserving and augmenting natural enemies and assessing the impact of biological control. Researchers have collected four species of predatory phytoseiid mites from Oregon apple and pear orchards: *Galendromus* (=*Typhlodromus* or *Metaseiulus*) *occidentalis* (Nesbitt), *Typhlodromus arboreus* Chant, *T. pyri* Scheuten, and *Amblyseius andersoni* (Chant) (Westigard and Berry 1970; AliNiazee 1979a,b; Hadam, AliNiazee, and Croft 1986). In southern Oregon *G. occidentalis* appears to be predominant, whereas *T. pyri* and *T. arboreus* are more common in the Willamette Valley and the Hood River Valley.

RESULTS

Objective C: The Conservation and Augmentation of Beneficial Organisms

Over the last 10 years, researchers in Oregon have worked on conserving the three predominant phytoseiid predators—*Galendromus occidentalis*, *Typhlodromus pyri*, and *T. arboreus*. They found that the use of insecticide had a detrimental effect on the predators. For example, M. T. AliNiazee (1979a) reported that the use of carbofuran on a test plot virtually eliminated *T. arboreus* populations, resulting in severe outbreaks of yellow mites—over 700 mites per leaf—that had, by the end of the season, left trees defoliated. The test plot did not regain its natural balance until after nearly 3 years without pesticide use. P. H. Westigard and D. W. Berry (1970) reported a buildup of yellow mites, and a concurrent decline of *G. occidentalis*, in azinphosmethyl-treated plots.

Scientists have pursued two approaches to conserving phytoseiid mites. First, they attempted to reduce the use of insecticide by pursuing alternative approaches for controlling major pests (Moffit and Westigard 1984). Second, they looked for a buildup of resistance in the beneficial phytoseiids. M. T. AliNiazee and J. E. Cranham (1979) and M. T. AliNiazee (1984) reported that *T. pyri* and *T. arboreus* were susceptible to different synthetic pyrethroid insecticides. Their data suggested that most of the commonly used insecticides were toxic to phytoseiids, but that by using these chemicals sparingly, enough predatory mites might survive to provide effective biological control. B. A. Croft and M. T. AliNiazee (1983) found that *T. arboreus* developed a resistance to azinphosmethyl and carbaryl, and *G. occidentalis* to diazinon and phosalone as well. J. J. Hadam, M. T. AliNiazee, and B. A. Croft (1986) reported that *T. pyri* developed resistance to azinphosmethyl, carbaryl, and parathion. The development of resistance in these phytoseiids could be very beneficial to conservation and augmentation programs.

Objective D: Impact

M. T. AliNiazee (1979a,b) reported that *T. arboreus*, when not exposed to pesticides, responded numerically to the increase of yellow mite populations and was able to reduce the pest densities to levels that did not cause economic injury. In a 6-year study of yellow mites on apples, the numbers of yellow mites declined to less than five per leaf, and the numbers of *T. arboreus* increased to over three per leaf. By the end of the study period yellow mites were scarce. *G. occidentalis* and *T. pyri* have also been shown to provide effective biological control.

RECOMMENDATIONS

The biological control of the yellow mite is highly effective under undisturbed conditions. The beneficial phytoseiid mites have developed some resistance to a few common pesticides, but are still affected by the repeated disruption from the indiscriminate use of broad-spec-trum insecticides. Researchers should attempt to develop a more selective pesticide-use methodology to conserve the natural enemies of yellow mites.

REFERENCES CITED

AliNiazee, M. T. 1979a. Role of a predatory mite, *Typhlodromus arboreus*, in biological control of spider mites on apples in western Oregon. In *Proc. 4th Internat. Congr. Acarology, 1974, Salfedan, Austria,* 637–42.

———. 1979b. Mite populations in western Oregon. In *Recent advances in acarology,* ed. J. G. Rodriguez, 71–76. New York: Academic Press.

———. 1984. Effect of two synthetic pyrethroids on the predatory mite, *Typhlodromus arboreus*, in the apple orchards of western Oregon. In *Acarology VI,* eds. D. A. Griffiths and C. E. Bowman, 655–58. New York: Wiley.

AliNiazee, M. T., and J. E. Cranham. 1979. Effect of four synthetic pyrethroids on a predatory mite, *Typhlodromus pyri*, and its prey, *Panonychus ulmi*, on apples in southeast England. *Environ. Entomol.* 9:436–39.

Croft, B. A., and M. T. AliNiazee. 1983. Differential resistance to insecticides in *Typhlodromus arboreus* Chant and associated phytoseiid mites in apples in the Willamette Valley, Oregon. *Environ. Entomol.* 12:1420–23.

Hadam, J. J., M. T. AliNiazee, and B. A. Croft. 1986. Phytoseiid mites (Parasitiformis: Phytoseiidae) of major crops in the Willamette Valley, Oregon, and pesticide resistance in *Typhlodromus pyri. Environ. Entomol.* 15:1255–63.

Moffitt, H. R., and P. H. Westigard. 1984. Suppression of the codling moth (Lepidoptera: Tortricidae) population on pears in southern Oregon through mating disruption with sex pheromone. *J. Econ. Entomol.* 77:1513–19.

Westigard, P. H., and D. W. Berry. 1970. Life history and control of the yellow spider mite on pear in southern Oregon. *J. Econ. Entomol.* 63:1433–37.

12 / GREENHOUSE THRIPS

J. A. McMURTRY

> **greenhouse thrips**
> *Heliothrips haemorrhoidalis* (Bouché)
> Thysanoptera: Thripidae

INTRODUCTION

Biology and Pest Status

In recent years greenhouse thrips has become an increasingly important avocado pest in California. Found mainly on the Haas variety, greenhouse thrips occur primarily on the fruit, causing brown scarring that results in the downgrading of the produce.

Historical Notes

The trichogrammatid egg parasitoid *Megaphragma mymaripenne* Timberlake is not an effective regulating agent, even when more than 50 percent of the thrips's egg blisters show parasitoid emergence holes (Hessein and McMurtry 1988, 1989). Predators, including the green lacewing, *Chrysoperla* (=*Chrysopa*) *carnea* (Stephens), and the vespiform thrips, *Franklinothrips vespiformis* (Crawford), sometimes cause significant thrips mortality, but their occurrence is generally sporadic (McMurtry unpub. data).

RESULTS

Objective A: The Identification and Introduction of Beneficial Organisms

A eulophid larval parasitoid, *Goetheana parvipennis* Gahan, was imported to California from Trinidad in 1962 and from the Bahamas in 1983 (Hessein and McMurtry 1989). In neither case did the parasitoid become established (McMurtry and Johnson unpub. data). A second eulophid larval parasitoid, *Thripobius semiluteus* Boucek, was introduced from Australia in 1986 and Brazil in 1988; it has become established on avocado in southern California (McMurtry, Johnson, and Newberger 1991). It lays eggs in either the first- or second-instar thrips larvae, and the host usually develops to the prepupa, after which the parasitoid pupates and emerges as an adult. *T. semiluteus* is only known to para-sitize thrips in the subfamily Panchaetothripinae (LaSalle and McMurtry 1989).

Between 1986 and 1990, more than 500,000 *T. semiluteus* parasitoids were released at 50 sites in five California counties. At most of the sites, the recoveries of parasitoids 2 to 3 months after release have been impressive. Researchers have also recorded parasitoid survival through the winter and dispersal from release sites throughout an orchard.

At some sites researchers have observed a rate of up to 70 percent parasitization of thrips larvae, and a subsequent decline of thrips. At other sites parasitoid activity has been low. One of the factors critical to the success of this parasitoid will be its ability to recover from severe reductions in thrips numbers, which result from the early picking of fruit, parasitization, or severe weather conditions.

RECOMMENDATIONS

1. Surveys are needed to assess the rate of spread of *T. semiluteus* within and between orchards.

2. Experiments should be conducted to determine the impact of *T. semiluteus* on greenhouse thrips populations, and on the actual reduction of fruit scarring, by comparing plots with and without parasitoids.

3. The feasibility and impact of augmentative releases of *T. semiluteus* should be determined.

REFERENCES CITED

Hessein, N. A., and J. A. McMurtry. 1988. Observations on *Megaphragma mymaripenne* Timberlake (Hymenoptera: Trichogrammatidae), an egg parasite of *Heliothrips haemorrhoidalis* (Bouché) (Thysanoptera: Thripidae). *Pan-Pacific Entomol.* 64:250–54.

———. 1989. Biological studies of *Goetheana parvipennis* (Gahan) (Hymenoptera: Eulophidae), an imported parasitoid, in relation to the host species *Heliothrips haemorrhoidalis* (Bouché) (Thysanoptera: Thripidae). *Pan-Pacific Entomol.* 65:25–33.

LaSalle, J., and J. A. McMurtry. 1989. The first record of *Thripobius semiluteus* (Hymenoptera: Eulophidae) in the New World. *Proc. Entomol. Soc. Wash.* 91:634.

McMurtry, J. A., H. G. Johnson, and S. J. Newberger. 1991. Imported parasite of greenhouse thrips established on California avocado. *Calif. Agric.* 45(6):31–32.

13 / MELON THRIPS

M. W. JOHNSON AND D. M. NAFUS

> **melon thrips**
> *Thrips palmi* Karny
> Thysanoptera: Thripidae

INTRODUCTION

Biology and Pest Status

The melon thrips, *Thrips palmi* Karny, is a polyphagous species with an extensive host range that includes vegetables and melons (cantaloupe, long beans, bush beans, lettuce, muskmelon, onion, squash, and Chinese spinach), field crops (cotton and cowpeas), tree crops (avocado, citrus, peach, and plum), and ornamentals (carnation and chrysanthemum) (Bournier 1983). Its wide host range, and the ineffectiveness of most insecticides applied for its control (Suzuki et al. 1982), have made this species a serious pest.

Studies in Hawaii and Guam have concentrated on monitoring pest populations, identifying which natural enemies affect *T. palmi* populations, and determining the influence of insecticides on both the pest and its natural enemies.

Historical Notes

Thrips palmi was first collected in 1925 from tobacco in Indonesia (Bournier 1983). Until 1976, its distribution was believed to be limited to Asia, Sumatra, and Java. However, during the last decade, this pest was found in India, Thailand, Malaysia, Pakistan, Taiwan, Japan, New Caledonia, Western and American Samoa, and the Philippines (Bhatti 1980; Bournier 1983; Waterhouse and Norris 1987).

T. palmi was introduced to Hawaii around 1982, but was not correctly identified until 1984 (Nakahara, Sakimura, and Heu 1984). In Guam it was first observed in 1983 at low densities on cucumber; researchers believe the thrips entered Guam on orchids or other flowers imported from Hawaii. By 1984 the thrips was severely damaging the island's watermelon, bean, and eggplant crops.

RESULTS

Objective A: The Identification and Introduction of Beneficial Organisms

In Hawaii. During surveys conducted between April and September 1984, *Thrips palmi* infestations were common on watermelon grown on Oahu. Most of the populations were composed of active immature stages; researchers counted less than 4 adults per leaf, compared with 1 to 50 immatures per leaf (Johnson 1986). During these surveys, an *Orius* sp. was observed feeding on the immature stages of *T. palmi*. A later survey of 43 watermelon plantings on Oahu and Molokai indicated that by 1986, *T. palmi* was present in plantings on more than 70 percent of all sample dates (Johnson et al. 1989). This indicated that it had become a primary pest of watermelon, surpassing nominal treatment thresholds on more than 16 percent of all survey dates.

In the summer of 1986 researchers conducted field studies on Oahu to determine which natural enemies were attacking *T. palmi* on eggplant, cucumber, and green beans. During the study, they took weekly foliage samples from the three plant hosts. *T. palmi* was the predominant thrips species (at 99 percent) infesting the various crop hosts; the highest recorded densities were on eggplant (185.7 individuals per leaf), followed by cucumber (84.3 individuals per leaf), and beans (16.8 individuals per leaf). Eggplant consistently supported the highest thrips densities and beans the lowest. The first- and second-most-common predators were *Franklinothrips vespiformis* (Crawford) (Thysanoptera: Aeolothripidae) and *Orius insidiosus* (Say) (Hemiptera: Anthocoridae). Other predators that were occasionally recorded included *Curinus coeruleus* (Mulsant) (Coleoptera: Coccinellidae), *Rhinacloa forticornis* Reuter (Hemiptera: Miridae), and *Paratriphleps laevisculus* Champion (Hemiptera: Anthocoridae). The highest recorded densities of *F. vespiformis* were 0.62, 0.17, and

The assistance of Wilmar Snell in collecting data on the natural enemies of thrips in Hawaii was greatly appreciated.

0.92 individuals per leaf on eggplant, cucumber, and beans, respectively. The higher densities on beans may have resulted from the presence of greenhouse whitefly, one of the predator's alternative hosts. The densities of *O. insidiosus* remained below 0.1 individuals per leaf on all crops throughout the survey period. These results indicated that the natural enemies that were established before the introduction of *T. palmi* have altered their habits to accept this prey. Although an *Orius* sp. was previously reported attacking *T. palmi* on watermelon, the association between the predatory *F. vespiformis* and *T. palmi* was not reported in the scientific literature.

Objective C: The Conservation and Augmentation of Beneficial Organisms

In Guam. Researchers tested insecticides in use on Guam with the thrips' natural enemies, including the minute pirate bug, *Orius niobe* Herring, and phytoseiid mites. They found that dimethoate, fenvalerate, malathion, and naled reduced the number of predatory mites and increased the number of thrips. In plots treated with dimethoate, fenvalerate, and malathion, the densities of predators averaged between 0.3 and 1.5 mites per 20 leaves, compared with 16.9 mites per 20 leaves in untreated plots. Naled had less effect, reducing the number of predatory mites by only about 35 percent.

RECOMMENDATIONS

The rapid spread of this previously unrecognized pest species and its subsequent rise to major pest status suggest that in its geographical area of origin, the melon thrips is probably under good biological control.

1. Studies should be carried out in Sumatra and Java, the suggested native range of the pest, to determine whether natural enemies regulate *T. palmi* in those regions.

2. If effective biological control agents do exist, these should be introduced where *T. palmi* causes problems. However, due to the possibility that general thrips predators will attack other predatory thrips (such as *Franklinothrips vespiformis*) or phytophagous species introduced for the control of weeds (namely, *Liothrips urichi* Karny on *Clidemia hirta*), caution should be exercised so that current levels of biological control are not disrupted.

3. Because *T. palmi* infestations can be worsened by pesticide applications, growers should restrict the use of pesticides against this thrips to times when population densities become economically significant.

REFERENCES CITED

Bhatti, J. S. 1980. Species of the genus *Thrips* from India (Thysanoptera). *Syst. Entomol.* 5:109–66.

Bournier, J. D. 1983. A polyphagous insect, *Thrips palmi* (Karny). Important cotton pests in the Philippines. *Cotton Fibres Trop.* 38:286–88.

Johnson, M. W. 1986. Population trends of a newly introduced species, *Thrips palmi* (Thysanoptera: Thripidae), on commercial watermelon plantings in Hawaii. *J. Econ. Entomol.* 79:718–20.

Johnson, M. W., R. F. L. Mau, A. P. Martinez, and S. Fukuda. 1989. Foliar pests of watermelon in Hawaii. *Trop. Pest Manage.* 35:90–96.

Nakahara, L. M., K. Sakimura, and R. A. Heu. 1984. New state record. *Hawaii Department of Agriculture, Hawaii Pest Report* 4(1):1–4.

Suzuki, H., S. Tamaki, and A. Miyara. 1982. Physical control of *Thrips palmi* Karny. *Proc. Assoc. Plant Protect. Kyushu* 28:134–37.

Waterhouse, D. F., and K. R. Norris. 1987. *Biological control: Pacific prospects.* Melbourne: Inkata Press.

14 / SOUTHERN GREEN STINK BUG

W. A. JONES, L. E. EHLER, M. P. HOFFMANN, N. A. DAVIDSON, L. T. WILSON, AND J. W. BEARDSLEY

> **southern green stink bug**
> *Nezara viridula* (L.)
> Heteroptera: Pentatomidae

INTRODUCTION

Biology and Pest Status

The southern green stink bug, *Nezara viridula* (L.), is one of the most important insect pests of food and fiber crops around the world. N. B. DeWitt and G. L. Godfrey (1972), J. W. Todd and D. C. Herzog (1980), A. R. Panizzi and F. Slansky (1985), W. A. Jones (1988), and J. W. Todd (1989) have reviewed various aspects of *N. viridula*. The stink bug was first recorded in the New World in 1798 and in Florida in 1885. It is now found in the United States in South Carolina and Florida and west along the Gulf States to Texas, where it is primarily a pest of soybeans and truck crops. After being discovered in Hawaii in 1961 (Mitchell 1965) and in California in 1986 (Hoffman, Wilson, and Zalom 1987a,b), the bug reached pest status almost immediately in those states.

The southern green stink bug deposits its pale yellow eggs in large masses on the undersides of host-plant leaves. It has five nymphal instars. Both the nymphs and adults damage crops by piercing the developing seeds or fruit, which results in direct losses or unmarketable produce. The stink bug usually has three to five generations annually. Adults prefer to overwinter in aboveground sites; the bug's distribution is therefore limited by winter temperatures (Jones and Sullivan 1981).

Historical Notes

W. A. Jones et al. (1983) and W. A. Jones (1988) have noted previous attempts to establish new parasitoid species in the southeastern states. Now that the stink bug is believed to have originated in the Ethiopian Region, researchers have identified previously unrecognized sources of natural enemies (Jones and Powell 1982; Hokkanen 1986; Jones 1988). Worldwide, they have recorded 57 species of parasitoids on *N. viridula* (Jones 1988).

RESULTS

Objective A: The Identification and Introduction of Beneficial Organisms

Researchers in other parts of the world have targeted this pest for biological control by introducing and establishing egg and adult parasitoids. The programs in Hawaii and Australia are considered landmark examples in classical biological control (Caltagirone 1981). In California, where *N. viridula* threatens a multimillion-dollar vegetable and nut industry, researchers have recently introduced and disseminated a parasitoid species; the results are discussed later in this chapter.

In Hawaii. After its initial appearance, in 1961, the southern green stink bug spread rapidly throughout the state. It caused major economic losses in several fruit, vegetable, and nut crops as well as in certain ornamentals. The Hawaii State Department of Agriculture initiated biological control efforts soon after the discovery of the pest, introducing, propagating, and releasing several egg parasitoids (Encyrtidae and Scelionidae) and two adult parasitoids (Tachinidae) (Davis and Krauss 1963). The scelionid egg parasitoid *Trissolcus basalis* (Wollaston) from Australia and Pakistan, and the tachinids *Trichopoda pennipes* (F.) from Florida and *Trichopoda pilipes* (F.) from the West Indies, became successfully established. By 1963, in many areas populations of the pest were

We thank J. R. Coulson, W. W. Harrison, and J. W. Todd for reviewing an early draft. We also acknowledge the support and assistance of the USDA-ARS, European Biological Control Laboratory in Montpellier, France, as well as the cooperation of the following in making parasitoid collections in Europe: T. Cabello, Centro de Investigación y Desarrollo Agrario in Granada, Spain; J. Voegele and N. Volkoff, Institut National de la Recherche Agronomique in Antibes, France; F. Bin and S. Colazza, University of Perugia in Perugia, Italy; D. Coutinot, EBCL; and L. B. Jones.

greatly reduced from the peak, postinvasion levels (Davis 1964). At present, *N. viridula* occasionally causes economic losses in Hawaii, particularly to beans and macadamia nuts, but is under satisfactory biological control in most agricultural environments (Davis 1967; Waterhouse and Norris 1987).

The W-84 project's involvement with stink bug biological control in Hawaii consisted primarily of biological and evaluation studies on the established parasitoids (Shahjahan 1968a,b; Shahjahan and Beardsley 1975).

In California. Following the 1986 discovery of a major infestation of *N. viridula* on processing tomatoes in the Sacramento Valley (Hoffmann, Wilson, and Zalom 1987a,b), entomologists with the University of California at Davis and the USDA-ARS European Parasite Laboratory (EPL) near Paris, France, developed a cooperative effort to collect the egg parasitoid *T. basalis* from the Mediterranean area. Using climate maps, researchers identified specific areas in Spain, France, and Italy as possessing the two subclimate types found in the Sacramento Valley. During the early summer of 1987, workers collected stink bug egg masses that had been parasitized in the three target areas: two egg masses from Spain (Andalusia and Granada), two from France (near Antibes), and seven from Italy (Umbria and Pontenuevo, Deruta). These were returned to the EPL, where the emerging parasitoids were confined with fresh egg masses. The parasitized eggs were then shipped to the quarantine laboratory at Davis.

The Spanish, French, and Italian races of *T. basalis*, after being maintained separately in quarantine, were evaluated by M. C. Awan, L. T. Wilson, and M. P. Hoffmann (1990). Of the approximately 7,300 parasitized eggs in quarantine, 187 parasitized egg masses were consigned to the rearing program. In the fall of 1987 researchers released about 2,500 parasitoids in Yolo County, where they became established.

During 1988, over 100,000 parasitoids were released in six northern California counties. By the end of the summer, *T. basalis* had become established at most release sites; at some locations more than 90 percent of field-collected southern green stink bug eggs had been parasitized. The parasitoid was also recovered at numerous release sites during 1989, indicating that it is now well established in the region. So far, *T. basalis* appears to be an effective biological control agent for the southern green stink bug in northern California. It should be noted that the founding population of *T. basalis* —11 parasitized egg masses—was relatively small, and may even represent the progeny of no more than 11 females.

RECOMMENDATIONS

1. Continue evaluating the impact of *T. basalis* on southern green stink bug populations in California.

2. If additional natural enemies are needed, consider introducing trichopodine tachinids that parasitize adults and larger nymphs, as well as egg parasitoids that may be compatible with *T. basalis*.

3. Evaluate the interactions between hosts and parasitoids to enhance parasitoid effectiveness.

REFERENCES CITED

Awan, M. S., L. T. Wilson, and M. P. Hoffmann. 1990. Comparative biology of three geographic populations of *Trissolcus basalis* (Wollaston) (Hymenoptera: Scelionidae). *Environ. Entomol.* 19:387–92.

Caltagirone, L. E. 1981. Landmark examples in classical biological control. *Annu. Rev. Entomol.* 26:213–32.

Davis, C. J. 1964. The introduction, propagation, liberation and establishment of parasites to control *Nezara viridula* variety *smaragdula* (Fabricius) in Hawaii (Heteroptera: Pentatomidae). *Proc. Hawaii. Entomol. Soc.* 18:369–75.

Davis, C. J. 1967. Progress in the biological control of the southern green stink bug, *Nezara viridula* variety *smaragdula* (F.) in Hawaii (Heteroptera: Pentatomidae). *Mushi (Tokyo)* 3 (Suppl.):9–16.

Davis, C. J., and N. L. H. Krauss. 1963. Recent introductions for biological control in Hawaii VIII. *Proc. Hawaii. Entomol. Soc.* 18:245–49.

DeWitt, N. B., and G. L. Godfrey. 1972. The literature of arthropods associated with soybeans. II: A bibliography of the southern green stink bug *Nezara viridula* (Linnaeus) (Hemiptera: Pentatomidae). Illinois Natural Historical Survey of Biology Notes 78.

Hoffmann, M. P., L. T. Wilson, and F. G. Zalom. 1987a. Control of stink bugs in tomatoes. *Calif. Agric.* 41(5–6):4–6.

Hoffmann, M. P., L. T. Wilson, and F. G. Zalom. 1987b. The southern green stink bug, *Nezara viridula* (L.) (Heteroptera; Pentatomidae): New location. *Pan-Pacific Entomol.* 63:333.

Hokkanen, H. 1986. Polymorphism, parasites, and the native area of *Nezara viridula* (Hemiptera, Pentatomidae). *Ann. Entomol. Fennici* 52:28–31.

Jones, W. A. 1988. World review of the parasitoids of the southern green stink bug *Nezara viridula* (L.) (Heteroptera: Pentatomidae). *Ann. Entomol. Soc. Am.* 81:262–73.

Jones, W. A., and J. E. Powell. 1982. Potential for biological control of the southern green stink bug on soybeans. In *Mississippi Entomol. Assoc. Proc.* 1:21–22.

Jones, W. A., and M. J. Sullivan. 1981. Overwintering habitats, spring emergence patterns and winter mortality of some South Carolina Hemiptera. *Environ. Entomol.* 10:409–14.

Jones, W. A., S. Y. Young, M. Shepard, and W. H. Whitcomb.

1983. Use of imported natural enemies against insect pests of soybean. In *Natural enemies of arthropod pests in soybean*, ed. H. Pitre, 63–77. Southern Cooperative Series Bulletin 285.

Mitchell, W. C. 1965. Status of the southern green stink bug in Hawaii. *USDA Agric. Sci. Rev.* 3:32–35.

Panizzi, A. R., and F. Slansky, Jr. 1985. Review of the phytophagous pentatomids (Hemiptera: Pentatomidae) associated with soybeans in the Americas. *Fla. Entomol.* 68:184–214.

Shahjahan, M. 1968a. Superparasitization of the southern green stink bug by the tachinid parasite *Trichopoda pennipes pilipes* and its effect on the host and parasite survival. *J. Econ. Entomol.* 61:1088–91.

———. 1968b. Effect of diet on the longevity and fecundity of the adults of the tachinid parasite *Trichopoda pennipes pilipes*. *J. Econ. Entomol.* 61:1102–3.

Shahjahan, M., and J. W. Beardsley, Jr. 1975. Egg viability and larval penetration in *Trichopoda pennipes pilipes* Fabricius (Diptera: Tachinidae). *Proc. Hawaii. Entomol. Soc.* 22:133–36.

Todd, J. W. 1989. Ecology and behavior of *Nezara viridula* (L.). *Annu. Rev. Entomol.* 34:273–92.

Todd, J. W., and D. C. Herzog. 1980. Sampling phytophagous Pentatomidae on soybean. In *Sampling methods in soybean entomology,* eds. M. Kogan and D. C. Herzog, 438–78. New York: Springer-Verlag.

Waterhouse, D. F., and K. R. Norris. 1987. *Biological control: Pacific prospects.* Melbourne: Inkata Press.

15 / SQUASH BUG

J. R. NECHOLS

> **squash bug**
> *Anasa tristis* DeGeer
> Heteroptera: Coreidae

INTRODUCTION

Biology and Pest Status

The squash bug, *Anasa tristis* DeGeer, is an important pest of squash, pumpkin, and other vine crops in the genus *Cucurbita* throughout most of the United States. R. L. Beard (1940), D. C. Elliot (1935), and H. N. Worthley (1923) have reported on the general biology of the squash bug. More recent investigations have dealt with aspects of the squash bug's reproduction (Al-Obaidi 1977; Nechols 1987; Page et al. 1989), development (Fargo and Bonjour 1988; Fielding and Ruesink 1988), dormancy (Nechols 1987, 1988; Fielding 1988a), and general population dynamics (Fargo et al. 1988; Fielding 1988b).

Adult squash bugs overwinter in and around cultivated fields; females do so in a reproductive diapause. This diapause ends by early spring (Nechols 1988), after which the squash bugs mate and oviposit on host plants. Each female lays an average of 200 to 300 tannish eggs over a 1- to 2-month period (Beard 1940; Nechols 1987). The eggs, which are deposited in masses of 12 to 14, darken to a brick red and hatch into green-and-black nymphs in 9 to 10 days at 27°C. Squash bugs pass through four additional grayish nymphal instars before emerging as adults. The entire life cycle takes about a month. Some adults are reproductively active until late summer, and squash bugs continue to develop and emerge into late autumn (Nechols 1987).

Historical Notes

Anasa tristis is considered indigenous to North America, although its native range includes Central America and northern South America, where it may have originated. The squash bug is attacked by a complex of natural enemies, including a tachinid fly, *Trichopoda pennipes* Fabricius, and various hymenopteran egg parasitoids in the families Scelionidae and Encyrtidae (see the Results section). *T. pennipes* is a nymphal-adult parasitoid that is widely distributed throughout the United States (Worthley l924a,b; Beard 1940, 1942; Dietrick and van den Bosch l957). Its biology was studied by R. L. Beard (1940, 1942) and H. N. Worthley (1924a,b).

In contrast, the biologies of the parasitoids of squash bug eggs are poorly understood. For example, the early literature includes just one paper on the bionomics of the scelionid *Hadronotus ajax* Girault (=*Gryon pennsylvanicum* [Ashmead]) (Schell 1943). In recent comparative studies of selected species, our research has somewhat improved this situation (Tracy and Nechols 1987, 1988; Nechols, Tracy, and Vogt 1989). L. Masner (1983) published a revision of *Gryon*, the genus to which all currently known scelionid egg parasitoids of *A. tristis* belong. However, the taxonomy of egg parasitoids in the family Encyrtidae still needs to be researched (Tracy and Nechols 1987; G. Gordh person. commun.).

RESULTS

Objective A: The Identification and Introduction of Beneficial Organisms

Surveys of squash and pumpkin fields in northeastern Kansas from 1984 to 1986 showed that squash bugs are attacked—but never at high rates—by a complex of naturally occurring enemies. These include the tachinid fly, *Trichopoda pennipes*, and nine species of hymenopteran egg parasitoids, seven of which had not been previously recorded on *A. tristis* eggs (Nechols unpub. data). We recovered three species of wasps in the family Scelionidae: *Gryon pennsylvanicum* (Ashmead), *G. floridanum* (Ashmead), and *G. longipenne* Masner. In the family Encyrtidae, we found *Ooencyrtus anasae* (Ashmead), *O. californicus* Girault, and what appear to be four undescribed *Ooencyrtus* species (G. Gordh person. commun.). However, only *O. anasae*, *G. pennsylvanicum*, and *G. floridanum* have been collected from squash bug eggs in more than one season. *G. pennsylvanicum* is the only egg parasitoid that has been collected annually in Kansas.

Objective B: Ecological and Physiological Studies

Because naturally occurring egg parasitoids are poorly synchronized with squash bugs, and are present at low densities throughout the season (see Objective D), we did comparative laboratory studies of the egg parasitoids *G. pennsylvanicum*, *O. anasae*, and two currently undescribed *Ooencyrtus* species to evaluate which species show the greatest potential for possible manipulation in an augmentation program. We quantified biological attributes and computed demographic statistics for various species at 27°C. In addition, we measured responses to temperature (such as development and survival), host quality (egg age), and host-plant quality (squash bug–resistant or –susceptible varieties). The developmental rate, proportion of female progeny, and intrinsic rate of population increase (r_{max}) were higher at 27°C for *O. anasae* than for *O.* sp. 'light form' (Tracy and Nechols 1987, 1988). However, *G. pennsylvanicum* had markedly higher reproductive and parasitization rates, as well as a higher r_{max} value, than did either of the *Ooencyrtus* species (Nechols, Tracy, and Vogt 1989).

All of the egg parasitoids that were tested developed and survived well when reared under a range of constant temperatures between 21 and 29.4°C. However, *G. pennsylvanicum* exhibited a considerably lower survivorship at the more extreme upper (32.7°C) and lower (18.3°C) temperatures than did the *Ooencyrtus* species (Nechols, Tracy, and Vogt 1989). Field temperatures in excess of 32.7°C are common during Kansas summers; hence, the thermal tolerance of *G. pennsylvanicum* under fluctuating laboratory and field conditions needs to be determined.

Under no-choice laboratory conditions, *G. pennsylvanicum* and the *Ooencyrtus* species parasitized, developed, and emerged at approximately equal rates in hosts ranging from newly deposited eggs to those that contained well-developed first-instar nymphs (Nechols, Tracy, and Vogt 1989). Thus, these parasitoids appear to be capable of using almost the full range of host egg ages in the field.

The preimaginal developmental rates, survival, sex ratio, oviposition rates, and fecundity of these egg parasitoids were unaffected by host eggs from squash bugs that had been reared on resistant plant varieties (Vogt 1992; Vogt and Nechols 1993). Thus, biological control and host-plant resistance can probably be used compatibly in a squash bug integrated pest management program.

Objective D: Impact

Surveys have indicated that the nymphal-adult parasitoid *T. pennipes* and all egg parasitoids are poorly synchronized with squash bugs, since none of these natural enemies was present in cucurbit fields until at least a month after the squash bugs had begun ovipositing.

Thus, squash bug egg populations cannot be suppressed by naturally occurring egg parasitoids. *T. pennipes* adults emerge early enough to parasitize overwintered squash bug adults, but they do not affect the much larger first summer generation that occurs when the adult parasitoids are no longer present. Furthermore, parasitization by the tachinid does not prevent squash bugs from mating, and adult females may deposit two-thirds of their eggs before dying. These findings suggest that natural biological control with the present enemy complex has little potential for maintaining *A. tristis* populations at acceptable levels.

RECOMMENDATIONS

1. Augmentative releases of one or more candidate species (for example, the scelionid egg parasitoid *Gryon pennsylvanicum*) should be considered during the spring to achieve early seasonal synchrony with hosts. This should be followed by a field evaluation to determine the impact that these releases have on squash bug populations throughout the season.

2. The comparative life history and demographic data at 27°C suggest that *G. pennsylvanicum* may have the greatest potential for controlling squash bug egg populations. However, high field temperatures may be a limiting factor. Thus, the thermal tolerance of this egg parasitoid needs to be evaluated further.

3. A taxonomic revision of the genus *Ooencyrtus* needs to be made; it should include the undescribed putative species of squash bug egg parasitoids found in Kansas.

REFERENCES CITED

Al-Obaidi, A. A. 1977. Reproductive bionomics of the squash bug, *Anasa tristis* (Heteroptera: Coreidae) as affected by temperature. Master's thesis, Oklahoma State University, Stillwater.

Beard, R. L. 1940. The biology of *Anasa tristis* DeGeer, with particular reference to the tachinid parasite, *Trichopoda pennipes* Fabr. *Conn. Agric. Exp. Sta. Bull.* 440:597–679.

————. 1942. On the formation of the tracheal funnel in *Anasa tristis* DeG. induced by the parasite *Trichopoda pennipes* Fabr. *Ann. Entomol. Soc. Am.* 35:68–72.

Dietrick, E. J., and R. van den Bosch. 1957. Insectary propagation of the squash bug and its parasite *Trichopoda pennipes* Fabr. *J. Econ. Entomol.* 50:627–29.

Elliot, D. C. 1935. The squash bug in Connecticut. *Conn. Agric. Exp. Sta. Bull.* 368:224–31.

Fargo, W. S., and E. L. Bonjour. 1988. Developmental rate of the squash bug, *Anasa tristis* (Heteroptera: Coreidae), at constant temperatures. *Environ. Entomol.* 17:926–29.

Fargo, W. S., P. E. Rensner, E. L. Bonjour, and T. L. Wagner. 1988. Population dynamics in the squash bug (Heteroptera: Coreidae)–squash plant (Cucurbitales: Cucurbitaceae) system in Oklahoma. *J. Econ. Entomol.* 81:1073–79.

Fielding, D. J. 1988a. Photoperiodic induction of diapause in the squash bug, *Anasa tristis*. *Entomol. Exp. Appl.* 48:187–193.

———. 1988b. Phenology and population dynamics of the squash bug, *Anasa tristis*. Ph.D. diss., University of Illinois, Urbana-Champaign.

Fielding, D. J., and W. G. Ruesink. 1988. Prediction of egg and nymphal developmental times of the squash bug *Anasa tristis*, in the field. *J. Econ. Entomol.* 81:1377–82.

Masner, L. 1983. A revision of *Gryon* Haliday in North America (Hymenoptera: Proctotrupoidea: Scelionidae). *Can. Entomol.* 115:123–74.

Nechols, J. R. 1987. Voltinism, seasonal reproduction, and diapause in the squash bug (Heteroptera: Coreidae) in Kansas. *Environ. Entomol.* 16:269–73.

———. 1988. Photoperiodic responses of the squash bug (Heteroptera: Coreidae): Diapause induction and maintenence. *Environ. Entomol.* 17:427–31.

Nechols, J. R., J. L. Tracy, and E. A. Vogt. 1989. Comparative ecological studies of indigenous egg parasitoids (Hymenoptera: Scelionidae; Encyrtidae) of the squash bug, *Anasa tristis* (Hemiptera: Coreidae). *J. Kansas Entomol. Soc.* 62:177–88.

Page, J. III, W. P. Morrison, J. K. Wangberg, and R. J. Whitworth. 1989. Ovipositional host association of *Anasa tristis* (DeGeer) and various *Cucurbita* cultivars on the Texas High Plains. *J. Agric. Entomol.* 6:5–8.

Schell, S. C. 1943. The biology of *Hadronotus ajax* Girault (Hymenoptera: Scelionidae), a parasite in the eggs of squash bug (*Anasa tristis* DeGeer). *Ann. Entomol. Soc. Am.* 36:625–35.

Tracy, J. L., and J. R. Nechols. 1987. Comparisons between the squash bug egg parasitoids *Ooencyrtus anasae* and *O.* sp. (Hymenoptera: Encyrtidae): Development, survival, and sex ratio in relation to temperature. *Environ. Entomol.* 16:1324–29.

———. 1988. Comparison of thermal responses, reproductive biologies, and population growth potentials of the squash bug egg parasitoids *Ooencyrtus anasae* and *O.* sp. (Hymenoptera: Encyrtidae). *Environ. Entomol.* 17:636–43.

Vogt, E. A. 1992. Compatibility of biological control and host plant resistance: The influence of three *Cucurbita* cultivars on the squash bug (*Anasa tristis* DeGeer) (Hemiptera: Coreidae) and its egg parasitoid *Gryon pennsylvanicum* (Ashmead) (Hymenoptera: Scelionidae). Ph.D. diss., Kansas State University, Manhattan.

Vogt, E. A., and J. R. Nechols. 1993. Responses of the squash bug (Hemiptera: Coreidae) and its egg parasitoid, *Gryon pennsylvanicum* (Hymenoptera: Scelionidae), to three *Cucurbita* cultivars. *Environ. Entomol.* 22:238–45.

Worthley, H. N. 1923. The squash bug in Massachusetts. *J. Econ. Entomol.* 16:73–79.

———. 1924a. The biology of *Trichopoda pennipes* Fab. (Diptera: Tachinidae), a parasite of the common squash bug. I. *Psyche* 31:7–16.

———. 1924b. The biology of *Trichopoda pennipes* Fab. (Diptera: Tachinidae), a parasite of the common squash bug. II. *Psyche* 31:57–77.

16 / LYGUS BUGS

C. G. JACKSON, J. W. DEBOLT, AND J. J. ELLINGTON

> **lygus bugs**
> *Lygus* spp.
> Heteroptera: Miridae

INTRODUCTION

Biology and Pest Status

In the United States three species of lygus bug are major pests of agricultural and horticultural crops. *Lygus lineolaris* (Palisot) occurs throughout the country but is a pest primarily in the eastern half, whereas *L. hesperus* Knight and *L. elisus* Van Duzee are western species (Kelton 1975). All three species are native to North America. *Lygus desertus* (=*desertinus*) Knight, previously considered to be a fourth western species, has recently been found to be a synonym of *L. elisus* (Lattin, Henry, and Schwartz 1992).

The lygus bug species are multivoltine and have two to five generations per year. Under summer conditions, *L. hesperus* and *L. elisus* develop from egg to adult in 20 to 24 days. The fall generation of adults overwinters and starts the first spring generation. In the desert of the Southwest, some development continues throughout the year, but adult *L. hesperus* undergo a reproductive diapause from late September until mid- to late December (Beards and Strong 1966).

Lygus bugs feed chiefly on developing reproductive structures—namely, buds, blooms, and seeds—or on the growing plant terminal. Their hosts include beans, peas, vegetable seed, alfalfa, cotton, safflower, stone and pome fruit, strawberries, and many flowers and nursery plants, including conifers (Scott 1977; Fye 1982; Young 1986). Lygus bugs are considered key pests on cotton because to control them it is sometimes necessary to use pesticides early in the season, which disrupts the natural enemy complex that keeps secondary pests under control. On other crops—for example, alfalfa seed—chemical control is necessary when the crop is pollinated and bees are active.

Historical Notes

North American efforts at biological control of the *Lygus* species began in the 1960s and have been summarized for the United States (Coulson 1987) and Canada (Craig and Loan 1987). Researchers have conducted surveys for exotic parasites in Europe, Turkey, Iran, Pakistan, India, Indonesia, and southern and eastern Africa. Several species of braconids in the genus *Peristenus* have been introduced, but with limited success.

Researchers have also completed a number of surveys for indigenous natural enemies in the United States and Canada. In the most substantial survey of crops and weeds in the western United States, D. W. Clancy and H. D. Pierce (1966) found substantial parasitism by two parasites: a braconid, *Leiophron uniformis* (Gahan), and a mymarid, *Anaphes ovijentatus*, renamed *A. iole* Girault by Huber and Rajakulendran (1988), which parasitize nymphs and eggs, respectively. A. Stoner and D. E. Surber (1969) also found *A. iole* to be a significant parasite of *L. hesperus* eggs in Arizona.

RESULTS

Objective A: The Identification and Introduction of Beneficial Organisms

Between 1970 and 1974, W-84 workers were involved in the introduction and release of *Peristenus rubricollis* (Thompson), *P. stygicus* (Loan), and *P. digoneutis* Loan in alfalfa in Arizona and California (Butler and Wardecker 1974; Van Steenwyk and Stern 1976, 1977). In California *P. stygicus* from France and Turkey were recovered for 2 years after the release, but no recoveries were made in Arizona.

A cooperative project (PL-480) between Egypt and the USDA-ARS laboratory in Tucson, Arizona, resulted in a survey in Egypt between 1978 and 1983 for mirid pests and their parasites. Researchers did not find any nymphal parasites from any mirids; they recorded only one egg parasite, *Telenomus* species, parasitizing *Taylorilygus pallidulus* (Blanchard), but it was not sent to the United States. During this project, researchers began a bibliography of the lygus bug literature (Graham, Negm, and Ertle 1984).

Biotypes of *P. stygicus* from Spain and Greece, collected by the European Parasite Laboratory (USDA-ARS), were released in alfalfa in Arizona between 1981 and 1983. During 1982 and 1983, parasitism levels reached 16.2 and 10.2 percent, respectively. However, researchers found no signs of overwintering and did not recover any parasites in the years after the releases ceased (Coulson 1987).

From 1981 to 1983, through an agreement between the USDA-ARS laboratory in Tucson, Arizona, and Texas A&M University in College Station, Texas, scientists conducted a search for mirid nymphal parasites in southern and eastern Africa. The braconids *Peristenus praeter* (Nixon) and *P. nigricarpus* (Szepligeti) were imported from the Republic of South Africa to the USDA-ARS quarantine laboratory in Stoneville, Mississippi, but sustained cultures were not established. *P. nigricarpus* and an unidentified *Leiophron* species were collected in Kenya, but also were not successfully cultured. The Commonwealth Institute of Biological Control (CIBC) station in Kenya sent further shipments of *Leiophron* species, which resulted in the establishment of cultures at the USDA-ARS laboratory in Stoneville, Mississippi. Although a few of these parasites were received in Arizona, none were released (Schuster 1987).

Objective B: Ecological and Physiological Studies

In surveying for the *Lygus* species in the southwestern United States, researchers have found that the three major species are associated with crop and weed hosts throughout the year (Graham, Jackson, and Butler 1982; Graham, Jackson, and Debolt 1986; Gordon et al. 1987). In the Pacific Northwest, *L. hesperus* and *L. elisus* occur on weeds and on field and tree crops (Fye 1980, 1982). Various researchers—including H. M. Graham and C. G. Jackson (1982), J. Ellington, K. Kiser, G. Fergusen, et al. (1984), J. Ellington, K. Kiser, M. Cardenas, et al. (1984), and H. M. Graham, C. G. Jackson, and K. R. Lakin (1984)—all developed or improved sampling procedures for the lygus bugs and their parasites.

Anaphes iole is the most common parasite of lygus bugs, occurring year-round in all the areas surveyed in Arizona and most of the common host plants of the *Lygus* species, especially when the plants are blooming or setting seed. All three *Lygus* species are parasitized at low (0 to 25 percent) to moderate (40 to 75 percent) levels (Graham and Jackson 1982; Jackson and Graham 1983; Graham, Jackson, and Debolt 1986). In the summer, when hay cutting is delayed in alfalfa, *Leiophron uniformis* reaches parasitization levels of 60 percent (Graham, Jackson, and Debolt 1986). A. Stoner and D. E. Surber (1971) and C. G. Jackson (1986, 1987) have

studied the biology of *A. iole*, and J. W. Debolt (1981) has studied that of *L. uniformis*. Western biotypes of *L. uniformis* readily and successfully parasitize *L. hesperus*, but are encapsulated by *L. lineolaris* (Debolt 1989a,b).

Through laboratory and field cage studies, researchers have shown that *Geocoris punctipes* (Say), *G. bullatus* (Say), *G. pallens* (Stal), *Nabis alternatus* (Parshley), and *N. americoferus* Carayon successfully prey on lygus bug nymphs (Butler 1967; Perkins and Watson 1972; Leigh and González 1976; Tamaki, Olsen, and Gupta 1978).

Objective C: The Conservation and Augmentation of Beneficial Organisms

After developing a meridic diet for *L. hesperus* (Debolt 1982) and a packaging system for the diet and oviposition substrate (Patana 1982), researchers were able to create a dependable large-scale rearing system for this species (Debolt and Patana 1985). The system ensures a supply of hosts for rearing the *L. hesperus* needed for field augmentation studies. J. W. Debolt (1987) has examined the potential of augmentative releases of parasites for lygus bug control. Currently, C. G. Jackson, J. W. Debolt, and J. Ellington, as well as other researchers, are working to determine the efficacy of these native parasites in alfalfa grown for seed and in strawberries.

RECOMMENDATIONS

1. Conduct foreign exploration for egg parasites and early-season nymphal parasites.

2. Identify *Lygus* predators and quantify their contribution to *Lygus* control under various host-population levels and host-species ratios.

3. Identify the differences in the biotypes of parasites—for example, the rates of development, fecundity, survival under environmental extremes, and pesticide resistance.

4. Make augmentative releases in different crop systems where lygus bugs are major pests.

REFERENCES CITED

Beards, G. W., and F. E. Strong. 1966. Photoperiod in relation to diapause in *Lygus hesperus* Knight. *Hilgardia* 37:345–62.

Butler, G. D. 1967. Big-eyed bugs as predators of *Lygus* bugs. *Progressive Agric. in Arizona* 19:13.

Butler, G. D., and A. L. Wardecker. 1974. Development of *Peristenus stygicus*, a parasite of *Lygus hesperus*, in relation to temperature. *J. Econ. Entomol.* 67:132–33.

Clancy, D. W., and H. D. Pierce. 1966. Natural enemies of some lygus bugs. *J. Econ. Entomol.* 59:853–58.

Coulson, J. R. 1987. Studies on the biological control of plant bugs (Heteroptera: Miridae): An introduction and history, 1961–1983. In *Economic importance and biological control of* Lygus *and* Adelphocoris *in North America,* eds. R. C. Hedland and H. M. Graham, 1–12. USDA-ARS-64.

Craig, C. H., and C. C. Loan. 1987. Biological control efforts on Miridae in Canada. In *Economic importance and biological control of* Lygus *and* Adelphocoris *in North America,* eds. R. C. Hedland and H. M. Graham, 48–53. USDA-ARS-64.

Debolt, J. W. 1981. Laboratory biology and rearing of *Leiophron uniformis* (Gahan) (Hymenoptera: Braconidae), a parasite of *Lygus* species (Hemiptera: Miridae). *Ann. Entomol. Soc. Am.* 74:334–37.

———. 1982. Meridic diet for rearing successive generations of *Lygus hesperus. Ann. Entomol. Soc. Am.* 75:119–22.

———. 1987. Augmentation: Rearing, release, and evaluation of plant bug parasites. In *Economic importance and biological control of* Lygus *and* Adelphocoris *in North America,* eds. R. C. Hedland and H. M. Graham, 82–87. USDA-ARS-64.

———. 1989a. Encapsulation of *Leiophron uniformis* by *Lygus lineolaris* and its relationship to host acceptance behavior. *Entomol. Exp. Appl.* 50:87–95.

———. 1989b. Host preference and acceptance by *Leiophron uniformis* (Hymenoptera: Braconidae): Effects of rearing on alternate *Lygus* (Heteroptera: Miridae) species. *Ann. Entomol. Soc. Am.* 82:399–402.

Debolt, J. W., and R. Patana. 1985. *Lygus hesperus.* In *Handbook of insect rearing,* eds. R. F. Moore and P. Singh, vol. 1, 329–38. Amsterdam: Elsevier.

Ellington, J., K. Kiser, G. Fergusen, and M. Cardenas. 1984. A comparison of sweepnet, absolute, and insectivac sampling methods in cotton ecosystems. *J. Econ. Entomol.* 77:599–605.

Ellington, J., K. Kiser, M. Cardenas, J. Duttle, and Y. Lopez. 1984. The Insectivac: A high-clearance, high-volume arthropod vacuuming platform for agricultural ecosystems. *Environ. Entomol.* 13:259–65.

Fye, R. E. 1980. Weed sources of lygus bugs in the Yakima Valley and Columbia Basin in Washington. *J. Econ Entomol.* 73:469–73.

———. 1982. Weed hosts of the *Lygus* (Heteroptera: Miridae) complex in central Washington. *J. Econ. Entomol.* 75:724–27.

Gordon, R., J. Ellington, G. F. Faubion, and H. Graham. 1987. A survey of the insect parasitoids from alfalfa and associated weeds in New Mexico. *Southwest. Entomol.* 12:335–50.

Graham, H. M., and C. G. Jackson. 1982. Distribution of eggs and parasites of *Lygus* species (Hemiptera: Miridae), *Nabis* species (Hemiptera: Nabidae), and *Spissistilus festinus* (Say) (Homoptera: Membracidae) on plant stems. *Ann. Entomol. Soc. Am.* 75:56–60.

Graham, H. M., C. G. Jackson, and G. D. Butler. 1982. Composition of the *Lygus* complex in some crop and weed habitats of Arizona. *Southwest. Entomol.* 7:105–10.

Graham, H. M., C. G. Jackson, and J. W. Debolt. 1986. *Lygus* spp. (Hemiptera: Miridae) and their parasites in agricultural areas of southern Arizona. *Environ. Entomol.* 15:132–42.

Graham, H. M., C. G. Jackson, and K. R. Lakin. 1984. Comparison of two methods of using the D-vac to sample mymarids and their hosts in alfalfa. *Southwest. Entomol.* 9:249–52.

Graham, H. M., A. A. Negm, and L. R. Ertle. 1984. *Worldwide literature of the* Lygus *complex (Hemiptera: Miridae), 1900–1980.* USDA-ARS Bibliogr. and Lit. Agric. 30.

Huber, J. T., and V. K. Rajakulendran. 1988. Redescription of and host-induced antennal variation in *Anaphes iole* Girault (Hymenoptera: Mymaridae), an egg parasite of Miridae (Hemiptera) in North America. *Can. Entomol.* 120:893–901.

Jackson, C. G. 1986. Effects of cold storage of adult *Anaphes ovijentatus* on survival, longevity, and oviposition. *Southwest. Entomol.* 11:149–53.

———. 1987. Biology of *Anaphes ovijentatus* (Hymenoptera: Mymaridae) and its host, *Lygus hesperus* (Hemiptera: Miridae), at low and high temperatures. *Ann. Entomol. Soc. Am.* 80:367–72.

Jackson, C. G., and H. M. Graham. 1983. Parasitism of four species of *Lygus* (Hemiptera: Miridae) by *Anaphes ovijentatus* (Hymenoptera: Mymaridae) and an evaluation of other possible hosts. *Ann. Entomol. Soc. Am.* 76:772–75.

Kelton, L. A. 1975. The lygus bugs (genus *Lygus* Hahn) of North America (Heteroptera: Miridae). *Mem. Entomol. Soc. Can.* 95.

Lattin, J. D., T. J. Henry, and M. D. Schwartz. 1992. *Lygus desertus* Knight, 1944, a newly recognized synonym of *Lygus elisus* Van Duzee, 1914 (Heteroptera: Miridae). *Proc. Entomol. Soc. Wash.* 94:12–25.

Leigh, T. F., and D. González. 1976. Field cage evaluation of predators for control of *Lygus hesperus* on cotton. *Environ. Entomol.* 5:948–52.

Patana, R. 1982. Disposable diet packet for feeding and oviposition of *Lygus hesperus* (Hemiptera: Miridae). *J. Econ. Entomol.* 75:668–69.

Perkins, P. V., and T. F. Watson. 1972. *Nabis alternatus* as a predator of *Lygus hesperus. Ann. Entomol. Soc. Am.* 65:625–29.

Schuster, M. F. 1987. Biological control of plant bugs in cotton. In *Economic importance and biological control of* Lygus *and* Adelphocoris *in North America,* eds. R. C. Hedland and H. M. Graham, 13–19. USDA-ARS-64.

Scott, D. R. 1977. An annotated listing of host plants of *Lygus hesperus* Knight. *Bull. Entomol. Soc. Am.* 23:19–22.

Stoner, A., and D. E. Surber. 1969. Notes on the biology and rearing of *Anaphes ovijentatus,* a new parasite of *Lygus hesperus* in Arizona. *J. Econ. Entomol.* 62:501–2.

———. 1971. Development of *Anaphes ovijentatus,* an egg parasite of *Lygus hesperus,* in relation to temperature. *J. Econ. Entomol.* 64:1566–67.

Tamaki, G., D. P. Olsen, and R. K. Gupta. 1978. Laboratory evaluation of *Geocoris bullatus* and *Nabis alternatus* as predators of *Lygus. J. Entomol. Soc. Brit. Columbia* 75:35–37.

Van Steenwyk, R. A., and V. M. Stern. 1976. The biology of

Peristenus stygicus (Hymenoptera: Braconidae), a newly imported parasite of lygus bugs. *Environ. Entomol.* 5:931–34.

———. 1977. Propagation, release and evaluation of *Peristenus stygicus*, a newly imported parasite of lygus bugs. *J. Econ. Entomol.* 70:66–69.

Young, O. P. 1986. Host plants of the tarnished plant bug, *Lygus lineolaris* (Heteroptera: Miridae). *Ann. Entomol. Soc. Am.* 79:747–62.

17 / ACACIA PSYLLID

J. W. BEARDSLEY, K. S. HAGEN, J. R. LEEPER, AND R. L. TASSAN

> **acacia psyllid**
> *Acizzia uncatoides* (Ferris & Klyver)
> **Homoptera: Psyllidae**

INTRODUCTION

Biology and Pest Status

The acacia psyllid, *Acizzia uncatoides* (Ferris & Klyver) (formerly *Psylla uncatoides*), feeds primarily on the young terminal growth of *Acacia* and *Albizia* species. Heavy infestations can severely defoliate and kill twigs and branches. J. A. Munro (1965) listed the known host species and ranked them according to the relative severity of psyllid attack. C. S. Koehler, M. E. Kattoulas, and G. W. Frankie (1966), and L. C. Madubunyi and C. S. Koehler (1974) have summarized the biology of the psyllid.

Historical Notes

G. F. Ferris and F. D. Klyver (1932) described *A. uncatoides* from specimens collected in New Zealand, where acacias are exotic. Although the psyllid was suspected of being of Australian origin, it was not found there until 1971, when J. W. Beardsley collected it near Melbourne (Leeper and Beardsley 1976).

In 1954 the acacia psyllid was discovered in California, where it became a serious pest of several introduced *Acacia* species widely used as ornamental and shade trees (Jensen 1957). In 1966 the psyllid was discovered in Hawaii; in 1970 it was found to be seriously damaging two native species of *Acacia*, *A. koa* Gray and *A. koaia* Hillebrand, at elevations of 3,000 to 6,600 feet on the island of Hawaii (Leeper and Beardsley 1976).

By 1971 it was apparent that biological control of the acacia psyllid was needed in both Hawaii (to protect native acacia stands important as timber, watershed, and wildlife resources) and California (to protect ornamental acacias, particularly those planted along highways). During the latter part of 1971, J. W. Beardsley and K. S. Hagen traveled to southeastern Australia where, as expected, they located the native home of the psyllid. Psyllid natural enemies were shipped from Melbourne and other locations in southeastern Australia to the University of California quarantine facility at Albany, where they were propagated.

RESULTS

Objective A: The Identification and Introduction of Beneficial Organisms

The three natural enemies shipped from southeastern Australia included two species of predaceous coccinellid beetles, *Diomus pumilio* Weise and *Harmonia conformis* (Boisduval), and an encyrtid parasitoid, *Psyllaephagus* species. Unfortunately, the *Psyllaephagus* could not be maintained in the insectary, and the culture was lost. However, researchers propagated and released both species of the coccinellids in California and in Hawaii. *D. pumilio* became established in California, but *H. conformis* apparently did not (Pinnock et al. 1978). In Hawaii, on the other hand, *H. conformis* became successfully established, but *D. pumilio* apparently did not (Leeper and Beardsley 1976).

Objective D: Impact

Researchers on the island of Hawaii carried out a detailed evaluation that included extensive prerelease studies of psyllid populations and their natural enemies in the areas where damage was most severe. They continued sampling for approximately 17 months after *H. conformis* was released.

H. conformis consumes mostly the eggs and small nymphs of *A. uncatoides*. It was recovered most readily on *Acacia koaia*, the most heavily infested and seriously damaged of the two affected *Acacia* species. On that host, the predator reduced populations of psyllid eggs and small nymphs by two-thirds to five-sixths (see fig. 17.1), eliminating the serious twig terminal dieback caused by psyllid damage (Leeper and Beardsley 1976). Occasional observations since 1976 have shown that *H.*

91

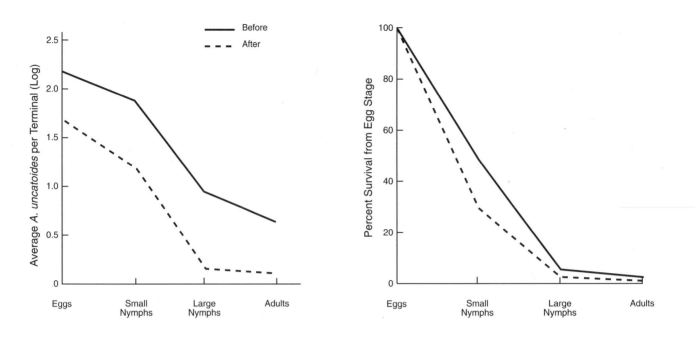

Figure 17.1. Graphs comparing the survival in logs (left) and percent (right) of the acacia psyllid, *Acizzia uncatoides*, before and after the predator *Harmonia conformis* was introduced in an *Acacia koaia* sanctuary in Hawaii.

conformis has continued to keep *A. uncatoides* populations below damaging levels (Beardsley unpub. data).

In California, by 1977 *D. pumilio* had distinctly reduced psyllid populations from the San Francisco Bay Area to San Diego, saving the California Department of Transportation over $55,000 per year by eliminating the need to spray acacias on freeways.

REFERENCES CITED

Ferris, G. F., and F. D. Klyver. 1932. Report upon a collection of Chermidae (Homoptera) from New Zealand. *Trans. R. Soc. New Zealand* 63:34–61.

Jensen, D. D. 1957. The albizia psyllid, *Psylla uncatoides* (Ferris & Klyver), in California. *Pan-Pacific Entomol.* 33:29–30.

Koehler, C. S., M. E. Kattoulas, and G. W. Frankie. 1966. Biology of *Psylla uncatoides*. *J. Econ. Entomol.* 59:1097–100.

Leeper, J. R., and J. W. Beardsley. 1976. The biological control of *Psylla uncatoides* (Ferris & Kylver) (Homoptera: Psyllidae) on Hawaii. *Proc. Hawaii. Entomol. Soc.* 22:307–21.

Madubunyi, L. C. 1967. Ecological investigations on the albizzia psyllid, *Psylla uncatoides* (Ferris & Klyver) (Homoptera: Psyllidae). Master's thesis, University of California, Berkeley.

Madubunyi, L. C., and C. S. Koehler. 1974. Development, survival, and capacity for increase of the Albizzia psyllid at various constant temperatures. *Environ. Entomol.* 3:1013–16.

Munro, J. A. 1965. Occurrence of *Psylla uncatoides* on *Acacia* and *Albizia*, with notes on control. *J. Econ. Entomol.* 58:1171–72.

Pinnock, D. E., K. S. Hagen, D. V. Cassidy, R. J. Brand, J. E. Milstead, and R. L. Tassan. 1978. Integrated pest management in highway landscapes. *Calif. Agric.* 32(2):33–34.

18 / LEUCAENA PSYLLID

J. W. BEARDSLEY, D. M. NAFUS, AND G. K. UCHIDA

leucaena psyllid
Heteropsylla cubana Crawford
Homoptera: Psyllidae

INTRODUCTION

Biology and Pest Status

The leucaena psyllid, *Heteropsylla cubana* Crawford, is often found in great numbers on *Leucaena leucocephala*. It lays its eggs, which hatch in 2 to 3 days, in the new growth of the uncurling leaves. The nymphs prefer to feed on new or developing leaves, but will accept older leaves when psyllid populations are high. The psyllid develops in 8 or 9 days and has five instars. Females live about 11 days, have a 1- to 3-day preovipositional period, and lay approximately 400 eggs (Waterhouse and Norris 1987).

Plants that have been severely attacked by the psyllid have a stunted, bunchy appearance. They may be defoliated or exhibit dead terminal growth and low pod production; in some circumstances, they may die. When this happens, the tree canopy is opened, allowing shifts in the plant community. *Leucaena* is widely used for forage or feed pellets, as a shade tree for cacao and other crops, as mulch or green fertilizer, and for firewood. In some areas it is considered a weed.

On Guam the W-84 project became involved with the control of the psyllid in 1986 with the release of the coccinellid *Curinus coeruleus* Mulsant. In 1987 the researchers initiated studies on the psyllid's population fluctuations and its associated natural enemies. Later they also began studies of tree growth in relation to psyllid attack, and of the impacts of defoliation on understory vegetation.

In Hawaii W-84 involvement focused on monitoring the establishment and evaluating the effectiveness of the purposely introduced encyrtid parasite, *Psyllaephagus yaseeni* Noyes, as well as evaluating the coccinellid predators.

Historical Notes

Leucaena psyllid is a relatively new pest in the Pacific, having been found in Hawaii first in 1984. By 1985 it was in Guam, Samoa, the Cook Islands, Fiji, the Solomon Islands, the Philippines, and Indonesia, as well as on numerous other islands. Apparently it originated in Central America and was spread with its host to the Caribbean, South America, and Florida. However, it did not become a serious pest until it was accidentally introduced into the Pacific region. Between 1986 and 1989, this pest extended its range to Taiwan and into Southeast Asia as far west as India and Sri Lanka.

RESULTS

Objective A: The Identification and Introduction of Beneficial Organisms

In 1986 *Curinus coeruleus* was imported from Hawaii and released in Guam. Researchers released a total of 308 beetles in lots of about 60 to 80 at four different sites (Nafus and Schreiner 1989). The beetle became established at only one of the sites, and spread very slowly. In 1988 it was still confined to within 1 kilometer of its initial release site.

Objective D: Impact

Researchers chose ten *Leucaena* stands distributed widely over the island of Guam for monitoring psyllid populations. They began monitoring in May 1987. Psyllid populations were highest during the wet season—July through November—and lowest in the dry season. At most sites—excluding those where *C. coeruleus* was present—psyllid populations fluctuated between 10 and 100 nymphs per gram of tip leaves (dry weight), although in three locations populations reached up to 200 nymphs per tip. The largest fluctuations occurred at the site where *C. coeruleus* had become established. In the 1987 and 1988 wet seasons, researchers noted a brief outbreak of psyllids: populations reached 350 to 400 nymphs per gram of terminal leaf. At other times populations were low, ranging between 0 and 50 nymphs. The *C. coeruleus* populations tracked the psyllid populations, but the number of beetles was not well correlated with that of psyllids.

The coccinellid *Olla v-nigrum* (Mulsant) has also been found feeding on the psyllid. Where the distributions of the coccinellids overlap, *C. coeruleus* is more abundant than *O. v-nigrum*, but the latter is more widely distributed. The bugs *Orius niobe* Herring (Anthocoridae) and *Campylomma lividicornis* Reuter (Miridae) have also been associated with psyllids, and may feed on psyllids or psyllid eggs.

In their evaluation studies on natural enemies of *H. cubana* in Hawaii, researchers showed that *C. coeruleus*, originally introduced from Mexico in 1922 to combat mealybugs, was the dominant predator. At most sites *O. v-nigrum* was also associated with the psyllid, but it was generally much less abundant. At sample sites on Oahu, psyllid populations fluctuated greatly during the year; the periodic population crashes seemed to be caused by a combination of coccinellid predation and the loss of suitable host foliage due to psyllid feeding and low rainfall. Following the rains, *Leucaena* plants put forth a new flush growth that allowed the psyllids to build up to damaging levels while coccinellid populations were low. However, psyllid populations then declined as the coccinellids increased. The addition of the encyrtid parasite of psyllid nymphs, *Psyllaephagus yaseeni*, did not noticeably reduce the psyllid populations. We found *P. yaseeni* to be heavily hyperparasitized by *Pachyneuron siphono-* *phorae* (Ashmead) (Pteromalidae) and two species of *Syrphophagus* (Encyrtidae). Hyperparasites may be limiting the effectiveness of *P. yaseeni* in Hawaii (Beardsley and Uchida 1990).

RECOMMENDATIONS

1. The efficacy of *C. coeruleus* as a biological control agent is questionable. It responds to populations of the psyllid but is unable to prevent explosive outbreaks and may even enhance their severity. On Guam, periodic defoliations occur whether the beetle is present or not. The impact of the beetle's presence on the growth of the plant needs to be determined.

2. Additional natural enemies of *H. cubana* should be obtained for evaluation and possible release.

REFERENCES CITED

Beardsley, J. W., and G. K. Uchida. 1990. Parasites associated with the leucaena psyllid, *Heteropsylla cubana* Crawford, in Hawaii. *Proc. Hawaii. Entomol. Soc.* 30:155–56.

Nafus, D., and I. Schreiner. 1989. Biological control activities in the Mariana Islands from 1911 to 1988. *Micronesica* 22:65-106.

Waterhouse, D. F., and K. R. Norris. 1987. *Biological control: Pacific prospects.* Melbourne: Inkata Press.

19 / PEAR PSYLLA

T. R. UNRUH, P. H. WESTIGARD, AND K. S. HAGEN

> **pear psylla**
> *Cacopsylla pyricola* (Förster)
> Homoptera: Psyllidae

INTRODUCTION

Biology and Pest Status

Cacopsylla pyricola (Förster) is a major pest of pear in northern Europe and North America. In the pear-growing areas of southern Europe a closely related species, *C. pyri* (L.), usually replaces *C. pyricola* as the dominant pest. Pear psylla causes damage when nymphs, feeding at high densities on leaves, produce enough honeydew to drip onto the fruit. A black, sooty mold fungus then grows in the honeydew, distorting and russeting the fruit surface, which substantially lowers its commercial value, especially on such clear-skinned cultivars as Williams, Comice, and D'Anjou. In western North America pear psylla is also the vector of the mycoplasma-like organism (MLO), which causes pear decline disease (Jensen et al. 1964). Between the late 1950s and late 1960s, this disease killed hundreds of thousands of pear trees, but was eventually controlled with resistant or tolerant rootstocks.

The pear psylla has three to four generations each year; each female lays an average of 400 eggs (Burts and Fisher 1967), although fecundity is a function of temperature and tree vigor (McMullen and Jong 1972). The egg stage is followed by five nymphal instars, the last referred to as the "hardshell" stage. *C. pyricola* overwinters in the adult stage as a specialized morph that is larger and darker than the summer-form adults. This winter form is produced in the fall in response to the exposure of nymphal stages to photoperiods shorter than 13.5 hours (McMullen and Jong 1976). The winter forms are highly dispersive and can be found outside of pear orchards on several plant species (Hodgson and Mustafa 1984a). In late winter the adults return to pear trees, and the females begin to oviposit on the wood at the base of the fruit buds. *C. pyricola* has a relatively low developmental temperature threshold—about 40°F—which is important to its potential for successful biological control.

Historical Notes

Cacopsylla pyricola was first reported in North America in 1832, when it was found in Connecticut (Pettit and Huston 1931); in 1939 it was found in the Pacific Northwest (Webster 1939). Its current distribution includes all of the commercial pear-growing areas of the United States, Canada, Europe, and Israel, but it is not yet found in Australia, New Zealand, or South Africa.

In western North America, the most abundant or most important predators of *C. pyricola* are several Anthocoridae and Miridae species (Madsen, Westigard, and Sisson 1963; McMullen and Jong 1967; Westigard, Gentner, and Berry 1968). The relative importance of individual mirid or anthocorid species varies geographically. For example, *Anthocoris antevolens* White and *A. melanocerus* Reuter are the most abundant species in British Columbia (McMullen and Jong 1967), *A. antevolens* is dominant in California (Madsen, Westigard, and Sisson 1963), and the mirid *Deraeocoris brevis* (Uhler) is dominant in Oregon (Westigard, Gentner, and Berry 1968). In much of western Europe, the most abundant species is *Anthocoris nemoralis* (Fabricius) (Hérard 1986). In general, these predators feed on a variety of prey from several host plants, and we believe that very few individuals overwinter on pear (but see Fye 1985). However, the predators tend to colonize pear from outside sources every year (Hodgson and Mustafa 1984b).

In North America, researchers have found seven parasitoid species on *C. pyricola* (Gutierrez 1965; Hansen

The following researchers have conducted foreign collections and shipments of psylla parasites that supplemented those of K. S. Hagen: J. Ball (University of California), from Afghanistan in 1974; H. Reidl (Oregon State University, Mid-Columbia Experiment Station), from Yugoslavia and Greece in 1980; and R. Messing (Oregon State University, Corvallis), from Austria, France, Hungary, and Italy in 1987 and 1988. M. Grbić brought insects from Yugoslavia in 1988. Also, F. Hérard of the USDA-ARS European Parasite Laboratory in Montpellier, France, sent parasites and predators from France in the early 1980s.

1975; Philogene and Chang 1978; Rathman and Brunner 1989). Of these, only two are known to be obligatory primary parasitoids: *Trechnites psyllae* Ruschka, which is abundant, and *Prionomitus mitratus* Dalman, which is quite rare. In western North America, scientists have observed *Trechnites insidiosus* Crawford (=*T. psyllae;* see Objective A) parasitizing late nymphal instars at rates as high as 70 to 90 percent (Nickel, Shimizu, and Wong 1965; Westigard, Gentner, and Berry 1968; Burts 1970; Unruh unpub. data). The relative importance of parasitoids in the suppression of pear psylla is unknown, but many researchers believe it to be less than that of the generalist predator complex. In contrast, V. I. Talitski (1966) has argued that *Trechnites psyllae* regulates the abundance of *Cacopsylla pyri* (L.) in Moldavia.

RESULTS

Objective A: The Identification and Introduction of Beneficial Organisms

Since the late 1960s, W-84 researchers have introduced numerous generalist predators and specialized parasitoids of the pear psylla in the western states. Those introductions are summarized in table 19.1. Unfortunately, the degree of specificity of the predators and parasitoids, and whether they have established, remains unknown.

The introduced parasitoids have come from two distinct geographic areas: Europe and, to a lesser extent, the Persian plateau. Because *Trechnites psyllae* from western Europe is morphologically indistinguishable from *T. insidiosus* (Hagen unpub. data), which is presumably endemic to North America, researchers made no postrelease evaluations for the various European races of *T. psyllae.* With the exception of an arrhenotokous race from Switzerland (also see Talitski 1966), all of the races introduced have probably been thelytokous, as is *T. insidiosus* in the western states (Hagen and Unruh unpub. data). It is likely that the *Trechnites* found in North America is *T. psyllae,* and that it became established here during the nineteenth century along with *C. pyricola,* as well as with at least one hyperparasitoid—*Syrphophagus* species (Hagen unpub. data). This scenario is consistent with the collection of *T. insidiosus* in California and British Columbia before its intentional release. However, the reproductive systems and other biological attributes of *T. psyllae* merit further study.

Researchers have also introduced a second group of *Trechnites* from the Indo-Persian region and Yugoslavia, but in quite limited numbers. This species or group of species is morphologically distinct from *T. psyllae* and displays arrhenotokous reproduction (Hansen 1975; Hagen and Unruh unpub. data). They may be promising candidates for continued importation and release.

Infrequent recoveries of the Yugoslavian *Trechnites* species were made at release sites in Washington in early 1991 (Unruh unpub. data), and recoveries of the Afghanistan biotype were made in California a few generations after release (Hagen unpub. data). However, researchers did not record any long-term establishment.

Prionomitus mitratus, like *Trechnites,* was recorded in North America before researchers began the introductions to control *C. pyricola* (Jensen 1957). It is likely that *P. mitratus* has a holarctic distribution with a diversity of psyllid hosts, or that it is made up of many host-associated races or sibling species. *P. mitratus* overwinters in the adult stage and, in Europe, attacks a diversity of psyllid species in and near pear orchards (Talitski 1966; Hérard 1986). In North America, particularly in the west, it must adapt to the seasonality of *C. pyricola,* the only psyllid in, or typically near, pear orchards. In Washington, *P. mitratus* has been reared from *C. pyricola* at very low frequencies—less than 1 percent of total parasitism (Burts 1970; Unruh unpub. data)—and is known to attack endemic, willow-infesting psyllids (Jensen 1957; McMullen 1971). From table 19.1 it is clear that European populations have been amply released in California with no significant effect (Hagen unpub. data). Since the pear ecosystem lacks alternative hosts, this species seems to hold little promise.

The introduced predators represent a group of generalist species, some of which were not originally derived from the pear ecosystem (*Calvia, Coccinella, Diomus, Leis,* and *Menochilus*). Of all the predators introduced, only *Anthocoris nemoralis* and perhaps *A. nemorum* L. have become established. In Wenatchee, Washington, *A. nemoralis* stems from populations that the Commonwealth Institute of Biological Control (CIBC) originally introduced into British Columbia and Nova Scotia (McMullen and Jong 1967), and that were imported to Wenatchee in the 1970s (E. C. Burts person. commun.). The origin of the population of *A. nemoralis* recently detected near San Francisco is unknown (Hagen and Dreistadt 1990). Recoveries of *A. nemorum* at very low levels have been reported in Yakima, Washington (R. Fye person. commun.).

RECOMMENDATIONS

The predators and parasitoids of *C. pyricola* are highly susceptible to most of the broad-spectrum pesticides used to control other pear pests, especially the codling moth, *Cydia pomonella* (Hagley and Simpson 1983; Hodgson and Mustafa 1984b); *C. pyricola* itself is highly resistant to most of these materials (Riedl et al. 1981; Burts et al. 1989). Some important pear cultivars are highly susceptible to fruit-marking injury at relatively low psylla densities (Westigard, Allen, and Gut 1981),

TABLE 19.1. Predators and parasites of the pear psylla released into the western United States between 1965 and 1990.

Species	Country of origin and date first released	Release area	Numbers	Dates of release
Prionomitus mitratus Dalman	Switzerland (1965)	Alameda Co., CA	2,092	1968–72
		Contra Costa Co., CA	254	1965–67
		El Dorado Co., CA	1,850	1969–71
		Napa Co., CA	2,219	1965–68
		Placer Co., CA	760	1969–72
		Sacramento Co., CA	2,829	1967–69
		Santa Clara Co., CA	1,121	1967–68
		Santa Cruz Co., CA	1,727	1967–68
		Solano Co., CA	8,920	1968–71
		Yolo Co., CA	2,641	1967–71
	Yugoslavia (1980)	Alameda Co., CA	538	1981
		Solano Co., CA	820	1981
		Jackson Co., OR	730	1981
		Yakima Co., WA	341	1981
	France (1983)	Klickitat Co., WA	40	1983
		Yakima Co., WA	94	1983
	France (1987)	Yakima Co., WA	108	1989–90
		Hood River Co., OR	70	1989
Trechnites psyllae Ruschka	Switzerland (1965)	Alameda Co., CA	1,747	1967–71
		Contra Costa Co., CA	492	1966
		Napa Co., CA	1,092	1967
		Placer Co., CA	25	1971
		Sacramento Co., CA	1,257	1967–72
		Santa Clara Co., CA	67	1967
		Santa Cruz Co., CA	404	1967
		Solano Co., CA	11,673	1967–72
		Yolo Co., CA	653	1967–72
	Greece (1987)	Contra Costa Co., CA	40	1987
		Hood River Co., OR	100	1989
		Yakima Co., WA	51	1989
	Italy (1988)	Hood River Co., OR	50	1989
		Yakima Co., WA	231	1989–90
	Yugoslavia (1989)	Yakima Co., WA	376	1990
Trechnites sp.	Iran (1974)	El Dorado Co., CA	59	1975
	Afghanistan (1974)	Santa Clara Co., CA	79	1976
		Contra Costa Co., CA	110	1978
		Mendocino Co., CA	552	1978
		Benton Co., OR	56	1977
		Jackson Co., OR	1,870	1977
		Cache, Utah Cos., UT	828	1977–78
		Chelan Co., WA	630	1976–78

(continued)

TABLE 19.1 (*continued*)

Species	Country of origin and date first released	Release area	Numbers	Dates of release
	Yugoslavia (1989)	Yakima Co., WA	14,153	1990–91
Anthocoris nemorum L.	France (1982–83)	Yakima Co., WA	457	1983
		Klickitat Co., WA	103	1983
Anthocoris nemoralis F.[*]	Europe (1962)	Chelan Co., WA	90	1976
		Yakima Co., WA	–	1979
Chrysoperla carnea (Stephens)	Greece (1982)	Klickitat Co., WA	170	1983
Calvia 14–guttata L.	Japan (1981)	Yakima Co., WA	52,820	1981–82
Propylaea 14–punctata (L.)	Yugoslavia (1980)	Klickitat Co., WA Yakima Co., WA	23,766	1981–82
Harmonia dimidiata (F.)	Pakistan (1974)	El Dorado Co., CA	30	1975
		Yakima Co., WA	204	1977
Harmonia conformis (Boisd.)	Australia (1975)	El Dorado Co., CA	108	1975
		Mendocino Co., CA	552	1978
		Jackson Co., OR	90	1978
		Colorado	500	1979
		Central WA[†]	45,429	1977–82
Harmonia axyridis (Pallas)	Japan (1978)	Central WA[†]	69,789	1978–82
Coccinella septempunctata L.	Japan (1978)	Klickitat Co., WA Yakima Co., WA	6,928	1979
Menochilus quadriplagiatus (Schoenkerr)	Hong Kong (1978)	Central WA[†]	151,387	1979–82
Oenopia conglobata (L.)	Yugoslavia (1980)	Alameda Co., CA	111	1981
		Solano Co., CA	251	1981
		Jackson Co., OR	221	1981
		Yakima Co., WA	2,489	1981–82
Diomus pumilio Weis	Australia (1975)	El Dorado Co., CA	20	1975
		Jackson Co., OR	130	1977

[*]The population was established in British Columbia from Switzerland, collected in British Columbia, then released in the Wenatchee area (McMullen 1971; Hagen and Dreistadt 1990; E. C. Burts person. commun.). Collections from established populations in Wenatchee were distributed throughout the Wenatchee and Yakima areas. The actual number released in Wenatchee and Yakima is unknown, but exceeds 100.

[†]See R. E. Fye (1981) for detailed localities; includes Wenatchee, White Salmon, Yakima, and vicinity.

and since the thermal developmental threshold for psylla is probably lower than that of its natural enemies, it can cause substantial early-season damage before the beneficial organisms can suppress its populations (Brunner 1975; Westigard, Allen, and Gut 1981). Although these factors do not lend themselves to biological control, research to ameliorate them is paying dividends.

Some chemical-management practices are more amenable to biological control than others. For example, using pyrethroid insecticides during the dormant phase of pear development suppresses *C. pyricola* at a period when the summer-colonizing predaceous species, such as *A. nemoralis,* are not adversely affected (Burts 1968; Hodgson and Mustafa 1984b; Solomon et al. 1989). However, even prebloom use of pyrethroids can disrupt spider mite populations later in the season (Westigard unpub. data). The problem caused by broad-spectrum, disruptive pesticides may be partly resolved by the recent development of selective chemicals, such as the insect growth regulator diflubenzeron for codling moth control (Burts 1983; Westigard, Gut, and Liss 1986). The tactic of using sex pheromones to disrupt the mating of codling moths has recently been registered in the United States; it is another promising approach that would further reduce insecticide use in pear orchards (Moffitt, Westigard, and Hathaway 1978; Howell et al. 1992).

Currently, even in orchards free of synthetic pesticides controlling *C. pyricola* with native or introduced natural enemies is tenuous. Successful biological control in these circumstances has been reported only as isolated cases in Oregon (Westigard 1973) and California (Madsen, Westigard, and Sisson 1963). But as the control paradigm changes from using broad-spectrum disruptive chemicals to using selective controls, the biological control of *C. pyricola* in the future looks more promising. Research emphasis should be placed on fostering currently established natural enemies, and foreign exploration for new beneficials should concentrate on poorly sampled parts of the Palearctic, such as Central Asia.

REFERENCES CITED

Brunner, J. F. 1975. Economic injury level of the pear psylla, *Psylla pyricola* Förster, and a discrete time model of a pear psylla–predator interaction. Ph.D. diss., Washington State University, Pullman.

Burts, E. C. 1968. An area control program for the pear psylla. *J. Econ. Entomol.* 61:261–63.

———. 1970. *The pear psylla in Central Washington.* Washington Agricultural Experiment Station Circular 516.

———. 1983. Planning programs ahead best strategy for early season pear pest control. *Ecofruit Grower* 34:6.

Burts, E. C., and W. R. Fisher. 1967. Mating behavior, egg pro-duction, and egg fertility in the pear psylla. *J. Econ. Entomol.* 60:1297–1300.

Burts, E. C., H. E. van de Baan, and B. A. Croft. 1989. Pyrethroid resistance in pear psylla, *Psylla pyricola* Förster (Homoptera: Psyllidae), and synergism of pyrethroids with piperonyl butoxide. *Can. Entomol.* 121:219–23.

Fye, R. E. 1981. Rearing and release of coccinellids for potential control of pear psylla. In *Advances in agricultural technology.* USDA-ARS, AAT-W-20. Washington, D.C.: U.S. Department of Agriculture.

———. 1985. Corrugated fiberboard traps for predators over-wintering in pear orchards. *J. Econ. Entomol.* 78:1511–14.

Gutierrez, A. P. 1965. The bionomics of two encyrtid parasites of *Psylla pyricola* Förster in northern California. Master's thesis, University of California, Berkeley.

Hagen, K. S., and S. H. Dreistadt. 1990. First California record for *Anthocoris nemoralis* (Fabr.) (Hemiptera: Anthocoridae), a predator important in the biological control of psyllids (Homoptera: Psyllidae). *Pan-Pacific Entomol.* 66:323–24.

Hagley, E. A. C., and C. M. Simpson. 1983. Effect of insecticides on predators of pear psylla, *Psylla pyricola* (Hemiptera: Psyllidae), in Ontario. *Can. Entomol.* 115:1409–14.

Hansen, R. R. 1975. Feeding rates and behavior of some predators of two *Psylla* spp. and a biology of a parasite of *Psylla pyricola* Förster. Master's thesis, University of California, Berkeley.

Hérard, F. 1986. Annotated list of the entomophagous complex associated with pear psylla, *Psylla pyri* (L.) (Homoptera: Psyllidae), in France. *Agronomie* 6:1–34.

Hodgson, C. J., and T. M. Mustafa. 1984a. The dispersal and flight activity of *Psylla pyricola* Förster in southern England. *Bull. Org. Int. Lutte Biol. Sect. Reg. Ouest Palearct.* 7(5):97–124.

———. 1984b. Aspects of chemical and biological control of *Psylla pyricola* Förster in England. *Bull. Org. Int. Lutte Biol. Sect. Reg. Ouest Palearct.* 7(5):330–53.

Howell, J. F., A. L. Knight, T. R. Unruh, D. F. Brown, J. L. Krysan, C. R. Sell, and P. A. Kirsch. 1992. Control of codling moth, *Cydia pomonella* (L.), in apple and pear with sex pheromone–mediated mating disruption. *J. Econ. Entomol.* 85:918–25.

Jensen, D. D. 1957. Parasites of the psyllidae. *Hilgardia* 27:71–99.

Jensen, W., H. Griggs, C. Q. Gonzales, and H. Schneider. 1964. Pear psylla proven carrier of pear decline virus. *Calif. Agric.* 18(3):2–3.

McMullen, R. D. 1971. *Psylla pyricola* Förster (Homoptera: Psyllidae). In *Biological control programmes in Canada, 1959–1968,* part 1, Agricultural insects, 33–38. Commonwealth Institute of Biological Control Technical Communication 4.

McMullen, R. D., and C. Jong. 1967. New records and discussion of predators of the pear psylla, *Psylla pyricola* Förster, in British Columbia. *J. Entomol. Soc. Brit. Columbia* 64:35–40.

———. 1972. Influence of temperature and host vigor on fecundity of the pear psylla (Homoptera: Psyllidae). *Can. Entomol.* 104:1209–12.

———. 1976. Factors affecting induction and termination of diapause in pear psylla (Homoptera: Psyllidae). *Can. Entomol.* 108:1001–5.

Madsen, H. F., P. H. Westigard, and R. L. Sisson. 1963. Observations on the natural control of the pear psylla, *Psylla pyricola* Förster in California. *Can. Entomol.* 95:837–44.

Moffitt, H. R., P. H. Westigard, and D. O. Hathaway. 1978. Pheromonal control of the codling moth and biological control of the pear psylla. *Proc. Oregon Hort. Soc.* 70:95–96.

Nickel, J. L., J. T. Shimizu, and T. T. Y. Wong. 1965. Studies on natural control of pear psylla in California. *J. Econ. Entomol.* 58:970–76.

Pettit, R. H., and R. Hudson. 1931. *Pests of apple and pear in Michigan.* Michigan State Agricultural Experiment Station Circular 137.

Philogene, B. J. R., and J. F. Chang. 1978. New records of parasitic chalcidoids of pear psylla (Homoptera: Psyllidae) in Ontario, Canada, with observations on the current world status of its parasitoids and predators. *Proc. Entomol. Soc. Ontario* 109:53–60.

Rathman, R. J., and J. F. Brunner. 1989. Observations on the biology of a new species of *Dilyta* (Hymenoptera: Charipidae) from Washington state. *Pan-Pacific Entomol.* 64:93–97.

Riedl, H., P. W. Westigard, R. S. Bethell, and J. E. DeTar. 1981. Problems with chemical control of pear psylla. *Calif. Agric.* 35(9–10):7–9.

Solomon, M. G., J. E. Cranham, M. A. Easterbrook, and J. D. Fitzgerald. 1989. Control of the pear psyllid, *Cacopsylla pyricola*, in Southeast England by predators and pesticides. *Crop Protect.* 8:197–205.

Talitski, V. I. 1966. Hymenoptera that are parasites of the pear sucker (*Psylla pyri* L.) in Moldavia (in Russian). *Trudy moldav. nauchno–issled.* Inst. Sadov. Vinogr. Vinod. 13:191–221.

Webster, R. L. 1939. Pear psylla survey. *Proc. Washington State Hort. Assoc.* 35:36–40.

Westigard, P. H. 1973. The biology of and effect of pesticides on *Deraeocoris brevis-piceatus* (Heteroptera: Miridae). *Can. Entomol.* 105:1105–11.

Westigard, P. H., R. B. Allen, and L. J. Gut. 1981. Pear psylla: Relationship of early season nymph densities to honeydew-induced fruit damage on two pear cultivars. *J. Econ. Entomol.* 74:532–34.

Westigard, P. H., L. G. Gentner, and D. W. Berry. 1968. Present status of biological control of the pear psylla in southern Oregon. *J. Econ. Entomol.* 61:740–43.

Westigard, P. H., L. J. Gut, and W. J. Liss. 1986. Selective control program for the pear pest complex in southern Oregon. *J. Econ. Entomol.* 79:250–57.

20 / ASH WHITEFLY

T. S. BELLOWS, JR., AND K. Y. ARAKAWA

> **ash whitefly**
> *Siphoninus phillyreae* (Haliday)
> Homoptera: Aleyrodidae

INTRODUCTION

Biology and Pest Status

Siphoninus phillyreae (Haliday) is a recently introduced, serious whitefly pest of fruit and ornamental trees and shrubs in the southwestern United States. Since its detection in 1988, damage from this pest has included 100 percent infestation of available foliage, defoliation, honeydew and sooty mold accumulation, severe yield reduction, and (in the case of young pear trees) repeated defoliation in a single season, leading to tree death.

No previously published studies are available on the biology of the whitefly, and the few references on its natural enemies are mostly taxonomic in nature (see, for example, Viggiani and Mazzone 1980).

Historical Notes

Siphoninus phillyreae (Haliday) is native to the Near East and the Mediterranean region of northern Africa and Europe; it has also been recorded on the Indian subcontinent (fig. 20.1). It has been recorded in countries representing a wider climatic range than that of any aleyrodid previously introduced into the United States.

The ash whitefly was first discovered in the United States in Los Angeles, in August 1988. Initial observations at that time showed heavy infestations on ash (*Fraxinus* spp.), pomegranate (*Punica granatum*), ornamental and bearing pear (*Pyrus* spp.), and apple (*Malus* spp.) trees. The very large whitefly populations and well-established infestations indicated that the pest was imported possibly as early as 1986. During 1989 this

Figure 20.1. Countries reporting *Siphoninus phillyreae* prior to its introduction into California.

Thanks to D. Gerling of Tel Aviv University and to G. Viggiani of the Laboratorio di Entomologia Agraria 'Filippo Silvestri', Portici, for their assistance in obtaining natural enemies. L. Bezark and J. Ball of California's Department of Food and Agriculture assisted in the development of the program described here. A. Polaszek of C.A.B. International Institute of Entomology identified the *Encarsia inaron*.

initial infestation substantially broadened in its range (fig. 20.2), which by 1991 included most of California and parts of Nevada, Arizona, and New Mexico. The spread has been more rapid and has incorporated more climatic zones in the western United States than was the case with any previously introduced whitefly.

In Europe this species is polyphagous, and it has been found on several new hosts since being introduced to the United States (see table 20.1). Such polyphagy is common among whiteflies achieving pest status (Byrne, Bellows, and Parrella 1990). It appears that some of the host-range expansion in the United States is due in part to the presence of new acceptable hosts and in part to the pest's adoption of host plants of marginal quality, perhaps because of its exceptionally large population densities.

The ash whitefly has a substantial and effective natural enemy fauna. Its parasitoids include several *Encarsia* species and at least one species of *Eretmocerus*; its predators include various coccinellid and drosophilid species (table 20.2). Further, some references in the literature on natural enemies have indicated that the whitefly, when first reported in a particular country—for example, Greece (Costacos 1963) and Bulgaria (Kolev 1973)—was a substantial pest of apple and pear, with population densities similar to those seen in California. The subsequent appearance of natural enemies led to a substantial amelioration of the problem. A similar situation existed in pesticide-disrupted orchards in Italy, where recommendations for pest management included avoiding highly toxic pesticides in order to retain the natural enemies (Tremblay 1969, 1973).

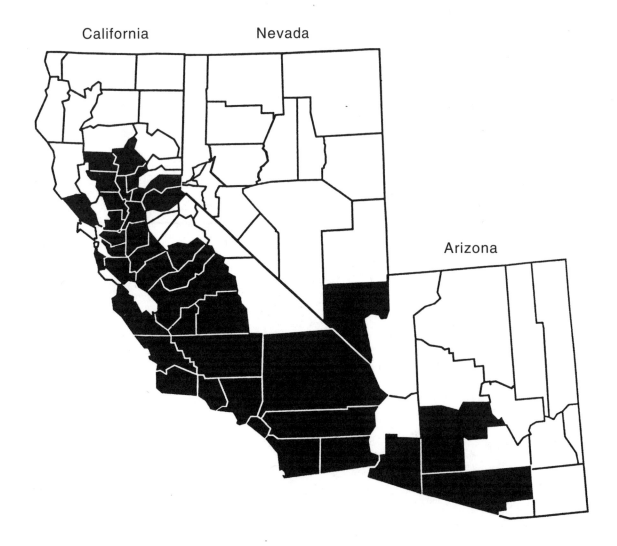

Figure 20.2. Counties (in black) in California, Nevada, and Arizona reporting *Siphoninus phillyreae* in November 1990.

Table 20.1. Hosts of *Siphoninus phillyreae* (Haliday) in Europe and in California.

Family	Genus and species	Common name	Family	Genus and species	Common name
Bignoniaceae	*Catalpa* sp. × *Chilopsis* sp.*	catalpa hybrid	Rosaceae (*cont.*)	*C. monogyna*[†]	hawthorn
				C. oxyacantha[†]	hawthorn
Leguminosae	*Afzelia* sp.[†]	pod mahogany		*Cydonia oblonga*[†]	quince
	*Cercis occidentalis*ered*	western redbud		*Eriobotrya deflexa*[*]	golden loquat
	*C. siliquastrum*ered*	Judas tree		*Heteromeles arbutifolia*ered*	California Christmas berry
Lythraceae	*Lagerstroemia indica*ered*	crape myrtle		*Malus domestica*ered*	apple
				*M. floribunda*ered*	Japanese flowering crabapple
Magnoliaceae	*Liriodendron tulipifera*ered*	tulip tree			
	*Magnolia stellata*ered*	star magnolia		*M. fusca*ered*	Oregon crabapple
				Malus sp. 'Scheideckeri'*ered*	Scheidecker crabapple
Oleacae	*Fraxinus excelsior*[†]	European ash			
	*F. latifolia*ered*	Oregon ash		*Malus* sp. 'Hopa'*ered*	crabapple
	*F. ornus*ered*[†]	flowering ash		*Malus* sp. 'Red Jade'*ered*	crabapple
	F. syriaca[†]	ash		*Mespilus* sp.[†]	
	F. velutina 'Modesto'*ered*	Modesto ash		*Prunus armeniaca*ered*	apricot
	*F. velutina glabra*ered*	Arizona ash		*P. persica*[†]	peach
	F. velutina var. *coriacea*ered*	Western ash		*P. blireiana*ered*	blue plum hybrid
	*F. uhdei*ered*	Shamel ash		*P. salicina*ered*	Santa Rosa plum
	F. uhdei 'Tomlinson'*ered*	Tomlinson ash		*P. virginiana* var. *melanocarpa*ered*	black chokecherry
	Ligustrum spp.*ered*	privet			
	Olea chrysophylla[†]	wild olive		*Pyracantha* sp.*ered*	pyracantha
	O. europa[†]	common olive		*Pyrus calleryana*ered*	ornamental pear
	Phillyreae latifolia[†]	phillyreae		*P. communis*[†]	pear
	P. media[†]	phillyreae		*P. kawakamii*ered*	evergreen pear
	*Syringa lanciniata*ered*	cut-leaf lilac		*P. pyrifolia*ered*	Japanese sand pear
	*S. vulgaris*ered*	common lilac		*P. sativa*[†]	
	*S. hyacinthiflora*ered*	common lilac			
			Rubiaceae	*Cephalanthus occidentalis* var. *californicus*ered*	buttonbush
Punicaciae	*Punica granatum*ered*[†]	pomegranate			
Rhamnaceae	*Rhamnus alaternus*[†]	buckthorn	Rutaceae	*Citrus sinensis*ered*	navel orange
	Ziziphus spina-christi[†]	crown of thorns		*C. sinenesis*[*]	Valencia orange
				*C. limonia*ered*	lemon
				C. paradisi 'Marsh'*ered*	grapefruit
Rosaceae	*Amelanchier dentiolata*ered*	serviceberry		*Citrus* sp.*ered*	tangerine
	*Chaenomeles speciosa*ered*	flowering quince		*Fortunella* sp.*ered*	kumquat
	Crataegus mollis[†]	downy hawthorn			

Source: Adapted from L. A. Mound and S. H. Halsey (1978) and from T. S. Bellows and coworkers (1990).
* Recorded in California.
[†] Recorded outside of the United States.

Table 20.2. Natural enemies reported in the literature for *Siphoninus phillyreae* (Haliday).

Natural enemy	Location	Literature reference
PARASITOIDS		
Hymenoptera, Aphelinidae		
Encarsia inaron (Walker)	Macedonia	Menzelos (1967)
(=*partenopea* Masi)	Egypt	Priesner and Hosny (1932)
	Italy	Tremblay (1969), Viggiani and Mazzone (1980)
E. dichroa (Mercet) (=*pseudopartenopea* Viggiani & Mazzone)	Italy	Viggiani and Mazzone (1980)
E. siphonini Silvestri	Italy	Viggiani and Mazzone (1980)
E. gautieri (Mercet)	Italy	Viggiani and Mazzone (1980)
E. aleyrodis (Mercet)	Italy	Viggiani and Mazzone (1980)
Eretmocerus siphonini Viggiani and Battaglia	Italy	Viggiani and Battaglia (1983)
Eretmocerus sp.	Israel	Gerling and Bellows (unpublished)
PREDATORS		
Coleoptera, Coccinellidae		
Clitostethus arcuatus (Rossi)	France	Gautier (1922)
	Macedonia	Mentzelos (1967)
	Italy	Tremblay (1969)
Diptera, Drosophilidae		
Acletoxenus formosus Loew	France	Gautier (1922)
	Thessalonica	Costacos (1963)
	Macedonia	Mentzelos (1967)
	Italy	Tremblay (1969)
Diptera, Syrphidae		
Syrphus sp.	Macedonia	Mentzelos (1967)
PATHOGENS		
Fungi		
Undocumented reference	Egypt	Priesner and Hosny (1932)

RESULTS

Objective A: The Identification and Introduction of Beneficial Organisms

Natural enemies were imported in 1989 from Italy and from Israel. The species or populations imported from Italy included the aphelinids *Encarsia inaron* (Walker), *E. gautieri* (Mercet), and *E. dichroa* (Mercet); those from Israel included the aphelinids *E. inaron* and *Eretmocerus* sp. and the coccinellids *Clitostethus arcuatus* (Rossi) and *Scymnus* sp. Of these, both populations of *E. inaron* reproduced in quarantine as did *C. arcuatus*. *Encarsia*

inaron developed from egg to adult in approximately 25 days at 25°C in the laboratory. *Clitostethus arcuatus* developed from egg to adult in approximately 20 days at 20°C.

Colonies of the two *E. inaron* populations became established at Riverside, and trial releases of between 30 and 100 females and males were made during the autumn and winter of 1989–90. The Italian population was released in San Diego County, the Israeli population in four other southern California counties. In several sites, progeny were recovered between 6 and 8 weeks after release, in numbers varying from approximately 50 to 350 percent of those released. The parasitoid's developmental time was approximately 6 weeks, reflecting the cool winter temperatures experienced during this period.

In May and June 1990, approximately 4,000 adult *E. inaron* from Israel were released in both Riverside and Los Angeles counties. The populations in Riverside were followed closely, on a weekly or semiweekly basis. The wasp populations established quickly; in 6 weeks significant parasitism was evident. Wasp populations subsequently expanded in both number and range, causing the ash whitefly population densities in the Riverside area to decline approximately a hundredfold by autumn 1990 (Gould, Bellows, and Paine 1992). Similar results were experienced from the releases in Los Angeles. By June 1991 ash whitefly densities in Riverside had declined up to a thousandfold and were nearly undetectable (Bellows et al. 1992). (See fig. 20.3.)

A large colony of the Israeli population of *E. inaron* was reared, and releases numbering several thousand wasps were subsequently made in most infested areas throughout California, and in Arizona and Nevada between 1990 and 1992. The wasp became established and provided control in all areas. In addition, *Clitostethus arcuatus* was released and became established in several California locations.

RECOMMENDATIONS

1. Continued dissemination of previously introduced natural enemies and evaluation of their effectiveness in all regions affected by ash whitefly will be necessary to maintain strategic planning during research and development aimed at achieving biological control for this species.

2. Biological control in the areas that received releases in 1990 appears to be complete and permanent. Additional releases in 1991 in climatically different regions (for example, in coastal, interior, and desert biomes) also provided complete control. Where the whitefly is known in the Mediterranean region it has a complex of natural enemies, indicating that several may play a role in its population suppression and regulation. Currently, in the United States the addition of the single wasp species appears to have been sufficient in all locations to which it has been introduced.

REFERENCES CITED

Bellows, T. S., Jr., T. D. Paine, K. Y. Arakawa, C. Meisenbacher, P. Leddy, and J. Kabashima. 1990. Biological control sought for ash whitefly. *Calif. Agric.* 44(1):4–6.

Bellows, T. S., Jr., T. D. Paine, J. R. Gould, L. G. Bezark, J. C. Ball,

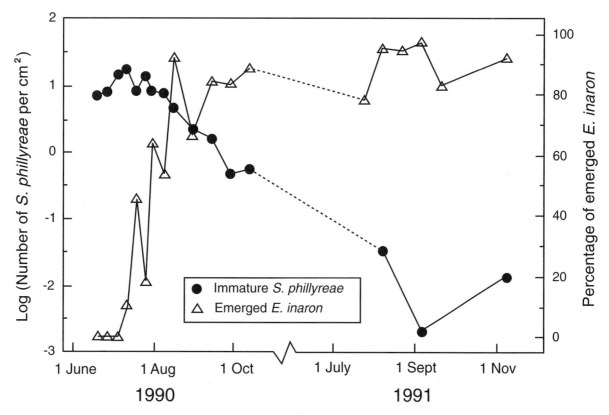

Figure 20.3. Decline in the whitefly *Siphoninus phillyreae* following introductions of *Encarsia inaron*. The percentage of *S. phillyreae* that were bearing parasitoid larvae is shown by the right-hand ordinate (adapted from Bellows et al. 1992).

W. Bentley, R. Coviello, J. Downer, P. Elam, D. Flaherty, P. Gouveia, C. Koehler, R. Molinar, N. O'Connell, E. Perry, and G. Vogel. 1992. Biological control of ash whitefly: A success in progress. *Calif. Agric.* 46(1):24,27–28.

Byrne, D. N., T. S. Bellows, Jr., and M. P. Parrella. 1990. Whiteflies in agricultural systems. In *Whiteflies: Their bionomics, pest status, and management,* ed. D. Gerling, 227–62. Andover, U.K.: Intercept Publications.

Costacos, T. A. 1963. On a severe attack by *Siphoninus phillyreae* (Haliday), subsp. *ineaqualis* Gautier, on fruit trees and its control (in Greek). *Geoponika* 105:3–7.

Gautier, C. 1922. Un aleurode parasite du poirer et du frêne, *Trialeurodes ineaqualis,* n. sp. (Hem.: Aleurodidae). *Ann. Soc. Entomol. France* 91:337–50.

Gould, J. R., T. S. Bellows, Jr., and T. D. Paine. 1992. Population dynamics of *Siphoninus phillyreae* (Haliday) in California in the presence and absence of a parasitoid, *Encarsia partenopea* (Walker). *Ecol. Entomol.* 17:127–34.

Kolev, K. 1973. A new pest of pear in this country (in Bulgarian). *Rastitelna Zaschita* 21:28.

Mentzelos, I. A. 1967. Contribution to the study of the entomophagous insects of *Siphoninus phillyreae* (Halid.) (=*ineaqualis* Gautier) (Aleyrodidae) on pear trees in central Macedonia (in Greek). *Report of the Plant Protection Agricultural Research Station* (Thessaloniki) 3:92–102.

Mound, L. A., and S. H. Halsey. 1978. *Whitefly of the world.* London: British Museum (Natural History).

Patti, I., and C. Rapisarda. 1981. Reperti morfo-biologici sugli aleirodidi nocivi alle piante coltivate in Italia. *Bolletino di Zoologia Agraria e di Bachicoltura* 16:135–90.

Priesner, H., and M. Hosny. 1932. Contributions to the knowledge of the whiteflies (Aleurodidae) of Egypt (I). Bulletin 121 of the Ministry of Agriculture (Egypt).

———. 1934. Contributions to the knowledge of the whiteflies (Aleurodidae) of Egypt (III). Bulletin 145 of the Ministry of Agriculture (Egypt).

Tremblay, E. 1969. Il controllo del *Siphoninus phillyreae* (Haliday) in Campania: Studi del gruppo di laboro del C.N.R. per lotta integrata contro i nemici animali delle piante: XL. *Bollettino del Laboratorio di Entomologia Agraria 'Filippo Silvestri' Portici* 27:161–76.

———. 1973. *Principi di lotta chimica razionale ai fitofagi.* Nota Divulgative 6. Naples: Instituto di Entomologia Agraria della Universita di Napole.

Viggiani, G., and D. Battaglia. 1983. Le specie italiane del genere *Eretmocerus* Hald. (Hymenoptera: Aphelinidae). *Bollettino del Laboratorio di Entomologia Agraria 'Filippo Silvestri' Portici* 40:97–101.

Viggiani, G., and P. Mazzone. 1980. *Encarsia pseudopartenopea* n. sp., parassita di *Siphoninus phillyreae* (Haliday) (Homoptera: Aleyrodidae). *Bollettino del Laboratorio di Entomologia Agraria 'Filippo Silvestri' Portici* 37:9–12.

21 / CITRUS WHITEFLY

T. S. BELLOWS, JR., AND K. Y. ARAKAWA

> **citrus whitefly**
> *Dialeurodes citri* (Ashmead)
> Homoptera: Aleyrodidae

INTRODUCTION

Biology and Pest Status

The citrus whitefly, *Dialeurodes citri* (Ashmead), is one of several whitefly species that are reported as serious pests of citrus worldwide (Byrne, Bellows, and Parrella 1990). It is known in most citrus-growing areas in the world, including Asia, the Orient, the Mediterranean region, South Africa, South America, Central America, Bermuda, and the United States. The genus *Dialeurodes*, the most numerous genus of aleyrodids, has the greatest number of species in Southeast Asia and the Orient, where this whitefly probably originated.

The citrus whitefly has been the object of limited research. A. W. Morrill and E. A. Back (1911) described its field biology in Florida. It is multivoltine, with three principal flights of adults each season; however, many nymphs of the second generation overwinter, as do the nymphs of the third generation. This phenology is also expressed in California (Bellows unpub. data). The species is oligophagous, and hosts in addition to *Citrus* spp. include Cape jasmine (*Gardenia jasminoides*) and other *Gardenia* species, privet (*Ligustrum* sp.), umbrella China tree (*Melia azedarach* cv. Umbraculifera), and prickly ash (*Zanthoxylum clava-Herculis*).

Historical Notes

The citrus whitefly was first reported in California in 1934, near the Sacramento area. Several years' eradication efforts had nearly met with success there by 1945; however, efforts stopped shortly thereafter. The pest was reported again in the 1960s and is now found in several counties in California. It is most serious in the southern coastal citrus-growing regions from San Diego to Los Angeles.

Limited work on biological control has been conducted on this species. Early exploration by R. S. Woglum (1913) revealed a few natural enemies on the Indian subcontinent; the whitefly was unknown in the Mediterranean region at that time. The introduction of *Encarsia lahorensis* (Howard) to Florida following that trip was unsuccessful. In the 1960s *E. lahorensis* (from Pakistan) and *Encarsia* sp. (from India) were introduced into California. These species are established in southern coastal California, but the whitefly continues to be a problem there. *Encarsia lahorensis* was also subsequently introduced into Florida and the Mediterranean region. In spite of these introductions, the citrus whitefly continues to pose an economic threat to citrus production, in many cases requiring treatments to avoid honeydew and sooty mold contamination of fruit.

RESULTS

Objective A: The Identification and Introduction of Beneficial Organisms

Since 1984 researchers have conducted several exploration trips for natural enemies of citrus whitefly. The areas searched include Malaysia (1984), Japan (1984, 1985), Taiwan (1985), Thailand (1985), India (1985), Pakistan (1985), Puerto Rico (1987), and Trinidad (1987). The whitefly was not found in Malaysia, Thailand, India, or Trinidad, but the species was readily located in the Lahore region of Pakistan, and 1,500 pupae of *E. lahorensis* were introduced into quarantine. However, the whitefly populations around Lahore were dense despite the presence of the parasitoid.

Exploration in Japan, where the whitefly may not be native, revealed three parasitoids: an *Eretmocerus* sp. and an *Encarsia* sp., from the Shimizu area in Shizuoka Prefecture, and *Encarsia strenua* (Silvestri), which was found in both northern and southern Kyushu. In the Shimizu area populations of the citrus whitefly were moderately high despite the presence of the parasitoids;

This work would not have been possible without the gracious and generous support of our colleagues in Pakistan, India, Thailand, Malaysia, Taiwan, Japan, Puerto Rico, and Trinidad. J. LaSalle conducted the exploration in Thailand, India, and Pakistan. A. Polaszek identified the *Encarsia strenua*.

however, in Kyushu whitefly populations were clearly at subeconomic levels. The citrus grown in both areas consisted of varieties of Mandarin orange. Search in Kyushu on other host plants of the whitefly did not reveal any significantly greater population densities. During 1984 and 1985 all three species were shipped to California, but only the *Eretmocerus* sp. successfully reproduced in quarantine. The *Encarsia* sp. from Kyushu (which did not establish in 1984) was shipped to California by colleagues in Japan in 1986 and was successfully colonized in quarantine.

The original shipment of *Eretmocerus* sp. yielded 9 males and 11 females. These were provided with citrus whitefly nymphs, and the F_1 progeny consisted of three females. Those were again placed with whitefly nymphs and gave rise to a deuterotokous colony of this species, with males present only rarely.

The *E. strenua* population in Kyushu consisted of both males and females; females were primary parasitoids of citrus whitefly nymphs, and males were secondary parasitoids of *E. strenua* females. Shipments received in 1986 yielded both males and females; attempts at rearing them included offering adult wasps different combinations of citrus whitefly nymphs, citrus whitefly nymphs bearing parasitoids (of *Eretmocerus* sp.), *Bemisia tabaci* (Gennadius) nymphs, and *B. tabaci* nymphs bearing parasitoids (of a different *Eretmocerus* sp.). The colony reproduced at very low levels for several weeks, requiring considerable manipulation to produce both male and female parasitoids. After approximately 12 weeks, we attempted to rear progeny from virgin females and succeeded in establishing a thelytokous population, which has since continued to reproduce without males.

Exploration in Taiwan revealed at least five species of *Encarsia*, a predaceous drosophilid, and pathogenic fungi attacking *Dialeurodes citri*, which is erroneously referred to in some literature there as *Bemisia hancocki* Corbett. Populations of citrus whitefly appeared generally to be at subeconomic levels. Shipments to California did not result in any colony establishment of natural enemies.

Exploration in Puerto Rico yielded another population of *E. strenua* attacking *Dialeurodes* sp. [?*citrifolii* (Morgan)]. Four pupae were introduced into quarantine, three of which produced adult females. These females reproduced readily on both *D. citri* and *D. citrifolii*, giving rise to a thelytokous population.

The *Eretmocerus* sp. from Japan was released in one southern California grove in 1985. Very few parasitoid exit holes were recovered from that grove in the spring of 1986, and once the grove was removed that spring, no further surveys were possible.

Subsequently, the *Eretmocerus* sp., *E. strenua* from Japan, and *E. strenua* from Puerto Rico have been released in several orchards in Orange and San Diego counties. Recoveries have been made during the season of release, but very limited overwinter recoveries have been made. The establishment of at least the *Eretmocerus* sp. appears to be hampered by the presence of *Encarsia lahorensis*, which parasitizes the *Eretmocerus* sp. pupae. When these pupae are present in small numbers, such hyperparasitism can destroy large portions of populations.

RECOMMENDATIONS

The citrus whitefly continues to be an economic problem for citrus producers worldwide. Successful biological control appears to exist in Taiwan and Japan, and to lesser degrees in Florida, the Caucasus region of Russia, and the Mediterranean. In many of these areas fungi play a role in reducing the populations, but in California (and probably throughout the continental Western Region) those fungi apparently are unable to establish and contribute to suppression.

1. Dissemination of the more recently available agents should be continued either until it is clear that establishment cannot be achieved or until establishment occurs. This would permit researchers to evaluate these agents' contributions. It may be wise to defer dissemination of *Encarsia lahorensis* until other agents have been introduced: its presence can make subsequent establishment of agents more difficult because of the initial differences in numbers between a small release colony of a new agent and a well-established field population of *E. lahorensis*.

2. If further agents are required to suppress the citrus whitefly, Taiwan appears to be an excellent location for additional exploration. It has the largest insect fauna of natural enemies known for this whitefly, including several species of parasitoids and predators that have not previously been employed elsewhere in attempts at biological control.

REFERENCES CITED

Byrne, D. N., T. S. Bellows, Jr., and M. P. Parrella. 1990. Whiteflies in agricultural systems. In *Whiteflies: Their bionomics, pest status and management,* ed. D. Gerling, 227–62. Andover, U.K.: Intercept Publications.

Morrill, A. W., and E. A. Back. 1911. *White flies injurious to citrus in Florida.* USDA Bureau of Entomology Bulletin 92. Washington, D.C.: U.S. Department of Agriculture.

Woglum, R. S. 1913. *Report of a trip to India and the Orient in search of the natural enemies of the citrus white fly.* USDA Bureau of Entomology Bulletin 120. Washington, D.C.: U.S. Department of Agriculture.

22 / CLOUDY-WINGED WHITEFLY

T. S. BELLOWS, JR., AND K. Y. ARAKAWA

> **cloudy-winged whitefly**
> *Dialeurodes citrifolii* (Morgan)
> Homoptera: Aleyrodidae

INTRODUCTION

Biology and Pest Status

The cloudy-winged whitefly, *Dialeurodes citrifolii* (Morgan), was first reported in California in 1984 in San Diego. It has since become widespread in the coastal and subcoastal citrus-growing areas between San Diego and Los Angeles. Although it is a less serious pest of citrus worldwide than the citrus whitefly, *Dialeurodes citri* (Ashmead), in several areas in California it has developed moderate to large populations. It damages citrus crops by producing honeydew, which encourages sooty mold contamination. In some cases the densities of the citrus whitefly have declined following the expansion of the cloudy-winged whitefly.

Little work has been published on the biology of this species. A. W. Morrill and E. A. Back (1911) reported its field biology in Florida. It has three principal periods of adult activity during the year and, thus, appears to be trivoltine.

Historical Notes

The origin of this pest is open to question, although the genus *Dialeurodes* is known primarily from tropical Asia and the Orient. At present, the cloudy-winged whitefly can be found in parts of Asia and in the Caribbean, Hawaii, Florida, and California.

Published information on natural enemies of this species is scarce. F. Silvestri (1927) reported *Encarsia perstrenua* (Silvestri) from it in Vietnam, and L. Fulmek (1943) reported *E. strenua* (Silvestri) from it in Indochina. Few other records of natural enemies exist. In California the species is not parasitized in the laboratory or the field by any of the available parasitoids of citrus whitefly: *Encarsia lahorensis* (Howard), *Encarsia* spp. from India, *E. strenua* from Japan, or *Eretmocerus* sp. from Japan.

RESULTS

Objective A: The Identification and Introduction of Beneficial Organisms

Researchers have searched for natural enemies in Japan (1985), Thailand (1985), India (1985), Pakistan (1985), Puerto Rico (1987), Trinidad (1987), and Hawaii (1989). Cloudy-winged whitefly was found only in Puerto Rico.

Collections of four parasitized pupae from a grapefruit orchard at an Agricultural Experiment Station in Puerto Rico yielded three *E. strenua* females. This parasitoid was reared on citrus and cloudy-winged whitefly and produced thelytokous colonies on both. This is the first parasitoid reported to attack both of these *Dialeurodes* species.

The parasitoid was released in San Diego on infested dooryard citrus. They were recovered in the season of release, but overwinter recoveries have not yet been made.

RECOMMENDATIONS

Encarsia strenua should be distributed widely and in sufficient numbers to allow its establishment. This will permit researchers to evaluate the effectiveness of this species, and assess the need for additional agents.

REFERENCES CITED

Fulmek, L. 1943. Wirtsindex der Aleyrodiden- und Cocciden-Parasiten. *Ent. Beih. Berl.-Dahlem* 10.

Morrill, A. W., and E. A. Back. 1911. *White flies injurious to citrus in Florida.* Bureau of Entomology Bulletin 92. Washington, D.C.: U.S. Department of Agriculture.

F. Bennett provided much helpful discussion at the beginning of this project. This work would not have been possible without the help of colleagues in Japan, Thailand, India, Pakistan, Puerto Rico, Trinidad, and Hawaii. A. Polaszek identified the *Encarsia strenua*.

Silvestri, F. 1927. Contribuzione alla conoscenza degli Aleurodidae (Inseta: Hemiptera) viventi su *Citrus* in Estremo Oriento e dei Loro parassiti. *Bollettino del Laboratorio di Zoologia e di gen. Agraria R. Scuola Agricultora Portici* 21:1–60.

23 / ORANGE SPINY WHITEFLY

D. M. NAFUS AND J. R. NECHOLS

---

orange spiny whitefly
Aleurocanthus spiniferus (Quaintance)
Homoptera: Aleyrodidae

INTRODUCTION

Biology and Pest Status

Aleurocanthus spiniferus (Quaintance) is a major pest of citrus in subtropical and tropical regions (Nakao and Funasaki 1979). In temperate regions it attacks roses, grapes, apples, and pears (Peterson 1955). The adults are gray, with white spots on the wings; their bodies are orange under the gray wax. The nymphs are black, egg shaped, and covered with spines; a fringe of white wax extends outward around the periphery of the body. The females lay 12 to 13 eggs in spiral masses on the lower leaf surface. Upon hatching, first-instar crawlers attach to the underside of the leaf and begin feeding. They molt in place through three sedentary nymphal stages before emerging as adults. The orange spiny whitefly has four to six generations a year; it develops from egg to adult in 1 to 2 months, depending on the temperature and, probably, the quality of the host plant. When immature whiteflies feed they produce a sticky honeydew, which allows the growth of a dense, black, sooty mold that is unsightly and impairs photosynthesis.

Historical Notes

Native to the Indo-Malay region, the orange spiny whitefly also occurs in southern China and the Philippines. In 1919 it was found in southern Japan (Clausen 1978). By the late 1940s and early 1950s, this pest had spread to the western Pacific islands of Chuuk and Guam in Micronesia (Peterson 1955). Currently, its distribution in Micronesia includes Kosrae (Nafus 1988), Pohnpei (Schreiner and Nafus 1986), and Yap in the Caroline Islands. It is also present in the Hawaiian Islands.

During the 1950s, researchers sent several parasitoids from Mexico to Guam, releasing them to control the orange spiny whitefly. Originally from western India and Pakistan, these parasitoids had been used in Mexico for the successful control of the citrus blackfly, *Aleurocanthus woglumi* Ashby (Clausen 1978). Two of the parasitoids, *Encarsia smithi* (Silvestri) and *Amitus hesperidum* Silvestri, became established (Peterson 1955), and have effectively controlled the orange spiny whitefly for over 30 years (Nafus and Schreiner 1989).

RESULTS

Objective A: The Identification and Introduction of Beneficial Organisms

In 1982, heavy infestations of *A. spiniferus* were discovered on oranges, tangerines, and other citrus on the island of Kosrae, damaging fruit production (Nafus 1988). W-84 project leaders in Guam were asked to train Kosraen agriculturalists and to ship parasitoids. Researchers in Guam sent two shipments of *E. smithi* to Kosrae in November 1983 (Nechols unpub. data). These included 240 females and 24 males, of which 142 wasps survived and were released (Nafus 1988).

Objective D: Impact

In 1984 and 1986 the establishment of *A. spiniferus* on Kosrae was monitored. In 1984 *E. smithi* was established at all survey sites, and parasitization rates of 50 to 90 percent were observed. The undersides of old leaves were almost completely covered with a combination of old whitefly pupal cases and living and dead immature whitefly stages. The upper surfaces were heavily encrusted with sooty mold. Newly matured leaves contained an average of about five nymphs and pupae per leaf. In contrast, the 1986 survey revealed no sooty mold, and orange spiny whiteflies or their remains were rarely found on the older leaves. Of the 3,600 leaves examined, only 0.5 percent were infested, with an average of less than 0.02 immatures per leaf. Parasitization rates ranged from 83 to 100 percent.

RECOMMENDATIONS

In Kosrae one species, *E. smithi*, provided virtually complete biological control. This parasitoid is the dominant natural enemy in Guam, and has probably been the primary agent responsible for the successful biological control of the orange spiny whitefly over the last 30 years. Based on these data, we recommend that this parasitoid be introduced to other tropical islands or continental regions where *A. spiniferus* becomes established.

REFERENCES CITED

Clausen, C. P. 1978. *Introduced parasites and predators of arthropod pests and weeds: A world review.* USDA-ARS Handbook 480. Washington, D.C.: U.S. Department of Agriculture.

Nafus, D. 1988. Establishment of *Encarsia smithi* on Kosrae for control of orange spiny whitefly, *Aleurocanthus spiniferus. Proc. Hawaii. Entomol. Soc.* 28:229–31.

Nafus, D., and I. Schreiner. 1989. History of biological control in the Mariana Islands. *Micronesica* 22:65–106.

Nakao, H. K., and G. Y. Funasaki. 1979. Introductions for biological control in Hawaii: 1975 and 1976. *Proc. Hawaii. Entomol. Soc.* 13:125–28.

Peterson, G. D. 1955. Biological control of the orange spiny whitefly in Guam. *J. Econ. Entomol.* 48:681–83.

Schreiner, I., and D. Nafus. 1986. Accidental introductions of insect pests to Guam, 1945–1985. *Proc. Hawaii. Entomol. Soc.* 27:45–52.

24 / SPIRALING WHITEFLY

J. R. NECHOLS AND D. M. NAFUS

spiraling whitefly
Aleurodicus dispersus Russell
Homoptera: Aleyrodidae

INTRODUCTION

Biology and Pest Status

The spiraling whitefly, *Aleurodicus dispersus* Russell, is a serious pest of a wide range of tropical and subtropical plants, including plumeria, mango, guava, coconut, and banana. Its host range includes over one hundred woody and herbaceous food and ornamental plants (Nakahara 1978).

A. dispersus has a life cycle typical of whiteflies: egg, four scalelike nymphal stages, and adult. Its total development takes about 1 month under field temperatures averaging 29°C. Females lay their eggs on lower leaf surfaces in a characteristic spiraling pattern that helps distinguish this species from other whiteflies. During heavy infestations, immature and adult whiteflies may completely cover the leaves and fruit of affected plants, giving them a powdery, white appearance. The feeding pests produce copious amounts of honeydew, promoting the growth of a black, sooty mold that is unsightly and impedes photosynthesis.

Historical Notes

A native of the New World tropics, the spiraling whitefly was accidentally introduced into Hawaii during the late 1970s (Lai, Funasaki, and Higa 1982). By 1981 it had spread throughout the major Hawaiian Islands (Kumashiro et al. 1983). In May 1981 the whitefly was discovered in Guam, where it quickly became a major pest of food and ornamental plantings throughout the island (Nechols 1981).

Currently, the spiraling whitefly has become established on other islands in the Pacific Basin, including the Northern Marianas, the Carolines (Belau and Pohnpei), the Marshalls, Fiji, and the Philippines. It is also present in American Samoa (Kumashiro et al. 1983) and parts of Southeast Asia. In the New World, this pest is distributed from southern Florida to Central and South America. It is also present in the West Indies (Kumashiro et al. 1983). L. M. Russell (1965) reported *A. dispersus* on the Canary Islands.

In the late 1970s Hawaii's Department of Agriculture introduced several natural enemies from the New World to Hawaii, including two coccinellids, *Nephaspis oculatus* (Blatchley) (=*N. amnicola* Wingo) and *Delphastus pusillus* (LeConte), and two aphelinids, *Encarsia ?haitiensis* Dozier and *Encarsia* sp. (Lai et al. 1982; Waterhouse and Norris 1989). Project W-84 scientists have been involved in introducing selected exotic enemies to Guam, monitoring their establishment, and shipping them to other areas where the spiraling whitefly has become established.

RESULTS

Objective A: The Identification and Introduction of Beneficial Organisms

The aphelinid *Encarsia ?haitiensis* and the coccinellid *Nephaspis oculatus* (=*amnicola*) were imported to Guam with the cooperation of the Hawaii Department of Agriculture (Nechols 1981). One of us (J. R. Nechols) released 4,000 to 5,000 *N. oculatus* adults between July and November of 1981. In November one shipment of 150 *E. ?haitiensis* was also released at a single locality. Both species became established (Nechols 1982).

In 1982, both of these exotic natural enemies were collected in Guam and released in the Northern Mariana Islands. Positive recoveries of both species were made in 1982 and 1983 (Nechols unpub. data). *E. ?haitiensis* was also released on the Caroline Islands of Belau and Pohnpei, and in Fiji; in all cases, the natural enemy became established and exerted good control (Nafus unpub. data).

Objective D: Impact

Whitefly counts on plumeria in Guam before, and soon after, the first natural enemy releases in 1981 showed that the population densities of fourth-instar spiraling whiteflies more than doubled between June and August (Nechols 1981, 1982). Average peak densities of about 90 nymphs per leaf were measured during this period. In contrast, approximately 1 year after the natural enemies were released, fewer than two emerged whiteflies per leaf were found (Nechols 1982).

N. oculatus was abundant at locations where whitefly infestations were heavy. However, after whitefly populations had declined, these beetles became scarce (Nechols unpub. data). The parasitoid *E. ?haitiensis* was found on all sample dates and at every locality, even when host densities were very low (Nechols 1982). These results are similar to those reported by B. R. Kumashiro et al. (1983) in Hawaii; they found that parasitoid populations increased rapidly on Oahu and, 9 months after their release, were widely distributed throughout the island. On Guam the rates of parasitization at most localities increased between July and September 1982, while host densities declined from low to very low levels. The coccinellid *Nephaspis roepkei* Fluiter and an unidentified chrysopid were also found in association with the spiraling whitefly. The impact of these enemies was not determined.

On Guam the spiraling whitefly has remained under substantial biological control (Nechols 1983; Nafus unpub. data). The parasitoid *E. ?haitiensis* appears capable of maintaining this pest at very low levels except during the annual "dry" season, which lasts approximately 5 months. During this period moderate but temporary resurgences are observed on guava, Indian almond, and other host plants (Nechols 1983). The coccinellid *N. oculatus* responds to pest increases, but not until whitefly densities become high. These results suggest that artificial control measures may be unnecessary because whitefly increases are mostly transient and subject to effective natural control.

RECOMMENDATIONS

1. Guam's classical biological control program of the spiraling whitefly should be considered a substantial success. Because exotic natural enemies may be exported to other Pacific islands, it would be useful to establish the relative importance of imported predators and parasitoids through experimental evaluation studies. However, in most cases, the best strategy may be the simultaneous release of *N. oculatus* and *E. ?haitiensis*, because it appears that the peak response of each natural enemy occurs at different whitefly densities. Thus, the action of the two enemies may be complementary.

2. Pest resurgences occur during Guam's dry season. However, the direct and indirect effects of rainfall and other moisture on the whitefly, its natural enemies, and its host plants are unknown. Research to establish these relationships would help entomologists make seasonal predictions of the pest's population dynamics.

REFERENCES CITED

Kumashiro, B. R., P. Y. Lai, G. Y. Funasaki, and K. K. Teramoto. 1983. Efficacy of *Nephaspis amnicola* and *Encarsia ?haitiensis* in controlling *Aleurodicus dispersus* in Hawaii. *Proc. Hawaii. Entomol. Soc.* 24:261–69.

Lai, P. Y., G. Y. Funasaki, and S. Y. Higa. 1982. Introductions for biological control in Hawaii: 1979 and 1980. *Proc. Hawaii. Entomol. Soc.* 24:109–113.

Nakahara, L. 1978. *Hawaii Cooperative Economic Pest Report, October 20.* Honolulu: Hawaii State Printing Office.

Nechols, J. R. 1981. *Entomology: Biological control,* 17. Annual Report of the Guam Agricultural Experiment Station.

———. 1982. *Entomology: Biological control,* 43–44,47–49. Annual Report of the Guam Agricultural Experiment Station.

———. 1983. *Entomology: Biological control,* 26–27. Annual Report of the Guam Agricultural Experiment Station.

Russell, L. M. 1965. A new species of *Aleurodicus* and two close relatives (Homoptera: Aleyrodidae). *Fla. Entomol.* 48:47–55.

Waterhouse, D. F., and K. R. Norris. 1989. *Biological control: Pacific prospects.* Supplement 1. Canberra: Australian Centre for International Agricultural Research.

25 / SWEETPOTATO WHITEFLY

T. J. HENNEBERRY AND T. S. BELLOWS, JR.

sweetpotato whitefly
Bemisia tabaci (Gennadius)
Homoptera: Aleyrodidae

INTRODUCTION

Biology and Pest Status

The sweetpotato whitefly, *Bemisia tabaci* (Gennadius), is a serious pest of cotton and a variety of other cultivated crops. In cotton it causes losses from feeding (Hussain and Trehan 1933), reduces yield (Mound 1965), transmits cotton leaf crumple virus (Dickson et al. 1954), and contaminates the lint with honeydew—making the cotton sticky—and with associated molds (Gerling et al. 1980). Sticky cottons are an increasing problem in textile mills worldwide, because they cannot be processed efficiently (Perkins 1983).

On vegetables, the sweetpotato whitefly's damage is compounded by its transmission of plant diseases that have become limiting factors in the production of several crops in desert areas of the southwestern United States (Flock and Mayhew 1981). Sweetpotato whitefly vectors viruses that infect cotton, lettuce, melons, cucurbits, and sugarbeets (Byrne, Bellows, and Parrella 1990). These diseases have caused annual losses in excess of $100 million (Duffus and Flock 1982).

Historical Notes

In the United States the sweetpotato whitefly was first reported on cotton in Arizona in 1927 and in California in 1928 (Russell 1957). Until the 1980s the pest caused only sporadic economic infestations in cotton, most frequently after fields had been treated with insecticides (Gerling 1967). However, since 1981 the sweetpotato whitefly has become an increasingly important pest of cotton and other crops (Butler, Henneberry, and Hutchison 1986). Worldwide, sweetpotato whitefly outbreaks also seem to have occurred most frequently after the widespread use of synthetic organic insecticides (Possibilities 1981). However, some other reasons that have been suggested for whitefly population increases include the development of resistance to insecticides, changing crop production practices, and interactions between insecticides and host plants that affect the pest's reproduction.

RESULTS

Objective B: Ecological and Physiological Studies

The biology of introduced and indigenous natural enemies. Researchers have evaluated two introduced biotypes of a predaceous phytoseiid mite, *Euseius scutalis* (Athias-Henriot), and an indigenous species, *E. hibisci* (Chant), as possible biological agents (Meyerdirk and Coudriet 1985, 1986). The introduced biotypes were collected from citrus in Morocco and from lantana in Jordan; the indigenous species is common in citrus in California. The most suitable hosts for both introduced biotypes were sweetpotato whitefly eggs, followed by first-instar nymphs. The Jordanian biotype appeared to be the more promising of the two, with higher fecundity, a longer oviposition period, and longer survival than the Moroccan biotype. *E. hibisci* also fed on sweetpotato whitefly stages; 75 percent of the predaceous mites developed to the adult stage when fed a mixture of sweetpotato whitefly eggs and first and second instars. Female *E. hibisci* consumed an average of 4.5 sweetpotato whitefly eggs per day, which suggested that the predator could be used for augmentative releases to complement the existing natural enemy complex.

The common green lacewing, *Chrysoperla* (=*Chrysopa*) *carnea* (Stephens), is present in many cultivated cropping systems, but little is known about its interaction with the sweetpotato whitefly. In studies researchers have shown that *C. carnea* larvae consumed sweetpotato whitefly pupae in 33 to 78 seconds (Butler and Henneberry 1988). Sweetpotato whitefly adults avoided cotton leaves when *C. carnea* were present, and for several days after the lacewing larvae were gone. This behavior not only resulted in lower sweetpotato whitefly populations, but suggested that *C. carnea*

115

releases a volatile chemical, or some other marking mechanism, that may be useful for control.

In many parts of the world, the aphelinid parasite *Eretmocerus mundus* (Mercet) has been reported as a sweetpotato whitefly parasite. To produce large numbers of a Jordanian biotype of this parasite for release, researchers studied the effects of temperature on developmental rates (Butler 1986). They found that the parasite developed from egg to adult at varying rates: 47.5 days at 17.5°C, but 14.0 days at 30°C. The regression equation for developmental time was $y = 4.763 + 0.00404x$. The parasite developed significantly more quickly than did the sweetpotato whitefly host.

Sweetpotato whitefly and parasite interactions on wild hosts. There is only limited information on the role of wild hosts in the population dynamics of the sweetpotato whitefly. The whitefly has been reported to reproduce on field bindweed, *Convolvulensis arvensis* L., and on wild sunflower, *Helianthus annus* L. (Coudriet et al. 1986). Both of these hosts are widely distributed in desert areas, present throughout the year, and are reservoirs of infectious yellows virus. The numbers of sweetpotato whitefly were low on both wild hosts from January through August, but increased dramatically between September and December. On both weed hosts, *Eretmocerus* species was the predominant parasite, but on one sample collected on field bindweed in February, researchers recorded *Encarsia* species. The highest rate of parasitism recorded on these hosts was 40 percent, which suggests that the parasites do not have enough of an effect to control the pest populations.

The effect of insecticides on sweetpotato whitefly parasites. Aggravating the sweetpotato whitefly problem has been its development of resistance to (Prabhaker, Coudriet, and Meyerdirk 1985), and reproductive stimulation by (Dittrich, Hassan, and Ernst 1985), pesticides. Also, the probable susceptibility of its natural enemies to pesticides is believed to contribute to whitefly outbreaks. Researchers found that two applications of Pydrin, Dylox, or malathion did not reduce sweetpotato whitefly populations, and that whitefly populations were, in fact, significantly higher in malathion-treated plots than in untreated plots (Dittrich, Hassan, and Ernst 1985). The numbers of parasites—both *Encarsia* species and *Eretmocerus* species—were generally higher in the untreated plots than in the insecticide-treated plots, but the differences were not significant. T. S. Bellows and K. Arakawa (1988) found that the rate of whitefly parasitization in cotton fields increased dramatically when spraying ceased late in the growing season. These results suggested that sweetpotato whitefly populations increase when insecticide treatments cause parasite populations to decline; however, other mechanisms—such as higher whitefly reproductive rates in the presence of the insecticides—cannot be ruled out.

RECOMMENDATIONS

1. Further evaluate the several predaceous mite species, the common green lacewing, and the aphelinid parasites discussed, as well as other exotic and native biological agents, for control of the sweetpotato whitefly through natural enemy conservation and augmentation.

2. Continue efforts to introduce additional exotic natural enemy species, in conjunction with a renewed effort to determine the geographical origin of the sweetpotato whitefly.

3. Further define the role of insecticides in elevating sweetpotato whitefly populations to economic pest status.

4. Investigate cultivated crop sequences and interactions between the sweetpotato whitefly and its natural enemies, and develop biological and cultural control approaches to reduce or eliminate insecticide use.

REFERENCES CITED

Bellows, T. S., Jr., and K. Arakawa. 1988. Dynamics of preimaginal populations of *Bemisia tabaci* (Homoptera: Aleyrodidae) and *Eretmocerus* spp. (Hymenoptera: Aphelinidae) in southern California cotton. *Environ. Entomol.* 17:483–87.

Butler, G. D., Jr. 1986. Time for development of *Eretmocerus mundus*, a parasite of the sweetpotato whitefly from Jordan. In *Cotton: A College of Agriculture report*, 229–31. University of Arizona, Tucson, Agricultural Experiment Station.

Butler, G. D., Jr., and T. J. Henneberry. 1988. Laboratory studies of *Chrysoperla carnea* predation on *Bemisia tabaci*. *Southwest. Entomol.* 13:165–70.

Butler, G. D., Jr., T. J. Henneberry, and W. D. Hutchison. 1986. Biology, sampling, and population dynamics of *Bemisia tabaci*. In *Agricultural zoology reviews*, ed. G. E. Russell, 167–95. Newcastle upon Tyne, U.K.: Intercept Press.

Byrne, D. N., T. S. Bellows, Jr., and M. P. Parrella. 1990. Whiteflies in agricultural systems. In *Whiteflies: Their bionomics, ecology, and control*, ed. D. Gerling, 227–61. Wimborne, U.K.: Intercept Press.

Coudriet, D. L., D. E. Meyerdirk, N. Prabhaker, and A. N. Kishaba. 1986. Bionomics of sweetpotato whitefly (Homoptera: Aleyrodidae) on weed hosts in the Imperial Valley, California. *Environ. Entomol.* 15:1179–83.

Dickson, R. C., M. McD. Johnson, and E. F. Laird. 1954. Leaf crumple, a virus disease of cotton. *Phytopathology* 44:479–80.

Dittrich, V. S., O. Hassan, and G. H. Ernst. 1985. Sudanese cotton and the whitefly: A case study of the emergence of a new primary pest. *Crop Protect.* 4:161–76.

Duffus, J. E., and R. A. Flock. 1982. The whitefly-transmitted disease complex of the desert Southwest. *Calif. Agric.* 36(11–12):4–6.

Flock, R. A., and D. Mayhew. 1981. Squash leaf curl, a new virus disease of cucurbits in California. *Plant Disease Reporter* 65:75–76.

Gerling, D. 1967. Bionomics of the whitefly complex associated with cotton in southern California (Homoptera: Aleyrodidae; Hymenoptera: Aphelinidae). *Ann. Entomol. Soc. Am.* 60:1306–21.

Gerling, D., U. Motro, and R. Horowitz. 1980. Dynamics of *Bemisia tabaci* (Gennadius) (Homoptera: Aleyrodidae) attacking cotton in the coastal plain of Israel. *Bull. Entomol. Res.* 70:213–19.

Hussain, M. A., and K. N. Trehan. 1933. The life-history, bionomics, and control of the white-fly of cotton (*Bemisia gossypiperda* M and L). *Indian J. Agric. Sci.* 3:701–53.

Meyerdirk, D. E., and D. L. Coudriet. 1985. Predation and developmental studies of *Euseuis hibisci* (Chant) (Acarina: Phytoseiidae) feeding on *Bemisia tabaci* (Gennadius) (Homoptera: Aleyrodidae). *Environ. Entomol.* 14:24–27.

————. 1986. Evaluation of two biotypes of *Euseuis scutalis* (Acari: Phytoseiidae) as predators of *Bemisia tabaci* (Homoptera: Aleyrodidae). *J. Econ. Entomol.* 79:659–63.

Mound, L. A. 1965. Effect of whitefly (*Bemisia tabaci*) on cotton in the Sudan Gezira. *Empire Cotton Growing Rev.* 42:290–94.

Perkins, H. H., Jr. 1983. Identification and processing of honeydew-contaminated cottons. *Textile Research Journal* 53:508–12.

Possibilities for use of biotic agents in the control of the whitefly, *Bemisia tabaci.* 1981. *Biocontrol News and Information,* Commonw. Inst. Biol. Contr. 2:1–7.

Prabhaker, H., D. L. Coudriet, and D. E. Meyerdirk. 1985. Insecticide resistance in the sweetpotato whitefly, *Bemisia tabaci* (Homoptera: Aleyrodidae). *J. Econ. Entomol.* 78:387–409.

Russell, L. M. 1957. Synonyms of *Bemisia tabaci* (Gennadius) (Homoptera: Aleyrodidae). *Bull. Brooklyn Entomol. Soc.* 52:122–23.

26 / WOOLLY WHITEFLY

D. M. NAFUS, J. W. BEARDSLEY, AND G. S. PAULSON

> **woolly whitefly**
> *Aleurothrixus floccosus* (Maskell)
> Homoptera: Aleyrodidae

INTRODUCTION

Biology and Pest Status

The woolly whitefly, *Aleurothrixus floccosus* (Maskell), attacks a wide range of crops but is predominantly a problem on citrus and guava. Heavy infestations can trigger an extensive buildup of sooty mold and cause premature leaf drop on citrus.

Historical Notes

The woolly whitefly is native to the tropical and subtropical parts of the Americas. In the mid-1960s it became established in California, spreading to Hawaii in 1981 and then to Guam in 1985. In 1981 Hawaii's Department of Agriculture introduced the parasitoids *Cales noacki* De Santis (Aphelinidae) and *Amitus spiniferus* (Brethes) (Platygasteridae) to control the whitefly; both species had already been used successfully in California. The two species became established and, although initially abundant, were later replaced by an undescribed species of *Eretmocerus* (Aphelinidae). The origin of the *Eretmocerus* species is unknown. G. S. Paulson (1983) suggested it may have switched from feeding on another whitefly, *Orchamoplatus mammaeferus* (Quaintance and Baker), which also occurs on citrus in Hawaii.

W-84 project researchers in Hawaii concentrated on determining the basic biology of the whitefly and on evaluating the effectiveness of *Eretmocerus* in controlling it. In Guam researchers have mainly focused on monitoring the populations of the woolly whitefly and *Eretmocerus* sp., which was apparently introduced with its host, to determine whether additional natural enemies were needed.

RESULTS

Objective B: Ecological and Physiological Studies

G. S. Paulson and J. W. Beardsley (1986) have studied the life cycle of the woolly whitefly. They found that at 22.5°C the whitefly took 27.4 days to reach adulthood from the time it hatched; it had a 1-day preovipositional period. The mean number of days required for each of the pest's stages were as follows: 9.9, egg; 4.9, first instar; 7.8, second instar; 8.1, third instar; and 6.6, fourth instar. The whiteflies emerged in the early morning, mostly between 6:00 and 9:00. Adults lived for an average of 36 days and laid an average of 53 eggs. The eggs were typically laid in concentric rings in groups of about 38, but small groups and single eggs were also laid. As do many other whitefly species, the woolly whitefly inserted egg pedicels into host-plant stomata (Paulson and Beardsley 1985). Crawlers settled within 20 minutes of hatching and remained on the undersides of the leaves on which they had hatched.

Objective D: Impact

In Hawaii researchers studied the population dynamics of the woolly whitefly and its natural enemies on citrus, guava, and plumeria in both wet and dry locations (Paulson 1983). On citrus and guava populations were high, reaching nearly 800 whiteflies per leaf on the former. Temperature, natural enemies, and interspecific competition all played important roles in the whitefly's population dynamics. *Eretmocerus* sp. exhibited a significant density-dependent response to changes in whitefly populations, and appeared to be able to check or reduce population growth when temperatures were above 24° to 25°C. Increasing temperature also reduced the abundance of the woolly whitefly. On citrus, the woolly whitefly appeared to be adversely affected by interspecific competition with the whitefly *O. mammaeferus*: populations of woolly whitefly were lower when populations of *O. mammaeferus* were high. On plumeria, a nonpreferred host, woolly whitefly numbers were always very low, and no conclusions about the effect of any parasitoids or predators could be drawn.

Other natural enemies associated with the woolly whitefly in Hawaii were *Allograpta obliqua* Say (Syrphidae), *Chrysopa comanche* Banks (Chrysopidae),

Delphastus pusillus (LeConte), *Serangium maculigerum* Blackburn, *Nephaspis oculatus* (Blatchley) (=*N. amnicola* Wingo), and *Coelophora pupillata* (Swartz) (Coccinellidae). However, researchers found that only at one site on the dry side of Oahu were the predators positively and significantly associated with woolly whitefly densities. Casual observations in Hawaii since 1983 (Beardsley unpub. data) indicate that the woolly whitefly is now under satisfactory biological control.

On Guam woolly whitefly populations were lower than those observed in Hawaii. At the start of the sampling program on tangerine, which took place near the end of the initial invasion peak, there was an average of 140 whiteflies per leaf. Populations later declined to less than 10 whiteflies per leaf, and except for brief outbreaks during the dry season—February to June—populations have remained at nonthreatening levels. *Eretmocerus* sp. is the principal parasitoid attacking the woolly whitefly on Guam; its parasitization rates have fluctuated between 15 and 85 percent, averaging around 60 percent. These rates were similar to those reported on Hawaii by G. S. Paulson (1983), although whitefly population levels on Guam were substantially lower. The warmer temperatures on Guam—average monthly temperatures fluctuate between 26° and 29°C—may be responsible for the better control; this is above the minimum temperature of 24° to 25°C that G. S. Paulson (1983) suggested as necessary for good biological control.

RECOMMENDATIONS

In Hawaii and Guam the woolly whitefly is under satisfactory biological control, and it does not appear necessary to introduce additional natural enemies. What is needed is additional information on the interaction between temperature, fecundity, and the survival of the woolly whitefly, as well as on the effectiveness of *Eretmocerus* sp.

REFERENCES CITED

Paulson, G. S. 1983. The biology and natural enemies of the woolly whitefly, *Aleurothrixus floccosus* (Maskell), in Hawaii. Master's thesis, University of Hawaii, Honolulu.

Paulson, G. S., and J. W. Beardsley. 1985. Whitefly (Hemiptera: Aleyrodidae) egg pedicel insertion into host plant stomata. *Ann. Entomol. Soc. Am.* 78:406–8.

———. 1986. Development, oviposition, and longevity of *Aleurothrixus floccosus* (Maskell) (Homoptera: Aleyrodidae). *Proc. Hawaii. Entomol. Soc.* 26:97–99.

27 / European Asparagus Aphid

K. M. DAANE, K. S. HAGEN, D. GONZÁLEZ, AND L. E. CALTAGIRONE

European asparagus aphid
Brachycorynella asparagi (Mordvilko)
Homoptera: Aphididae

INTRODUCTION

Biology and Pest Status

The European asparagus aphid, *Brachycorynella asparagi* (Mordvilko), is a major pest of garden asparagus. It is a small ($^1/_{16}$ inch) blue-green to gray aphid, covered with a powdery coating. It can be readily identified by the very short cornicles protruding from the posterior of the abdomen, which, along with parallel-sided cauda, form a tail-like appendage (Castle et al. 1987). The aphid's complex life cycle has only recently been understood (Tamaki, Gefre, and Halfhill 1983; Wright and Cone 1988a). In temperate climates, it overwinters in the egg stage on old fern (asparagus plants after the harvestable shoot stage) or in the soil. Eggs hatch in the spring, and the nymphs undergo four instars, the last of which becomes the "stem" mother. During the summer, these wingless females produce several asexual generations. The winged asexual aphids can be induced by environmental conditions or by crowding (Tamaki, Gefre, and Halfhill 1983). In the fall, sexual females and winged males are produced and mate; the females lay small, black eggs on ferns or in the soil. In subtropical regions, the overwintering cycle is unknown; instead, the aphid undergoes continuous anholocyclic development.

Infested plants have a tufted appearance marked by a blue-gray-green color. In large numbers, the European asparagus aphid stunts local growth, injects a toxin that seems to interfere with the storage of nutrients in the roots, and may cause plants to be more susceptible to fusarium wilt (Capinera 1974; Castle et al. 1987). After serious infestations, the following season's crop can have smaller and less marketable spears.

Currently, growers seek to control the aphids by applying insecticides and through sanitation practices.

In California malathion and Disyston are registered for aphid control; both are applied after the harvest, when the plant is in the fern stage. Sanitation, which consists of cutting and burning ferns in November and December after fern dieback, significantly reduces overwintering eggs (Halfhill, Gefre, and Tamaki. 1984). However, biological control offers a more permanent and less costly remedy.

Historical Notes

Garden asparagus, *Asparagus officinalis* L., is an important perennial crop in California, ranking 10th among vegetables produced. It was relatively free of major pests until 1984, when the European asparagus aphid, *B. asparagi*, was observed in California's Coachella Valley (Castle et al. 1987). The aphid was first found in the United States on Long Island, New York, in 1969 (Angalet and Stevens 1977). Over the next 10 years it spread west: in 1975 it was reported in British Columbia, in 1979 in Washington State (Tamaki, Gefre, and Halfhill 1983; Wright and Cone 1988a). It is now considered the major insect pest of asparagus in the western United States. Researchers began importing the aphid's natural enemies in 1986.

RESULTS

Objective A: The Identification and Introduction of Beneficial Organisms

In May and June 1985, L. E. Caltagirone searched for natural enemies of *B. asparagi* in the asparagus-growing regions of Spain and southern France, but found no trace of the aphid. In July and August 1986, K. S. Hagen searched in France, Switzerland, Italy, and Greece, and found some aphids in northern Greece and Italy—main-

We thank P. Starý and L. Polgar for shipping natural enemies from Czechoslovakia and Hungary; B. Rousch of Fresno State University; G. S. Sibbett, Tulare County Farm Advisor, for locating grower collaborators; and the McNabb, Knoll, Maitre, Wysinger, and Lovett asparagus farms.

ly old mummies from which parasites had emerged. In Switzerland, K. S. Hagen found aphids parasitized by *Diaeretiella rapae* (M'Intosh); these were sent to the University of California Biological Control Quarantine Facility at the Kearney Agricultural Center. A colony was established, and 115 of the parasites were released in Contra Costa County.

Since the European asparagus aphid is widespread in Europe yet seldom reaches pest status, researchers believe that improved biological control is possible. In 1989 scientists in Czechoslovakia (P. Starý) and Hungary (L. Polgar) collected natural enemies and sent them to the Albany Quarantine Facility at the Division of Biological Control. Researchers reared five species of natural enemies from this material. Three of these natural enemies were from Czechoslovakia:*Trioxys brevicornis* Haliday, *Aphidius colemani* Viereck, and *Diaeretiella rapae* M'Intosh (all three Hymenoptera: Aphidiidae). The remaining two were from Hungary: *Hippodamia variegata* Guerin-Meneville (Coleoptera: Coccinellidae), and *Chrysoperla* (=*Chrysopa*) *carnea* (Stephens) (Neuroptera: Chrysopidae).

The most common parasite reared was *T. brevicornis,* which had been received from the laboratory of P. Starý of the Czechoslovakian Academy of Sciences. This parasite had never before been released in California.

W-84 researchers established quarantine colonies and obtained release permits. They chose release sites—based on sufficient populations of the aphid and no pesticide treatments—in Contra Costa, Monterey, San Joaquin, Fresno, and Tulare counties. *T. brevicornis* releases began in September and ended when the European asparagus aphid density dropped with the onset of plant dormancy in December. In total, 14,090 parasites were released, mostly in Fresno and Tulare counties.

The researchers then conducted field and laboratory studies to determine whether *T. brevicornis* overwintered successfully and to develop possible conservation measures. Because *T. brevicornis* populations are reduced by sanitation practices, which help remove the aphid's overwintering eggs, growers in four release sites in Fresno and Tulare counties left some asparagus plants standing or, after cutting them, placed old ferns on the outskirts of the field. These plants were sampled during the winter months for aphid mummies; those found were taken to the laboratory for parasite emergence.

From this material we collected four primary parasites—*T. brevicornis, D. rapae, Aphelinus nigritus* Howard, and *Aphelinus asychis* Walker—and four secondary parasites—*Syrphophagus aphidivorus* (Mayr), *Pachyneuron siphonophorae* (Ashmead), *Asaphes lucens* (Provancher), and *Dendrocerus* sp. nr. *laticepi* (Hedicke). In spring we sampled new spear and fern growth and again collected *T. brevicornis.* As there had been no releases since

December, we concluded that *T. brevicornis* was able to overwinter (Daane et al. 1992).

We also conducted two laboratory tests. In the first, four potted asparagus plants were inoculated with the aphid in November and the cages were moved outside after exposure to *T. brevicornis.* Our results showed two parasite generations between December and March, each with a development period of 9 to 10 weeks. A third generation in April developed in only 3 to 4 weeks. In the second test, laboratory-produced *T. brevicornis* mummies were isolated and held at 41±3°F, 53±3°F, 63±3°F, and ambient winter temperatures. We found that *T. brevicornis* could not survive sustained periods at low temperatures; however, development continued above 50°F (Daane et al. 1992).

Objective D: Impact

In some asparagus fields in the eastern United States, natural enemies have reportedly provided sufficient control (Angalet and Stevens 1977); however, in the western states they have only provided inadequate or sporadic control (Wright and Cone 1988a). In California several natural enemies are present. Surveys conducted between 1986 and 1990 found the predators *Coccinella novemnotata franciscana* (Herbst), *Olla abdominalis* (Say), *Scymnus* (*Pullus*) *loewii* Mulsant, *Scymnus* species, *Hippodamia convergens* Guerin-Meneville, *C. carnea,* and syrphid flies. Parasites found included *D. rapae, Aphelinus semiflavus* Howard, *Aphelinus* species (probably a dark form of *semiflavus*), *A. asychis, Lysiphlebus testaceipes* (Cresson), and *Aphidius colemani.* Hyperparasites included *A. aphidivorus, Alloxysta* species, and *P. siphonophorae.*

D. rapae, imported from Switzerland, was morphologically identical to resident strains. This made it impossible to determine in the field whether the Swiss material was a different biotype.

Researchers sampled *T. brevicornis* release sites in Contra Costa, Monterey, and San Joaquin counties for parasitized aphids before and after release. At release sites in Fresno and Tulare counties, they took samples every other week before the parasites were released and weekly thereafter. Because the European asparagus aphid populations have a varying distribution within fields (Wright and Cone 1986) and on each plant (Wright and Cone 1983, 1988b), researchers used sequential sampling. Plants were chosen randomly, and one primary branch from the lower third was selected from each. Infested branches were taken to the laboratory, where mummies were isolated in gelatin capsules. The number of aphids, parasitized aphids, and parasite species was recorded.

No recoveries were made in Contra Costa, Monterey, or San Joaquin counties. However, in Fresno and Tulare counties the parasite was recovered at each release site. The establishment and performance of *T. brevicornis* was especially good in comparison to the well-established *D. rapae*. However, the most common parasite was *Aphelinus* (probably *semiflavus*). Release and recovery efforts were hampered by the late-season arrival of *T. brevicornis* from Czechoslovakia, which came just as aphid densities were beginning to decline. Another factor that could have affected recovery was the high number of secondary parasites common at the end of the season.

RECOMMENDATIONS

1. Develop a sampling program to be used by growers and Farm Advisors. The program should include economic threshold levels.

2. Continue *T. brevicornis* releases in California and, after evaluating their success, disseminate the parasite to other asparagus-producing regions in the western United States.

3. Develop conservation measures to improve the overwintering of *T. brevicornis*.

REFERENCES CITED

Angalet, G. W., and N. A. Stevens. 1977. The natural enemies of *Brachycolus asparagi* in New Jersey and Delaware. *Environ. Entomol.* 6:97–100.

Capinera, J. L. 1974. Damage to asparagus seedlings by *Brachycolus asparagi. J. Econ. Entomol.* 67:447–48.

Castle, S. J., T. M. Perring, C. A. Farrer, and A. N. Kishaba. 1987. Asparagus aphid is spreading fast. *Calif. Agric.* 41(9–10):13–14.

Daane, K. M., G. Y. Yokota, R. F. Gill, L. E. Caltagirone, K. S. Hagen, D. González, P. Starý, and W. E. Cheney. 1992. Imported parasite may help control European asparagus aphid. *Calif. Agric.* 46(6):12–14.

Halfhill, J. E., J. A. Gefre, and G. Tamaki. 1984. Cultural practices inhibiting overwintering survival of *Brachycolus asparagi* Mordvilko (Homoptera: Aphididae). *J. Econ. Entomol.* 77:954–56.

Tamaki, G., J. A. Gefre, and J. E. Halfhill. 1983. Biology of morphs of *Brachycolus asparagi* Mordvilko (Homoptera: Aphididae). *Environ. Entomol.* 12:1120–24.

Wright, L. C., and W. W. Cone. 1983. Extraction from foliage and within-plant distribution for sampling of *Brachycolus asparagi* (Homoptera: Aphididae) on asparagus. *J. Econ. Entomol.* 76:801–5.

———. 1986. Sampling plan for *Brachycorynella asparagi* (Homoptera: Aphididae) in mature asparagus fields. *Environ. Entomol.* 79:817–21.

———. 1988a. Population dynamics for the asparagus aphid, *Brachycorynella asparagi* (Homoptera: Aphididae), on different ages of asparagus foliage. *Environ. Entomol.* 17:699–703.

———. 1988b. Population dynamics of *Brachycorynella asparagi* (Homoptera: Aphididae) on undisturbed asparagus in Washington State. *Environ. Entomol.* 17:878–86.

28 / FILBERT APHID

M. T. ALINIAZEE AND R. H. MESSING

filbert aphid
Myzocallis coryli (Goetze)
Homoptera: Aphididae

INTRODUCTION

Biology and Pest Status

The filbert aphid, *Myzocallis coryli* (Goetze), is an important pest of cultivated hazelnuts—also known as filberts—in the northwestern United States. This monoecious and holocyclic aphid overwinters as a diapausing egg. Oviparous adults lay their black, shiny eggs in rows on twigs and branches during late fall. The eggs hatch in early spring—late February and early March—and the young nymphs move onto the swelling buds. They begin feeding as soon as the leaves start to unfold. Due to the cool temperatures at this time of year, the nymphs develop slowly: no adults appear until late April to early May. The aphids complete six to eight parthenogenetic generations throughout the summer. Males appear in late fall, and fertilized eggs are laid soon after. Filbert aphids cause damage throughout the season, although their population usually peaks in June and early July (AliNiazee 1983b).

Historical Notes

The filbert aphid is of Palearctic origin, and was probably introduced into this country during the late 1800s on infested plant material. Now widely distributed throughout the filbert-growing areas of the Pacific Northwest, it causes serious losses (AliNiazee 1980). Most growers apply one to three insecticide sprays per year to control this pest; as a result, in many commercial orchards populations of 200 to 300 aphids per leaf are not uncommon. Like many other aphids, the filbert aphid is an induced pest problem in most orchards (AliNiazee 1983a,b): it appears that the more pesticide used, the greater the aphid density. In unsprayed orchards, aphids are kept in check by their natural enemies and are rarely a problem.

Resistance to insecticides is common for this aphid. Since 1970 it has developed tolerance to almost all insecticides used on filberts. Resistance has developed rapidly (in 1 to 3 years) in each case, further reducing the effective life of the insecticide. These problems make biological control a highly desirable alternative.

RESULTS

Objective A: The Identification and Introduction of Beneficial Organisms

From the results of natural enemy surveys, researchers concluded that the filbert aphid was a suitable candidate for a classical biological control program based on the introduction of a host-specific parasitoid. They identified the following reasons: (1) *Myzocallis coryli* is an introduced pest and its distribution is well known; (2) in its native area, the aphid is rarely a pest and is generally controlled by a number of natural enemies; (3) previous studies (Starý 1978; Viggiani 1983) reported the effective parasitization of *M. coryli* in Europe by three species of *Trioxys*—*T. pallidus* Haliday, *T. curvicaudus* Mackauer, and *T. tenuicaudus* Starý—and (4) other closely related aphid species such as the walnut aphid, *Chromaphis juglandicola* (Kaltenbach), have been successfully controlled by importing and establishing host-specific parasitoids from Europe (van den Bosch et al. 1979).

An early attempt to import and release *T. tenuicaudus* failed, because it did not effectively parasitize the aphid either in the laboratory or in the field. In June 1984, researchers searched filbert orchards in France, Spain, and Italy for parasitoids of *M. coryli*; they found *T. pallidus* mummies in almost all sites (Messing and AliNiazee 1989). The collected mummies were shipped to the University of California Biological Control Insectaries in Albany for quarantine. About 30 percent of the 2,230 mummies shipped yielded *T. pallidus* adults. After one generation in quarantine, the F_1 parasitoids were transferred to Oregon, mass-reared, and released in many orchards. Researchers released the first parasitoids on June 25, 1984, and within a 2-week period, about one hundred mummies were recovered in the

Table 28.1. Field releases and the rate of establishment of *Trioxys pallidus* in Oregon filbert orchards from 1984 to 1990.

Orchards	Location	Year first released	Number released[*]	Establishment[†]
Abraham	Albany	1984	15A	No
Adelman	Brooks	1985	300A	Yes (1989)
Bestwick	Newberg	1987	245A	Yes (1989)
			160M	
Blake	Salem	1988	100A	Yes (1989)
Buchanan-A	Bellfountain	1984	150A	Yes (1989)
Buchanan-B	Bellfountain	1985	6,300A	Yes (1989)
Bush	Fern Ridge Dam	1984	1,850A	Yes (1988)
Calef	Coburg	1985	300A	Yes (1989)
Cohn	Bellfountain	1987	80A	Yes (1988)
Downing	Gaston	1988	150A	Yes (1990)
Duncan	Sherwood	1987	70A	Yes (1989)
			100M	
Gingerich	Canby	1987	140A	Yes (1990)
			75M	
Gratt	Woodburn	1986	60A	Yes (1990)
			200M	
Gray	Albany	1985	400A	Yes (1989)
Huffman	Newberg	1987	435A	Yes (1989)
			260M	
Knittel	Kiger Island	1985	4,000A	Yes (1989)
Lemert	Junction City	1984	2,350A	Yes (1989)
MacDonald	Wilsonville	1987	185A	Yes (1989)
			360M	
Malone	Amity	1984	400A	Yes (1990)
Mitchell	Newberg	1987	235A	Yes (1990)
			210M	
Newton	Dever Conner	1987	185A	Yes (1989)
			25M	
Nofziger	Dever Conner	1987	250A	Yes (1989)
			100M	
Pierce	Dayton	1987	170A	Yes (1990)
			350M	
Pierce	Newberg	1987	355A	Yes (1989)
			160M	
Pierce	Newberg	1987	95A	Yes (1990)
			100M	
Schrapel	Yamhill	1986	6A	Yes (1989)
			700M	
Simonson	Bellfountain	1985	300A	Yes (1989)
Smith	Monroe	1987	200A	Yes (1989)
Twedt	Corvallis	1984	6,800A	Yes (1989)
Wennerstrom	McMinnville	1986	600M	Yes (1990)

[*]A = adults; M = mummies.
[†]Including the year establishment was noticed.

field. Mass-rearing and releasing were continued throughout 1984 to 1988. During these years researchers made releases at 32 different sites. The parasite became established at nearly half of these sites within a year and at most of the remaining sites within 2 to 3 years (see table 28.1). So far, nearly 33,000 parasitoid wasps and mummies have been released in the filbert orchards of Oregon's Willamette Valley.

Objective B: Ecological and Physiological Studies

The biosystematics of the parasitoid. Researchers found the biotype of *T. pallidus* imported for filbert aphid control to be morphologically indistinguishable from the biotype previously imported to control the walnut aphid in California. In addition, reciprocal hybridization tests between the two biotypes showed no evidence that the biotypes were reproductively incompatible in either the parental or filial generations (Messing and AliNiazee 1988). Thus, they concluded that the two populations were conspecific. Interestingly, the biotype imported to control the walnut aphid, although well established in Oregon for over 10 years, was never observed to parasitize the filbert aphid, even in orchards where walnut and filbert trees were interplanted. Laboratory and field-cage tests confirmed that the "walnut aphid biotype" had greater reproductive success on walnut than on filbert aphids, and vice-versa. Thus, although conspecific, the two populations of parasites have differentiated along host lines through some combination of selection or conditioning.

Objective D: Impact

R. H. Messing and M. T. AliNiazee (1985) reported 55 species of aphidophagous predators in Willamette Valley's filbert orchards. The most important predaceous species among these were *Adalia bipunctata* (L.) and *Cyloneda polita* Casey (Coleoptera: Coccinellidae); *Deraeocoris brevis* (Uhler) (Hemiptera: Miridae); and species of *Hemerobius* and *Chrysopa* (Neuroptera: Hemerobiidae and Chrysopidae). Also found attacking filbert aphids were a parasitoid, *Mesidiopsis* species (Hymenoptera: Aphelinidae), and a pathogenic fungus, *Triplosporium fresenii* (Nowakowski) (Entomophthorales: Entomophthoraceae). However, the fungal disease was only seen during very high densities of the aphid, and the *Mesidiopsis* species was never found in numbers large enough to provide effective biological control.

Among the major predators, *A. bipunctata* was active in the orchards from April through October. Its populations peaked in early July with a second, smaller peak in October, indicating two generations per year. *Cyloneda polita* was found in large numbers at midseason. The mirid, *D. brevis*, was active from April through November. It overwinters as an adult and is active early enough in the spring to somewhat suppress aphids before they reach outbreak densities. In one survey (Messing and AliNiazee 1985), *D. brevis* accounted for 20 to 60 percent of all predators collected. Also a seemingly major factor in reducing the numbers of aphids were four species of *Chrysopa* and *Hemerobius*. In another study, R. H. Messing and M. T. AliNiazee (1986) used chemical and mechanical exclusion methods to show that the predator complex found in the filbert orchards was an important factor in regulating the aphid population. Under laboratory conditions the coccinellid *A. bipunctata* and the mirids *D. brevis*, *Heterotoma meriopterum* (Scopoli), and *Compsidolon salicellum* (Herrich-Schaeffer) fed voraciously, consuming 50 to 65 aphids per day. Although the coccinellids consumed more aphids than did the mirids, the mirids could survive much longer in the absence of aphids.

M. T. AliNiazee (1983b) and R. H. Messing and M. T. AliNiazee (1986) concluded that biological control based on predators alone, though effective, usually occurred too late in the season to prevent aphid populations from growing large enough to cause economic harm. Because using insecticides early in the season generally disrupted this predator-based biological control, an integrated control approach was needed. Researchers expected that a host-specific parasitoid would be more finely attuned to aphid phenology and would show better synchrony early in the season.

Early studies (Messing and AliNiazee 1989) on the efficacy of the imported *T. pallidus* showed it to be very effective in controlling filbert aphids. One season after the releases, aphid populations dropped by 11 to 33 percent. The parasitization rate was found to be between 26 and 28 percent. Further studies during 1987 and 1988 showed that the parasitoid was effective beyond expectation: the number of aphids had continuously declined, and mean percent parasitization had increased to levels of 25 to 50 percent (see table 28.2). *Trioxys pallidus* appears to survive commercial filbert management practices, including pesticide usage, and responds rapidly to increases in aphid density. It has also spread to many orchards where no releases were made. About 40 to 60 percent of the entire filbert acreage in the Willamette Valley is now influenced by this parasitoid. At the present rate of dispersal, within 5 to 10 years almost all of Oregon's filbert orchards could contain the parasitoid and derive benefits from it. The aphid control costs of nearly $600,000 per year have now been reduced by about on-half, and could be nearly eliminated in another 10 years. The initial cost of this biological control project—only $8,900—may return more than $1 million in benefits in 5 to 10 years, in addition to substantially reducing the environmental pollution caused by applying pesticides.

Table 28.2. The impact of *T. pallidus* in five selected filbert orchards in the Willamette Valley, Oregon, in 1989.

Orchards	Mean aphids/count		Mean % parasitism	
	Release site	Nonrelease site	Release site	Nonrelease site
SPRING 1989				
Blake	9.14	2.34	23.00	3.96
Buchanan	15.44	1.96	9.72	3.68
Bush	2.32	5.78	14.15	5.77
Calef	2.74	3.88	15.64	9.29
Twedt	11.00	8.10	21.99	5.72
FALL 1989				
Blake	0.54	1.42	25.00	2.29
Buchanan	0.20	0.72	56.67	15.74
Bush	0.58	2.00	6.67	2.25
Calef	5.16	2.16	6.04	4.26
Twedt	1.46	1.92	27.57	4.68

RECOMMENDATIONS

Data suggest that *T. pallidus* is an effective parasitoid of the filbert aphid; it is well synchronized with its host and is capable of maintaining pest densities below economically harmful levels. Further studies on the parasitoid's insecticide resistance and orchard dispersal would be beneficial.

REFERENCES CITED

AliNiazee, M. T. 1980. *Filbert insect and mite pests.* Oregon State University Agricultural Experiment Station Bulletin 643.

———. 1983a. Carbaryl resistance in the filbert aphid (Homoptera: Aphididae). *J. Econ. Entomol.* 76:1002–4.

———. 1983b. Pest status of filbert (hazelnut) insects: A ten-year study. *Can. Entomol.* 115:1155–62.

Messing, R. H., and M. T. AliNiazee. 1985. Natural enemies of *Myzocallis coryli* (Homoptera: Aphididae) in hazelnut orchards of western Oregon. *J. Entomol. Soc. Brit. Columbia* 82:14–18.

———. 1986. Impact of predaceous insects on filbert aphid, *Myzocallis coryli* (Homoptera: Aphididae). *Environ. Entomol.* 15:1037–41.

———. 1988. Hybridization and host-suitability for two biotypes of *Trioxys pallidus* (Haliday). *Ann. Entomol. Soc. Am.* 81:6–9.

———. 1989. Introduction and establishment of *Trioxys pallidus* (Hym.: Aphidiidae) in Oregon, U.S.A., for control of filbert aphid, *Myzocallis coryli* (Homoptera: Aphididae). *Entomophaga* 34:153–63.

Starý, P. 1978. Parasitoid spectrum of the arboricolous callaphidid aphids in Europe (Hymenoptera: Aphidiidae; Homoptera: Aphidoidea, Callaphididae). *Acta Entomol. Bohem.* 75:164–77.

van den Bosch, R., R. Hom, P. Matteson, B. D. Frazer, P. S. Messenger, and C. S. Davis. 1979. Biological control of the walnut aphid in California: Impact of the parasite *Trioxys pallidus*. *Hilgardia* 47:1–13.

Viggiani, G. 1983. Natural enemies of filbert aphids in Italy. In *Aphid antagonists,* ed. R. Cavalloro, 109–13. Rotterdam, Holland: Balkema.

29 / GREEN PEACH APHID

K. D. BIEVER

green peach aphid
Myzus persicae (Sulzer)
Homoptera: Aphididae

INTRODUCTION

Biology and Pest Status

The green peach aphid has a complex life cycle, with 10 to 25 generations a year. Most generations consist only of females that produce live nymphs without mating. Sexual reproduction does not occur in areas with mild winters; in cold-winter areas, a generation of sexual aphids occurs in the fall and eggs are laid on a winter host, usually peach. The overwintering eggs hatch between late January and March. Usually three aphid generations live on the peach trees or other plant hosts in the orchards; then, spring migrants gradually move to such secondary hosts as weeds, crop plants, and ornamentals. A single tree can produce up to 40,000 spring migrants. Although the winged aphids that colonize potato plants may come directly from peach trees or other winter hosts, they generally come from secondary hosts. In the fall, winged migrants appear and return to the winter hosts.

The green peach aphid occurs throughout the world on several hundred species of host plants. Its feeding causes many crops serious damage; it also transmits more than 100 viruses to cultivated crops (Bishop et al. 1982). It is a vector of potato leafroll virus, which causes leafroll and tuber net necrosis, resulting in smaller yields and lower-quality potatoes. The aphid is also an important vector of beet western yellows virus and beet yellows virus, both of which reduce sugarbeet yields.

Historical Notes

Growers control the green peach aphid primarily by applying systemic insecticides at planting or emergence and following up later with one to several foliar sprays. Replacing the orchard floor flora with nonhost grass can greatly limit the number of aphids migrating from orchards to crops (Tamaki 1975). Researchers generally believe that the low levels of naturally occurring parasitoids seldom reduce the number of aphids in commercial crops. A study in 10 different countries determined that parasitism seldom exceeded 1 percent (Mackauer and Way 1976). So far researchers have made no specific effort to introduce exotic species to control the aphid; neither has using biological control agents on noncrop or alternative crop plants been exploited.

In the past 20 years growers have been able to suppress the aphid with insecticides. This has probably created a false sense of security and contributed to the failure to develop biologically based management approaches, including the use of biological control agents.

RESULTS

Objective B: Ecological and Physiological Studies

In the peach ecosystem, predators are more effective than parasitoids against aphids (Tamaki, Landis, and Weeks 1967; Tamaki and Weeks 1968). Several studies have shown that native predators can reduce spring migrants produced on the weeds of the orchard floor by as much as 95 percent (Tamaki 1972, 1973; Tamaki, Annis, and Weiss 1981). Syrphids and geocorids are particularly effective. The response of several beneficial species to the aphid varied with the plant host (Tamaki, Annis, and Weiss 1981). Researchers also found that tree bands on peach trees can provide overwintering sites for aphid predators (Tamaki and Halfhill 1968).

In a number of studies, researchers have established threshold levels for the green peach aphid on potatoes. The models they developed, from surveys of commercial fields in Idaho, predicted that net necrosis caused by the potato leafroll virus would cause economic losses if there were more than 10 aphids per 50 leaves in 2 consecutive weeks prior to August (Byrne and Bishop 1979a). In commercial potato production in California, spraying is recommended when 5 percent of the leaves are infested (Bacon, Burton, and Wyman 1978).

Researchers have evaluated trapping procedures to monitor the activity of aphids, particularly the green peach aphid (Byrne and Bishop 1979b). Studies have also identified a number of nonpotato and nonsugarbeet hosts for the green peach aphid and related aphids (Bishop and Guthrie 1964; Tamaki 1975; Annis, Tamaki, and Berry 1981; Tamaki and Fox 1982). Recent work has also shown that two weed crucifers—tumble mustard (*Sisymbrium altissimum* L.) and shepherd's purse (*Capsella bursa-pastoris*)—can serve as overwintering reservoir hosts of potato leafroll virus, which can then be acquired and transmitted by the green peach aphid (Biever unpub. data).

RECOMMENDATIONS

1. Develop a methodology to use exotic and native natural enemies as classical and augmentative biological control agents.

2. Introduce and evaluate exotic species of parasitoids and fungi.

3. Develop procedures and evaluate the natural enemies in the peach- and nectarine-weed ecosystem.

4. Acquire additional information on the population ecology of the green peach aphid in the peach- and nectarine-weed ecosystem and on the dynamics involved in the aphid's movement into crop systems.

REFERENCES CITED

Annis, B., G. Tamaki, and R. E. Berry. 1981. Seasonal occurrence of wild secondary hosts of the green peach aphid, *Myzus persicae* (Sulzer), in agricultural systems in the Yakima Valley. *Environ. Entomol.* 10:307–12.

Bacon, O. G., V. E. Burton, and J. A. Wyman. 1978. Management of insect pests on potatoes. *Calif. Agric.* 32(2):26–27.

Bishop, G. W., and J. W. Guthrie. 1964. Home gardens as a source of the green peach aphid and virus diseases in Idaho. *Am. Potato Journal* 41:28–34.

Bishop, G. W., H. W. Homan, L. E. Sandvol, and R. L. Stoltz. 1982. *Management of potato insects in the western states.* Western Regional Extension Publication 64.

Byrne, D. N., and G. W. Bishop. 1979a. Relationships of green peach aphid numbers to spread of potato leafroll virus in southern Idaho. *J. Econ. Entomol.* 72:809–11.

———. 1979b. Comparison of water trap pans and leaf counts as sampling techniques for green peach aphids on potatoes. *Am. Potato Journal* 56:237–41.

Mackauer, M., and M. J. Way. 1976. *Myzus persicae* (Sulzer), an aphid of world importance. In *Studies in biological control. The international biological programme,* ed. V. L. Delucchi, chap. 9, 51–119. Cambridge: Cambridge University Press.

Tamaki, G. 1972. The biology of *Geocoris bullatus* inhabiting orchard floors and its impact on *Myzus persicae* on peaches. *Environ. Entomol.* 1:559–65.

———. 1973. Spring populations of the green peach aphid on peach trees and the role of natural enemies in their control. *Environ. Entomol.* 2:186–91.

———. 1975. Weeds in orchards as important alternative sources of green peach aphids in late spring. *Environ. Entomol.* 4:958–60.

Tamaki, G., and L. Fox. 1982. Weed species hosting viruliferous green peach aphids, vector of beet western yellows virus. *Environ. Entomol.* 11:115–17.

Tamaki, G., and J. E. Halfhill. 1968. Bands on peach trees as shelters for predators of the green peach aphid. *J. Econ. Entomol.* 61:707–11.

Tamaki, G., and R. E. Weeks. 1968. Use of chemical defoliants on peach trees in an integrated program to suppress populations of green peach aphids. *J. Econ. Entomol.* 61:431–35.

Tamaki, G., B. Annis, and M. Weiss. 1981. Response of natural enemies to the green peach aphid in different plant cultures. *Environ. Entomol.* 10:375–78.

Tamaki, G., B. J. Landis, and R. E. Weeks. 1967. Autumn populations of green peach aphid on peach trees and the role of syrphid flies in their control. *J. Econ. Entomol.* 60:433–36.

30 / PEA APHID AND BLUE ALFALFA APHID

D. GONZÁLEZ, K. S. HAGEN, P. STARÝ, G. W. BISHOP, D. W. DAVIS, AND K. S. PIKE

pea aphid *Acyrthosiphon pisum* (Harris) Homoptera: Aphididae	blue alfalfa aphid *Acyrthosiphon kondoi* Shinji Homoptera: Aphididae

INTRODUCTION

Biology and Pest Status

Although the blue alfalfa aphid, *Acyrthosiphon kondoi* Shinji, closely resembles the pea aphid, *Acyrthosiphon pisum* (Harris), subtle morphological differences exist (Kono 1975). In California both species have several generations per year, but the blue alfalfa aphid is more abundant in the spring, the pea aphid in the fall (V. Marble person. commun.). Each species has four nymphal instars; however, the blue alfalfa aphid produces far more winged forms than does the pea aphid.

Pea aphid mated females lay eggs that overwinter and produce parthenogenetic viviparous females in the spring. This morph is present during most of the growing season (Hutchison and Hogg 1985). At northern latitudes, males and females are produced in late summer and fall. An extensive bibliography on the pea aphid is available in A. Harper et al. (1978).

When it first became established in California, the blue alfalfa aphid was reported to competitively displace the pea aphid where they occurred together (Stern, Sharma, and Summers 1980; Summers and Coviello 1984a). The blue alfalfa aphid was identified as an early-season pest, affecting mainly the first and second alfalfa cuttings (Summers and Coviello 1984a). V. M. Stern, R. Sharma, and C. Summers (1980) reported that the blue alfalfa aphid cannot tolerate temperatures above 27°C. These field reports coincided with laboratory studies, in which a temperature of 25°C had the strongest effect on the aphid's generation time and intrinsic rate of increase (Kodet, Nielson, and Keuhl 1982). Both photoperiod and a temperature-photoperiod interaction had a relatively minor effect on the aphid's development at 20° to 25°C. However, photoperiod did influence the aphid's rate of development at 10° to 15°C. In a separate study, C. G. Summers and R. L. Coviello (1984b) determined that the maximum intrinsic rates of increase (r_m), and the minimum lower developmental threshold temperatures (t) and mean generation times (T), occurred at between 10.0° and 12.8°C.

V. M. Stern, R. Sharma, and C. Summers (1980) reported that, in the absence of natural enemies, densities of 10 to 12 blue alfalfa aphids per stem on new growth from January to March would significantly reduce alfalfa yields. They concluded that in such circumstances growers had no recourse but to treat infested fields. However, C. G. Summers and R. L. Coviello (1984a) pointed out that the impact of the blue alfalfa aphid on alfalfa growth and development is the result of a complex interaction among aphid numbers, the timing and duration of infestation, and the plant age and compensation factors. They concluded that chemical applications were not essential until aphid levels reached 50 to 60 per stem. When densities exceeded 80 to 90 per stem for more than a few days, the aphid caused substantial losses.

With the pea aphid, researchers have reported a similarly wide range of damage thresholds in relation to environmental factors. G. W. Cuperus et al. (1982) reported a damage threshold of 1.2 aphids per stem in nonirrigated alfalfa. G. A. Hobbs et al. (1961), on the other hand, reported that in irrigated alfalfa, yield was not significantly reduced by a population of 1,400 to 1,800 per sweep. Obviously, damage thresholds change with conditions and need to be interpreted with great caution. As F. L. Poston et al. (1983) pointed out, damage thresholds depend on such variables as the stage of plant growth during the initial infestation, the duration of infestation, the cultivar, the impact of natural enemies, interaction with other pests, and control and application costs.

Historical Notes

The pea aphid is believed to have been introduced to North America in the late 1800s (Davis 1915; Campbell

We appreciate the assistance of W. White, L. Etzel, and J. Quesada of the University of California in the rearing, release, and establishment phases of this study.

and Davidson 1924; Campbell 1926). By the early 1950s it was reported throughout the continental United States and was a serious pest in the Midwest and far west (Dudley and Cook 1952). In 1958 the pea aphid was first found in southern Idaho; since the late 1970s, it has become more widespread and increasingly abundant throughout the Boise and Magic valleys (Hahn and Bishop 1986). In Washington state, the pea aphid is present every year, but it only infrequently reaches high densities.

The blue alfalfa aphid was first collected in Manchukuo in 1937 and was described in 1938 by Shinji and Kondo (Dickson 1975). In the United States it was first collected in California in 1974 (Stern, Sharma, and Summers 1980). In 1975 it caused damage in the Imperial and San Joaquin valleys and then spread east and north. Although it was discovered in 1978 in Washington state, it has never been a serious pest there. In Idaho, the earliest infestations (in 1980) were moderate to heavy, but they occurred late in the season and caused no serious damage. In those same fields, the pea aphid predominated over the blue alfalfa aphid. In 1981, the blue alfalfa aphid was not found during spring or summer surveys; an extended warm period the previous fall is believed to have contributed to the low levels. In Utah, the blue alfalfa aphid was found in 1980, and by 1990 it was widely distributed throughout the southern half of the state.

The principal parasite of the two aphids is *Aphidius ervi* Haliday. *A. ervi* is part of a species complex, literature on which includes (reportedly Nearctic) *A. ervi pulcher* Baker, (reportedly Palearctic) *A. ervi ervi,* and (Nearctic) *A. pisivorus* Smith (Mackauer and Finlayson 1967). M. Mackauer and T. Finlayson synonymized *A. pisivorus* with *A. pulcher,* but this proved controversial. P. Starý (1974) believes that the true Nearctic *Aphidius* parasite of the pea aphid in North America was probably *A. pisivorus,* a valid species that was present before the 1959 introduction of *A. ervi* from France. P. Starý also believes that *A. pulcher* is a separate species, distinct from *A. pisivorus,* and is not normally a parasite of the pea aphid. We agree, and indicate this by adding *A. pisivorus* (in parentheses) after literature citations to Nearctic *Aphidius* "ervi" or to an "ervi pulcher." This designation was reaffirmed when T. Unruh et al. (1989) reviewed allozyme patterns for *A. ervi* and *A. pisivorus,* they concluded that the species are closely related but distinct. P. Marsh (1977) concluded that (1) *A. pisivorus* is a distinct species, not a synonym of *A. pulcher,* and it apparently attacks only the pea aphid; (2) *A. ervi* was probably present in North America before being introduced from Europe, and is not restricted to the pea aphid; and (3) *Aphidius nigripes* Ashmead is a senior synonym of *A. pulcher,* and it is not a parasite of the pea aphid.

G. W. Angalet and R. Fuester (1977) pointed out that in the early 1900s several authors reported an *Aphidius* parasite attacking the pea aphid in different areas of the United States. These two researchers concluded that it was *A. ervi* (=*A. pisivorus*), that it must have been widely distributed before 1900, and that it may have become established at the same time as (or soon after) the arrival of the pea aphid. Although *A. ervi pulcher* (*A. pisivorus*) was widely distributed in North America before the 1958 releases of exotic *Aphidius* species, most published reports on the pea aphid's natural enemies indicated that the parasite did not effectively control the aphid (Campbell 1926; Fluke 1929; Mackauer 1971).

Aphidius ervi (=*A. ervi ervi,* according to Mackauer and Finlayson [1967]) was first introduced into the United States in 1959, when about 1,000 individuals were sent to the USDA-ARS Beneficial Insects Introduction Laboratory in New Jersey from the USDA-ARS European Parasite Laboratory in France. The parasites had been collected from alfalfa near Paris (Angalet and Fuester 1977). From the laboratory colony, researchers made releases in New Jersey and Delaware in 1959. They made additional releases in Arizona and Maine between 1961 and 1963, and in New Jersey and Delaware during 1967 to 1968 (Halfhill, Featherston, and Dickie 1972). Between 1961 and 1963, researchers released about 11,000 *A. ervi* (*ervi*) in Washington, Idaho, and Oregon.

Other species introduced into the United States are *Aphidius smithi* Sharma & Subba Rao and the morphologically similar *Aphidius eadyi* Starý, González & Hall (Starý, González, and Hall 1980), the *Aphidius urticae* Haliday group, and *Aphidius staryi* Chen, González, & Luhman. All three species have been introduced in California. *Aphidius smithi* was introduced to the United States in 1958 from India; M. J. P. Mackauer and H. E. Bisdee (1965) summarized the *A. smithi* releases in eastern North America. Between 1958 and 1960, researchers reared and released approximately 220,000 *A. smithi* in California (Hagen and Schlinger 1960). Less than 10 years after the initial 1958 releases, *A. smithi* was established throughout the entire range of the pea aphid in North America, from central Mexico to Ontario, Canada (Angalet and Fuester 1977). In 1966 to 1967, G. W. Angalet and R. Fuester found *A. smithi* to be the dominant parasite of the pea aphid in the Deep South and in the Midwest. Between 1959 and 1964, researchers released about 256,000 *A. smithi,* 250,000 of them near Yakima, Washington, in 1964 (Halfhill, Featherston, and Dickie 1972). Except in Oregon, none of the parasites appeared to control the pea aphid. Between 1965 and 1969, an estimated 500,000,000 *A. smithi* were released near Walla Walla, Washington (Halfhill and Featherston 1973).

G. W. Angalet and R. Fuester (1977) have pointed out that in the United States competitive displacement of *Aphidius* parasites of the pea aphid on alfalfa has occurred twice. First, *A. smithi* displaced the native *A. pulcher* (*A. pisivorus*) in the southern and midwestern states; later, *A. smithi* was itself displaced by introduced *A. ervi* (*ervi*). Additional evidence for a more recent displacement of *A. smithi* by *A. ervi* in the western United States is presented in the next section.

RESULTS

Objective A: The Identification and Introduction of Beneficial Organisms

Exploration and introduction. When the blue alfalfa aphid was accidentally introduced in California in 1974, subsequently spreading and damaging alfalfa in the western states, researchers began a biological control program. Some of the results of that program are reported here.

The blue alfalfa aphid caused widespread damage throughout areas with greatly different climatic and habitat patterns, including fields in low and high deserts; in southern, central, and northern interior valleys; and in coastal valleys. For this reason, one of the researchers' major objectives was to obtain aphid parasites from areas with very diverse climates and habitats.

Parasites were collected from 16 countries on 4 continents (González et al. 1975, 1978; Starý and González 1978; González, Miyazaki, et al. 1979). The collection areas differed from one another in climate (winter and summer); habitat (crop types, crop mixtures, and wild plants); and management practices (large monocultural areas and small farms with diverse plantings, fertilization, irrigation, cultivation, and pest control). These parasites were then released in several areas of the western United States having large differences in climate and habitat.

Biosystematic analyses. A major focus of the work was to obtain biotypes of one or more parasite species, to study the potential of using biotypes in biological control programs. Studies of the biotypes that were introduced into the western United States included analyses of genetic variation and also some life table analyses (see Objective B). The term *biotype* can be defined as "a population with a potentially distinct and unique behavior, or having a capacity to regulate its pest host's density or the host's capacity to damage the crop" (González, Gordh, et al. 1979; Caltagirone 1985; González 1988).

The parasites collected in this study included approximately 57 potential biotypes representing seven species. More than 60 percent of those biotypes were *Aphidius ervi* Haliday. *A. ervi* was the predominant parasite in all countries where parasites were collected, regardless of location, weather, hosts, or cultural practices.

T. Unruh et al. (1989) evaluated populations—that is, potential biotypes—of *Aphidius ervi* from nine geographically isolated areas. They concluded that (1) *A. ervi* in California and *A. ervi* from western Europe and the Mediterranean region, introduced to the United States by D. González et al. (1978), were closely related; (2) *A. ervi* from Japan and Pakistan and *A. pisivorus* from northern California are distinct from each other as well as from the western European populations; and (3) *A. ervi* populations from central Asia probably exist as a complex of sibling species or semispecies, and these areas are promising sources of new races and species of *Aphidius* nr. *ervi*.

T. Unruh et al. (1989) concluded that systematic studies of biological control agents can be enhanced by multilocus electrophoretic studies, which are useful in clarifying relationships among groups above the species level. For example, results from T. Unruh et al. (1985, 1989) support separate species status for *A. ervi, A. smithi, A. pisivorus,* and *A. eadyi.* The allozyme electrophoresis work of T. Unruh et al. (1985) also clearly showed that *Aphidius eadyi* (Starý, González, and Hall 1980) and a population later described as *Aphidius staryi* (Chen, González, and Luhman 1990) were closely related to, but distinct from, *A. smithi.* Because all three species are so similar morphologically, the allozyme work done by T. Unruh et al. (1985) stimulated the taxonomic research of J. H. Chen, D. González, and J. Luhman (1990).

Releases. After 1975 most of the parasites released in alfalfa fields in the western states were *A. ervi.* This reflected that species' predominance in the overseas alfalfa field collections and in U.S. insectary colonies after 1975. Of more than 188,000 parasites reared at Riverside and released in California between 1976 and 1986, more than 80 percent were *A. ervi.* These releases occurred over a period of time and in various localities. In California, *A. ervi* were not released in the same fields at the same time with other species. The greatest numbers of *A. ervi* were released in California's inland and coastal valleys and, to a lesser extent, in the low desert. Besides *A. ervi,* more than 14,000 *A. eadyi* were released in southern coastal and inland valleys, and more than 20,000 *A. smithi* were released in central California valleys. More than 2,000 individuals of a new species, *Aphidius staryi* Chen & Luhman, which closely resembles *A. smithi,* were released in the inland valleys of southern California; and more than 4,000 *Ephedrus plagiator* (Nees) and more than 2,900 *Praon barbatum* Mackauer were released in the coastal and inland valleys of southern and central California.

Between 1979 and 1983 J. Quezada, L. Etzel, and K. Hagen released *A. ervi* from the insectary of the Division of Biological Control at Berkeley, California. Of a total of 931,000, they released more than 850,000 in Fresno, Kings, and Tulare counties. Other releases were made in Alameda, Colusa, Glenn, Kings, Monterey, Modoc, San Joaquin, Siskiyou, and Yolo counties. In 1980, *A. ervi* (Iran) was released near Gooding, Idaho, and *A. ervi* (USSR) was released at Kimberly, Idaho. *A. ervi* (USSR) was also released in southern Idaho in 1981 and 1982.

In Utah, no parasite releases have been made against the blue alfalfa aphid. Parasite releases against the pea aphid were made before 1954.

Establishment. Beginning in 1975 and continuing through 1987, researchers sampled alfalfa fields in California to examine the parasite complex on *Acyrthosiphon* species. In areas sampled from 1975 to 1980, *Aphidius smithi* was the predominant species. However, beginning in 1980 and increasing progressively thereafter, *A. ervi* was found in greater numbers and at more localities, while the numbers of *A. smithi* decreased correspondingly. *A. smithi* was last collected in 1983 in the Santa Maria and Santa Ynez valleys. It seems that *A. ervi* has displaced *A. smithi*, which was first introduced to California in 1958 and was originally the pea aphid's dominant parasite. *A. ervi* appeared to take 3 to 5 years to become established—except in Escondido, where *A. ervi* were recovered the year following their release.

D. González or W. White sampled fields in Arizona (1987), Torreón, Mexico (1985), Baja California (1985 to 1986), Colorado (1985 to 1986), New Mexico (1986, 1987), South Dakota (1977, 1978), and Utah (1986, 1987). All parasites emerging from the pea aphid and the blue alfalfa aphid in these samples were *A. ervi*; no other parasite species was recovered.

In 1980 researchers did not find *A. ervi* in surveys in Idaho, but by 1982 they confirmed that it was established. Of nearly 1,300 primary parasites reared from the pea aphid, 54 percent were *A. ervi*, 38 percent *A. smithi*, and 4 percent *Praon pequodorum* Viereck (=*simulans* [Provancher]) (Hahn and Bishop 1986).

In Utah the blue alfalfa aphid occurs throughout the southern half of the state, where it is commonly parasitized by *A. ervi*, a species that has never been released in Utah.

Objective B: Ecological and Physiological Studies

Researchers have emphasized developing techniques and procedures for identifying and comparing the attributes of biotypes and clarifying the systematic relationships among aphidiids that attack the pea aphid or the blue alfalfa aphid (see Objective A). T. Unruh et al. (1983) developed and tested a series of buffer/pH isozyme techniques on more than 27 enzyme loci of *A.* *ervi*. These techniques were used initially to measure the decline of allozyme variability in seven laboratory populations. They were also developed for future studies using markers for identifying biotypes, and for measuring heterozygosity and its loss in founder colonies and after shipment, culture, and colonization. T. Unruh et al. (1985) demonstrated the use of multilocus electrophoretic studies for systematic studies in biological control. T. Unruh et al. (1989) used enzyme electrophoresis to evaluate the genetic relationships among 17 populations of *Aphidius* Nees reared from the pea aphids collected throughout the Holarctic region.

E. Botto, D. González, and T. Bellows (1988) measured and compared the biological traits of two populations of *A. ervi*. They used the results to evaluate the biological differences among biotypes and the implication of those differences in applied biological control programs. T. Chua, D. González, and T. Bellows (1990) compared the competitiveness of *A. ervi* and *A. smithi* against the pea aphid in laboratory experiments. They found that separately *A. smithi* parasitized more hosts than did *A. ervi*. However, when the two species competed for the same hosts, significantly more *A. ervi* emerged. The authors suggested that this may be one reason that *A. ervi* displaced *A. smithi*.

Objective D: Impact

Aphidius ervi appears to have displaced *A. smithi* in all sampled areas, including Arizona, California, Colorado, Idaho, Nebraska, South Dakota, Utah, and Coahuila and Baja California, Mexico. In states where the blue alfalfa aphid had become a serious pest, it is no longer a pest. In fact, the blue alfalfa aphid is rare or nonexistent in most of the areas where it was previously established (G. Bishop, H. Homan, S. Halbert, University of Idaho person. commun.; D. Davis, University of Utah person. commun.; K. Pike, Washington State University person. commun.). In many of these areas—especially in California, Idaho, and Washington—researchers released substantial numbers of exotic parasites (predominantly *A. ervi*) against the blue alfalfa aphid and the pea aphid. Throughout the western states, the establishment and abundance of *A. ervi* was highly correlated with the corresponding decline of the blue alfalfa aphid. In addition, low levels of the blue alfalfa aphid are usually highly parasitized by *A. ervi*. Also, in California all of the newer, nondormant, alfalfa varieties have moderate to high levels of resistance to the blue alfalfa aphid and the pea aphid (V. Marble person. commun.). There is insufficient evidence to form a clear analysis of the impact of *A. ervi* alone on the blue alfalfa aphid and the pea aphid. It seems likely that, together, *A. ervi* and resistant alfalfa varieties have reduced the blue alfalfa aphid to levels that do not cause economic injury.

Although the reasons for *A. ervi's* apparent success against the aphids is not clear, we strongly suspect that its polyphagous habit is an important contributing factor. P. Starý (1972) and P. I. Cameron and G. P. Walker (1989) similarly believe that polyphagy can be an important attribute in biological control programs. M. Mackauer (1971), after lengthy laboratory studies of pea aphid parasites, concluded that *A. smithi* was superior to *A. ervi* and to several other species in searching behavior, fecundity, developmental time, and other parameters. The complete displacement of *A. smithi* by *A. ervi* is strong evidence that laboratory and greenhouse studies are of limited use in predicting which of several species—to say nothing of biotypes—will be the most effective under field conditions. Weather is also a factor in maintaining low levels of the blue alfalfa aphid, which cannot tolerate field temperatures above 27°C.

The effect of natural enemies on the pea aphid, measured by correlative evidence similar to that applied to the blue alfalfa aphid and *A. ervi*, is less significant, though still substantial. In California, Idaho, and Washington, sporadic outbreaks of the pea aphid still occur. In Idaho H. Homan, the University of Idaho Alfalfa Extension Coordinator (person. commun.) estimated that growers use chemical applications against the pea aphid about one year out of ten. In California the frequency is estimated to be one out of three or four years (C. Summers person. commun.). Pea aphid outbreaks are unpredictable and are likely to be associated with climatic and possibly agronomic variables. The pea aphid has a much greater alternative host range than the blue alfalfa aphid in terms of numbers of both crops and wild hosts. Nevertheless, G. W. Angalet and R. Fuester (1977) reported a marked reduction in pea aphid damage in all regions after the mid-1960s. They suggested that *A. ervi (ervi)* and *A. smithi* may have contributed to the decline in pea aphid populations in many parts of the United States.

Successful aphid biological control does more than reduce the costs and residues that accompany chemical applications. Reducing chemical use also preserves the natural enemies that are normally effective against a sequence of alfalfa pests—including three principal aphid species, several species of lepidopterous larvae, and alfalfa weevils. Destroying natural enemies in the West's large acreage of alfalfa is equivalent to destroying a vast "field insectary" of natural enemies that regulate many potential pests in both alfalfa and surrounding crops. When this array of native natural enemies is missing, there are more outbreaks of secondary pests, resurgences of primary pests, and an increased pest resistance to chemicals, which leads to an ever-increasing need for applications.

RECOMMENDATIONS

1. Foreign exploration: Natural enemies need to be collected from areas having different climates, habitats, and host plants. The center of origin of the pest species should be defined, and collections should include samples from the origin as well as the edges of the known distribution to obtain and compare species or biotypes that are effective throughout the pest's range. Also, data should be collected on factors that are associated with low pest densities, including climate (especially seasonal temperatures), host-plant varieties, cultivated and wild hosts, and natural enemy species and potential biotypes. An ongoing computerized data bank should be established for every major introduced exotic pest to document the factors associated with low pest densities over a wide range of conditions in the pests' areas of origin, edges of distribution, and newly invaded areas. Correlation, regression, and multifactor analyses of the variables associated with low pest densities can reduce the inventory of natural enemy species or biotypes selected for further studies.

2. Identification: Greater emphasis on the biosystematics of natural enemies—using morphological, behavioral, and genetic techniques—is essential. Voucher specimens should be preserved as a part of all foreign collections. Correctly identifying natural enemies is the first and most important component of biological control programs. This is particularly true of species complexes in which morphologically similar natural enemy populations are available from areas differing widely in climate and habitat, as has occurred with the alfalfa aphid parasites. Tests to detect biotypes should use several techniques (including allozyme electrophoresis) and morphological and biological studies (including fecundity, development rates, and sex ratio).

3. Rearing: Colonies should be monitored for loss of heterozygosity and to identify potential biotypes that can then be successfully colonized. Also, colonies should be released in the shortest time possible to prevent inbreeding under laboratory conditions.

4. Release: Natural enemy species or biotypes that are highly correlated with low pest densities should be released at the earliest date possible into areas newly invaded by the pest, with each natural enemy species or biotype released into several areas differing in climate and habitat. Where the pest has a wide distribution, several natural enemy species or biotypes may need to be released under different environmental conditions. Also, concurrent genetic, morphological, and biological studies should accompany the release program. Long-term preintroduction studies are not justified for two major reasons. First, there is no evidence

that preintroductory studies predict which natural enemies will be the most effective (González and Gilstrap 1992). Second, delay leads to the continued widespread use of insecticides, with the attendant economic, ecological, and social costs.

5. Establishment and evaluation: The two most neglected aspects of introducing exotic natural enemies against introduced pests are identification and evaluation. Identifying successful populations accurately is especially important when several biotypes are released into several areas. Equally essential in this phase are ongoing genetic, morphological, and biological studies, as described above. Voucher specimens need to be collected and preserved. Also, a computerized data bank with input from successful biological control programs, relating known correlative variables, needs to be established on a national level and continued as a long-term project. At the national level, emphasis is also needed on specific studies aimed at defining the key mortality factors and their limits.

REFERENCES CITED

Angalet, G. W., and R. Fuester. 1977. The *Aphidius* parasites of the pea aphid, *Acyrthosiphon pisum,* in the eastern half of the United States. *Ann. Entomol. Soc. Am.* 70:87–96.

Botto, E., D. González, and T. Bellows. 1988. Effects of temperature on some biological parameters of two populations of *Aphidius ervi.* In *Advances in parasitic Hymenoptera research, Proceedings of the 2nd Conference on the Taxonomy and Biology of Parasitic Hymenoptera, 19–21 November 1987, University of Florida, Gainesville,* ed. V. Gupta, 367–77. Leiden, Holland: E. J. Brill.

Caltagirone, L. 1985. Identifying and discriminating among biotypes of parasites and predators. In *Biological control in agricultural IPM systems,* eds. M. Hoy and D. Herzog, 189–200. Orlando, Florida: Academic Press.

Cameron, P. I., and G. P. Walker. 1989. Release and establishment of *Aphidius* spp. (Hymenoptera: Aphidiidae), parasitoids of pea aphid and blue-green aphid in New Zealand. *New Zealand J. Agric. Res.* 32:281–90.

Campbell, R. 1926. *The pea aphid in California,* 2nd ed. USDA Bureau of Entomology Circular 43. Washington, D.C.: U.S. Department of Agriculture.

Campbell, R. E., and W. M. Davidson. 1924. Notes on aphidophagous syrphidae of southern California. *Bull. So. Calif. Acad. Sci.* 23:3–9.

Chen, J. H., D. González, and J. Luhman. 1990. A new species of *Aphidius* attacking the pea aphid, *Acyrthosiphon pisum. Entomophaga* 35:509–14.

Chua, T., D. González, and T. Bellows. 1990. Searching-efficiency and multiparasitism in *Aphidius smithi* and *A. ervi,* parasites of pea aphid, *Acyrthosiphon pisum. J. Appl. Entomol.* 110:101–6.

Cuperus, G. W., E. B. Radcliffe, D. K. Barns, and G. C. Martin. 1982. Economic injury levels and economic thresholds for pea aphid, *Acyrthosiphon pisum* (Harris), on alfalfa. *Crop Protect.* 1:453–63.

Davis, J. J. 1915. The pea aphid in relation to forage crops. USDA Bulletin 276. Washington, D.C.: U.S. Department of Agriculture.

Dickson, R. C. 1975. Identity, origin, and host range of the blue alfalfa aphid. In *Proceedings of the 5th California Alfalfa Producers Symposium, Fresno, California, 22–23.* Berkeley: University of California Cooperative Extension Service, Division of Agricultural Sciences.

Dudley, J., and W. Cook. 1952. The pea aphid. In *Insects, Yearbook of Agriculture 1952,* 538–43. Washington, D.C.: U.S. GPO.

Fluke, C. L. 1929. *The known predaceous and parasitic enemies of the pea aphid in North America.* Wisconsin Agricultural Experiment Station Bulletin 93.

González, D. 1988. Biotypes in biological control. Examples with populations of *Aphidius ervi, Trichogramma pretiosum,* and *Anagrus epos.* In *Advances in parasitic Hymenoptera research, Proceedings of the 2nd Conference on the Taxonomy and Biology of Parasitic Hymenoptera, 19–21 November 1987, University of Florida, Gainesville,* ed. V. Gupta, 475–82. Leiden, Holland: E. J. Brill.

González, D., and F. E. Gilstrap. 1992. Foreign exploration: Assessing and prioritizing natural enemies and consequences of preintroduction studies. In *Selection criteria and ecological consequences of importing natural enemies,* eds. W. C. Kauffman and J. R. Nechols, 53–70. Entomological Society of America, Thomas Say Symposia Proceedings Series. Hyattsville, Md.: ESA.

González, D., G. Gordh, S. N. Thompson, and J. Adler. 1979. Biotype discrimination and its importance to biological control. In *Genetics in relation to insect management: Working papers,* eds. M. Hoy and J. McKelvey, Jr., 129–37. New York: Rockefeller Foundation.

González, D., W. White, R. Dickson, and R. van den Bosch. 1975. The potential for biological control of the blue alfalfa aphid. In *Proceedings of the 5th California Alfalfa Producers Symposium, Fresno, California, 35–38.* Berkeley: University of California Cooperative Extension Service, Division of Agricultural Sciences.

González, D., W. White, J. Hall, and R. C. Dickson. 1978. Geographical distribution of Aphidiidae (Hymenoptera) imported to California for biological control of *Acyrthosiphon kondoi* and *Acyrthosiphon pisum* (Homoptera: Aphididae). *Entomophaga* 23:239–248.

González, D., M. Miyazaki, W. White, H. Takada, R. C. Dickson, and J. C. Hall. 1979. Geographical distribution of *Acyrthosiphon kondoi* Shinji (Homoptera: Aphididae) and some of its parasites and hyperparasites in Japan. *Kontyu* (Tokyo) 47:1–7.

Hagen, K. S., and E. I. Schlinger. 1960. Imported Indian parasite of pea aphid established in California. *Calif. Agric.* 14(5):5–6.

Hahn, J. D., and G. W. Bishop. 1986. The hymenopterous parasites of alfalfa aphids in southern Idaho. *J. Idaho Acad. Sci.* 22:40–44.

Halfhill, J. E., and P. E. Featherston. 1973. Inundative releases of *Aphidius smithi* against *Acyrthosiphon pisum*. *Environ. Entomol.* 2:469–72.

Halfhill, J. E., P. E. Featherston, and A. G. Dickie. 1972. History of the *Praon* and *Aphidius* parasites of the pea aphid in the Pacific Northwest. *Environ. Entomol.* 1:402–5.

Harper, A., J. Miska, G. Manglitz, B. Irwin, and E. Armbrust. 1978. *The literature of arthropods associated with alfalfa. III. A bibliography of the pea aphid,* Acyrthosiphon pisum. Illinois Agricultural Experiment Station Special Publication 50.

Hobbs, G. A., N. D. Holmes, G. E. Swailes, and N. S. Church. 1961. Effect of the pea aphid, *Acyrthosiphon pisum* (Harr.) (Homoptera: Aphididae), on yields of alfalfa hay on irrigated land. *Can. Entomol.* 93:801–4.

Hutchison, W. D., and D. B. Hogg. 1985. Time-specific life tables for the pea aphid, *Acyrthosiphon pisum* (Harris), on alfalfa. *Res. Popul. Ecol.* 27:231–53.

Kodet, R., M. Nielson, and R. Keuhl. 1982. *Effect of temperature and photoperiod on the biology of the blue alfalfa aphid,* Acyrthosiphon kondoi *Shinji.* USDA Technical Bulletin 1660. Washington, D.C.: U.S. Department of Agriculture.

Kono, T. 1975. Distribution and identification of the blue alfalfa aphid. In *Proceedings of the 5th California Alfalfa Producers Symposium, Fresno, California,* 24. Berkeley: University of California Cooperative Extension Service, Division of Agricultural Sciences.

Mackauer, M. 1971. *Acyrthosiphon pisum,* pea aphid. In *Biological control programs against insect pests and weeds in Canada, 1959–68,* 1–10. Technical Communication 4. Trinidad: Commonwealth Institute of Biological Control.

Mackauer, M., and H. E. Bisdee. 1965. *Aphidius smithi* Sharma & Subba Rao (Hymenoptera: Aphidiidae) a parasite of the pea aphid new in southern Ontario. *Proc. Entomol. Soc. Ont.* 95:121–24.

Mackauer, M., and T. Finlayson. 1967. The hymenopterous parasites of the pea aphid in eastern North America. *Can.*

Entomol. 99:1051–82.

Marsh, P. 1977. Notes on the taxonomy and nomenclature of *Aphidius* species parasitic on the pea aphid in North America. *Entomophaga* 22:365–72.

Poston, F. L., L. E. Pedigo, and S. M. Welch. 1983. Economic injury levels: Reality and practicality. *Bull. Entomol. Soc. Am.* 29:49–53.

Starý, P. 1972. Host range of parasites and ecosystem relations, a new viewpoint in multilateral control concept. *Ann. Soc. Entomol. France* (n.s.) 8:351–58.

———. 1974. Taxonomy, origin, distribution, and host range of *Aphidius* species in relation to biological control of the pea aphid in Europe and North America. *Zeit. angew. Entomol.* 77:141–71.

Starý, P., and D. González. 1978. Parasitoid spectrum of *Acyrthosiphon* aphids in Central Asia (Hymenoptera: Aphidiidae). *Entomol. Scandinavica* 9:140–45.

Starý, P., D. González, and J. C. Hall. 1980. *Aphidius eadyi* n. sp. (Hymenoptera: Aphidiidae), a widely distributed parasitoid of the pea aphid, *Acyrthosiphon pisum* (Harris), in the Palearctic. *Entomol. Scandinavica* 1:473–80.

Stern, V. M., R. Sharma, and C. Summers. 1980. Alfalfa damage from *Acyrthosiphon kondoi* and economic threshold studies in southern California. *J. Econ. Entomol.* 73:145–48.

Summers, C. G., and R. L. Coviello. 1984a. Impact of *Acyrthosiphon kondoi* (Homoptera: Aphididae) on alfalfa: Field and greenhouse studies. *J. Econ. Entomol.* 77:1052–56.

———. 1984b. Influence of constant temperatures on the development and reproduction of *Acyrthosiphon kondoi. Environ. Entomol.* 13:236–42.

Unruh, T., W. White, D. González, and R. Luck. 1985. Electrophoretic studies of parasitic Hymenoptera and implications for biological control. *Misc. Publ. Entomol. Soc. Am.* 61:150–63.

Unruh, T., W. White, D. González, and J. Woolley. 1989. Genetic relationships among 17 *Aphidius* populations, including six species. *Ann. Entomol. Soc. Am.* 82:754–68.

Unruh, T. R., W. White, D. González, G. Gordh, and R. F. Luck. 1983. Heterozygosity and effective size in laboratory populations of *Aphidius ervi* (Hym.: Aphididae). *Entomophaga* 28:245–58.

31 / RUSSIAN WHEAT APHID

R. M. NOWIERSKI AND J. B. JOHNSON

> **Russian wheat aphid**
> *Diuraphis noxia* (Mordvilko)
> Homoptera: Aphididae

INTRODUCTION

Biology and Pest Status

The Russian wheat aphid, *Diuraphis noxia* (Mordvilko), is a small (less than 2 mm long), lime green aphid with a long, spindle-shaped body and prominent dark eyes. It can be distinguished from other common grain aphids by its short antennae, the absence of prominent cornicles, and the presence of a supracaudal process that gives the posterior of the abdomen a forked appearance when viewed laterally (Johnson 1988).

Its preferred hosts are mainly Triticeae, including wheat, barley and, to lesser degrees, rye and oats. The aphids usually occur at the base of the two or three most recently developed leaves (Walters 1984), but can occupy any aerial portion of the plant (especially when populations are high), including the inflorescence and developing grain (Alfaro 1947).

Fecundity varies from 31.9 to 81.5 nymphs per female at 13°C and 17°C, respectively (Aalbersberg et al. 1987; Webster and Starks 1987). The aphid can reproduce at 5°C (Michels and Behle 1988) and probably at even lower temperatures. Individuals require four to six—usually five—instars to mature (Aalbersberg et al. 1987; Webster and Starks 1987). Nymphs develop in 7.5 to 56.5 days at constant temperatures of 27°C and 5°C, respectively (Webster and Starks 1987; Michels and Behle 1988). However, the nymph is able to develop at temperatures as low as 0.54°C (Aalbersberg et al. 1987).

The Russian wheat aphid may overwinter in winter cereals or infest spring crops early in the season, usually by apterae moving into the field from nearby grasses (Kriel et al. 1984). At first population growth is slow, but it increases during tillering and stem elongation and peaks from the boot stage on (Kriel et al. 1984; Hughes

1988). Winged forms, produced when the crop senesces, are distributed by wind currents, which accounts for the rapid spread of the Russian wheat aphid in the United States (Peairs 1987). In autumn the population can increase a second time on early-planted winter grains (Halbert and Johnson unpub. data). The aphids' damage is caused by toxins in their saliva, which induce leaf curling and destroy chloroplasts, producing, in warm temperatures, white to yellow streaks on the lamina or, in cooler temperatures, pink to purple streaks (Hughes 1988). Young plants may die (Grossheim 1914). Infestations in older plants result in reduced tillering, fewer fertile tillers (Hughes 1988), and less grains per ear (Aalbersberg 1987), as well as a drop in photosynthetic capacity (Walters 1981).

The Russian wheat aphid has been a major threat to grain production in North America since it was first detected in Texas in 1986 (Stoetzel 1987; Burton 1990; Kovalev et al. 1991). Direct and indirect economic losses due to the aphid from 1987 to 1991 have been estimated at nearly $670 million (Burton and Webster 1993). In South Africa the aphid has also been reported to transmit brome mosaic virus, barley yellow dwarf virus, and barley stripe mosaic virus (von Wechmar 1984), but the extent to which the transmission occurs under field conditions is unknown.

Historical Notes

Diuraphis noxia appears to have originated in the Ukraine and central Asia (Balachowsky and Mesnil 1935). It was first recorded in the New World in 1980 in Mexico (Gilchrist, Rodríguez, and Burnett 1984). Since then it has been reported in Argentina (Blackman and Eastop 1984), in the United States via the Texas Panhandle (Webster, Starks, and Burton 1987), and in Canada and Chile (Stoetzel 1987; Zerene, Caglevic, and

Appreciation is extended to the following for contributing information to this chapter: M. G. Feng, F. Gilstrap, S. E. Halbert, G. D. Johnson, S. E. Lajeunesse, J. L. Littlefield, J. Miller, F. B. Peairs, and G. Piper. Support was provided in part by a grant from the Montana Wheat and Barley Committee and the Idaho Wheat Commission.

Ramírez 1988). In the United States and Canada, its range now includes 15 central and western states and 3 provinces (APHIS-NPPSDS 1988; *Russian Wheat Aphid News* 1990).

RESULTS

Objective A: The Identification and Introduction of Beneficial Organisms

One of the major challenges to the biological control of this aphid is finding natural enemies that are effective both early and late in the growing season, when the aphid is most damaging, as well as during periods of relatively low aphid density. A premium should be placed on natural enemies that are cold adapted, such as the neuropteran family Hemerobiidae (Neuenshchwander 1976), or that have an excellent searching capacity.

Researchers have found that although many aphidophagous insects feed on the Russian wheat aphid, they have a limited impact on aphid populations in cropland (Berest 1980; Aalbersberg, van der Westhuizen, and Hewitt 1988). The greatest contribution of these insects to control may be on noncrop hosts, especially between harvest and fall plantings. Heat-tolerant natural enemies could be important during this period of low aphid populations, helping reduce the aphid's invasion of winter cereals. The leaf curling that results from the aphid's feeding reduces the effectiveness of some natural enemies—particularly the larger species—because they cannot gain access to the aphids.

R. D. Hughes (1988) lists 8 species of parasitoids, 5 hyperparasitoids, and 14 or more predators of the Russian wheat aphid. The following hymenopteran genera all contain species that parasitize the aphid: *Aphelinus* (Aphelinidae: Aphelininae); *Aphidius, Diaretus, Diaeretiella, Ephedrus, Lysiphlebus,* and *Praon* (Braconidae: Aphidiinae). Aphidiinae appear to be the most effective controls on exposed Russian wheat aphids, but *Ephedrus plagiator* (Nees) will probe and presumably oviposit through the gap between the margins of rolled wheat leaves (Johnson and Halbert unpub. data). The smaller, more compact *Aphelinus varipes* (Foerster) will enter rolled leaves and can attack numerous aphids within a colony (Lajeunesse person. commun.), but its efficiency may be limited when honeydew is abundant. It appears likely that rapidly reproducing parasitoids such as *Aphidius rhopalosiphi* De Stefani—which attacks exposed aphids—will be nicely complemented by species such as *A. varipes*, which attacks the aphids within the rolled leaves. Another parasitoid that has shown promise is *Diaeretiella rapae* (M'Intosh). In an endemic population of this parasitoid in Washington state, researchers have observed occasional high levels of

aphid parasitism—up to 90 percent late in the season, following the aphid's population crash (K. Pike person. commun.).

In 1988, researchers began to acquire exotic parasitoid species for release against the Russian wheat aphid. Scientists at the USDA-ARS European Parasite Laboratory (EPL) in Paris, France, and at the International Institute of Biological Control (IIBC) in Délemont, Switzerland, as well as other international and U.S. state and federal cooperators, have collected potential parasitoids from different species of cereal aphids (including *D. noxia*) and plant hosts in Turkey, France, Spain, Poland, and parts of the former Soviet Union. Most of the parasitoids were processed by the EPL and sent to the Texas A&M Quarantine Facility.

Five species of parasitoids—*A. varipes, Aphidius matricariae* Haliday, *A. rhopalosiphi, E. plagiator,* and *Praon gallicum* Starý (collected in Turkey and France)—were mass-reared at the quarantine facility and then sent to state and federal cooperators. In 1988 researchers made field releases in Washington, Oregon, Idaho, and Montana. Although *A. rhopalosiphi* appeared to be the easiest to rear in the laboratory, in the field only *A. varipes* and *E. plagiator* showed evidence of mummy formation (Johnson, Halbert, Lajeunesse, Nowierski, and Littlefield unpub. data). However, the next spring researchers did not recover any of the parasitoids released in 1988. Extremely cold temperatures in February 1989 probably had an adverse effect on the aphid's host plants and on the aphid itself; in Idaho and Montana, researchers found a 99 percent mortality rate (Feng and Johnson unpub. data; G. D. Johnson person. commun.).

More parasitoid and predator material—collected during fall 1988 in Spain and spring 1989 in Czechoslovakia, Pakistan, Iraq, and Jordan—was sent to the Texas A&M Quarantine Facility, where it was reared and released to cooperators. However, because of the extremely harsh temperatures in such northern states as Montana, aphid populations during the 1989 growing season were too low to justify releasing parasitoids (Nowierski unpub. data).

The following genera all contain predators of the Russian wheat aphid: *Aeolothrips* (Thysanoptera: Aeolothripidae); *Orius* and *Nabis* (Hemiptera: Anthocoridae and Nabidae); *Chrysoperla* and *Chrysopa* (Neuroptera: Chrysopidae); *Adonia, Coccinella, Hippodamia, Propylaea,* and *Scymnus* (Coleoptera: Coccinellidae); *Malachius* (Coleoptera: Melyridae); *Tachyporus* (Coleoptera: Staphylinidae); *Aphidoletes, Leucopis,* and *Syrphus* (Diptera: Cecidomyiidae, Chamaemyiidae, and Syrphidae). Exposed aphids appear to be more vulnerable to attack by predators. However, researchers have observed that smaller species, such as chamaemyiids and cecidomyiids, can attack the aphid in the curled leaves, as can the smaller instars of some of the larger predatory species—for example, coccinellids and syrphids (Lajeunesse, Littlefield, and Johnson unpub. data).

In southwestern Idaho, researchers have recovered six species of entomopathogenic fungi from the Russian wheat aphid (Feng et al. 1990a,b), including *Beauveria bassiana* (Balsamo) Vuillemin, *Verticillium lecanii* (Zimmermann), *Conidiobolus obscurus* (Hall and Dunn) Remaudiere and Keller, *Pandora neoaphidis* (Remaudiere and Hennebert) Baltko, *Pandora radicans* (Brefeld) Baltko, and *Neozygites frieseni* (Nowakowski) Remaudiere and Keller. Most of these species, plus *Entomophthora planchoniana* Cornu, are known to attack the Russian wheat aphid in Turkey (Burton 1988). These fungi appear to have a limited effect: in irrigated wheat in southwestern Idaho, fungal-induced mortality rarely exceeds 25 percent (Feng person. commun.). The potential of these fungi, however, remains unknown.

RECOMMENDATIONS

1. Surveys should be conducted in small grain and Conservation Reserve Program fields for entomophagous species that could be used in the biological control of the Russian wheat aphid.

2. Exotic predators, parasitoids, and pathogens of the aphid should continue to be obtained, mass-reared, and released.

3. Aphid mummies should be obtained from small grain and Conservation Reserve Program fields throughout the growing season for taxonomic analysis before and after the release of exotic parasitoids.

4. The impact of natural enemies should be evaluated in the field. This will require traditional methods of natural enemy exclusion, as well as more modern techniques that make it possible to discriminate among different strains of a natural enemy or host species.

5. Selective insecticides, key periods of pesticide application, and minimal dosages should be identified that will control the aphid with minimal impact on the natural enemies.

6. Wheat and barley varieties that are aphid resistant and compatible with entomophagous species should be identified.

7. Semiresistant Conservation Reserve Program grasses should be identified; these are needed to maintain a residual aphid population, providing a reservoir for the natural enemy community after pesticide applications or harvest.

REFERENCES CITED

Aalbersberg, Y. K. 1987. Ecology of the wheat aphid *Diuraphis noxia* (Mordvilko) in the Eastern Orange Free State. Master's thesis, University of Orange Free State, Bloemfontein, Republic of South Africa.

Aalbersberg, Y. K., F. du Toit, M. C. van der Westhuizen, and P. H. Hewitt. 1987. Development rate, fecundity, and life span of apterae of the Russian wheat aphid, *Diuraphis noxia* (Mordvilko) (Hemiptera: Aphididae), under controlled conditions. *Bull. Entomol. Res.* 77:629–35.

Aalbersberg, Y. K., M. C. van der Westhuizen, and P. Hewitt. 1988. Natural enemies and their impact on *Diuraphis noxia* (Mordvilko) (Hemiptera: Aphididae) populations. *Bull. Entomol. Res.* 78:111–20.

Alfaro, A. 1947. Notes on *Brachycolus noxius* Mordw., a new pest for our wheats and barleys (in Spanish). *Boletín de Patología Vegetal y Entomología Agrícola* 15:125–30.

APHIS National Plant Pest Survey and Detection Service. 1988. Geographic distribution of *Diuraphis noxia*. Hyattsville, Md.: APHIS.

Balachowsky, A., and L. Mesnil. 1935. *Les insectes nuisibles aux plantes cultivées*. Paris: N.p.

Berest, Z. L. 1980. Parasites and predators of the aphids *Brachycolus noxius* and *Schizaphis graminum* in crops of barley and wheat in the Nikolayev and Odessa regions (in Russian). *Vestnik Zoologii* 5:84–87.

Blackman, R. L., and V. F. Eastop. 1984. *Aphids on the world's crops: An identification and information guide.* New York: Wiley.

Burton, R. L. 1988. The Russian wheat aphid. First Annual Report of the USDA-ARS. Stillwater, Okla.: U.S. Department of Agriculture.

———. 1990. The Russian wheat aphid. Second Annual Report of the USDA-ARS. Stillwater, Okla.: U.S. Department of Agriculture.

Burton, R. L., and J. A. Webster. 1993. The Russian wheat aphid. Fifth Annual Report of the USDA-ARS. Stillwater, Okla.: U.S. Department of Agriculture.

Feng, M. G., J. B. Johnson, and L. P. Kish. 1990a. Survey of entomopathogenic fungi naturally infecting cereal aphids (Homoptera: Aphididae) of irrigated grain crops in southwestern Idaho. *Environ. Entomol.* 19:1534–42.

———. 1990b. Virulence of *Verticillium lecanii* (Zimm.) and a new isolate of *Beauvaria bassiana* (Bals.) to six species of cereal aphids (Homoptera: Aphididae). *Environ. Entomol.* 19:815–20.

Gilchrist, L. I., R. Rodríguez, and P. A. Burnett. 1984. The extent of Freestate streak and *Diuraphis noxia* in Mexico. In *Barley yellow dwarf, proceedings of the workshop,* 157–63. Mexico City: International Maize and Wheat Improvement Center (CIMMYT).

Grossheim, N. A. 1914. The aphid *Brachycolus noxius* Mordwilko. *Memoirs of the Natural History Museum of the Zemstvo of the government of Taurida, Simferopol* (in Russian), vol. 3, 35–78.

Hughes, R. D. 1988. A synopsis of information on the Russian wheat aphid, *Diuraphis noxia* (Mordwilko). Division of Entomology Technical Paper 28, 1–39. Melbourne: CSIRO.

Johnson, G. D. 1988. The Russian wheat aphid: A threat to Montana grain production. *MontGuide* Feb. issue. Bozeman: Montana State University Extension Service.

Kovalev, O. C., T. J. Poprawski, A. V. Stekolshchikov, A. B. Vereshchagina, and S. A. Gandraabur. 1991. *Diuraphis* Aizenberg (Homoptera: Aphididae): Key to apterous viviparous females, and review of Russian language literature on the natural history of *Diuraphis noxia* (Kurdjumov) 1913. *J. Appl. Entomol.* 112:425–36.

Kriel, C. F., P. H. Hewitt, J. de Jager, M. C. Walters, A. Fouché, and M. C. van der Westhuizen. 1984. Aspects of the ecology of the Russian wheat aphid, *Diuraphis noxia,* in the Bloemfontein district: II. Population dynamics. In *Progress in Russian wheat aphid* (Diuraphis noxia Mordv.) *research in the Republic of South Africa,* ed. M. C. Walters, 14–21. Technical Communication 191. Johannesburg: Republic of South Africa Department of Agriculture.

Michels, G. J., Jr., and R. J. Behle. 1988. Influence of temperature on reproduction, development, and intrinsic rate of increase of Russian wheat aphid, greenbug, and bird cherry–oat aphid (Homoptera: Aphididae). *J. Econ. Entomol.* 82:439–44.

Neuenschwander, P. 1976. Biology of the *Hemerobius pacificus. Environ. Entomol.* 5:96–100.

Nordlund, D. A., R. L. Jones, and W. J. Lewis. 1981. *Semiochemicals: Their role in pest control.* New York: Wiley.

Peairs, F. B. 1987. Aphids in small grains. *Service in Action* 5:568. Fort Collins: Colorado State University Cooperative Extension Service.

Russian Wheat Aphid News. 1990. 4(1):1–6. Colorado State University, Fort Collins.

Stoetzel, M. B. 1987. Information on and identification of *Diuraphis noxia* (Homoptera: Aphididae) and other aphid species colonizing leaves of wheat and barley in the United States. *J. Econ. Entomol.* 80:696–704.

von Wechmar, M. B. 1984. Russian aphid spreads gramineae viruses. In *Progress in Russian wheat aphid* (Diuraphis noxia Mordv.) *research in the Republic of South Africa,* ed. M. C. Walters, 38–41. Technical Communication 191. Johannesburg: Republic of South Africa Department of Agriculture.

Walters, S. S. 1981. The Russian wheat aphid in the winter rainfall region (in Afrikaans). *Winterreënnuusbrief* Feb. issue.

———. 1984. Grain aphids in the winter rainfall region (in Afrikaans). *Winter Rain* special edition, Feb.

Webster, J. A., and K. J. Starks. 1987. Fecundity comparison of *Schizaphis graminum* (Rondani) and *Diuraphis noxia* (Mordvilko) at three temperature regimes. *J. Kansas Entomol. Soc.* 60:580–82.

Webster, J. A., K. J. Starks, and R. L. Burton. 1987. Plant-resistance studies with *Diuraphis noxia* (Homoptera: Aphididae), a new United States wheat pest. *J. Econ. Entomol.* 80:944–49.

Zerene, Z. M., D. M. Caglevic, and A. I. Ramírez. 1988. A new cereal aphid detected in Chile (in Spanish). *Agric. Técnica* (Santiago) 48:60–61.

32 / WALNUT APHID — PART 1

M. T. ALINIAZEE AND K. S. HAGEN

walnut aphid
Chromaphis juglandicola (Kaltenbach)
Homoptera: Aphididae

INTRODUCTION

Biology and Pest Status

In the United States, commercial walnuts are produced mainly in the Pacific states of California, Oregon, and Washington. California is by far the largest producer, and the walnut aphid, *Chromaphis juglandicola* (Kaltenbach), is one of the crop's major pests. The aphid was probably introduced into the United States during the latter part of the nineteenth century on infested nursery stocks (Essig 1909); by the turn of the century, it had invaded most of the walnut-growing areas of California (Davidson 1914). In addition to damaging leaves, long-term aphid feeding can devastate tree vigor. Under extreme conditions, crops can be reduced by 50 percent (Michelbacher and Oatman 1956).

Historical Notes

A number of studies have documented the biological control of the walnut aphid using generalist predators (Michelbacher, Middlekauff, and Wagenek 1950; Sluss 1967; Hagen and van den Bosch 1968). The aphid's most common predators were *Chrysoperla* (=*Chrysopa*) *carnea* (Stephens), *C. nigricornis* Burmeister, *Hemerobius ovalis* Carpenter, *Hippodamia convergens* Guerin-Meneville, and *Zelus* species, as well as many species of coccinellids and spiders (Frazer and van den Bosch 1973; Nowierski 1979). However, these several generalist predators did not provide a satisfactory level of control.

RESULTS

Objective A: The Identification and Introduction of Beneficial Organisms

As early as 1959 R. van den Bosch searched for natural enemies of *C. juglandicola*. Because in its Old World habitat the insect was nearly always kept in check by parasitoids, it appeared that it could be controlled by a host-specific natural enemy. In southern France, R. van den Bosch found an aphidiid parasitoid, *Trioxys pallidus* Haliday, parasitizing the walnut aphid (van den Bosch, Schlinger, and Hagen 1962). A later search in other parts of Europe and Asia showed that *T. pallidus* was prevalent throughout the geographical range of *C. juglandicola*. In 1959 specimens of *T. pallidus* were collected in France and shipped to California. Nearly 12,000 individuals were reared in insectaries and released in about a dozen California counties. The parasite became established the same year (Schlinger, Hagen, and van den Bosch 1960). A population buildup occurred in walnut groves in both southern and northern California. Later studies showed that the parasitoid was well suited to the mild coastal areas of southern California, where it spread rapidly (van den Bosch et al. 1970).

However, in hotter and drier areas of California, such as the Central Valley, the French strain of *T. pallidus* did poorly. This suggested two options. One was to wait for an unknown period of acclimatization, hoping that successive generations would slowly adapt. The second option was to search for *T. pallidus* in locations with climates resembling that of the Central Valley. Researchers eventually collected a strain from areas of Iran that had a summer climate nearly identical to that of the Central Valley (van den Bosch et al. 1970). These wasps were imported in the spring of 1968 and released in several areas of central California. They immediately established vigorous populations and spread into many nonrelease sites. In approximately 2 years, researchers noticed a moderate to heavy parasitization of walnut aphids. Researchers found that *T. pallidus* from Iran, when not disturbed by applications of pesticides, could reduce walnut aphid populations to very low numbers (van den Bosch et al. 1979). One of the greatest threats to the successful biological control of the aphid has been the use of insecticides to control such pests as the codling moth, the walnut husk fly, and spider mites.

Current status and future outlook. Both the French and Iranian strains of *T. pallidus* are now well established in California and Oregon. The Oregon population is relatively small and appears to be mostly the Iranian strain, which can withstand summer heat. In California, the San Joaquin Valley and other interior valleys seem to be dominated by the Iranian strain, whereas the milder coastal areas are probably dominated by the French strain. The unique geographic distribution of these two strains demonstrates how important adaptability to climate is when identifying and importing natural enemies. Together, the two strains appear to provide effective biological control of the walnut aphid.

Objective C: The Conservation and Augmentation of Beneficial Organisms

Areas where insecticides are used indiscriminately have experienced repeated outbreaks of walnut aphids. When the parasitoids were killed, the aphid numbers increased to damaging levels. In some areas hyperparasitoids have been a limiting factor. On the positive side, after repeated exposure, *T. pallidus* has also developed low levels of resistance to certain compounds. In another W-84 project, M. A. Hoy used laboratory selection to develop insecticide-resistant populations of *T. pallidus*. Preliminary results (Hoy and Cave 1988) suggested that the parasitoid has the ability to develop resistance to many insecticides (see chap. 33).

The biological control of the walnut aphid by *T. pallidus* has been a highly successful example of a classical biological control program. California walnut growers have reaped substantial economic benefits from this program, much of whose research was conducted under the auspices of Western Regional Research Project W-84.

RECOMMENDATIONS

Due to the effectiveness of *T. pallidus*, the walnut aphid is essentially no longer a pest in many walnut orchards. However, aphid outbreaks still do occur when the parasitoid is disrupted by pesticides. To minimize the dosage and the number of sprays, management strategies need to be identified for other walnut pests. Also beneficial would be genetic improvements that would help *T. pallidus* tolerate pesticides.

REFERENCES CITED

Davidson, W. M. 1914. *Walnut aphid in California.* USDA Bulletin 100.

Essig, E. O. 1909. Aphididae of southern California II. *Pomona J. Entomol.* 1:47–52.

Frazer, B. D., and R. van den Bosch. 1973. Biological control of the walnut aphid in California: The interrelationship of the aphid and its parasite. *Environ. Entomol.* 2:561–67.

Hagen, K. S., and R. van den Bosch. 1968. Impact of pathogens, parasites, and predators on aphids. *Annu. Rev. Entomol.* 13:325–84.

Hoy, M. A., and F. E. Cave. 1989. Toxicity of pesticides used on walnuts to a wild and azinphosmethyl-resistant strain of *Trioxys pallidus* (Hymenoptera: Aphidiidae). *J. Econ. Entomol.* 82:1585–92.

Michelbacher, A. E., and E. Oatman. 1956. Walnut aphid studies in 1956. *Calif. Agric.* 10(3):9–10.

Michelbacher, A. E., W. W. Middlekauff, and E. Wagenek. 1950. The walnut aphid in northern California. *J. Econ. Entomol.* 43:448–56.

Nowierski, R. M. 1979. The field ecology of the walnut aphid, *Chromaphis juglandicola*, and its introduced parasitoid, *Trioxys pallidus*. Ph.D. diss., University of California, Berkeley.

Schlinger, E. I., K. S. Hagen, and R. van den Bosch. 1960. Parasite of walnut aphid. *Calif. Agric.* 14(10):3–4.

Sluss, R. R. 1967. Population dynamics of the walnut aphid, *Chromaphis juglandicola*, in northern California. *Ecology* 48:41–58.

van den Bosch, R., E. I. Schlinger, and K. S. Hagen. 1962. Initial field observations in California on *Trioxys pallidus*, a recently introduced parasite of the walnut aphid. *J. Econ. Entomol.* 55:857–62.

van den Bosch, R., B. D. Frazer, C. S. Davis, P. S. Messenger, and R. Hom. 1970. *Trioxys pallidus*: An effective walnut aphid parasite from Iran. *Calif. Agric.* 24(11):8–10.

van den Bosch, R., R. Hom, P. Matteson, B. D. Frazer, P. S. Messenger, and C. S. Davis. 1979. Biological control of the walnut aphid in California: Impact of the parasite *Trioxys pallidus*. *Hilgardia* 47:1–13.

33 / WALNUT APHID — PART 2

M. A. HOY

---

walnut aphid
Chromaphis juglandicola (Kaltenbach)
Homoptera: Aphididae

INTRODUCTION

Biology and Pest Status

(See chap. 32, "Walnut Aphid—Part 1," by M. T. AliNiazee and K. S. Hagen.)

Historical Notes

In commercial walnut orchards in California, the exotic parasite *Trioxys pallidus* Haliday can be an effective parasite of the walnut aphid *Chromaphis juglandicola* (Kaltenbach) (Schlinger, Hagen, and van den Bosch 1960; van den Bosch, Schlinger, and Hagen 1962; van den Bosch et al. 1970; Frazer and van den Bosch 1973; Riedl, Barnes, and Davis 1979; van den Bosch et al. 1979). Unfortunately, two other key pests of walnuts— codling moth, *Cydia pomonella* (L.), and navel orangeworm, *Amyelois transitella* Walker—are commonly controlled with azinphosmethyl (Guthion), which is harmful to *T. pallidus*. Although applications of this pesticide have been associated with secondary outbreaks of the walnut aphid (Riedl, Barnes, and Davis 1979; Sibbett, Bettiga, and Bailey 1981), growers still use Guthion because it provides more effective control of codling moth and navel orangeworm than do other pesticides. Thus, developing and implementing a Guthion-resistant strain could improve the usefulness of *T. pallidus* in integrated pest management programs in California walnut orchards.

RESULTS

Objective B: Ecological and Physiological Studies

Using a time-response technique, a laboratory study evaluated the mode of inheritance of Guthion resistance in a laboratory-selected population of *T. pallidus*. An evaluation of the reciprocal F_1 and backcross progeny indicates that Guthion resistance in this strain is determined by more than one gene (Brown et al. 1992).

Although the resistance appears to be polygenic, the F_1 progeny are more like their resistant parents than like their susceptible parents; thus, resistance is inherited in a semidominant manner. The estimated dominance value (D) is 0.33 (a value of 0 indicates intermediate inheritance, +1 a complete dominant, and −1 a complete recessive). This means that the F_1 progeny should survive relatively well in the field.

To determine how long Guthion resistance will persist in mixed populations of resistant and susceptible *T. pallidus* held without selection with Guthion, we conducted replicated population cage studies. The results indicate that the resistance allele(s) are not lost rapidly (Hoy, Cave, and Caprio 1991). In addition, we compared the ability of the pesticide-resistant and susceptible *T. pallidus* colonies to enter diapause under controlled temperature and day-length conditions as an indicator of their potential for successful overwintering (Hoy, Cave, and Caprio 1991). We also evaluated the mechanism of resistance through a joint project with G. Georghiou of the University of California, Riverside.

Objective C: Conservation and Augmentation

After surveying colonies collected from California walnut orchards for variability, we selected *Trioxys pallidus* for resistance to Guthion in the laboratory (Hoy and Cave 1988, 1991). Although the selection responses were slow, we achieved a 10-fold increase in resistance. The Guthion-resistant strain is cross-resistant to Supracide (methidathion), Lorsban (chlorpyrifos), Thiodan (endosulfan), and Zolone (phosalone), which should enhance this strain's usefulness in an IPM program (Hoy and Cave 1989a,b).

In 1988 approximately 75,000 Guthion-resistant parasites were released into five orchards (Hoy et al. 1989, 1990). Colonies of *T. pallidus* collected from each orchard before the releases were tested with a discriminating dose of Guthion, to estimate resistance levels of the wild population. Samples of *T. pallidus* were collected once or twice from each block after the releases, and from adjacent walnut blocks in two sites. These samples were tested with Guthion. The high survival rates of the parasites indicated that the Guthion-resistant strain of *T. pallidus* established and persisted through the growing season in four of the five walnut blocks and dispersed to nearby nonrelease blocks in two of two sites sampled. Counts of aphids and mummies support the conclusion that the laboratory-selected strain survived applications of Guthion or Supracide, parasitized aphids, and persisted within the release and nearby nonrelease sites during the 1988 growing season. In 1989 and 1990, parasites were collected from the 1988 release sites and tested with Guthion; the results indicate that the released strain persisted, particularly during the 1989 field season, although at low frequencies.

Implementing the Guthion-resistant strain of *T. pallidus* is a challenge for several reasons. We have two different options for implementation: inoculative or augmentative releases. If the releases are inoculative the strain must not only establish, persist, overwinter, and compete with the widespread "native" population, but, to be successful, it must either replace the native population or its resistance allele(s) must be incorporated into the wild population. We have assumed that inoculative releases are necessary because the rearing and release of resistant *T. pallidus* on a yearly basis—the augmentation model—is not economically viable; however, this assumption could be false. Although sources that can produce the immense numbers of *T. pallidus* needed for releases into several hundred walnut orchards are currently lacking, the strain could be released augmentatively if commercial insectaries decided to produce these parasites, or if walnut growers banded together to mass-produce the parasites.

Together with M. Caprio and B. Tabashnik of the University of Hawaii, we developed a computer model to test different methods for maximizing the likelihood of establishing Guthion-resistant *T. pallidus* in walnut orchards in the San Joaquin Valley of California (Caprio, Hoy, and Tabashnik 1991; Hoy, Cave, and Caprio 1991). The model incorporates as much of the known biology, ecology, and genetics of *T. pallidus* as possible and is potentially useful in predicting the establishment and dispersal of both genetically manipulated and nonmanipulated arthropod natural enemies. The model, along with data obtained from current experiments, will allow us to recommend specific options for implementing the Guthion-resistant strain of *T. pallidus*.

REFERENCES CITED

Brown, E., F. E. Cave, and M. A. Hoy. 1992. Mode of inheritance of azinphosmethyl resistance in a laboratory-selected strain of *Trioxys pallidus* Haliday (Hymenoptera: Aphidiidae). *Entomol. Exp. Appl.* 63:229–36.

Caprio, M. A., M. A. Hoy, and B. E. Tabashnik. 1991. A model for implementing a genetically improved strain of the parasitoid *Trioxys pallidus* Haliday (Hymenoptera: Aphidiidae). *Am. Entomol.* 37(4):232–39.

Frazer B. D., and R. van den Bosch. 1973. Biological control of the walnut aphid in California: The interrelationship of the aphid and its parasite. *Environ. Entomol.* 2:561–68.

Hoy, M. A. 1990. Pesticide resistance in arthropod natural enemies: Variability and selection responses. In *Pesticide resistance in arthropods,* eds. R. T. Roush and B. Tabashnik, 203-36. New York: Chapman & Hall.

Hoy, M. A., and F. E. Cave. 1988. Guthion-resistant strain of walnut aphid parasite. *Calif. Agric.* 40(4):4–5.

———. 1989a. Parasite tolerates other pesticides. *Calif. Agric.* 43(5):4–26.

———. 1989b. Toxicity of pesticides used on walnuts to a wild and azinphosmethyl-resistant strain of *Trioxys pallidus* (Hymenoptera: Aphidiidae). *J. Econ. Entomol.* 82:1585–92.

———. 1991. Genetic improvement of a parasitoid: Response by *Trioxys pallidus* to laboratory selection with azinphosmethyl. *Biocontrol Science and Techn.* 1:31–41.

Hoy, M. A., F. E. Cave, R. H. Beede, J. Grant, W. H. Krueger, W. H. Olson, K. M. Spollen, W. W. Barnett, and L. C. Hendricks. 1989. Guthion-resistant walnut aphid parasite: Release, dispersal, and recovery in orchards. *Calif. Agric.* 43(5):21–23.

———. 1990. Release, dispersal, and recovery of a laboratory-selected strain of the walnut aphid parasite *Trioxys pallidus* (Hymenoptera: Aphidiidae) resistant to azinphosmethyl. *J. Econ. Entomol.* 83:89–96.

Riedl, H., M. M. Barnes, and C. S. Davis. 1979. Walnut pest management: Historical perspective and present status. In *Pest management programs for deciduous tree fruits and nuts,* eds. D. M. Boethel and R. D. Eikenbary, 15–80. New York: Plenum Press.

Schlinger, E. I., K. S. Hagen, and R. van den Bosch. 1960. Imported French parasite of walnut aphid established in California. *Calif. Agric.* 14(11):3–4.

Sibbett, G. S., L. Bettiga, and M. Bailey. 1981. Impact of summer infestation of walnut aphid on quality. *Sun-Diamond Grower* June–July:8–9, 50.

van den Bosch, R., E. I. Schlinger, and K. S. Hagen. 1962. Initial field observations in California on *Trioxys pallidus* (Haliday), a recently introduced parasite of the walnut aphid. *J. Econ. Entomol.* 55:857–62.

van den Bosch, R., B. D. Frazer, C. S. Davis, P. S. Messenger, and R. Hom. 1970. *Trioxys pallidus:* An effective new walnut aphid parasite from Iran. *Calif. Agric.* 24(11):8–10.

van den Bosch, R., R. Hom, P. Matteson, B. D. Frazer, P. S. Messenger, and C. S. Davis. 1979. Biological control of the walnut aphid in California: Impact of the parasite *Trioxys pallidus*. *Hilgardia* 47(1):1–13.

34 / BLACK SCALE

K. M. DAANE, C. E. KENNETT, L. E. CALTAGIRONE, AND M. S. BARZMAN

> **black scale**
> **Saissetia oleae (Olivier)**
> **Homoptera: Coccidae**

INTRODUCTION

Biology and Pest Status

The black scale, *Saissetia oleae* (Olivier), is the most serious insect pest of olives in California. It damages the tree directly through its feeding and indirectly by producing honeydew upon which black, sooty molds—most commonly *Cladosporium* species—then grow. To control the scale, growers often rely on annual insecticide applications. Although this can control the pest, it is often unnecessary (Shoemaker, Huffaker, and Kennett 1979) and disrupts regulation by natural enemies (Paraskakis, Neuenschwander, and Michelakis 1980).

Black scale outbreaks are unpredictable: it is not uncommon for one orchard to have repeated invasions and outbreaks while neighboring orchards are free of the scale. As a result, little work has been done to develop and implement a reliable pest-management program. K. M. Daane (1988) found that cultural practices such as pruning could cause local populations to fluctuate, and that the Central Valley's regional fluctuations were influenced by superior natural enemy activity in the northern orchards compared with the southern orchards. These findings suggested that biological control could be improved by manipulating cultural practices or by establishing natural enemies in the San Joaquin Valley.

However, the scale's development pattern of one generation per year impedes the establishment of parasites in the San Joaquin Valley, because for long periods the proper host stage cannot be found (Daane and Caltagirone 1989). Many of the initial attempts to establish parasites in this region failed because of inadequate biological information (Bartlett 1978). For example,

until the 1930s many primary parasites with obligatory male hyperparasitic habits were discarded as dangerous secondaries. Racial differences may also influence establishment. A. Panis and J. P. Marro (1978) found that two biotypes of *Metaphycus lounsburyi* (Howard) had pronounced physiological and behavioral differences, which allowed them to be used in the same orchard for biological control.

Historical Notes

Black scale is believed to be native to South Africa, where many of its natural enemies are found (De Lotto 1976). The scale was first reported in California in the 1870s. As a result of its wide host range and high reproductive capacity, it has spread throughout the state and is now endemic to California's Central Valley, where it causes recurrent infestations.

In California a major biological control campaign has been directed against *S. oleae*. Scientists began importing natural enemies in 1891; since then, species have been imported from Asia, Africa, Australia, Central and South America, Europe, and the Middle East (Bartlett 1978). These efforts have controlled the scale quite successfully in coastal regions; however, the results in the Central Valley have been disappointing. Of approximately 50 natural enemies introduced there, less than 15 have become established, and these have not provided satisfactory control (Daane, Barzman, Kennett, et al. 1991).

In other countries the natural enemies of the black scale have had better success: in Greece, Italy, and Libya, population dynamic studies have attributed the significant black scale mortality to natural enemies (Argyriou and Katsoyannos 1976; Monaco 1976; Paraskakis, Neuenschwander, and Michelakis 1980; Roberti 1981).

This research was funded in part by the California Olive Committee, the University of California Statewide Integrated Pest Management Project, the USDA-ES Smith-Lever IPM Project Development, the W-84 Regional Project, and the Division of Biological Control. We thank M. Martin of Olberti Olives; G. Martin and L. Ferguson of the University of California, Davis; S. Sibbett of Tulare County Cooperative Extension; and K. Hagen, D. Rowney, and D. Brombeger of the University of California, Berkeley.

RESULTS

Objective A: The Identification and Introduction of Beneficial Organisms

Since W-84 funding began, researchers have introduced several new parasites. These introductions, made between 1979 and 1986, include *Metaphycus inviscus* Compere, *Metaphycus* species A, *Prococcophagus probus* Annecke & Mynhardt, *Prococcophagus saissetia* Annecke & Mynhardt, *Aloencyrtus saissetia* Compere, and *Coccophagus rusti* Compere (Kennett 1986; Daane and Caltagirone 1989).

A 4-year survey of parasites in Central Valley olive orchards revealed that *P. probus* and *C. rusti* have become established. Since its release in 1986, *M.* species A has been continually recovered from release sites near the coast and is now considered to be established in California (Daane unpub. data).

Objective B: Ecological and Physiological Studies

Scale outbreaks have been linked to the temperature and relative humidity of the olive orchard (Rosen, Harpaz, and Samish 1971; Morillo 1975; Podoler, Bar-Zacay, and Rosen 1979). Crawlers typically hatch in May and June and are immediately subjected to hot, dry summer conditions. This is the time of the greatest mortality.

The cultural practices used in olive orchards can affect the microclimate, which in turn can influence black scale development and survival (Daane 1988). For example, closed canopies and high-volume sprinkler systems lower the ambient temperature and raise humidity, producing an environment more conducive to scale outbreaks. We have produced guidelines to help growers maximize the natural abiotic mortality (Daane, Caltagirone, Ferguson, Krueger, et al. 1989); these include pruning strategies that either increase summer mortality or influence scale development patterns that favor parasite establishment.

Previous sampling programs have ranged from periodic observations and collections (Kennett 1980) to random collections of branches of specified lengths (Rosen, Harpaz, and Samish 1971; Argyriou and Katsoyannos 1976) to more intricate sampling procedures involving fixed sampling trees or units (Podoler, Bar-Zacay, and Rosen 1979; Neuenschwander and Paraskakis 1980; Paraskakis, Neuenschwander, and Michelakis 1980). However, such intricate monitoring procedures are too cumbersome for growers to use.

To help growers predict scale outbreaks, we have developed a simple monitoring program (Daane and Caltagirone 1990), having determined that fecundity would vary from 400 to over 1,200 eggs per scale, depending on cultural practices and temperatures.

Growers can predict mortality and the level of the next season's scale population from adult scale counts, summer temperatures, and olive canopy type. These estimates will help them decide on such management techniques as parasite release or pruning.

Objective C: The Conservation and Augmentation of Beneficial Organisms

Black scale develops at different rates on different plant species. Because of this, alternative hosts can provide refuges for the parasite in the orchard. Researchers have followed the scale's development on oleander (*Nerium oleander*), pepper tree (*Schinus molle*), coyote bush (*Baccharis pilularis consanguinea*), Japanese boxwood (*Buxus japonica*), and toyon (*Heteromeles arbutifolia*), under controlled conditions of temperature and light.

At all temperatures tested—65°, 70°, 80°, and 90°F—oleander produced the most rapid scale development. Scale infestations are found in the Central Valley, where temperatures often exceed 100°F for many days in a row; however, at a constant temperature, the upper development threshold for early instars on all host plants was 90°F.

For many years augmentative releases of *Metaphycus helvolus* (Compere) have been used for black scale control in coastal citrus. Recently, we have investigated this technique on olives (Daane, Barzman, and Caltagirone 1991). In 1989 and 1990, data from paired plots with and without *M. helvolus* release consistently showed that release plots had significantly greater numbers of *M. helvolus*, as well as, in most cases, a significantly greater black scale mortality. However, in plots whose scale populations were at outbreak levels, damage remained above the level of economic injury during the first year; this indicates that where scale populations are very high, *M. helvolus* release cannot be used in place of chemical control.

Objective D: Impact

As part of their biological control efforts, scientists have distributed a number of parasitic and predaceous insects in California. Some have become established and are able, under certain circumstances, to provide satisfactory control of the scale (Kennett 1986; Daane 1988). Most of the natural enemies that became established were released in the 1930s. Soon afterward, natural regulation was disrupted when the olive industry was hit by a new and far more damaging insect pest, the olive scale, *Parlatoria oleae* (Colvée). After biological control efforts were successful with the olive scale (Huffaker and Kennett 1966), black scale again rose to prominence, and importation efforts were resumed.

Since then several parasites have been introduced successfully. The most important is *Metaphycus bartletti*

Annecke & Mynhardt, which was introduced in 1958 and has now spread throughout the state (Kennett 1986). It accounts for the greatest percentage of parasitism of preovipositional and adult scales (Kennett 1980; Daane 1988). Between 1979 and 1983 *C. rusti*, *P. probus*, and *P. saissetia* were released. Although recoveries have not been high, *C. rusti* and *P. probus* both have become established (Daane, Barzman, Kennett, et al. 1991). *M. inviscus* was released initially in 1958 and again in 1980. In coastal sites it has been recovered continually since soon after release (Kennett 1986). Finally, in 1985 *Metaphycus* species A was imported from Spain. Although it has been recovered from northern, southern, and coastal release sites (Daane and Caltagirone 1989), field surveys have shown its effect to be negligible.

RECOMMENDATIONS

1. Develop an economical production method for the black scale parasite *M. helvolus* for use in augmentative release programs.

2. Develop a taxonomic key of black scale parasites found in California for pest control advisors and Cooperative Extension personnel.

3. Determine economic injury levels for black scale. Also, the damage caused to the tree's nutrient resources and photosynthetic capabilities by sooty mold should be determined.

4. Examine the use of ground cover to increase natural enemies, and to influence soil fertility, erosion, water loss, and olive diseases.

REFERENCES CITED

Argyriou, L. C., and P. Katsoyannos. 1976. Establishment and dispersion of *Metaphycus helvolus* Compere in Kerkyra (Corfu) on *Saissetia oleae* (Oliver). *Ann. Inst. Phytopath.* Benaki (N. S.) 11:200–8.

Bartlett, B. R. 1978. Coccidae. In *Introduced parasites and predators of arthropod pests and weeds: A world review*, ed. C. P. Clausen, 57–74. USDA Agricultural Reserve Service Handbook 480. Washington, D.C.: U.S. Department of Agriculture.

Daane, K. M. 1988. The effect of cultural practices on the population dynamics of black scale, *Saissetia oleae* (Olivier), and its natural enemies in California olive orchards. Ph.D. diss., University of California, Berkeley.

Daane, K. M., and L. E. Caltagirone. 1989. Biological control in California olive orchards: Cultural practices affect biological control of black scale. *Calif. Agric.* 43(1):9–11.

———. 1990. Black scale distribution in the olive canopy:

Development of a sampling plan. Proceedings of the First International Symposium on Olive Growing, Córdoba, Spain. *Acta Horticulturae* 286:347–50.

Daane, K. M., M. S. Barzman, and L. E. Caltagirone. 1991. Augmentative release of *Metaphycus helvolus* for control of black scale, *Saissetia oleae*, in olives. *Plant Protection Quarterly* 1:6–9.

Daane, K. M., M. S. Barzman, C. E. Kennett, and L. E. Caltagirone. 1991. Parasitoids of black scale in California: Establishment of *Prococcophagus probus* Annecke & Mynhardt, and *Coccophagus rusti* Compere (Hymenoptera: Aphelinidae) in olive orchards. *Pan-Pacific Entomol.* 67:99–106.

Daane, K. M., L. E. Caltagirone, L. Ferguson, B. Krueger, and S. Sibbett. 1989. *Black scale in olive orchards: Control techniques.* California Olive Committee.

De Lotto, G. 1976. On the black scales of southern Europe (Homoptera: Coccoidea: Coccidae). *J. Entomol Soc. So. Afr.* 39:147–49.

Huffaker, C. B., and C. E. Kennett. 1966. Biological control of *Parlatoria oleae* (Colvée) through the compensatory action of two introduced parasites. *Hilgardia* 37:283–335.

Kennett, C. E. 1980. Occurrence of *Metaphycus bartletti* Annecke & Mynhardt, a South African parasite of black scale, *Saissetia oleae* (Olivier), in central and northern California (Hymenoptera: Encyrtidae; Homoptera: Coccidae). *Pan-Pacific Entomol.* 56:107–10.

———. 1986. A survey of the parasitoid complex attacking black scale, *Saissetia oleae* (Olivier), in central and northern California (Hymenoptera: Chalcidoidea; Homoptera: Coccidae). *Pan-Pacific Entomol.* 62:363–69.

Monaco, R. 1976. Nota su *Metaphycus lounsburyi* (How.) (Hymenoptera: Encyrtidae) parassita di *Saissetia oleae* (Oliv.). *Entomol. Bari.* 12:143–51.

Morillo, C. 1975. Regulación de las poblaciones de *Saissetia oleae* (Olivier, 1791): Factores de mortalidad. (Hem.: Coccidae). *Graellsia* 29:221–31.

Neuenschwander, P., and M. Paraskakis. 1980. Studies on the distribution and population dynamics of *Saissetia oleae* (Oliv.) (Homoptera: Coccidae) within the canopy of the olive tree. *Zeit. angew. Entomol.* 90:366–78.

Panis, A., and J. P. Marro. 1978. Variation du comportement chez *Metaphycus lounsburyi* (Hymenoptera: Encyrtidae). *Entomophaga* 23:9–18.

Paraskakis, M., P. Neuenschwander, and S. Michelakis. 1980. *Saissetia oleae* (Oliv.) (Homoptera: Coccidae) and its parasites on olive trees in Crete, Greece. *Zeit. angew. Entomol.* 90:-450–64.

Podoler, H., I. Bar-Zacay, and D. Rosen. 1979. Population dynamics of the Mediterranean black scale, *Saissetia oleae* (Olivier), on citrus in Israel: I. A partial life table. *J. Entomol. Soc. So. Afr.* 42:257–66.

Roberti, D. 1981. Osservazioni sulla dinamica di popolazione e sulla parassitizzazione della *Saissetia oleae* (Oliv.) su olivo in Puglia. *Entomologica, 16, Bari.* 25:113–20.

Rosen, D., I. Harpaz, and M. Samish. 1971. Two species of *Saissetia* (Homoptera: Coccidae) injurious to olive in Israel and their natural enemies. *Israel J. Entomol.* 6:35–53.

Shoemaker, C. A., C. B. Huffaker, and C. E. Kennett. 1979. A systems approach to the integrated management of a complex of olive pests. *Environ. Entomol.* 8:182–89.

35 / CITRICOLA SCALE

C. E. KENNETT, K. S. HAGEN, AND K. M. DAANE

citricola scale
Coccus pseudomagnoliarum (Kuwana)
Homoptera: Coccidae

INTRODUCTION

Biology and Pest Status

A native of Asia, the citricola scale, *Coccus pseudomagnoliarum* (Kuwana), has been an important pest of citrus in California since the early 1900s (Quayle 1938). Although the scale is univoltine, its uniparental nature and high reproductive capacity enable it to increase rapidly in a single season. High temperatures (35+°C) during the reproductive and postreproductive periods from June to August take a severe toll on the young scales, but usually enough survive to cause economic damage.

Citricola scale, like most soft scales, excretes a honeydew on which sooty-mold fungi develop. This reduces the citrus's photosynthesis and results in the downgrading or cullage of discolored—that is, blackened—fruits. Extreme scale densities can also damage leaves and twigs to such an extent that flowering and fruit set are reduced in subsequent years.

Historical Notes

The natural range of citricola scale is apparently limited to Japan and southern China. Until the 1950s, the only other places the scale was found were California and possibly Arizona. Since then it has been found in a number of Near East and Mediterranean countries.

In California, attempts at biological control began in 1922, when several species of parasitoids were imported from Japan (Compere 1924). Only one of these was colonized in the field, and it failed to become established. Between 1936 and 1953 five additional parasitoids were colonized, but again, none became established (Flanders 1942; Gressitt, Flanders, and Bartlett 1954; Flanders and Bartlett 1964).

During the latter part of these efforts, researchers found that in southern California the scale had come under good biological control. This control was credited to a parasitoid complex consisting of native, cosmopolitan, and exotic species (Bartlett 1953); the latter having been introduced for control of the black scale, *Saissetia oleae* (Olivier). In central and northern California, however, this parasitoid complex has rarely affected scale population levels—a failure that has been attributed to various biotic and abiotic factors (Bartlett 1953; Kennett 1988).

In 1981 researchers revived this work, driven by the need for biological control in northerly citrus districts.

RESULTS

Objective A: The Identification and Introduction of Beneficial Organisms

Between 1981 and 1986 researchers explored Japan, collecting all the natural enemies (parasitoids) known to occur on citricola scale in that country. They discovered no new species. However, they found that only nonhost-specific parasitoids attacked the closely related brown soft scale *Coccus hesperidum* L. on citrus, whereas host-specific parasitoids were highly dominant on the citricola scale collected from hackberry, *Celtis sinensis* var. *japonica*. Both scales were rarely found on trifoliate orange, *Poncirus trifoliata*, a supposedly preferred host in Japan.

Researchers only propagated the host-specific encyrtid species, *Metaphycus orientalis* (Compere) and *Microterys okitsuensis* (Compere). These were colonized in 1985, and, though the former species was recovered in the field soon after release, later samplings failed to show that either had become established.

Recent efforts with *Metaphycus flavus* (Howard), introduced from Italy in 1987, gave similar results: the parasitoid showed only temporary establishment. Efforts with this species are continuing. *M. flavus* is a nonhost-specific parasitoid that apparently adapted to citricola scale after the scale became established in Italy in the 1970s.

RECOMMENDATIONS

To improve the level of control, which is presently negligible, additional parasitoid species need to be introduced. These should include the following: (1)

Nonhost-specific species that attack other soft scales, particularly those able to reproduce on the closely related brown soft scale. (2) Parasitoids that attack a broad range of host sizes—those in which larval development is solitary in smaller hosts (about 1.0 to 1.5 mm long) and gregarious in larger hosts (1.75 to 4.0 mm long). Such habits promote parasitoid survival when the host scale is univoltine and only one—usually an evenbrooded—stage is present at a time. (3) Parasitoids from the Mediterranean Basin, where the scale is a recent, post-1950s invader. For example, a number of species have been reported on the citricola scale in Turkey.

REFERENCES CITED

Bartlett, B. R. 1953. Natural control of citricola scale in California. *J. Econ. Entomol.* 46:25–28.

Compere, H. 1924. A preliminary report on the parasitic enemies of the citricola scale, *Coccus pseudomagnoliarum* (Kuwana), with descriptions of two new chalcidoid parasites. *Bull. So. Calif. Acad. Sci.* 24:113–23.

Flanders, S. E. 1942. Biological observations on the citricola scale and its parasites. *J. Econ. Entomol.* 35:830–33.

Flanders, S. E., and B. R. Bartlett. 1964. Observations on two species of *Metaphycus* (Hymenoptera: Encyrtidae) parasitic on citricola scale. *Mushi* (Fukuoka) 38:39–42.

Gressitt, J. L., S. E. Flanders, and B. R. Bartlett. 1954. Parasites of citricola scale in Japan, and their introduction into California. *Pan-Pacific Entomol.* 30:5–9.

Kennett, C. E. 1988. Results of exploration for parasitoids of citricola scale, *Coccus pseudomagnoliarum* (Homoptera: Coccidae), in Japan and their introduction in California. *Kontyu* (Tokyo) 56:445–57.

Quayle, H. J. 1938. *Insects of citrus and other subtropical fruits.* Ithaca, N.Y.: Comstock.

36 / ICEPLANT SCALES

R. L. TASSAN AND K. S. HAGEN

iceplant scales
Pulvinariella mesembryanthemi (Vallot) and *Pulvinaria delottoi* Gill
Homoptera: Coccidae

INTRODUCTION

Biology and Pest Status

The biologies of *Pulvinariella mesembryanthemi* (Vallot) and *Pulvinaria delottoi* Gill have been intensively studied under both field and laboratory conditions in California (Washburn 1984). Field studies of the biology of *P. mesembryanthemi* have also been made in Argentina (Quintana 1956). Briefly, both species reproduce parthenogenetically; although males have been observed in *P. mesembryanthemi* populations in California, they do not appear to be functional. Reproductive potential is quite high; females have been recorded laying in excess of 2,400 eggs. However, their realized reproductive capacity is considerably lower, ranging from 350 to 800 crawlers per female. *P. delottoi* is univoltine in northern California; *P. mesembryanthemi* is bivoltine in northern California, but has three to four generations per year in southern California, where temperatures are generally higher. Each species also differs in host use; *P. delottoi* tends to settle on mature plant parts, while *P. mesembryanthemi* colonizes the new terminal growth. The two scales damage the plant by feeding in vascular tissues; they also excrete honeydew, upon which sooty molds then grow, decreasing photosynthesis and contributing to a loss of plant vigor.

Previously free of insect pests, iceplant in California was first attacked in 1971, when *P. mesembryanthemi* (Vallot) was discovered infesting *Carpobrotus* species at private residences in Napa. In 1973, the scale was discovered in Alameda County; a second exotic scale species, later described as *P. delottoi* (Gill 1979), was found infesting iceplants in freeway landscaping in Alameda and Santa Cruz counties.

The California Department of Transportation (CALTRANS) estimated that, statewide, it was responsible for maintaining about 6,000 acres of various iceplant species in freeway landscaping. An equal acreage was estimated to exist on residential, commercial, and other public lands. One reason for the extensive use of iceplants was they were relatively free of insect damage (Donaldson et al. 1978), though they did support populations of spittlebugs, aphids, and a few lepidopterans.

Between 1971 and 1983, *P. mesembryanthemi* was found in 19 coastal counties, from Napa in northern California to San Diego in the south. In 1979 it was detected in the Central Valley, in Sacramento and Yolo counties. Recently, it has been found in Fresno County (Washburn 1984; Gill person. commun.). *P. delottoi* has a much more restricted distribution, having been found only between the San Francisco Bay Area and Monterey County.

By 1978 CALTRANS was finding it difficult to control the scales with insecticides. Even after repeated applications, iceplant scales were causing dead areas in freeway landscapes. The agency's chemical control costs had escalated—they came to over $50,000 per year in the San Francisco Bay Area alone—and the cost of replacing the iceplant with another groundcover was estimated at $20 million. CALTRANS was also concerned about its workers' increased exposure to pesticides, and about public and political pressures to use environmentally compatible control measures (Cassidy person. commun.).

Historical Notes

In the last 50 years in California, native and introduced iceplants in the families Aizoaceae and Crassulaceae have been used extensively as groundcover for ornamental landscaping, dune stabilization, and erosion control. The most widely used are *Carpobrotus edulis* (=*Mesembryanthemum edulis*), an imported species from South Africa, and the hybrid *C. edulis* × *C. aequilateralis*, a cross between *C. aequilateralis* (=*M. chilense*), a native species, and *C. edulis*. Iceplant scales were reported on ornamental plantings of *Carpobrotus* species and *Lampranthus* (Donaldson et al. 1978) and on a native species of *Dudleya* (Crassulaceae) in Monterey County (Shoener person. commun.). Laboratory tests of 49 species of Aizoaceae indicated that an additional 34 iceplant species in 18 genera would support iceplant scales (Washburn 1984).

Researchers believe that both *P. mesembryanthemi* and *P. delottoi* originated in southern Africa (Brain 1920; De Lotto 1967; Gill 1979), even though *P. mesembryanthemi* was first described in southern France and *P. delottoi* in California. *P. mesembryanthemi* has also been reported in southern Europe along the Mediterranean, on the Canary and Scilly Islands, in Germany, Chile, Argentina, and Australia (Quintana 1956; De Lotto 1967; Hodgson 1967; Collins and Scott 1982). *P. delottoi* has been found only in South Africa and California (Gill 1979).

RESULTS

Objective A: The Identification and Introduction of Beneficial Organisms

In April and May 1978, researchers searched for natural enemies of iceplant scales in South Africa, southern France, and Monaco (Hagen and Tassan 1982). Additional collections in South Africa were carried out by S. Nesser of the Republic of South Africa's Department of Agricultural Technical Services, Plant Protection Research Institute, in Stellenbosch. All of the natural enemies eventually introduced into California came from South Africa (see table 36.1). These consisted of five encyrtid parasitoids and two coccinellid predators, which were released after being successfully propagated in the laboratory (Tassan, Hagen, and Cassidy 1982). Subsequent natural enemy releases were made from both laboratory rearings and collections at the initial field colonization sites.

Metaphycus funicularis Annecke, *M. stramineus* Compere, and *Encyrtus saliens* Prinsloo & Annecke successfully established permanent populations and, either singly or in concert, significantly reduced iceplant scale populations throughout California. At the time these species were imported, there was little information on their biological attributes other than host associations. Since then, C. Y. Kitayama (1983) has investigated the biology of *M. funicularis*; R. L. Tassan (1988) that of *M. stramineus*, and E. J. Wright (1983) that of *E. saliens*. The following summary of the parasitoid's biologies is drawn from those studies.

All three species are primary, internal parasitoids of both scale species. The *Metaphycus* species are facultatively gregarious, whereas *E. saliens* is an obligate, solitary parasitoid. *M. funicularis* and *M. stramineus* attack all stages of host scales, from late first instar to nearly mature females. *E. saliens*, whose females are apterous, has a much more restricted range of host size, preferring

Table 36.1. South African natural enemies of iceplant scales introduced into California.

Natural enemy (origin)	Parent generation	Number released	Number of releases	Years released	Current status
HYMENOPTERA					
Encyrtidae					
Metaphycus funicularis (Pretoria, Transvaal Province)	15 F, 1 M*	555,600	265	1978–1983	established
Metaphycus stramineus (Masbaai, Cape Province)	21 F, 1 M	221,700	191	1978–1983	established
Metaphycus "A" (Jeffrey's Bay, Cape Province)	1 F, 1 M	100	1	1980	not recovered
Coccophagus cowperi (Beaufort West, Cape Province)	15 F, 0 M	29,300	34	1978–1981	recovered in first 2 years
Encyrtus saliens (Jeffrey's Bay, Cape Province)	1 F, 3 M	1,600	21	1981–1984	established
COLEOPTERA					
Coccinellidae					
Exochomus flavipes (Pretoria, Transvaal Province)	24 undet. sex	20,500	148	1978–1983	recovered in first 3 years
Hyperaspis senegalensis hottentotta (Shelly Beach, Natal Province)	93 undet. sex	52,100	192	1978–1983	recovered in first 2 years

Source: In part, R. L. Tassan (1988).
* M = male; F = female.

third instar and early adult scales for oviposition. Although the parasitoid species all develop slightly faster in *P. mesembryanthemi* than in *P. delottoi*, there do not appear to be significant differences in how they use either scale species. All three parasitoids have similar lower temperature thresholds but different immature developmental times: for *M. funicularis*, 198 degree-days above 11.4°C; for *M. stramineus*, 181 degree-days above 12.6°C; and for *E. saliens,* 363 degree-days above 12.9°C. At a constant 20°C, their immature development would take about 22, 25, and 52 days, respectively. In comparison, at the same temperature *P. mesembryanthemi* would require more than 3 months to complete immature development. Life-table studies have indicated that the approximate total fecundities were 160 eggs for *M. stramineus*, 180 eggs for *M. funicularis*, and 235 eggs for *E. saliens*.

Once the parasitoids cleared quarantine, researchers established laboratory colonies as sources for field releases. When field populations of *M. funicularis* and *M. stramineus* had become established in northern California, large numbers of these species were reared from field-collected material and then released in other areas of the state.

Researchers made the first field releases in northern California, from Napa south to Monterey County. As other counties reported infestations of iceplant scales, they were included in the release program. By the end of the project, releases had been made in every coastal county south to San Diego, as well as in Sacramento and Yolo counties in the Central Valley. CALTRANS personnel were active both in direct releases and in redistributing the natural enemies. Between 1978 and 1984, almost 900,000 individuals from the seven species had been reared and released (see table 36.1). CALTRANS personnel also collected scale-infested iceplant cuttings from release sites where the natural enemies had become established and distributed them to nonrelease sites.

Metaphycus funicularis, *M. stramineus*, and *E. saliens* became established soon after they were first released, often after only a single release. Each of these species has been observed continuously in the field since the initial releases.

The other introduced species failed to establish permanent populations. Although *Exochomus flavipes* was quick to establish in northern California, after the winter of 1981—an unusually cold one—it was no longer detectable. Researchers made several subsequent releases in northern California, but in each case there were no recoveries.

Objective D: Impact

Between February 1982 and August 1983, researchers systematically monitored the densities of natural enemies and scales in freeway landscaping at four northern California sites (Tassan and Hagen unpub. data); an additional eight sites were monitored from April through December 1982 (Frankie and Hagen 1985). Although the relative numbers of natural enemies and scales varied considerably from site to site, as did the seasonal timing of events such as population peaks, the data from one site where both *Metaphycus* species occurred illustrate the general patterns (Tassan and Hagen unpub. data). In a scale-infested planting at the intersection of Interstate 880 and Thornton Avenue in Newark (Alameda County), 25 cups, each 11 centimeters in diameter, were filled with yellow antifreeze and arranged in a 5-by-5 grid on 2-meter centers. Figure 36.1 shows the average numbers of scale crawlers and adult parasitoids found in the cups in each 2-week interval.

At the start of the sampling, scale densities were high throughout the plot, and some plants were dying. The densities of the scales, reflected by the numbers of crawlers found in the traps, decreased by about 50 percent with each successive generation: June 1982, October 1982, and June 1983. In February 1982, few *Metaphycus* were recovered, but by the summer of 1982 *Metaphycus* populations began to increase substantially. Between June and October 1982, *M. funicularis* slightly outnumbered *M. stramineus*. During the late fall and early winter of 1982 to 1983, nearly equal numbers were recovered. By early 1983, *M. stramineus* became the more numerous of the two species, more than doubling in number at each peak (February, May, and August). In 1983 very few *M. funicularis* were recovered. A sample taken from this site in the summer of 1989 revealed that both *M. stramineus* and *M. funicularis* were still present, even though scale-infested plants were difficult to locate.

This pattern of ever-decreasing scale densities and a concurrent increase in either or both *M. funicularis* and *M. stramineus* was observed at almost all the intensively studied sites, as well as at other locations throughout the state that were only irregularly inspected. The effectiveness of *M. funicularis*, *M. stramineus,* and *E. saliens* in maintaining iceplant scale populations at low levels has nearly eliminated the use of insecticides in freeway landscaping (CALTRANS landscapers person. commun.).

RECOMMENDATIONS

1. Periodic year-long sampling programs, at 3- to 5-year intervals, should be conducted at northern California sites to document the long-term impact of the introduced natural enemies. Additional studies of the dynamics of this system may contribute to the general theory of biological control.

2. An effective means of maintaining high visibility for this project needs to be developed. Urban landscapers

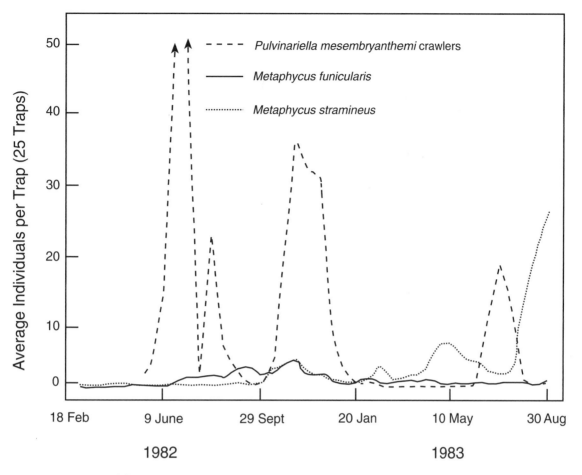

Figure 36.1. Seasonal fluctuations in the densities of iceplant scale (*Pulvinariella mesembryanthemi*) crawlers, and two adult parasitoids (*Metaphycus funicularis* and *M. stramineus*), in a northern California freeway landscaped with iceplant, February 1982 to August 1983.

are faced with a bewildering array of information that is often out of date. As a result, they tend to intervene when an iceplant scale infestation is found, often making unnecessary insecticide applications.

REFERENCES CITED

Brain, C. K. 1920. The Coccidae of South Africa. *Bull. Entomol. Res.* 11:1–41.

Collins, L., and J. K. Scott. 1982. Interaction of ants, predators, and the scale insect, *Pulvinariella mesembryanthemi*, on *Carpobrotus edulis*, an exotic plant naturalized in Australia. *Aust. Entomol. Mag.* 8:73–78.

De Lotto, G. 1967. The soft scales (Homoptera: Coccidae) of South Africa. *So. Afr. J. Agric. Sci.* 10:781–810.

Donaldson, D. R., W. S. Moore, C. S. Koehler, and J. L. Joos. 1978. Scales threaten iceplant in the Bay Area. *Calif. Agric.* 32(10):4, 7.

Frankie, G. W., and K. S. Hagen. 1985. Ecology and biology of iceplant scales, *Pulvinaria* and *Pulvinariella*, in California. Final Report FHWA–CA–HM–85. Sacramento: State Department of Transportation.

Gill, R. J. 1979. A new species of *Pulvinaria* Targioni-Tozzetti (Homoptera: Coccidae) attacking ice plant in California. *Pan-Pacific Entomol.* 55:241–50.

Hagen, K. S., and R. L. Tassan. 1982. Ecology and biological control of ice plant scales, *Pulvinariella mesembryanthemi* and *Pulvinaria delottoi*, in California. Subproject 2: Biological control of ice plant scales in California. Interim Report FHWA–CA–UCB–E79-HM-08. Sacramento: State Department of Transportation.

Hodgson, C. J. 1967. Some *Pulvinaria* species (Homoptera: Coccidae) from the Ethiopian region. *J. Entomol. Soc. So. Afr.* 30:198–211.

Kitayama, C. Y. 1983. The biology and ecology of *Metaphycus funicularis* (Hymenoptera: Encyrtidae). Ph.D. diss., University of California, Berkeley.

Quintana, F. J. 1956. *Pulvinaria mesembryanthemi* (Vallot) (Homoptera: Sternorrhyncha), nueva cochinilla para la fauna argentina y sus zooparásitos. *La Plata Universidad Nacional, Facultad de Agronomía Rev.* 32:75–110.

Tassan, R. L. 1988. Some biological attributes of *Metaphycus stramineus* (Hymenoptera: Encyrtidae): A parasitoid imported into California for biological control of iceplant scales. Ph.D. diss., University of California, Berkeley.

Tassan, R. L., K. S. Hagen, and D. V. Cassidy. 1982. Imported natural enemies established against ice plant scales in California. *Calif. Agric.* 36(9–10):16–17.

Washburn, J. O. 1984. The biology and ecology of iceplant scales, *Pulvinariella mesembryanthemi* and *Pulvinaria delottoi*, in California. Ph.D. diss., University of California, Berkeley.

Wright, E. J. 1983. The biology of *Encyrtus saliens* (Hymenoptera: Encyrtidae), an imported parasite of ice plant scales in California. Ph.D. diss., University of California, Berkeley.

37 / CALIFORNIA RED SCALE

R. F. LUCK

California red scale
Aonidiella aurantii (Maskell)
Homoptera: Diaspididae

INTRODUCTION

Biology and Pest Status

The California red scale is an insect pest of citrus in arid and semiarid regions of the world (Bodenheimer 1951; Ebeling 1959). In California it is one of the two key pests of citrus (Luck 1981). The scale infests all above-ground portions of the plant, but thrives on the leaves and fruits and, to a lesser extent, on the twigs (Quayle 1938). When the scale population is not abundant enough to reduce plant growth, infested fruit may be culled according to industry standards and market conditions. At higher densities, the scale may defoliate portions of a tree, inhibit fruit production, and kill branches or even sections of a tree.

The California red scale is multivoltine and develops year-round, producing two to three generations a year depending on geographic location and temperature. All of its stages overwinter, though the earlier stages—particularly the molt stages and the male pupae—are frequently killed by cooler winter weather (Abdelrahman 1974a). Mature females give birth to fully developed crawlers that walk to suitable feeding sites and settle within 2 to 6 hours (Bodenheimer 1951; Willard 1973; Hare unpub. data). The crawler inserts its stylets into plant tissues, secretes a waxy scale cover, and becomes immobile.

After molting to the second instar, male and female scales differentiate morphologically: males becoming oblong, whereas females remain round. Females molt once again to the third instar, and, about 2 days later, begin emitting a sex pheromone. Males undergo two additional nonfeeding stages after the second instar and emerge as winged adults at the same time as females begin producing pheromones. Adult males do not feed, and they live no more than a few hours. After mating, the female increases in size, becomes sclerotized and attached to the scale cover, and secretes a waxy sheath beneath her body; she then begins to form eggs. Females produce 2 or 3 crawlers per day, totaling 100 to 150 over their lifetime. At 27°C it takes a female scale approximately 47 days to develop from a crawler to a mature, crawler-producing female.

Historical Notes

The scale first appeared in southern California between 1868 and 1875, arriving on shipments of citrus nursery stock from Australia (Quayle 1938). In their control efforts, researchers have introduced 52 species or strains of natural enemies, 8 of which became established. These include three coccinellid predators—*Rhyzobius lophanthae* (Blaisdell), *Orcus chalybeus* (Boisduval), and *Chilocorus similis* (Rossi)—and five parasitoids—*Comperiella bifasciata* Howard, *Habrolepis rouxi* Compere, *Aphytis lingnanensis* Compere, *Encarsia* (=*Prospaltella*) *perniciosi* Tower, and *Aphytis melinus* DeBach (Rosen and DeBach 1978). But it was not until *A. melinus* was introduced in 1956 and 1957 that the scale came under control in southern California (Rosen and DeBach 1978, 1979).

In contrast, control has been unsuccessful in the San Joaquin Valley and the desert areas of southern California, even though *R. lophanthae, C. bifasciata,* and *A. melinus* are indigenous to those areas. Growers in the San Joaquin Valley have thus relied exclusively on broad-spectrum synthetic organic insecticides (Flaherty, Pehrson, and Kennett 1973). My colleagues and I have sought to understand why suppression of the scale by natural enemies differs between the two regions, in order to identify the general attributes of an effective natural enemy or predator-prey, parasitoid-host interactions, and to improve the biological control of scale in the San Joaquin Valley.

RESULTS

Objective B: Ecological and Physiological Studies

Aphytis melinus is a facultatively gregarious ectoparasitoid that lays one to several eggs—how many depends on the host's size—beneath the cover on the outside of the scale body (Luck, Podoler, and Kfir 1982). The scale is paralyzed when the egg is laid, which halts scale growth; thus, the only food available to the wasp's developing offspring is that contained in the scale at the time of attack (Rosen and DeBach 1979; Luck, Podoler, and Kfir 1982).

The parasitoid only attacks male and female second-instar scales larger than 0.15 square millimeters (body area equals length times width and is an index of scale size) and third-instar female scales before they reach the gravid stage (Luck and Podoler 1985; Opp and Luck 1986); it is unable to parasitize the molt or gravid female scales (Abdelrahman 1974b; Rosen and DeBach 1979). The parasitoid can control the sex of its eggs by determining whether or not they are fertilized at the time of oviposition; usually it allocates female eggs only to scales larger than 0.39 square millimeters; smaller scales receive male eggs 91 percent of the time (Abdelrahman 1974b; Luck and Podoler 1985; Opp and Luck 1986). This size-dependent sex allocation is not a result of a greater mortality of females on small hosts; instead, it is an active "choice" by the female wasp at the time the egg is laid (Luck and Podoler 1985; Opp and Luck 1986). Thus, to persist in sufficient numbers *A. melinus* needs a continuous supply of scales larger than 0.39 square millimeters.

Scale size also affects wasp size: the larger the scale, the larger the wasp (Opp and Luck 1986). Wasp size in turn affects *A. melinus*'s lifetime fecundity (Luck unpub. data). Moreover, *A. melinus* prefers to parasitize the largest scales (Yu 1986). Thus, the size of scales at the time adult *A. melinus* are searching largely determines (1) the likelihood that a scale will be attacked (the probability increases with its size); (2) how many eggs will be laid (gregarious parasitism is uncommon on scales smaller than 0.55 square millimeters), and (3) which sex will be allocated (mostly male eggs are allocated to scales less than or equal to 0.39 square millimeters, larger scales are mostly—70 percent of the time—allocated female eggs).

Two important elements must be kept in mind, however, when considering scale size in relation to the risk of attack: (1) the size of the scale at its transition from the third instar to the gravid female stage, at which it becomes immune to attack, and (2) the developmental time and size range of scales during their growth from 0.39 square millimeters to the transition size. Because *A. melinus* cannot parasitize the gravid stage, the density of scales that are between 0.39 square millimeters and the transition size is crucial to the wasp's persistence in the field. The developmental time of the scale during this period determines the period when it is at risk of attack.

Several factors influence how large a scale is at the time of transition. Scales that grow during the summer are smaller than those that grow in cooler seasons—namely, autumn and spring (Yu and Luck 1988; Hare,

Yu, and Luck 1990; Hare and Luck 1991). Furthermore, scale size varies depending on the citrus cultivar and the substrate—fruit, leaves, twigs, or branches—on which it grows (Luck and Podoler 1985; Walde et al. 1989; Hare, Yu, and Luck 1990; Hare and Luck 1991). Scales that grow on fruit are the largest; those on the branches are the smallest; and those on the other substrates are of intermediate size (Luck and Podoler 1985; Yu 1986).

Moreover, an interaction occurs between scale size and the cultivar, the cultivar substrate, and the climate in which it grows (Luck and Podoler 1985; Hare, Yu, and Luck 1990; Hare and Luck 1991). Scales on certain cultivars and substrates (such as grapefruit branches) vary more in size, seasonally, than do those on leaves. Cultivar and substrate also influence the survival and fecundity of scales. For example, scales that settle as crawlers on the branches of oranges are more likely to die or to produce fewer offspring than are scales that settle on grapefruit or lemon branches (Hare, Yu, and Luck 1990). Because *A. melinus* is less likely to parasitize scales that grow on the branches (Walde et al. 1989; Yu, Luck, and Murdoch 1990), the branches of grapefruit and lemon serve as a refuge for the scale. The crawlers from these sheltered scales can contribute substantially to the scale population in a tree, influencing the seasonal population dynamics (Luck, Yu, Murdoch, and Walde unpub. data).

We suspect that biological control of the scale in the San Joaquin Valley fails because scales of the size that *A. melinus* needs in order to produce daughters are periodically scarce or absent in the late spring. At least three interacting factors appear to cause this periodic scarcity, or resource bottleneck: (1) the differential winter mortality of the younger scale stages synchronizes the scale's age structure in the late spring and early summer and leads to an absence or scarcity of third-instar scales; (2) the hot summer temperatures of the San Joaquin Valley foster rapid developmental rates and hence small third-instar scales; and (3) the cultivar and substrates within the cultivar provide refuge for the scales. As a result, in spring the scale population is dominated by one or two age classes that move as a wave through each succeeding stage. Such an age structure leads to periods when the hosts that *A. melinus* needs in order to produce daughters are abundant, followed by periods when those hosts are absent or scarce (Luck 1986). In the inland coastal valleys of southern California, such synchronization appears to be less pronounced (Luck 1986).

RECOMMENDATIONS

1. Test the resource-bottleneck hypothesis by conducting augmentative releases of *Aphytis melinus* timed to coincide with susceptible scale stages in the field.

2. Compare the seasonal change in the red scale's age structure in the San Joaquin Valley with that of the scale in the inland coastal valleys of southern California.

3. Foreign exploration for more biological control agents—especially for *Pteropterix* and other species from southern China—should be undertaken.

REFERENCES CITED

Abdelrahman, I. 1974a. The effects of extreme temperatures on California red scale, *Aonidiella aurantii* (Mask.) (Hemiptera: Diaspididae) and its natural enemies. *Aust. J. Zool.* 22:203–12.

———. 1974b. Studies in the ovipositional behavior and control of sex in *Aphytis melinus* DeBach, a parasite of California red scale, *Aonidiella aurantii* (Mask.). *Aust J. Zool.* 22:231–47.

Bodenheimer, F. S. 1951. *Citrus entomology in the Middle East.* The Hague: Junk.

Ebeling, W. 1959. *Subtropical fruit pests.* Berkeley: University of California Division of Agricultural Science.

Flaherty, D. L., J. E. Pehrson, and C. E. Kennett. 1973. Citrus pest management studies in Tulare County. *Calif. Agric.* 27(11):3–7.

Hare, J. D., and R. F. Luck. 1991. Indirect effects of citrus cultivars on life history parameters of a parasitic wasp. *Ecology* 72:1576–85.

Hare, J. D., D. S. Yu, and R. F. Luck. 1990. Variation in survival, ultimate size, and fecundity of the California red scale on different citrus cultivars. *Ecology* 71:1451–60.

Luck, R. F. 1981. Integrated pest management in California citrus. *Proc. Internat. Soc. Citriculture* 2:630–35.

———. 1986. The role of scale size in the biological control of California red scale. In *Commission of the European Communities Experts Meeting,* 355–64. Integrated Pest Control in Citrus Groves, March 26–29, 1985, Acireale, Italy.

Luck, R. F., and H. Podoler. 1985. Competitive exclusion of *Aphytis lingnanensis* by *A. melinus*: Potential role of host size. *Ecology* 66:904–13.

Luck, R. F., H. Podoler, and R. Kfir. 1982. Host selection and egg allocation behavior by *Aphytis melinus* and *A. lingnanensis*: Comparison of two facultatively gregarious parasitoids. *Ecol. Entomol.* 7:397–408.

Opp, S. B., and R. F. Luck. 1986. Effects of host size on selected fitness components of *Aphytis melinus* DeBach and *A. lingnanensis* Compere (Hymenoptera: Aphelinidae). *Ann. Entomol. Soc. Am.* 79:700–4.

Quayle, H. J. 1938. *Insects of citrus and other subtropical fruit.* Ithaca, N.Y.: Comstock.

Rosen, D., and P. DeBach. 1978. Diaspididae. In *Introduced parasites and predators of arthropod pests and weeds: A world review,* ed. C. P. Clausen, 78–128. USDA-ARS Handbook 480. Washington, D.C.: U.S. Department of Agriculture.

———. 1979. *Species of Aphytis of the World.* The Hague: Junk.

Walde, S. J., R. F. Luck, D. S. Yu, and W. W. Murdoch. 1989. A refuge for red scale: The role of size-selectivity by a parasitoid wasp. *Ecology* 70:1700–6.

Willard, J. R. 1973. Wandering time of the crawlers of California red scale, *Aonidiella aurantii* (Mask.) (Homoptera: Diaspididae). *Aust. J. Zool.* 21:217–29.

Yu, D. S. 1986. The interactions between California red scale, *Aonidiella aurantii* (Maskell), and its parasitoids in citrus groves of inland southern California. Ph.D. diss., University of California, Riverside.

Yu, D. S., and R. F. Luck. 1988. Temperature dependent size and development of California red scale (Homoptera: Diaspididae) and its effect on host availability for the ectoparasitoid, *Aphytis melinus. Environ. Entomol.* 17:154–61.

Yu, D. S., R. F. Luck, and W. W. Murdoch. 1990. Competition, resource partitioning, and coexistence of an endoparasitoid, *Encarsia perniciosi,* and an ectoparasitoid, *Aphytis melinus,* of the California red scale. *Ecol. Entomol.* 15:469–80.

38 / Coconut Scale

J. W. BEARDSLEY, A. M. VARGO, AND C. H. CHIU

> coconut scale
> *Aspidiotus destructor* Signoret
> Homoptera: Diaspididae

INTRODUCTION

Biology and Pest Status

The coconut scale, *Aspidiotus destructor* Signoret, is a polyphagous armored scale pest that is widely distributed in tropical and subtropical regions. Some of the more important crops it attacks are avocado, banana, breadfruit, cacao, cassava, coconut, cotton, guava, oil palm, papaya, rubber, sugarcane, and tea. Since the 1920s biological control measures have been applied against this pest in many tropical regions (Clausen 1978; Waterhouse and Norris 1988).

Historical Notes

Aspidiotus destructor first appeared in Hawaii in 1968 (Beardsley 1970). It caused damage primarily to coconut and other palms, guava, and ornamentals. In 1983 the scale was discovered in American Samoa, where most of the damage has been to coconut palms, breadfruit, and ornamentals (Vargo 1986). In 1969 researchers initiated a classical biological program in Hawaii.

RESULTS

Objective A: The Identification and Introduction of Beneficial Organisms

Between 1969 and 1975 the Hawaii State Department of Agriculture introduced several predaceous coccinellid beetles to combat the coconut scale. *Crytognatha nodiceps* Marshall, introduced from Trinidad in 1969, failed to become established, but two species that were introduced from Guam in 1970—*Chilocorus nigritus* (Fabricius) and *Pseudoscymnus anomalus* Chapin—became successfully established. Several other natural enemies—either previously introduced to combat other armored scales or fortuitous accidental introductions—were also found to attack the coconut scale.

In 1984 Hawaii's Department of Agriculture shipped several natural enemies to American Samoa to combat the coconut scale there. These included the coccinellids *Pseudoscymnus anomalus*, *Rhyzobius satelles* Blackburn, and *Telsimia nitida* Chapin; two *Aphytis* species; and the predatory mite *Hemisarcoptes* species. Two coccinellids—*Chilocorus nigritus* and *P. anomalus*—were obtained from Vanuatu in 1984 and were also released.

Objective D: Impact

Between 1982 and 1985, as part of the W-84 project, researchers conducted an evaluation study of the natural enemies attacking *Aspidiotus destructor* infesting guava on the island of Oahu, Hawaii (Chiu 1986). This study showed that on guava, the scale was being controlled by a complex of natural enemies made up of the coccinellid beetles *Pseudoscymnus anomalus*, *Rhizobius satelles,* and *Telsimia nitida*; the aphelinid parasitoids *Aphytis* species (tentatively identified as *A. chrysomphali* Mercet and *A. lingnanensis* Compere), and *Encarsia agilior* (Berlese); and a predatory mite, *Hemisarcoptes* species.

Researchers found that at higher mean temperatures *A. lingnanensis* parasitized more scales and *A. chrysomphali* parasitized fewer scales. *Hemisarcoptes* mites—which had higher populations in humid, wet conditions—killed up to 43 percent of scales in the study populations. Of the three coccinellid beetles, *P. anomalus* proved the most effective.

In American Samoa, a late 1985 survey of natural enemies showed that *C. nigritus*, *P. anomalus,* and *T. nitida* had become established and that scale populations on the island had decreased markedly.

RECOMMENDATIONS

In many parts of the tropics, the biological control of the coconut scale by introduced natural enemies has been successful. Additional introductions should be carried out wherever and whenever new infestations are discovered (Waterhouse and Norris 1988).

REFERENCES CITED

Beardsley, J. W. 1970. *Aspidiotus destructor* Signoret, an armored scale pest new to the Hawaiian Islands. *Proc. Hawaii. Entomol. Soc.* 20:505–8.

Chiu, C. H. 1986. Impact of natural enemies and climate on population dynamics of coconut scale (*Aspidiotus destructor* Signoret). Ph.D. diss., University of Hawaii, Honolulu.

Clausen, C. P. 1978. *Introduced parasites and predators of arthropod pests and weeds: A world review.* USDA-ARS Handbook 480. Washington, D.C.: U.S. Department of Agriculture.

Vargo, A. 1986. Biological control and its uses in American Samoa. *Food and Farm Fair Jour.* 1:60–66.

Waterhouse, D. F., and K. R. Norris. 1988. *Biological control: Pacific prospects.* Melbourne: Inkata Press.

39 / OBSCURE SCALE

L. E. EHLER

> obscure scale
> *Melanaspis obscura* (Comstock)
> Homoptera: Diaspididae

INTRODUCTION

Biology and Pest Status

The obscure scale, *Melanaspis obscura* (Comstock), is native to the eastern half of the United States, where it is commonly associated with plants in the genera *Quercus*, *Castanea*, and *Carya*. The scale was originally described by John Henry Comstock (as *Aspidiotus obscurus*), based on material taken from the willow oak (*Quercus phellos* L.) in Washington, D.C. (Comstock 1881). During the early part of this century the scale was known as *Chrysomphalus obscurus* (Comstock); for the last 50 or so years it has been known as *Melanaspis obscura* (Comstock). In its native range, the obscure scale is both an urban and an agricultural pest. In the eastern United States, it is a major pest of the pin oak (*Q. palustris* Muenchhausen) in home and park plantings (Stoetzel and Davidson 1971, 1973), whereas in the southeastern states and Texas it is a minor pest of pecan (*Carya illinoensis* [Wangenheim] K. Koch). The scale evidently does not commonly attain pest status on its native hosts in natural stands.

The biology and ecology of the obscure scale have been investigated on pecan in Louisiana (Baker 1933) and on pin oak in Maryland (Stoetzel and Davidson 1971) and Kentucky (Potter, Jensen, and Gordon 1989). Its developmental biology is typical for a bisexual species of armored scale: for females three instars (with the last instar neotenic) and for males two instars, plus prepupal, pupal, and adult stages. The scale is univoltine; however, the seasonal distribution of its developmental stages may vary considerably according to the host plant (Stoetzel and Davidson 1973). As would be expected for a native species, the obscure scale is exploited by a complex of natural enemies, including parasites, predators, and a pathogen. Preliminary observations have indicated that in natural or undisturbed habitats—but not necessarily in urban or agricultural systems—these natural enemies keep scale populations at relatively low levels.

Historical Notes

Since the 1930s there have been three documented infestations of the obscure scale in California. In March 1933 the first infestation was detected on pecan and English walnut in the San Fernando Valley of Los Angeles County. In 1933 and 1934 the application of oil sprays eradicated the scale (Mackie 1937). In March 1934 the second infestation was found, this time on pecan in San Diego County. Eradication was more difficult, but it was eventually achieved by applying oil and fumigating with HCN (Mackie 1943). The third infestation was discovered in Sacramento County in early 1962; at least four oak trees in Sacramento's Capitol Park were infested. Trunk injections of Bidrin failed to eradicate the scale (Harper 1966). At present the obscure scale is widespread in Capitol Park, where it infests both introduced oaks, such as *Q. palustris* and *Q. rubra* L., and native ones, such as *Q. agrifolia* Nee and *Q. wislizenii* A.DC. Except for an occasional *Aphytis* species, the scale is free of parasites. Beginning in 1981, the Sacramento population of obscure scale became a target for classical biological control.

RESULTS

Objective A: The Identification and Introduction of Beneficial Organisms

My exploration for natural enemies was directed primarily toward parasites associated with the obscure scale on pecan in Texas. Twelve species in eight genera were recovered (see table 39.1). Four species were generally dominant and are briefly discussed here. *Ablerus clisiocampae* (Ashmead) is evidently an obligate sec-

I thank M. E. Schauff, G. Viggiani, and J. B. Woolley for confirming my taxonomic determinations.

Table 39.1. Parasites associated with the obscure scale on pecan in Texas.

Family	Genus	Species and authority
Aphelinidae	*Ablerus*	*clisiocampae* (Ashmead)
	Aphytis	*melanostictus* Compere
		species A
	Physcus	*varicornis* (Howard)
	Coccophagoides	*fuscipennis* (Girault)
	Encarsia	*aurantii* (Howard)
		species A
		species B
	Marietta	*pulchella* (Howard)
Signiphoridae	*Signiphora*	nr. *flavopalliata* Ashmead
		species A
	Thysanus	*ater* Haliday

ondary parasite; however, this has not been confirmed experimentally, despite some attempts in the quarantine laboratory at the University of California, Davis. *Physcus varicornis* (Howard) is a nearctic primary parasite of several diaspidid scales, and apparently occurs in California already. *Coccophagoides fuscipennis* (Girault) is a nearctic primary parasite that is evidently host-specific to the obscure scale; it is apparently one of the few if not the only member of the parasite complex that has had a close coevolutionary relationship with the scale. Because culturing obscure scale in the laboratory is very difficult, I have been unable to culture either *P. varicornis* or *C. fuscipennis* on this host. Attempts to culture these species on oleander scale (*Aspidiotus nerii* Bouché) were also unsuccessful. The fourth parasite, *Encarsia aurantii* (Howard), was probably accidentally introduced into North America. It is a widespread primary parasite of a number of diaspidid scales and occurs in California, where it is associated primarily with the yellow scale, *Aonidiella citrina* (Coquillett). This parasite has been misidentified in the past, and I suspect that certain records in the literature identifying *E. berlesei* (Howard) as a dominant parasite of the obscure scale should actually name *E. aurantii*.

Because the obscure scale infestation in Capitol Park was localized and posed no immediate threat to either native oak stands or deciduous fruit and nut trees, it proved a good opportunity to pursue a more experimental approach to classical biological control. Instead of releasing all of the available species of primary parasites (an empirical approach), I chose to release a single species and determine its impact free of competing species. I selected *Encarsia aurantii* because it was easily reared on oleander scale, and because it possessed many of the attributes of an effective natural enemy: it has a relatively short generation time (about 45 days on olean-

der scale under laboratory conditions); it is thelyotokous (although arrhenotokous forms may also exist); and it has a dominant position in the parasite complex in the pest's native home. In many ways *E. aurantii* is an opportunistic species, so it appeared that, when released free of other members of the parasite complex, it might provide the desired level of biological control. However, the first step in this experiment was to get *E. aurantii* established.

Imported *E. aurantii* from Texas were released from quarantine in mid-1987, and I released about 50 females that fall. Over 11,000 more females were released during the late summer and fall of 1988. In both years the parasites were the progeny of a founder population collected from a scale-infested pecan on the Brazos County campus of Texas A&M University. The releases were timed to coincide with the presence of first- and second-instar scales, which are the stages in which the parasite oviposits. (The second-instar scale is eventually killed by the parasite.) The establishment of *E. aurantii* was confirmed in 1990, and its population density has increased steadily since then. Evaluation of the parasite's impact on the scale population is currently under way.

RECOMMENDATIONS

1. Continue to evaluate the ecological impact of *E. aurantii* on the obscure scale.

2. Introduce *C. fuscipennis* if *E. aurantii* fails to control the obscure scale satisfactorily.

REFERENCES CITED

Baker, H. 1933. *The obscure scale on the pecan and its control.* USDA Circular 295.

Comstock, J. H. 1881. Report of the entomologist. In *USDA Annual Report of the Commissioner, 1880,* 235–373.

Harper, R. W. 1966. Bureau of Entomology. *Bull. Calif. Dept. Agric.* 55(2):92–98.

Mackie, D. B. 1937. Entomological service. *Bull. Calif. Dept. Agric.* 26(4):418–38.

———. 1943. Bureau of Entomology and Plant Quarantine. *Bull. Calif. Dept. Agric.* 32(4):240–74.

Potter, D. A., M. P. Jensen, and F. C. Gordon. 1989. Phenology and degree-day relationships of obscure scale (Homoptera: Diaspididae) and associated parasites on pin oak in Kentucky. *J. Econ. Entomol.* 82:551–55.

Stoetzel, M. B., and J. A. Davidson. 1971. Biology of the obscure scale, *Melanaspis obscura* (Homoptera: Diaspididae), on pin oak in Maryland. *Ann. Entomol. Soc. Am.* 64:45–50.

———. 1973. Life history variations of the obscure scale (Homoptera: Diaspididae) on pin oak and white oak in Maryland. *Ann. Entomol. Soc. Am.* 66:308–11.

40 / Pineapple Mealybugs and Pineapple Field Ants

J. W. Beardsley, N. J. Reimer, and T. S. Su

pineapple mealybugs *Dysmicoccus* spp. Homoptera: Pseudococcidae	pineapple field ants *Pheidole megacephala* (Fabricius) Hymenoptera: Formicidae

INTRODUCTION

Biology and Pest Status

Pineapple mealybug wilt disease occurs in most areas where pineapples are grown commercially (Carter 1973). The disease is associated with the presence of certain mealybug species (Illingsworth 1931; Carter 1932). In Hawaii the principal species spreading the disease are the gray pineapple mealybug, *Dysmicoccus neobrevipes* Beardsley, and the pink pineapple mealybug, *Dysmicoccus brevipes* (Cockerell) (Carter 1932; Zimmerman 1948; Beardsley 1959; Beardsley et al. 1982). Although the cause of mealybug wilt is a matter of controversy (Carter 1951, 1963; Rohrback et al. 1988), commercial control has been achieved primarily by eliminating the pineapple field ants that clean up the mealybugs' honeydew secretions and interfere with the mealybugs' natural enemies (Carter 1973).

The big-headed ant, *Pheidole megacephala* (Fabricius), is the dominant ant species in most of Hawaii's pineapple fields (Phillips 1934). It is a tropicopolitan species apparently native to Africa (Wilson and Taylor 1967). Although it is the dominant ground-nesting species in most of Hawaii's lowland (below 700 meters in elevation) agricultural areas, above 600 meters it is sometimes replaced by the Argentine ant, *Iridomyrmex humilis* Mayr (Fluker & Beardsley 1970). In very dry, open lowland areas the fire ant, *Solenopsis geminata* Fabricius, may be dominant (Phillips 1934).

The big-headed ant is a polydomous species, characterized by mobile, ramifying, multiqueen colonies (Phillips 1934). Nuptial flights take place, but mated queens must rejoin established colonies to survive. The same is true of the Argentine ant, the second-most-important species in Hawaiian pineapple fields. The inability of newly mated queens to found new colonies has an important bearing on ant-control practices. (Newly mated fire ant queens, however, can initiate new colonies.)

Historical Notes

Pineapple growers formerly controlled the field ants by applying mirex bait or broadcast sprays of heptachlor. These insecticides were so effective that mealybug infestations and mealybug wilt became extremely rare. However, after the Environmental Protection Agency banned those materials in 1977 and 1982, respectively, the pineapple industry was left without effective formicides. W-84 project researchers focused on developing pest management strategies that would reduce the need for pesticides, and on developing efficacy data and application methods for environmentally acceptable formicides. A major part of this research has been composed of basic studies on the field ecology of ants and mealybugs, which are necessary to understand the interrelationships between these organisms and mealybug wilt disease (Beardsley et al. 1982).

RESULTS

Objective B: Ecological and Physiological Studies

Our research reconfirmed the symbiotic relationship between pineapple mealybugs and their attendant ants. In the control plots where ants were allowed to flourish, mealybugs reached damaging levels within a few months, and mealybug wilt resulted. In plots where ants were controlled before the mealybugs became established, mealybug populations remained at very low or undetectable levels. Where ants were eliminated, mealybug populations rapidly declined (Beardsley et al. 1982). At present we are investigating the interrelationship between ants and mealybugs on pineapple. We have identified a number of the pineapple mealybugs' natural enemies—coccinellid beetles, predaceous Diptera larvae, and parasitic encyrtid wasps—and are evaluating the ants' impact on them. We also are seeking to clarify other ways that ants may affect the establishment and growth of mealybug populations (such as the movement of mealybugs by ants) and the importance, if any, of ant-mediated sanitation (such as honeydew removal).

In devising strategies to combat pineapple mealybug wilt, we developed a technique to detect invading ants on field margins, using a system of lath stakes baited with a mixture of peanut butter and soy oil. By detecting and eliminating incipient infestations, it is possible to avoid large-scale formicide applications.

After performing various tests we found formicidal bait formulations to be superior to broadcast spray applications because of their minimal impact on nontarget organisms and the relatively small amounts of an effective toxicant required to achieve control. AMDRO bait, or hydomethylnon, proved highly effective against big-headed ants. This material is registered for noncrop use and can be applied to noncrop areas to prevent ants from moving into uninfested fields. We have also found other promising bait formulations and are testing them for efficacy against big-headed and Argentine ants.

RECOMMENDATIONS

1. Studies on the ecology and behavior of pineapple field ants should be continued. At present there are insufficient data on the migration rates of ants into fields, in-field survival after tillage, food sources in fallow fields, foraging behavior, and population seasonality. Such information might reveal vulnerable points in the ant's biology that could lead to better control.

2. The biological control of mealybugs needs to be reconsidered. Because of the longtime use of insecticidal sprays that not only destroyed mealybugs but also reduced natural enemy populations, this tactic was probably was never given a fair trial. Because the biological control of mealybugs will not be effective without the concurrent control of ant populations, a program that integrates the management of the two should be developed.

3. Ant predators and parasites need to be studied. Researchers have discovered a number of natural ene-

mies of fire ants, but very little work has been done on the natural enemies of big-headed ants.

REFERENCES CITED

Beardsley, J. W. 1959. On the taxonomy of pineapple mealybugs in Hawaii, with a description of a previously unnamed species (Homoptera: Pseudococcidae). *Proc. Hawaii. Entomol. Soc.* 17:29–37.

Beardsley, J. W., T. H. Su, F. L. McEwen, and D. Gerling. 1982. Field investigations on the interrelationships of the big-headed ant, the gray pineapple mealybug, and pineapple mealybug wilt disease in Hawaii. *Proc. Hawaii. Entomol. Soc.* 24:51–67.

Carter, W. 1932. Studies of populations of *Pseudococcus brevipes*, and wilt of pineapple. *Phytopathology* 23:207–42.

———. 1951. The feeding sequence of *Pseudococcus brevipes* (Ckl.) in relation to mealybug wilt of pineapples in Hawaii. *Phytopathology* 41:769–80.

———. 1963. Mealybug wilt of pineapple: A reappraisal. *Ann. New York Acad. Sci.* 105:741–64.

———. 1973. The systemic phytotoxemias. In *Insects in relation to plant disease,* 2nd ed., 274–308. New York: Wiley.

Fluker, S. S., and J. W. Beardsley. 1970. Sympatric associations of three ants: *Iridomyrmex humilis, Pheidole megacephala,* and *Anoplolepis longipes,* in Hawaii. *Ann. Entomol. Soc. Am.* 63:1290–96.

Illingsworth, J. F. 1931. Preliminary report on evidence that mealybugs are an important factor in pineapple wilt. *J. Econ. Entomol.* 24:877–89.

Phillips, S. S. 1934. The biology and distribution of ants in Hawaiian pineapple fields. *Experiment Station Pineapple Producers Cooperative Association Bulletin* 15:1–57.

Rohrback, K. R., J. W. Beardsley, T. L. German, N. J. Reimer, and W. C. Sanford. 1988. Mealybug wilt, mealybugs, and ants on pineapple. *Phytopathology* 72:558–65.

Wilson, E. O., and R. W. Taylor. 1967. *The ants of Polynesia (Hymenoptera: Formicidae).* Pacific Insects Monograph 14. Honolulu: Bishop Museum.

Zimmerman, E. C. 1948. *Insects of Hawaii,* vol 5. Honolulu: University of Hawaii Press.

41 / SPHERICAL MEALYBUG

J. R. NECHOLS

> **spherical mealybug**
> *Nipaecoccus viridis* (Newstead)
> Homoptera: Pseudococcidae

INTRODUCTION

Biology and Pest Status

The spherical mealybug, *Nipaecoccus viridis* (Newstead) (=*N. vastator* [Maskell]), is a serious tropical pest of many important food, forage, and ornamental woody plants, including citrus, mango, guava, cotton, and various leguminous shrubs and trees (Al-Rawy, Kaddou, and Al-Omar 1977; Nechols unpub. data).

After the winged males mate with the sessile females, the female mealybugs secrete a white, cottony, spherical ovisac that envelops most of the body. As batches of eggs accumulate, the ovisac increases. Newly deposited eggs are yellow but turn purplish as the embryos develop. A female can lay 400 to 700 eggs during her adult life (Ali 1957). Because *N. viridis* typically lives in aggregations on host plants, large masses of coalesced, honeydew-impregnated egg sacs are commonly observed. First-instar nymphs hatch in the ovisac masses, emerge, disperse to suitable feeding sites, and begin feeding on the plant's vascular tissues. The females pass through three nymphal instars; males have an additional pupalike resting stage. Female preimaginal development takes about 3 weeks at 32°C, and a generation—from egg to egg—is about 1 month (Ali 1957).

Historical Notes

Except in the New World, *N. viridis* is widely distributed throughout the tropics, including many of the islands in the Pacific Basin (Sharaf and Meyerdirk 1987). Although the date of its entry into Guam and the Northern Mariana Islands is unknown, outbreaks of the bug occurred between 1977 and 1980. The principal host plant was the woody legume, *Leucaena leucocephala* (Lam.) DeWit., a dominant shrub-tree used as an energy and forage source. Because this legume and other ornamental and fruit trees are very important to Guam's economy, in 1980 W-84 researchers began to identify and evaluate naturally occurring biological control agents.

RESULTS

Objective A: The Identification and Introduction of Beneficial Organisms

In surveys done between 1981 and 1983, researchers found that the spherical mealybug was attacked by a complex of naturally occurring enemies, including the dipteran predators *Kalodiplosis* sp. (Cecidomyiidae) and ?*Cacoxenus* sp. (Drosophilidae), the beetle predators *Scymnus roepkei* (Fluiter) (=*Nephus roepkei* Fluiter) and *Cryptolaemus montrouzieri* (Mulsant) (Coccinellidae), and the gregarious hymenopteran parasitoid *Anagyrus indicus* Shafee, Alam, & Angarwal (Encyrtidae). The most abundant and consistently encountered natural enemy was *A. indicus* (Nechols 1981, 1982, 1983; Nechols and Seibert 1985). This parasitoid was particularly effective at finding mealybugs at very low densities, and it exhibited high rates of parasitization—over 90 percent—at both high and low pest densities (Nechols 1982).

Objective B: Ecological and Physiological Studies

In host selection experiments in the laboratory J. R. Nechols and R. S. Kikuchi (1985) found that, when each host stage was exposed individually, *A. indicus* oviposited in all three nymphal instars and in the adult female mealybug. However, the larger (and thus older) the host, the higher the rates of parasitization, numbers of eggs laid per host, and proportion of female progeny. Moreover, when given a choice, female parasitoids preferred adult female hosts to third-instar nymphs, and they did not oviposit in first- or second-instar nymphs. By discriminating among host stages (that is, sizes), *A. indicus* should increase its fitness, because host survivorship is greater in the older stages. Larger hosts also provide more, and possibly better, food for the developing parasitoid. Finally, because larger hosts can support the development of multiple progeny of this gregarious parasitoid, oviposition efficiency is increased. Studies of the reproductive biology of *A. indicus* showed that female parasitoids mate within 1 hour of adult eclosion and deposit their first eggs about 6 hours after mating. Larger females are significantly more fecund, and lay considerably more eggs per day, than smaller females (Nechols unpub. data).

Objective D: Impact

When we conducted ant-exclusion experiments on natural stands of *Leucaena leucocephala* in Guam, we discovered that ants rarely interfere with the mealybug and its natural enemies (Nechols 1982; Nechols and Seibert 1985), and that in fact the natural enemies—*A. indicus* and various predators—have a major suppressive effect on mealybug populations (Nechols and Seibert 1985). Islandwide surveys on Guam and the Northern Marianas also showed that *A. indicus* almost always parasitizes even extremely low densities of mealybugs (Nechols 1982). These data provide indirect evidence that this parasitoid has an important impact on mealybug populations.

RECOMMENDATIONS

Field and laboratory data indicate that, in Guam, the naturally occurring parasitoid *Anagyrus indicus* is an extremely effective biological control agent. This natural enemy should be considered a prime candidate for introduction to other tropical areas where *N. viridis* becomes a pest.

REFERENCES CITED

Ali, S. M. 1957. Some bio-ecological studies on *Pseudococcus vastator* Mask. (Coccidae: Hemiptera). *Indian J. Entomol.* 19:54–58.

Al-Rawy, M. A., I. K. Kaddou, and M. A. Al-Omar. 1977. The present status of the spherical mealybug, *Nipaecoccus vastator* (Maskell) (Homoptera: Pseudococcidae), in Iraq. *Bull. Biol. Res. Centre* 8:3–15.

Nechols, J. R. 1981. *Entomology: Biological control*, 16–18. Annual Report of the Guam Agricultural Experiment Station.

———. 1982. *Entomology: Biological control*, 33–38, 40–42, 45–46. Annual Report of the Guam Agricultural Experiment Station.

———. 1983. *Entomology: Biological control*, 26. Annual Report of the Guam Agricultural Experiment Station.

Nechols, J. R., and R. S. Kikuchi. 1985. Host selection of the spherical mealybug (Homoptera: Pseudococcidae) by *Anagyrus indicus* (Hymenoptera: Encyrtidae): Influence of host stage on parasitoid oviposition, development, sex ratio, and survival. *Environ. Entomol.* 14:32–37.

Nechols, J. R., and T. F. Seibert. 1985. Biological control of the spherical mealybug, *Nipaecoccus vastator* (Homoptera: Pseudococcidae): Assessment by ant exclusion. *Environ. Entomol.* 14:45–47.

Sharaf, N. S., and D. E. Meyerdirk. 1987. A review on the biology, ecology, and control of *Nipaecoccus viridis* (Homoptera: Pseudococcidae). *Misc. Publ. Entomol. Soc. Am.* 66:1–18.

42 / ALFALFA WEEVIL

D. W. DAVIS

alfalfa weevil
Hypera postica (Gyllenhal)
Coleoptera: Curculionidae

INTRODUCTION

The alfalfa weevil, *Hypera postica* (Gyllenhal), was first recorded in the United States near Salt Lake City, Utah, in 1904 by E. G. Titus (1908). It spread rather rapidly; by 1925 it had become established in most regions from Colorado westward. It was most serious as a pest in areas with higher elevations and short seasons. Between 1911 and 1913, scientists made a major effort to introduce parasites (Chamberlin 1924). *Bathyplectes curculionis* (Thomson) became established and could soon be found in all weevil-infested areas.

Between 1939 and 1960, two other forms of alfalfa weevil became pests in the United States: the Egyptian alfalfa weevil, *H. brunneipennis* (Boheman), predominated in the hotter areas of the west and southwest, whereas the eastern alfalfa weevil biotype was dominant from the Great Plains to the east. The initial population, *H. postica* (Gyllenhal), became known as the western alfalfa weevil.

During the 1950s, chemical control of alfalfa weevils was highly effective, with growers commonly using low rates of heptachlor early in the season. However, about 1960 the weevil started showing high levels of resistance to the cyclodiene insecticides. This was followed by the banning of those chemicals on forages. None of the insecticides then available was an adequate replacement.

By the time the W-84 project began in 1964, there had been a rebirth of interest in biological control of the alfalfa weevil complex. In 1957 the USDA Beneficial Insects Research Laboratory had started a second major effort to introduce alfalfa weevil parasites (Dysart and Day 1976). Concentrating on the eastern weevil biotype, they released seven parasite species during the ensuing years. In the West, W-84 workers released parasites reared by the USDA and the state of California. Unfortunately, the parasite species adapted to the eastern alfalfa weevil did not adapt to the western biotype. During this period, the alfalfa weevil complex had become a major pest in most of the contiguous states plus several Canadian provinces. One published figure placed the mean loss of alfalfa forage at 15 percent, but the source of that figure is unknown. In any case, losses in some regions were negligible yet in others so high that alfalfa was no longer a viable crop. USDA estimates for 1982 valued forage alfalfa in the United States at about $7 billion annually.

Since about 1980 nationwide losses due to the alfalfa weevil complex have declined. In the eastern United States, several weevil parasite species and a fungal pathogen commonly keep weevil populations below economically injurious levels. In the western states, losses have been reduced by integrated pest management approaches involving cultural practices, a better understanding of the biology of both the weevil and *B. curculionis*, and a more precise use of insecticides.

A large portion of the alfalfa weevil research in the western United States was interrelated with the W-84 project: between 1964 and 1982, approximately 15 theses and dissertations and 20 journal publications were based on alfalfa weevil research under that project. In many aspects of the work, especially those concerning parasite introductions, no clear-cut distinction was made between work on the western weevil and on the Egyptian weevil. Between 1964 and 1970, most of the research was centered in Colorado, Idaho, and Utah, with California involved in supplying parasites. From 1971 to 1980 most of the research took place in Colorado and Utah. By about 1982, W-84 attention had subsided, though scattered activity on the weevil persists.

RESULTS

Objective A: The Identification and Introduction of Beneficial Organisms

Researchers made repeated efforts to introduce new alfalfa weevil parasites—some obtained through the W-84 project and some from USDA-APHIS (Animal and Plant Health Inspection Services)—into populations of the western alfalfa weevil. With the exception of *Bathyplectes stenostigma* (Thomson), which became established in Colorado, those new introductions were unsuccessful. In contrast, there have been quite a few successful recoveries from Egyptian weevil populations. In southwestern Utah, where both western and Egyptian weevils occur, the larval parasite *Tetrastichus incertus* (Ratzeburg) (Eulophidae) has become established.

In northern Utah, researchers also found an egg parasite, *Patasson* (=*Anaphes*) *luna* (Girault) (Mymaridae). It is apparently an unrecognized holdover from the 1911 to 1913 introductions, and not a result of recent work. Unfortunately, it occurs too late in the season to be of much value.

Objective B: Ecological and Physiological Studies

Throughout the period of primary alfalfa weevil studies in Colorado, Idaho, and Utah, a lot of attention went to ecological and physiological studies. Most contributions fall into three categories: alfalfa weevil biology, parasite and predator behavior and biology, and hyperparasites and other factors adversely affecting *B. curculionis*.

Major attention was given to alfalfa weevil phenology, especially degree-day determinations (Eklund and Simpson 1977). In addition, researchers studied flight patterns and seasonal activity (Southwick and Davis 1968), feeding by adult weevils (Bjork and Davis 1984), and oviposition in weed stems (Ben Saad and Bishop 1969).

Most of the research relating to parasites and predators has centered on *B. curculionis,* though several predators (Ouayogode and Davis 1981) and *B. stenostigma* have also been studied. In Utah research centered on *B. curculionis* diapause and the effect of that parasite on weevil hosts (Duodu and Davis 1974a,b,c; Parrish and Davis 1978; Hama and Davis 1983). In Idaho and Colorado major efforts were devoted to field behavior and the seasonal development of *B. curculionis* (Ben Saad and Bishop 1969; Foster and Bishop 1970; Eklund and Simpson 1977). Of particular interest were studies in Colorado that showed that nectar from various weeds extended the longevity of adult parasites beyond that attained by parasites fed on alfalfa nectar. These findings suggest that weed-produced nectar is superior.

The mortality factors affecting *B. curculionis* are extensive and varied. In addition to pesticides, which were studied in all three states (Davis 1970), hyperparasites received major attention in Colorado (Best and Simpson 1975; Coseglia and Simpson 1977; Ellsbury and Simpson 1978; Simpson et al. 1979), and the effects of high temperatures were studied in Utah (Hama and Davis 1983). All three mortality categories proved highly important.

Objective C: The Conservation and Augmentation of Beneficial Organisms

Researchers have also studied how different agronomic practices relate to the total IPM programs in alfalfa ecosystems. These practices include burning (Simpson and Clark 1967; Hansen and Simpson 1968); early cutting, green chopping, and weevil sampling (Jech 1988); preserving parasites and predators (Davis 1970); and using partially selective pesticides. Most of these studies included work on other insects in the alfalfa fields, as well as the alfalfa weevil and its natural enemies.

RECOMMENDATIONS

1. Future work with parasite introductions should stress biotype matching. There is no question that many of the failures to get parasites established can be explained by the nature of the weevil itself and the climatic adaptations of the parasites.

2. Despite the availability of a wide range of effective IPM procedures—including sampling, using predictive models, and altering cultural practices—neither growers nor Cooperative Extension Service personnel have made much use of this information for alfalfa weevil control. These management practices must be promoted to a much greater degree.

3. Serious consideration should be given to planting selected nectar sources to improve the effectiveness of *B. curculionis.*

4. Several species of alfalfa weevil parasites currently established to control the eastern alfalfa weevil are ineffective against the western weevil; this problem needs fuller study.

5. Research attention should be directed toward reducing the adverse effects of hyperparasites.

REFERENCES CITED

Ben Saad, A., and G. W. Bishop. 1969. Egg laying by the alfalfa weevil in weeds. *J. Econ. Entomol.* 62:1225–27.

Best, R. L., and R. G. Simpson. 1975. Biology of *Eupteromalus americanus:* A hyperparasite of *Bathyplectes curculionis.* *Ann. Entomol. Soc. Am.* 68:1117–20.

Bjork, C. D., and D. W. Davis. 1984. Consumption of alfalfa by adult alfalfa weevils (Coloeoptera: Curculionidae). *Environ. Entomol.* 13:432–38.

Chamberlin, T. R. 1924. *Introduction of parasites of the alfalfa weevil into the United States.* USDA Circular 301.

Coseglia, A. F., and R. G. Simpson. 1977. Biology of *Mesochorus nigripes,* a hyperparasite of *Bathyplectes* spp. *Ann. Entomol. Soc.*

Am. 70:695–98.

Davis, D. W. 1970. Insecticidal control of the alfalfa weevil in northern Utah and some resulting effects on the weevil parasite (*Bathyplectes curculionis*). *J. Econ. Entomol.* 63:119–25.

Duodu, Y. A., and D. W. Davis. 1974a. Effects of *Bathyplectes curculionis* (Thomson) on the development, morphological appearance, and activity of alfalfa weevil larvae. *Environ. Entomol.* 3:396–98.

———. 1974b. Selection of alfalfa weevil larvae by, and mortality due to, the parasite *Bathyplectes curculionis* (Thomson). *Environ. Entomol.* 3:549–52.

———. 1974c. A comparison of growth, food consumption, and food utilization between unparasitized alfalfa weevil larvae and those parasitized by *Bathyplectes curculionis* (Thomson). *Environ. Entomol.* 3:705–10.

Dysart, R. J., and W. H. Day. 1976. *Release and recovery of introduced parasites of the alfalfa weevil in eastern North America.* USDA-ARS Production Research Report 167.

Eklund, L. R., and R. G. Simpson. 1977. Correlation of activities of the alfalfa weevil and *Bathyplectes curculionis* with alfalfa height and degree-day accumulation in Colorado. *Environ. Entomol.* 6:69–71.

Ellsbury, M. M., and R. G. Simpson. 1978. Biology of *Mesochorus agilis*: An indirect hyperparasite of *Bathyplectes curculionis*. *Ann. Entomol. Soc. Am.* 71:865–68.

Foster, D., and G. W. Bishop. 1970. *Distribution and life history of an alfalfa weevil parasite,* Bathyplectes curculionis (*Thompson*) (*Ichneumonidae: Hymenoptera*), *in northern Idaho*. Idaho Experiment Station Research Bulletin 78.

Hama, N. N., and D. W. Davis. 1983. Lethal effects of high temperatures on nondiapausing pupae of *Bathyplectes curculionis* (Hym: Ichneumonidae). *Entomophaga* 28:295–302.

Hansen, R. W., and R. G. Simpson. 1968. Studies on alfalfa weevil control in Colorado: Development of flaming gauges. In *Proceedings of the 5th Annual Symposium on Thermal Agriculture,* 26–28.

Jech, Larry E. 1988. Alfalfa weevil, *Hypera postica* (Gyllenhal) (Coleoptera: Curculionidae), response to environmental factors in alfalfa fields in northern Utah. Ph.D. diss., Utah State University, Logan.

Ouayogode, B. V., and D. W. Davis. 1981. Feeding by selected predators on alfalfa weevil larvae. *Environ. Entomol.* 10:62–64.

Parrish, D. S., and D. W. Davis. 1978. Inhibition of diapause in *Bathyplectes curculionis,* a parasite of the alfalfa weevil. *Ann. Entomol. Soc. Am.* 71:103–7.

Simpson, R. G., and S. J. Clark. 1967. Effects of flaming to control the alfalfa weevil in Colorado. In *Proceedings of the 4th Annual Symposium on Thermal Agriculture,* 34–36.

Simpson, R. G., J. E. Cross, R. L. Best, M. M. Ellsbury, and A. F. Coseglia. 1979. Effects of hyperparasites on population levels of *Bathyplectes curculionis* in Colorado. *Environ. Entomol.* 8:96–100.

Southwick, J. W., and D. W. Davis. 1968. Behavior patterns of the adult alfalfa weevil in Cache County, Utah. *Ann. Entomol. Soc. Am.* 61:1224–28.

Titus, E. G. 1908. The alfalfa-leaf weevil. *Deseret Farmer* 4(8):3, 15.

43 / Coconut Flat Beetles

D. M. NAFUS AND A. M. VARGO

> **coconut flat beetles**
> *Brontispa* spp.
> Coleoptera: Chrysomelidae

INTRODUCTION

Biology and Pest Status

The genus *Brontispa* contains several species that are pests of coconut, betelnut, oil palm, and other palms. In their native ranges these beetles are not particularly damaging, but they are serious pests on the islands they have invaded. The larvae and adults feed in unopened leaves, grazing on the inner side of the leaf and leaving brown tracks parallel to the midrib. Heavy infestations give the leaf a brown, scorched appearance. Although they attack all ages of palms, the beetles are more common and damaging in young palms, where they can severely curtail growth and occasionally kill the plant. Heavy feeding also weakens older plants, increasing their susceptibility to disease, drought, or other stresses (Waterhouse and Norris 1987).

Brontispa longissima (Gestro) lays sets of one to four eggs in narrow tracks chewed into the leaf and then covers them with excreta. The eggs hatch in about 5 days and the larvae take 30 to 40 days to develop. The beetle has a 3-day prepupal period and a 6-day pupal period. Adults can live for over 200 days. The preovipositional period lasts 1 to 2 months, and females can lay over 100 eggs. The beetle can go through three generations per year (Waterhouse and Norris 1987). The biologies of the Micronesian species are less well known, but are probably similar to that of *B. longissima*.

Historical Notes

The region covered by the W-84 project contains four species of *Brontispa*: *B. longissima* in American Samoa; *B. mariana* Spaeth on Saipan, Tinian, and Rota of the Mariana Islands and on Chuuk and Yap of the Caroline Islands; *B. palauensis* (Esaki & Chujo) on Guam and on Belau of the Caroline Islands; and *B. chalybeipennis* (Zacher) on Kosrae and Pohnpei of the Caroline Islands, on the Marshall Islands, and in Hawaii. *Brontispa longissima* is native to the Indonesian–Papua New Guinea

region, but has now invaded much of the South Pacific. *Brontispa mariana*, *B. palauensis*, and *B. chalybeipennis* are all native to the Carolines and have only recently spread to the other islands in their current distribution. On Guam and American Samoa the W-84 project has been involved in releasing and evaluating the parasitoids and pathogens that attack these beetles.

Tetrastichus brontispae (Ferriere) has been widely used to control several species of *Brontispa*; success rates have depended on the species and race of the beetle it was released against. Different parasitoids have been released in other locations, but in general they are less effective (Waterhouse and Norris 1987).

RESULTS

Objective A: The Identification and Introduction of Beneficial Organisms

In 1974 researchers in Guam released strains of *Tetrastichus brontispae* from Saipan, New Caledonia, the Solomon Islands, and Vanuatu. The parasitoid became established, and by 1975 it was being collected and redistributed to other areas. In an islandwide survey in 1980, researchers found parasitization rates ranging from 2 to 75 percent; typical rates were about 30 percent (Muniappan, Duenas, and Blas 1980). Periodic beetle outbreaks still occur.

In American Samoa *T. brontispae* was released on five occasions. Researchers confirmed its establishment in 1987, but it appears to have had little impact, since it is rarely found (Vargo unpub. data). In 1986, after *B. chalybeipennis* was discovered in Honolulu, *Tetrastichus brontispae* was introduced into Hawaii. Although it has become established, its impact has not been evaluated (Funasaki, Nakahara, and Kumashiro 1988).

Also in American Samoa the entomopathogenic fungus *Metarrhizium anisopliae* var. *anisopliae* has been used to control *B. longissima*. K. J. Marshall found a naturally occurring strain on field-collected specimens of *Brontispa* and subsequently developed and mass-produced a laboratory strain. Although four applications of this fungus are recommended, researchers found that a single application controlled *B. longissima* for up to 4 years (Marshall and Vargo unpub. data).

The earwig *Chelisoches morio* Fabricius and an undescribed species of mite in the family Heterocoptidae also prey on *B. longissima* in American Samoa. *C. morio* is also found on Guam.

RECOMMENDATIONS

1. Follow-up studies on the release of *T. brontispae* on Guam are needed, including information on the strains of the wasp that have become established, the degree of seasonal fluctuation of parasitoid and beetle populations, the seasonal changes in parasitization rates, and the extent of damage to coconut.

2. Introductions of new species of parasitoids are suggested for both Guam and American Samoa. In Western Samoa a eulophid wasp, *Chrysonotomyia* species, parasitizes an average of 96 percent of the fourth-instar larvae (Waterhouse and Norris 1987). This wasp should be considered for introduction into both Guam and American Samoa unless prerelease surveys indicate that it is already present. Surveys are under way in American Samoa.

REFERENCES CITED

Funasaki, G. Y., L. M. Nakahara, and B. R. Kumashiro. 1988. Introductions for biological control in Hawaii: 1985 and 1986. *Proc. Hawaii. Entomol. Soc.* 28:101–4.

Muniappan, R., J. G. Duenas, and T. Blas. 1980. Biological control of the Palau coconut beetle, *Brontispa palauensis* (Esaki & Chujo), on Guam. *Micronesica* 16:359–60.

Waterhouse, D. F., and K. R. Norris. 1987. *Biological control: Pacific prospects*. Melbourne: Inkata Press.

44 / COCONUT RHINOCEROS BEETLE

A. M. VARGO

coconut rhinoceros beetle
Oryctes rhinoceros L.
Coleoptera: Scarabaeidae

INTRODUCTION

Biology and Pest Status

The coconut rhinoceros beetle, *Oryctes rhinoceros* L., is a major pest of coconuts. Its life cycle begins when females burrow into moist, decomposing vegetation such as rotting coconut or banana stumps, standing dead palms, or compost piles to lay their eggs. The eggs, which are ovoid and 3 to 4 millimeters long, hatch in 8 to 12 days. The newly emerged grubs consume their eggshells and feed on the frass left in the tunnel by the burrowing female. The larvae are typical scarab grubs—whitish on emergence, with the head capsule turning reddish brown within the first 24 hours. The three larval instars last 10 to 21, 12 to 21, and 60 to 165 days, respectively (Waterhouse and Norris 1987). Third-instar larvae construct a cocoon either by boring into more solid wood and lining the cavity with a coating of liquid fecal material, or—in compost or other soft media—burrowing at least 15 centimeters before forming a cocoon. During this final instar, which may reach a size of 105 millimeters in length and 20 millimeters in diameter, the larvae change from bluish gray to a creamy coloration as they enter the prepupal stage. The prepupal stage lasts 8 to 13 days after cocoon construction; the pupal period lasts 17 to 28 days.

Adult beetles are black and about 40 millimeters long. They remain in the pupal cell for 17 to 22 days while completing full sclerotization. The male has a large horn on its head and a smooth and shiny abdomen. The female has a smaller horn but is distinguished by long, erect, reddish hairs that are visible at the tip of her abdomen. Adults live at least 3 months (Lever 1969); insectary studies have shown that, on average, males live 6.4 months and females 9.1 months. Females have a mean fecundity of 51 eggs (Bedford 1976).

Adults fly at night to feeding sites; their flight is usually limited to a few hundred meters, but if necessary they can fly farther. Rain and moonlight can inhibit flight. The beetles fly to the crowns of palms, enter the side of an upper or middle frond, crawl down the surface of the petiole, wedge their way between that petiole and the one growing more centrally, and then bore into the center, where the immature fronds and inflorescence are crowded together. They then macerate the plant tissue, leaving a characteristic V-shaped notch when the fronds mature. The beetles do not ingest the solid plant material, but consume the sap that extrudes from the lacerated surfaces; they also eat their own, and each other's, frass. The beetles then exit the plant through the entrance hole (Waterhouse and Norris 1987).

The coconut rhinoceros beetle attacks 31 genera of palms (Gressitt 1953), with the coconut palm being its most important host. Other hosts are pandanus, sugarcane, pineapple, banana, and possibly taro (Waterhouse and Norris 1987). Trees on the edges of plantations and taller trees that offer a prominent silhouette are those most prone to beetle attack.

In areas where the copra industry is important, the beetle is a severe threat. Several studies have attempted to quantify the relationship between frond-area reduction and nut yield. In Papua New Guinea, researchers found that 40 percent defoliation resulted in a 40 percent loss in nut yield (Bailey, O'Sullivan, and Perry 1977). In Western Samoa, a 25 percent drop in leaf area resulted in a 25 percent drop in nut yield (Zelazny 1979). Feeding tunnels also provide an entrance for plant pathogens and make it easier for various species of palm weevils, including *Rhynchophorus bilinceatus* (Montrouzier), to invade the trunk.

Historical Notes

Oryctes rhinoceros is notorious as one of the most serious pests of the coconut palm. It is found throughout Southeast Asia, and its natural range is thought to include the Indian subcontinent, Indonesia, Malaysia, the Philippines, Hong Kong, Sri Lanka, and Taiwan.

The beetle was first discovered in the Pacific in Samoa in 1909, having purportedly arrived from Sri Lanka—

formerly Ceylon—with rubber seedlings. It also reached Fiji, Manus Island, New Britain, New Ireland, Palau, Pearl Island, Tokelau, Tuvalu, and Wallis Island. It arrived in Mauritius around 1962 but does not yet occur in Africa, where related pest species attack coconuts and other palms (Waterhouse and Norris 1987).

Under the auspices of the W-84 program, researchers are conducting surveillance for outbreaks of the rhinoceros beetle, which frequently occur after hurricanes, on the five major islands of American Samoa; they are also consulting with the local government on appropriate relief measures.

RESULTS

Objective A: The Identification and Introduction of Beneficial Organisms

In 1965 the United Nations Development Project and the South Pacific Commission initiated a project for research on coconut rhinoceros beetle control. In 1963 researchers had discovered a virus in Malaysia that kills larvae, pupae, and adults of the genus *Oryctes* and a few related genera. In the early stages, the virus exclusively infects the midgut without affecting other organs. Therefore, infected beetles can still fly and mate, spreading the virus orally and through their feces.

In 1967 the virus was introduced into Western Samoa, and in 1969 and 1970 researchers noticed a dramatic reduction in the numbers of adult beetles and larvae, as well as damaged palms.

In American Samoa, the release of virus-infected beetles began in 1972. In 33 months the percentage of damaged fronds fell from 89 to 47 percent. Studies showed that the virus spread from 0.8 to 1.6 kilometers per month. Within 3 kilometers of the release site, the damage to coconuts fell from 13 percent to 2 percent in 33 months, and within 8 to 13 kilometers it fell from 16 percent to 7 percent (Swan 1972).

A fungal pathogen, *Metarrhizium anisopliae*, has also been used with some success in the control of the rhinoceros beetle. However, the fungus must be periodically applied to breeding places and is not as self-sustaining as the virus.

Objective C: The Conservation and Augmentation of Beneficial Organisms

In Western Samoa the damage to palms began to increase after 1970. Consequently, the Samoan–German Beetle Project undertook another release program in 1975. After this second release of the virus, palm damage again decreased markedly (Marschall and Ioane 1982).

In American Samoa researchers conducted a baseline survey of rhinoceros beetle damage in 1988, when the W-84 project was initiated there. Because damage levels were low—less than 7 percent—augmentation was not necessary.

RECOMMENDATIONS

1. A second release of the virus may be advisable in American Samoa when outbreaks occur. Collecting and dissecting beetles will confirm the presence of the virus and also indicate virus levels.

2. Where rhinoceros beetles are a threat, continuous monitoring and release of the virus should be made to prevent severe outbreaks.

3. Other cultural controls that could be implemented include plantation maintenance to eliminate breeding places and planting cover crops such as the legume *Pueratia phaseoloides*.

REFERENCES CITED

Bailey, P., D. O'Sullivan, and C. Perry. 1977. Effect of artificial defoliation on coconut yields in Papua New Guinea. *Papua New Guinea Agric. J.* 26:39–44.

Bedford, G. O. 1976. Observations on the biology and ecology of *Oryctes rhinoceros* and *Seapanes australis* (Coleoptera: Scarabaeidae): Dynastinaeli pests of coconut palms in Melanesia. *J. Aust. Entomol. Soc.* 15:241–51.

Gressitt, J. L. 1953. *The coconut rhinoceros beetle* (Oryctes rhinoceros) *with particular reference to the Palau Islands*. Bishop Museum Bulletin 212. Honolulu: Bishop Museum.

Lever, R. J. A. 1969. *Pests of the coconut palm.* Rome, Italy: Food and Agriculture Organization of the United Nations.

Marschall, J. K., and I. Ioane. 1982. The effect of re-release of *Oryctes rhinoceros* baculovirus in the biological control of rhinoceros beetles in Western Samoa. *J. Invert. Pathol.* 39:267–76.

Swan, D. I. 1972. *UN/SPC Rhinoceros Beetle Project, Annual Report,* 166–69. Noumea, New Caledonia: South Pacific Commission.

Waterhouse, D. F., and K. R. Norris. 1987. *Biological control: Pacific prospects.* Melbourne: Inkata Press.

Zelazny, B. 1979. Loss of coconut yield due to *Oryctes rhinoceros* damage. *Food and Agriculture Organization Plant Protection Bulletin* 27:65–70.

45 / COLORADO POTATO BEETLE

K. D. BIEVER, M. J. TAUBER, AND C. A. TAUBER

> **Colorado potato beetle**
> *Leptinotarsa decemlineata* (Say)
> Coleoptera: Chrysomelidae

INTRODUCTION

Biology and Pest Status

The adult Colorado potato beetle, *Leptinotarsa decemlineata* (Say), overwinters in the soil and emerges in the spring about the time that volunteer potatoes appear. Although the potato is its primary host, the beetle also attacks tomato, eggplant, nightshade, and ground cherry. Over a 4- to 8-week period, the adult lays up to 500 eggs, which hatch in 4 to 9 days. Larvae pass through four growth stages in 2 to 5 weeks, pupating in the soil. The adults emerge about a week later, feed, and, in most areas, begin laying eggs after a few days. In cooler areas there is only one generation; in warmer areas there may be as many as three. The larvae and the first generation of newly emerged adults cause most of the damage.

The beetle occurs in several western states and is an important pest in Idaho, Montana, Oregon, Washington, and Wyoming. Larvae and adults feed on potato foliage, and heavy infestations can severely damage the vines. In the western region, beetle populations are usually suppressed by the insecticides applied to control the green peach aphid; however, since these insecticides may become ineffective or unavailable, new management strategies are needed.

Historical Notes

The Colorado potato beetle—a major pest of potato and several other solanaceous plants worldwide—is believed to have originated in Mexico and to have gradually moved north, where it fed on *Solanum rostratum*. It was first reported as a pest of potatoes in the Midwest, and by the 1880s was established throughout most potato-growing areas of the United States. In the northeastern and mid-Atlantic regions, its resistance to insecticides is already a serious problem; in other areas of the United States and Canada it is a developing problem (Forgash 1985; Johnson and Sandvol 1986; Boiteau, Parry, and Harris 1987).

Before 1980 the majority of published work on the biological control of the beetle came from eastern Europe (the Colorado potato beetle was accidentally introduced into Europe from North America). The beetle's more common natural enemies include lady beetles and stinkbugs, which attack the eggs and larvae. In some areas tachinid flies, which attack the larvae, are common parasitoids. However, natural enemies are seldom populous enough to suppress the beetles.

RESULTS

Objective A: The Identification and Introduction of Beneficial Organisms

In an attempt to introduce biological control of the beetle, U.S. researchers imported the eulophid egg parasitoid *Edovum puttleri* Grissell from Colombia and established its potential through augmentative field releases (Puttler and Long 1983). However, neither the Colombian strain nor another strain from Mexico can survive the harsh winter conditions of North America's potato-producing areas, so annual releases would be required (Obrycki et al. 1987). Although the two strains of *E. puttleri* are morphologically indistinguishable, the Mexican females survive slightly better under low temperatures (Obrycki et al. 1987; Ruberson, Tauber, and Tauber 1988), and the Colombian females accept a broader age range of beetle eggs (Ruberson, Tauber, and Tauber 1987). The two biotypes also differ in their response to pesticides (Obrycki, Tauber, and Tingey 1986). A high proportion of the mortality inflicted by *E. puttleri* results from host stabbing and feeding (Ruberson, Tauber, and Tauber 1987; Ruberson, Tauber, Tauber, et al. 1991). Currently the parasitoid is not used on potatoes, but it has been useful against the Colorado potato beetle on eggplant in the eastern United States.

Researchers have investigated the interactions among three trophic levels in the potato system: the parasitoid *E. puttleri*, the herbivorous beetle, and resistant potato plants (*Solanum berthaultii* accessions). In the laboratory, foliage from resistant potato plants was able to kill *E. puttleri* (Obrycki et al. 1985); however, in the field, relatively high levels of parasitism occurred on potato plants with certain types of resistance mechanisms, but not on others (Ruberson et al. 1989). Researchers also found that the negative effects of the resistant plants were limited and could be overcome by additional releases. Other studies of tritrophic interactions on resistant potatoes are described in chapter 2.

Objective B: Ecological and Physiological Studies

In several field and laboratory studies of two beetle populations—one from a cool, inland area (upstate New York) and one from a warm, coastal area (Long Island)—researchers quantified the beetle's developmental and reproductive responses to temperature; voltinism and the induction of estival diapause; and geographical variation in response to photoperiod and temperature during and after dormancy. Both populations produced one full plus a partial second generation each year, but a large proportion of first-generation adults entered estival-autumnal-hibernal diapause without ovipositing, or after ovipositing only briefly (Tauber, Tauber, Obrycki, et al. 1988a). Thus, the overwintering population was composed of adults from both the first and second summer generations. In general, the upstate population had a greater propensity to enter estivation without ovipositing. Although the two populations did not differ significantly in their thermal requirements (Tauber, Tauber, Gollands, et al. 1988), females from the coastal population generally oviposited fertile eggs at lower temperatures. The intensity of diapause had a distinct seasonal pattern in both populations, but diapause was less intense in the Long Island beetles (Tauber, Tauber, Obrycki, et al. 1988b).

A northwestern beetle population has exhibited a remarkable ability for prolonged dormancy. Field cage studies established that some of this population has a prolonged diapause that is at least threefold greater—amounting to two years longer—than previously reported; for one adult, diapause lasted through five winters (Biever and Chauvin 1990).

The phenological studies described here provide a basis for the development of accurate descriptive and ultimately predictive models of beetle development in the field. These models will be useful in forecasting emergence, development, and damage, and in predicting the occurrence of stages during which the pest is susceptible to attack by natural enemies.

Researchers also evaluated the impact of *Perillus bioculatus* (Fabr.), a pentatomid predator, by studying the biology, constructing life tables, and determining the feeding potential of its different life stages on the potato beetle's eggs and larval stages (Tamaki and Butt 1978). They concluded that this predator was ineffective in suppressing Colorado potato beetle populations in central Washington.

Objective C: The Conservation and Augmentation of Beneficial Organisms

One of the beetle's most common natural enemies is the tachinid fly *Myiopharus* (=*Doryphorophaga*) *doryphorae* (Riley), a larval parasitoid. Under typical field conditions, this fly's parasitism can exceed 75 percent. However, this occurs late in the season—in August and September—after the beetles have already damaged the crop (Tamaki, Chauvin, and Burditt 1983; Gollands, Tauber, and Tauber 1991). The parasitoid's effectiveness is limited by its low numbers during the first generation. Researchers have therefore developed procedures to rear this parasitoid (Tamaki, Chauvin, and Hsiao 1982), which is a step toward manipulating it to increase its effectiveness. Future work with *M. doryphorae* may include inoculative or mass-release programs in the field. Unfortunately, neither the overwintering site nor the overwintering stage of this parasitoid is known (Tamaki, Chauvin, and Burditt 1983; Gollands, Tauber, and Tauber 1991). Several tests have demonstrated that the parasitoid does not overwinter in an immature stage in the soil; in fact, it appears that adults emerge and leave potato fields by mid-November, suggesting that the overwintering stage is the adult (Biever unpub. data). A second species of *Myiopharus*, *M. aberrans*, also attacks Colorado potato beetle larvae. This species overwinters within hibernating beetles in the soil, and researchers have identified habitat manipulation as a possible method for increasing that parasitoid's effectiveness (Gollands, Tauber, and Tauber 1991).

Other research teams have conducted augmentative release studies by inoculating cage and field plots with the laboratory-produced predaceous stinkbugs *Perillus bioculatus* and *Podisus maculiventris* (Say) (Biever and Chauvin 1992a). Based on cage tests, when beetle populations were higher than 150 per plant, *P. bioculatus* was more effective than *P. maculiventris*. When beetle populations were about 100 per plant, a release of 8 third-stage *P. bioculatus* stinkbugs per plant reduced beetle populations by 68 percent and significantly reduced foliage damage. In small plot tests, *P. bioculatus* released at 3 per plant reduced beetle larval populations by 80 percent, significantly reduced foliage damage, and increased yield significantly over that of the untreated plots. These tests also demonstrated that although both species of stinkbugs were from laboratory colonies maintained on cabbage looper larvae, in the field they were effective against the Colorado potato beetle. Recent studies have established that delayed colonization of potatoes by the beetle had a negative effect on predation by *P. bioculatus* (Biever and Chauvin 1992b).

RECOMMENDATIONS

1. Study aspects of overwintering, host-plant interactions, and dispersal behavior to provide the fundamental information needed to develop beetle management strategies.

2. Continue to evaluate the potential of introduced and native parasitoids, pathogens, and predators as both classical and augmentative biological control agents.

REFERENCES CITED

Biever, K. D., and R. L. Chauvin. 1990. Prolonged dormancy in a Pacific Northwest population of the Colorado potato beetle, *Leptinotarsa decemlineata* (Say) (Coleoptera: Chrysomelidae). *Can. Entomol.* 122:175–77.

———. 1992a. Suppression of the Colorado potato beetle (Coleoptera: Chrysomelidae) with augmentative releases of predaceous stinkbugs (Hemiptera: Pentatomidae). *J. Econ. Entomol.* 85:720–26.

———. 1992b. Impact of time of colonization by the Colorado potato beetle (Coleoptera: Chrysomelidae) on the suppressive effect of field released stinkbugs (Hemiptera: Pentatomidae) in the Pacific Northwest. *Environ. Entomol.* 21:1212–19.

Boiteau, G., R. H. Parry, and C. R. Harris. 1987. Insecticide resistance in New Brunswick populations of the Colorado potato beetle (Coleoptera: Chrysomelidae). *Can. Entomol.* 119: 459–63.

Forgash, A. J. 1985. Insecticide resistance in the Colorado potato beetle. In *Proceedings of the Symposium on the Colorado Potato Beetle, 17th Congress of Entomology*, eds. D. N. Ferro and R. H. Voss, 33–52. Massachusetts Agricultural Experiment Station Bulletin 704.

Gollands, B., M. J. Tauber, and C. A. Tauber. 1991. Seasonal cycles of *Myiopharus aberrans* and *M. doryphorae* (Diptera: Tachinidae) parasitizing Colorado potato beetle in upstate New York. *Biol. Control* 1: 153–63.

Johnson, R. L., and L. E. Sandvol. 1986. Susceptibility of Idaho populations of Colorado potato beetle to four classes of insecticides. *Am. Potato J.* 63:81–85.

Obrycki, J. J., M. J. Tauber, and W. M. Tingey. 1986. Comparative toxicity of pesticides to *Edovum puttleri* (Hymenoptera: Eulophidae), an egg parasitoid of the Colorado potato beetle (Coleoptera: Chrysomelidae). *J. Econ. Entomol.* 79:948–51.

Obrycki, J. J., M. J. Tauber, C. A. Tauber, and B. Gollands. 1985.

Edovum puttleri (Hymenoptera: Eulophidae), an exotic egg parasitoid of the Colorado potato beetle (Coleoptera: Chrysomelidae): Responses to temperate zone conditions and resistant potato plants. *Environ. Entomol.* 14:48–54.

———. 1987. Developmental responses of the Mexican biotype of *Edovum puttleri* (Hymenoptera: Eulophidae) to temperature and photoperiod. *Environ. Entomol.* 16:1319–23.

Puttler, B., and S. H. Long. 1983. Host specificity tests of an egg parasite, *Edovum puttleri* (Hymenoptera: Eulophidae), of the Colorado potato beetle, *Leptinotarsa decemlineata* (Coleoptera: Chrysomelidae). *Proc. Entomol. Soc. Wash.* 85:384–87.

Ruberson, J. R., M. J. Tauber, and C. A. Tauber. 1987. Biotypes of *Edovum puttleri* (Hymenoptera: Eulophidae): Responses to developing eggs of the Colorado potato beetle (Coleoptera: Chrysomelidae). *Ann. Entomol. Soc. Am.* 80:451–55.

———. 1988. Reproductive biology of two biotypes of *Edovum puttleri*, a parasitoid of Colorado potato beetle eggs. *Entomol. Exp. Appl.* 46:211–19.

Ruberson, J. R., M. J. Tauber, C. A. Tauber, and W. M. Tingey. 1989. Interactions at three trophic levels: *Edovum puttleri* Grissell (Hymenoptera: Eulophidae), the Colorado potato beetle, and insect-resistant potatoes. *Can. Entomol.* 121:841–51.

Ruberson, J. R., M. J. Tauber, C. A. Tauber, and B. Gollands. 1991. Parasitism by *Edovum puttleri* (Hymentoptera: Eulophidae) in relation to host density in the field. *Ecol. Entomol.* 16:81–89.

Tamaki, G., and B. A. Butt. 1978. Impact of *Perillus bioculatus* on the Colorado potato beetle and plant damage. USDA Technical Bulletin 1581.

Tamaki, G., R. L. Chauvin, and T. Hsiao. 1982. Rearing *Doryphorophaga doryphorae,* a tachinid parasite of the Colorado potato beetle, *Leptinotarsa decemlineata.* USDA-ARS, Advances in Agricultural Technology, Western Series 21.

Tamaki, G., R. L. Chauvin, and A. K. Burditt, Jr. 1983. Field evaluation of *Doryphorophaga doryphorae* (Diptera: Tachinidae), a parasite and its host, the Colorado potato beetle. *Environ. Entomol.* 12:386–89.

Tauber, C. A., M. J. Tauber, B. Gollands, R. J. Wright, and J. J. Obrycki. 1988. Preimaginal development and reproductive responses to temperature in two populations of the Colorado potato beetle (Coleoptera: Chrysomelidae). *Ann. Entomol. Soc. Am.* 81:755–63.

Tauber, M. J., C. A. Tauber, J. J. Obrycki, B. Gollands, and R. J. Wright. 1988a. Voltinism and the induction of aestival diapause in the Colorado potato beetle, *Leptinotarsa decemlineata* (Coleoptera: Chrysomelidae). *Ann. Entomol. Soc. Am.* 81:748–54.

———. 1988b. Geographical variation in responses to photoperiod and temperature by *Leptinotarsa decemlineata* (Coleoptera: Chrysomelidae) during and after dormancy. *Ann. Entomol. Soc. Am.* 81:764–73.

46 / EGYPTIAN ALFALFA WEEVIL

L. K. ETZEL, K. S. HAGEN, D. GONZÁLEZ, AND J. J. ELLINGTON

> **Egyptian alfalfa weevil**
> *Hypera brunneipennis* (Boheman)
> Coleoptera: Curculionidae

INTRODUCTION

Biology and Pest Status

The Egyptian alfalfa weevil, *Hypera brunneipennis* (Boheman), is the most serious insect pest of alfalfa in California, where it typically attacks the first seasonal alfalfa growth. The California Department of Food and Agriculture, based on its most recently available figures, estimated that alfalfa losses due to weevils were $21 million in 1974 and $13.5 million in 1977. The situation has not improved since. Alfalfa has typically been one of the three most valuable crops in the state, and therefore plays a key role in California's agricultural economy—not only in the dairy and livestock industries, but also in crop rotation. Alfalfa weevils have also devastated rangeland bur clover, the plant that ranchers most prize as forage for livestock.

In Fresno County the weevil larval population usually peaks between the middle and end of March. It occurs earlier in Imperial County and later in the upper Central Valley. Emerging adults feed for several weeks to accumulate body fat reserves before migrating from the alfalfa or bur clover to nearby estivation sites, where they can be protected from unfavorable environmental conditions. Estivating weevils can be found under loose tree bark, in bark fissures, or wherever there are cracks or crevasses, such as in buildings. Between the end of October and the middle of December, the adult weevils migrate back to the fields. There they feed and lay eggs in plant stems throughout the winter, as weather conditions permit. In Fresno County a population rise of hatching larvae occurs between the middle and end of February.

Historical Notes

The Egyptian alfalfa weevil was first discovered in the United States near Yuma, Arizona, where it had been accidentally introduced in the mid-1930s (van den Bosch and Marble 1971). From there it spread into the nearby Imperial Valley in southern California and on to

San Diego County in 1950, San Bernardino County in 1954, San Luis Obispo County in 1956, the Salinas Valley, and, by the mid-1960s, the upper and lower Central Valley (van den Bosch and Marble 1971). The Egyptian alfalfa weevi—or its close relative the alfalfa weevil, *H. postica* (Gyllenhal)—is now found throughout the Southwest, including the alfalfa-growing regions of California. It also attacks bur clover (*Medicago hispida* Haertner) in California rangelands.

Researchers have been engaged in releasing parasites against the Egyptian alfalfa weevil since it was first discovered in Arizona. In the Yuma Valley they released a European strain of the larval parasite *Bathyplectes curculionis* (Thomson), obtained from Utah, where it had been successfully established against *H. postica*. Once it became established, the parasite spread into California along with the weevil (van den Bosch, Finney, and Lagace 1971). In California, the University of California initiated a biological control program in the early 1960s. During the 1960s and 1970s researchers procured and released several species of parasites, though usually in limited numbers. Most of the California releases were made by R. van den Bosch until his untimely death in 1978. The USDA also did some work in California (Clancy 1969).

Parasites other than *B. curculionis* that have been released in California include an egg parasite, *Patasson* species; the larval parasites *Bathyplectes anurus* (Thomson), *B. stenostigma* (Thomson), and *Tetrastichus incertus* (Ratzeburg) (=*T. erdoesi* Domenichini); a parasite of the prepupae and pupae, *Dibrachoides dynastes* (Forester) (=*D. druso* Walker); and the adult weevil parasites *Microctonus aethiopoides* Loan—previously misidentified as *M. aethiops* (Needs)—*M. colesi* Drea, and an unidentified *Microctonus* (van den Bosch, Finney, and Lagace 1971). Another larval parasite released was *Habrocytus* species (Fisher, Schlinger, and van den Bosch 1961; Clancy 1969). R. van den Bosch's releases mostly emphasized *B. anurus* and *M. aethiopoides* (van den Bosch, Finney, and Lagace 1971).

Although *B. curculionis* can now be found wherever the weevil is collected in California, it cannot control the pest by itself. As a result, researchers have continued their efforts. They released *B. anurus* in southern California in 1967, but it was not detected later (Clancy 1969). However, in 1973 *B. anurus* were recovered in the San Joaquin Valley after releases had been made there, and distribution was being attempted (Hagen et al. 1976). *D. druso* became established in southern California, but at a low and ineffective level of parasitization (González, van den Bosch, and Dawson 1969). D. W. Clancy (1969) released large numbers of European *Tetrastichus incertus* in Arizona and California, but did not recover it. In central and northern California in the late 1960s and early 1970s, an Iranian strain of *T. incertus* was also colonized in large numbers, but was only recovered in Albany in 1970 at up to 70 percent parasitization (Hagen et al. 1976). In 1975, 150 specimens of a second Iranian strain were released in California in Colusa County, and in 1976, another 200 individuals of the same strain were liberated in Fresno County. The USDA released *M. aethiopoides* in Arizona and southern California in 1965, but recovery attempts were unsuccessful (Clancy 1969). Although it was also released widely in northern California in the 1960s, the European and Iranian strains were not recovered until 1973 and 1974, when they were detected at montane release sites in northeastern California (Hagen et al. 1976).

In continued efforts between 1977 and 1981, researchers released 5,257 *M. aethiopoides* (plus 800 parasitized adult weevils), 1,375 *B. anurus*, and 90 *T. incertus*. *M. aethiopoides* was recovered annually from a university alfalfa plot in Albany, but it, and also *B. anurus*, were rarely recovered elsewhere. *T. incertus* is now established in many localities in the San Joaquin Valley (Pitcairn and Gutierrez 1989). In 1982, at the Kearney Horticultural Center and the West Side Field Station, *T. incertus* appeared in mid-May and achieved a 25 percent parasitization rate by the end of June. However, such late-season parasitization is of little value, because the weevil larval populations collapse and do no significant damage beyond mid-April.

RESULTS

Objective A: The Identification and Management of Beneficial Organisms

The difficulty of procuring or producing enough weevil parasites for field release was a key factor in the failure to establish permanent populations. However, a renewed effort to establish parasites began in 1982 and expanded with the approval of a new Experiment Station Project in 1983. Production problems were alleviated by improved rearing techniques and by a new source of supply: parasites field-collected in the eastern United States by the USDA Animal and Plant Health Inspection Services (APHIS).

In 1982 researchers altered the method of dispersing *B. anurus*, releasing 1,600 in field cages with field-collected larvae. In addition, 3,754 *M. aethiopoides* were released that year.

Between 1983 and 1989, totals of 98,300 *B. anurus*, 53,321 *M. aethiopoides*, 26,235 *D. dynastes*, 3,181 *M. colesi*, and 113 *B. stenostigma* were released in California. The releases were made in bur clover rangeland and alfalfa in counties throughout California, including Imperial, Riverside, Kern, Santa Barbara, Monterey, San Benito, Fresno, Madera, Merced, Glenn, Colusa, Butte, Siskiyou, and Lassen. Most of the releases, however, took place in Glenn, Colusa, and Fresno counties.

The releases of *B. anurus* received from the USDA-APHIS were most intensive at experimental alfalfa plots at the Kearney Agricultural Center and the West Side Field Station, both in California's Central Valley; it has become established in both areas. Close monitoring of these two alfalfa fields between 1984 and 1989 revealed that the peak weevil larval population occurred in March or the first week of April, depending on weather conditions. There was no consistent pattern of peak population dates between the two sites, which are 50 miles apart.

The mean peak population at Kearney decreased from 308 per sweep (with a standard insect net) in 1985 to 184 in 1986, 125 in 1987, 80 in 1988, and 62 in 1989; the University of California has set the economic injury level at 20 per sweep. Concurrently, between 1984 and 1987 the annual peak percentage of parasitism by *B. anurus* gradually increased to 2.4 percent, 4.2 percent, 5.1 percent, and 7.3 percent, respectively. Although in 1988 parasitism collapsed to 1.8 percent, it rebounded in 1989 to 7.7 percent.

At the West Side Field Station, the mean peak larval populations per sweep for 1984 to 1989 were 37, 44, 43, 263, 80, and 77, respectively. The dramatic weevil population increase in 1987 was directly attributable to a *Eucalyptus* grove that had been coincidentally planted adjacent to the alfalfa field in 1985. The trees grew rapidly in 1986, providing migrating adult weevils a significant increase in estivation shelter, which caused a decrease in mortality. The annual peak percentage of parasitism for *B. anurus* for those years also reflected this phenomenon. It gradually increased for the first three years—to 2.2, 3.9, and 5.3 percent—and then dropped to 2.3 percent for 1987. The actual number of parasites obtained per 100 sweeps in 1986 and 1987 remained the same, however. In 1988 the peak percentage of parasitism collapsed, as it did at Kearney—in this instance, to 0.3 percent. Likewise there was a rebound in 1989, to 1.9 percent.

At both Kearney and West Side, the last release of *B. anurus* against weevil larvae occurred in 1986. It should be noted that the Kearney alfalfa field was adjacent to orchards, so the high weevil populations there can also be attributed to the presence of superior shelters for the estivating weevils. Insecticides were not used on either the Kearney or West Side fields.

More encouraging results have been obtained from releases of a strain of *M. aethiopoides* originally collected in Iran. This strain had initially become established at a small alfalfa plot at the University of California's Gill Tract Experiment Station in Albany, California. Enough parasites were recovered there to annually re-establish an insectary colony. Since 1983 researchers have released *M. aethiopoides* from these insectary colonies in bur clover rangeland in Glenn and Colusa counties, where it is now well established. In one field in 1986, field-collected overwintering weevils were parasitized in a range between 37.9 and 53.8 percent; summer-estivating adults were parasitized at 28.9 percent. Furthermore, releases of *M. aethiopoides* in 300 acres of alfalfa next to bur clover rangeland resulted in a good initial parasitization of 8.6 percent, despite insecticide applications to the alfalfa.

In 1988 *M. aethiopoides* parasitism at the best rangeland site in Glenn County, where small releases—totaling 758 parasites—had been made in 1984 and 1986, was 35 percent of the 2,789 estivating adults collected. The fact that 20 percent of 1,556 weevils collected about a third of a mile away were also parasitized indicated that the parasite had dispersed well. The same was true at a site in Colusa County, where only one small release had been made—of 290 parasites in 1984. Parasitism of 620 estivating weevils collected in 1988, more than a half-mile away from that release point, was 10 percent. Similar success has not been achieved in alfalfa. At the best location, where 5,661 *M. aethiopoides* had been released over 4 years, the mean percentage of parasitism in 1988 was only 2.5 percent of 2,973 estivating weevils.

In 1989 *M. aethiopoides* parasitism continued to be very high in bur clover rangeland in Glenn and Colusa counties. In Glenn County between Elk Creek and Stonyford, a distance of about 20 miles, where releases in the 1980s had been made at two sites (the last in 1986), *M. aethiopoides* was found at each of seven sites sampled. The evidence indicated that the parasite had spread up to 5 miles from the nearest release site. At these sites it parasitized diapausing adult weevils of the 1989 to 1990 generation at rates from 16.7 percent to 44.1 percent, with an average of 38.7 percent.

These results with *M. aethiopoides* in California are similar to those R. G. van Driesche and G. G. Gyrisco (1979) reported for New York. These two researchers found, in New York alfalfa fields between 1971 and 1973, that the peak parasitization rate of overwintering alfalfa weevils, *Hypera postica*, ranged between 20 and 88 percent; the peak parasitization rate for summer adults ranged from 3 to 27 percent. This parasitization was accompanied by a 47 percent mean reduction in weevil egg laying.

The similarity of the New York and California results is very promising, since W. H. Day (1981) has indicated that by 1979 biological control of the weevil in New York had saved growers an annual $4.6 million. W. H. Day further noted that the most effective parasites in the Northeast were *M. aethiopoides* and *B. anurus*.

Methods were developed to sample for the establishment of *M. aetheiopoides* and to field-collect it for propagation and distribution. Adult weevils are generally difficult to collect by sweep-netting alfalfa, because they are frequently in the duff on the ground. They are even more difficult to collect in bur clover because of its recumbent growth habit. An effective sampling method takes advantage of the fact that new adult weevils migrate to estivation sites after they have fed and accumulated fat reserves. The easiest way to sample them is by collecting estivating adults in corrugated cardboard bands stapled to trees adjacent to the release sites. These collections recover the most adults when done in the late spring or early summer.

RECOMMENDATIONS

1. Recent California progress in field parasitization of the Egyptian alfalfa weevil by *M. aethiopoides* indicates that there is a distinct possibility that the Northeast's successful biological control can be repeated in California. In fact, some ranchers in Glenn County believe there is an increase in bur clover in the release areas. It is important to verify this by initiating an assessment program and following the changes that may occur in the rangeland.

2. A much larger field-collection, propagation, and distribution program to establish *M. aethiopoides* should be inaugurated in other alfalfa and bur clover areas as quickly as possible.

3. Future success with *B. anurus* might be achieved by introducing new strains of the parasite from Mediterranean and Middle Eastern regions.

REFERENCES CITED

Clancy, D. W. 1969. Biological control of the Egyptian alfalfa weevil in California and Arizona. *J. Econ. Entomol.* 62:209–13.

Day, W. H. 1981. Biological control of the alfalfa weevil in the northeastern United States. In *Biological control in crop production,* ed. G. C. Papavizas, 361–74. BARC Symposium 5. Totowa, N.J.: Allanheld, Osmun.

Fisher, T. W., E. I. Schlinger, and R. van den Bosch. 1961. Biological notes on five recently imported parasites of the Egyptian alfalfa weevil, *Hypera brunneipennis. J. Econ. Entomol.* 54:196–97.

González, D., R. van den Bosch, and L. H. Dawson. 1969. Establishment of *Dibrachoides druso* on the Egyptian alfalfa weevil in southern California. *J. Econ. Entomol.* 62:1320–22.

Hagen, K. S., G. A. Viktorov, K. Yasumatsu, and M. F. Schuster. 1976. Biological control of pests or range, forage, and grain crops. In *Theory and practice of biological control,* eds. C. B. Huffaker and P. S. Messenger, 397–442. New York: Academic Press.

Pitcairn, M. J., and A. P. Gutierrez. 1989. Biological control of *Hypera postica* and *Hypera brunneipennis* (Coleoptera: Curculionidae) in California, with reference to the introduction of *Tetrastichus incertus* (Hymenoptera: Eulophidae). *Pan-Pacific Entomol.* 65:420–28.

van den Bosch, R., and V. L. Marble. 1971. Egyptian alfalfa weevil: The threat to California alfalfa. *Calif. Agric.* 25(5):3–4.

van den Bosch, R., G. L. Finney, and C. F. Lagace. 1971. Egyptian alfalfa weevil: Biological control possibilities. *Calif. Agric.* 25(5):6–7.

van Driesche, R. G., and G. G. Gyrisco. 1979. Field studies of *Microctonus aethiopoides*, a parasite of the adult alfalfa weevil, *Hypera postica*, in New York. *Environ. Entomol.* 8:238–44.

47 / ELM LEAF BEETLE

D. L. DAHLSTEN AND S. H. DREISTADT

---

elm leaf beetle
Xanthogaleruca luteola (Müller)
Coleoptera: Chrysomelidae

INTRODUCTION

Biology and Pest Status

The elm leaf beetle, *Xanthogaleruca luteola* (Müller), was introduced to the eastern United States from Europe in the 1830s. First reported in California in the 1920s, the beetle is considered the third-most-important urban forest insect pest in the western United States. In most areas of California it has two or more generations annually. Overwintering adults emerge in the spring to lay eggs on the foliage of various *Ulmus* species. After the three larval instars have fed in the canopy, the mature larvae crawl down the trunk to pupate near the base of the tree.

Historical Notes

In the eastern United States researchers began introducing natural enemies in 1908 (Clausen 1978). In California introductions—mostly focusing on the egg parasitoid *Oomyzus* (=*Tetrastichus*) *gallerucae* (Fonscolombe)—were carried out from 1933 to 1990; three primary parasitoids and one secondary parasitoid have become established.

Researchers have employed a biotype approach to classical biological control, introducing *O. gallerucae* biotypes from Europe, North Africa, and the Middle East in a range of climatic areas throughout California. In 1933, 7,005 European *O. gallerucae* were released in Fresno (Berry 1938a; Clausen 1978); in 1934 and 1936, 14,000 *Tetrastichus* sp.—an egg parasitoid of *Pyrrhalta maculicollis* (Mots.)—were introduced from Japan (Flanders 1936; Clausen 1978). Neither biotype became established. Undocumented attempts to establish *O. gallerucae* continued through the 1950s and 1960s, apparently without success.

RESULTS

Objective A: The Identification and Introduction of Beneficial Organisms

In the late 1970s R. F. Luck and G. T. Scriven introduced *O. gallerucae* from Morocco, and it became established in Snow Creek Village in Riverside County. *Oomyzus gallerucae* collected from Snow Creek have subsequently been introduced throughout California, including in Stockton, the northernmost point at which this strain has become established (Ehler et al. 1987; Dahlsten et al. 1990). The Stockton population may also derive from earlier releases of parasitoids collected in Israel.

Between 1983 and 1988 researchers released over 135,000 insectary-reared *O. gallerucae* from five strains (Clair, Dahlsten, and Hart 1987) at 17 sites in 12 northern California counties. The introduced strains were from France, one each from Châteauneuf le Rouge and the Rhône area; from Snow Creek and Stockton, California; and from Columbus, Ohio. The Ohio strain is apparently of European origin (Berry 1938a; Hall and Johnson 1983). *O. gallerucae* was not known to have become established in the eastern United States until its discovery in Ohio in 1982 (Hall and Johnson 1983), approximately 50 years after its last known introduction (Clausen 1978).

In 1934 about 18,600 adults of *Oomyzus* (=*Tetrastichus*) *brevistigma* (Gahan) from the eastern United States became established in central California (Berry 1938b). *O. brevistigma*, which emerges from host pupae, was thought to be limited to the United States (Burks 1979); it may be native to the East Coast, where it possibly moved onto the elm leaf beetle from an unknown native host (Clausen 1956; Clausen 1978). However, in the mid-1970s, *O. brevistigma* was imported from Iran

---

This research was funded in part by the California Departments of Forestry and Fire Protection, Transportation, and Food and Agriculture. R. F. Luck and L. E. Ehler, with the University of California in Riverside and Davis, respectively, provided data on their research and helpful discussions. D. J. Clair, L. E. Caltagirone, S. M. Tait, D. L. Rowney, and G. Y. Yokota of the Division of Biological Control, UC Berkeley, and E. R. Hart of Iowa State University, Ames, were particularly helpful in this research. R. W. Hall of Ohio State University, Columbus, and R. F. Luck provided parasitoids.

(Olkowski et al. 1986; van den Bosch unpub. data). Reportedly, it was also collected from Greece and introduced in northern California in the 1970s, but these records are vague (Olkowski et al. 1986; Olkowski and Olkowski unpub. data). Nothing is published on this parasitoid in Europe, and there is no mention of it in Silvestri's extensive review of the elm leaf beetle and its natural enemies (Silvestri 1910).

In 1939, 31 European females of *Erynniopsis antennata* (Rondani) (Diptera: Tachinidae), a larval and larval-adult parasitoid, were introduced at five sites near Stockton, in central California, and became established (Clausen 1978). Introductions of this parasitoid, under the names *Erynnia nitida* and *Erynniopsis rondanii*, were made from France between 1955 and 1972 (anonymous unpub. data).

Objective B: Ecological and Physiological Studies

In laboratory studies during 1986 researchers found that of the *O. gallerucae* strains then in culture, the one from Châteauneuf le Rouge, France, had the highest fecundity and shortest developmental time. In the field this strain achieved the highest level of parasitism during the season of its release.

Insectary tests by L. A. Strong (1935) indicated that *O. gallerucae* "passes the winter in the adult stage, and this may explain the difficulty of securing establishment in the United States." Our laboratory studies between 1987 and 1989 demonstrate that under cool conditions (10°C), *O. gallerucae* can survive for 6 months or longer as adults or immature parasitoids in host eggs. However, a low proportion of parasitoids survived this nearly host-free period.

Objective D: Impact

In northern California, the release-season parasitism by *O. gallerucae* was as high as 95 percent. Before 1989 only one or two parasitized beetle egg clusters were found at 3 of 17 sites in northern California in years following *O. gallerucae*'s release. In 1989, however, 88 percent of the egg clusters in a Marysville park were parasitized at the end of summer; *O. gallerucae* had last been released there in 1987.

Although L. E. Ehler et al. (1987) reported that *O. gallerucae* "would appear to have considerable potential when utilized in an inundative release program," there is limited data on its efficacy. However, S. H. Dreistadt (1988) has provided evidence that inoculative parasitoid releases in 1986 significantly reduced the beetle damage to Siberian elm (*Ulmus pumila*) in Fall River Mills and Susanville, California. And R. F. Luck and G. T. Scriven (person. commun.) stated that "where it [*O. gallerucae*] has been established the longest (Snow Creek), it appears to have significantly reduced defoliation by the elm leaf beetle."

O. brevistigma, on the other hand, is patchily distributed in northern California, having been found in only 7 of 12 cities during 1986 and 1987. Its populations in California are low and apparently do little to control the beetle in either the northern or southern parts of the state (Luck and Scriven 1976; Dreistadt and Dahlsten 1990).

During 1986 and 1987, researchers also found *E. antennata* in 11 of 12 northern California cities; in four of them, it had a maximum apparent parasitism rate of over 40 percent (Dreistadt and Dahlsten 1990). *E. antennata* can play a significant role in helping control the elm leaf beetle, but its effectiveness is limited by a secondary parasitoid, *Baryscapus* (=*Tetrastichus*) *erynniae* (Domenichini), and by a lack of synchrony with its host's first generation (Luck and Scriven 1976).

RECOMMENDATIONS

1. The field ecology of overwintering *O. gallerucae* should be studied in locations where this parasitoid is well established.

2. The ability of established populations of *O. gallerucae* to reduce elm leaf beetle populations in California should be assessed through pre- and postintroduction comparisons.

3. The biotype approach to *O. gallerucae* introductions should be continued, based on the long-term successes with this parasitoid in the eastern United States and in southern California.

4. Additional parasitoid species, such as *Aprostocetus celtidis* (Erdos) from Europe, should be introduced.

5. Sampling methods should be improved, and the biologies of *Erynniopsis antennata* and *Oomyzus brevistigma* should be investigated as a prerequisite for quantifying the regulating ability of these parasitoids.

REFERENCES CITED

Berry, P. A. 1938a. Laboratory studies on *Tetrastichus xanthomelaenae* Rond. and *Tetrastichus* sp., two hymenopterous egg parasites of the elm leaf beetle. *J. Agric. Res.* 57:859–63.

———. 1938b. Tetrastichus brevistigma *Gahan, a pupal parasite of the elm leaf beetle.* USDA Circular 485. Washington, D.C.: U.S. Department of Agriculture.

Burks, B. D. 1979. Family Eulophidae. In *Catalog of Hymenoptera in America north of Mexico*, eds. K. V. Krombein, P. D. Hurd, D. R. Smith, and B. D. Burks, vol. 1, 967–1022. Washington, D.C.: Smithsonian Institute.

Clair, D. J., D. L. Dahlsten, and E. R. Hart. 1987. Rearing *Tetrastichus gallerucae* (Hymenoptera: Eulophidae) for biological control of elm leaf beetle, *Xanthogaleruca luteola*. *Entomophaga* 32:457–61.

Clausen, C. P. 1956. *Biological control of insect pests in the continental United States*. USDA Technical Bulletin 1139. Washington, D.C.: U.S. Department of Agriculture.

———. 1978. Chrysomelidae. In *Introduced parasites and predators of arthropod pests and weeds: A world review,* ed. C. P. Clausen, 255–57. USDA-ARS Handbook 480. Washington, D.C.: U.S. Department of Agriculture.

Dahlsten, D. L., S. H. Dreistadt, J. R. Geiger, S. M. Tait, D. L. Rowney, G. Y. Yokota, and W. A. Copper. 1990. *Elm leaf beetle biological control and management in northern California*. Final Report to the California Department of Forestry and Fire Protection. Sacramento: State GPO.

Dreistadt, S. H. 1988. Ecology and management of the elm leaf beetle, *Xanthogaleruca luteola* (Müller) (Coleoptera: Chrysomelidae), in northern California. Ph.D. diss., University of California, Berkeley.

Dreistadt, S. H., and D. L. Dahlsten. 1990. Distribution and abundance of *Erynniopsis antennata* (Diptera: Tachinidae) and *Tetrastichus brevistigma* (Hymenoptera: Eulophidae), two introduced elm leaf beetle parasitoids in northern California.

Entomophaga 35:527–36.

Ehler, L. E., R. L. Bugg, M. B. Hertlein, J. P. Sauter, and K. Thorarinsson. 1987. Patch-exploitation patterns in an egg parasitoid of elm leaf beetle. *Entomophaga* 32:233–39.

Flanders, S. E. 1936. Japanese species of *Tetrastichus* parasitic on eggs of *Galerucella xanthomelaena* (Schrank). *J. Econ. Entomol.* 29:1024–25.

Hall, R. W., and N. F. Johnson. 1983. Recovery of *Tetrastichus gallerucae* (Hymenoptera: Eulophidae), an introduced egg parasitoid of the elm leaf beetle (*Pyrrhalta luteola*) (Coleoptera: Chrysomelidae). *J. Kansas Entomol. Soc.* 56:297–98.

Luck, R. F., and G. T. Scriven. 1976. The elm leaf beetle, *Pyrrhalta luteola*, in southern California: Its pattern of increase and its control by introduced parasites. *Environ. Entomol.* 5:409–16.

Olkowski, W., S. Daar, M. Green, D. Anderson, and J. Hyde. 1986. Update: New IPM methods for elm leaf beetle. *The IPM Practitioner* 8:1–7.

Silvestri, F. 1910. Contribuzioni alla conoscenza degli insetti dannosi e dei Loro simbionta. *Bollettino del Laboratorio di Zoologia Generale e Agraria.* 4(1909):246–89.

Strong, L. A. 1935. Report of the Chief of the Bureau of Entomology and Plant Quarantine, 60. Washington, D.C.: U.S. Department of Agriculture.

48 / New Guinea Sugarcane Weevil

J. W. BEARDSLEY, J. R. LEEPER, M. TOPHAM, AND S. L. WAGGY

New Guinea sugarcane weevil
Rhabdoscelis obscurus (Boisduval)
Coleoptera: Curculionidae

INTRODUCTION

Biology and Pest Status

The New Guinea sugarcane weevil, *Rhabdoscelis obscurus* (Boisduval), is a major pest of sugarcane in Hawaii. Throughout the tropical Pacific, in addition to sugarcane it also infests pandanus, coconut, and other palms. Adult females oviposit in cracks, crevices, or holes they have drilled with their mandibles in the outer rind of mature cane stalks or palm trunks. The larvae bore within the living stalk tissue, producing frass-filled tunnels that weaken the stalks and permit fungal pathogens to enter. Mature grubs pupate in cocoons of plant fibers within the infested stalks.

In the 1960s and 1970s, there was a resurgence of *R. obscurus* damage to sugarcane in Hawaii, but the factors responsible are not well documented. However, we believe that the planting of susceptible cultivars of cane—that is, those with relatively soft rinds or rinds subject to cracking—a general lengthening of the period from planting until harvest, and other changes in cultural practices (particularly the use of chemical herbicides to eliminate all in-field and field margin weeds) probably caused the increase in damage.

Historical Notes

Rhabdoscelis obscurus, apparently native to New Guinea and the adjacent islands, is now widely distributed in the tropical Pacific. It reached Hawaii about 1854, possibly in sugarcane from Tahiti. The first damage to cane was observed at Lahaina, Maui, in 1865. Subsequently, the weevil developed into a major pest of sugarcane throughout the Hawaiian Islands (Muir and Swezey 1916).

In 1910 F. Muir successfully introduced a tachinid parasite, *Lixophaga sphenophori* (Villeneuve), into Hawaii from New Guinea. This parasite larviposits into openings in weevil galleries, and the first-stage maggots enter the grubs they encounter; from one to as many as six or eight flies may develop in a single grub (Muir and Swezey 1916). After this introduction, weevil damage dropped markedly throughout the islands, though some plantations still experienced minor losses (Muir and Swezey 1916). However, during the 1960s and early 1970s, weevil damage increased substantially, particularly in plantations in windward Kauai. This worsening situation caused the Hawaiian Sugar Planters' Association and the State of Hawaii to seek to improve the biological control of this pest. Researchers again explored New Guinea and surrounding areas for natural enemies. They also began studies on the field ecology of the weevil and its parasite, in order to determine which factors were responsible for the increases in weevil damage. As part of the W-84 project, we investigated the food requirements and feeding habits of *L. sphenophori* adults in Hawaii (Leeper 1974; Topham and Beardsley 1975), searching for ways to improve the parasite's efficiency. We also carried out cross-mating studies with *L. sphenophori* and a newly introduced race of *Lixophaga* from the highlands of New Guinea (Waggy and Beardsley 1974), which ultimately proved to be an undescribed sibling species of *L. sphenophori* (Hardy 1981).

RESULTS

Objective A: The Identification and Introduction of Beneficial Organisms

In cross-breeding experiments with *Lixophaga* populations obtained from New Guinea in 1968 and a population from Hawaii derived from flies collected in 1910 in the New Guinea lowlands, near Port Moresby, researchers found indications that Hawaiian *L. sphenophori* were reproductively isolated from the newly introduced flies from Wau and Garaina in the eastern highlands. *Lixophaga* from the two highland localities were interfertile and appeared to represent a sibling species, possibly a highland homologue of the lowland *L. sphenophori* (Waggy and Beardsley 1974). D. E. Hardy (1981) described this new highland species as *Lixophaga beardsleyi*.

Although *L. beardsleyi* was reared and released in Hawaii for more than one year, it has never been recovered and apparently failed to become established.

Objective C: The Conservation and Augmentation of Beneficial Organisms

Studies on the feeding habits and foraging behavior of adult *Lixophaga sphenophori* in Hawaiian sugarcane fields showed that the nectar from certain weed species of the family Euphorbiaceae was a major food source. J. R. Leeper (1974) reported that these flies commonly used garden spurge (*Euphorbia hirta* L.) and castor bean (*Ricinus communis* L.) growing on ditch banks and field margins, as sources of nectar. Graceful spurge, *E. glomerifera* (Millsp.), was used less extensively. M. Topham and J. W. Beardsley (1975) found that the flies also readily used wild spurge, *E. geniculata* Ortega, and Mexican fire plant, *E. heterophylla* L., when these plants grew in and around cane fields. These authors showed that nectar source plants may also serve as mating sites where males congregate.

Most importantly, these investigators demonstrated that periodic applications of herbicides on nectar-source plants in the ditch banks and field margins greatly reduced adult *L. sphenophori* populations. In cane fields where nectar source plants were eliminated, the parasitization by *L. sphenophori* of *R. obscurus* grubs declined from around 95 percent before treatment, to 10 percent 35 days after treatment (Topham and Beardsley 1975).

RECOMMENDATIONS

As a result of the research described here, Hawaiian sugar plantations have modified their herbicide practices to retain *Euphorbia* species on field margins and ditch banks. The damage to sugarcane by *R. obscurus* appears to have declined, although follow-up studies are needed to document this.

REFERENCES CITED

Hardy, D. E. 1981. Diptera: Cyclorrahapa IV. In *Insects of Hawaii*, vol. 14. Honolulu: University of Hawaii Press.

Leeper, J. R. 1974. Adult feeding behavior of *Lixophaga sphenophori*, a tachinid parasite of the New Guinea sugarcane weevil. *Proc. Hawaii. Entomol. Soc.* 21:403–12.

Muir F., and O. H. Swezey. 1916. *The cane borer beetle in Hawaii and its control by natural enemies*. Experiment Station, Hawaiian Sugar Planters' Association, Entomological Series Bulletin 13.

Topham, M., and J. W. Beardsley. 1975. Influence of nectar source plants on the New Guinea sugarcane weevil parasite, *Lixophaga sphenophori* (Villeneuve). *Proc. Hawaii. Entomol. Soc.* 22:145–54.

Waggy, S. L., and J. W. Beardsley. 1974. Biological studies on two sibling species of *Lixophaga* (Diptera: Tachinidae), parasites of the New Guinea sugarcane weevil, *Rhabdoscelus obscurus* (Boisduval). *Proc. Hawaii. Entomol. Soc.* 21:485–94.

49 / Omnivorous Looper and *Amorbia cuneana* Walsingham

E. R. OATMAN

omnivorous looper Sabulodes aegrotata (Guenée) Lepidoptera: Geometridae	Amorbia cuneana Walsingham Lepidoptera: Tortricidae

INTRODUCTION

The omnivorous looper, *Sabulodes aegrotata* (Guenée), and the tortricid *Amorbia cuneana* Walsingham are the most common lepidopterous pests on avocado in southern California (McKenzie 1935; Ebeling and Pence 1953). Apparently indigenous species, both are distributed throughout the avocado-growing areas. The larvae of both species feed primarily on foliage, usually causing light damage; however, high populations can severely defoliate plants and scar fruit. The omnivorous looper is generally the more serious of the two pests, although both species are usually kept at low densities by a complex of natural enemies (Fleschner, Ricker, and Johnson 1957; Oatman et al. 1983).

RESULTS

Objective A: The Identification and Introduction of Beneficial Organisms

Fifteen primary parasites were reared from the omnivorous looper and 15 from *Amorbia cuneana* (Oatman et al. 1983). Seven were parasitic on both pests (see table 49.1). In addition to its parasite complex, the omnivorous looper is commonly infected with a granulosis virus.

Objective C: The Conservation and Augmentation of Beneficial Organisms

Of all the parasites, *Trichogramma platneri* Nagarkatti appeared the most suitable for an augmentative program. Accordingly, between 1979 and 1983 researchers conducted studies to determine whether mass releases of this egg parasite would provide biological control of both *A. cuneana* and the omnivorous looper. They also investigated parasite-release rates, the timing of releases, and the number of release sites per acre that would be required for effective control.

Preliminary release studies in an experimental avocado orchard in 1981 showed that *T. platneri* preferred *A. cuneana* egg masses over omnivorous looper egg clusters (parasitizing 87 percent of the masses versus 58 percent of the clusters), and that it was more effective against *A. cuneana* eggs on leaves inside of the tree canopy. Most of the eggs within the *A. cuneana* masses were parasitized by *T. platneri*, whereas most of those within the omnivorous looper egg clusters were not.

Release studies in commercial avocado orchards in 1983 showed that both *A. cuneana* and the omnivorous looper (but especially the former) can be effectively controlled by releases of 50,000 *T. platneri* in each of four uniformly spaced trees per acre (Oatman and Platner 1985). At least three weekly releases were required to control the looper, whereas only two were required for *A. cuneana*.

RECOMMENDATIONS

Since *A. cuneana* and the omnivorous looper are not serious problems every year, their adult populations should be monitored daily each year with pheromone or blacklight traps to determine when and if mass releases of *T. platneri* are needed (Bailey, Hoffman, et al. 1988; Bailey, Olsen, et al. 1988). Releases to control the egg populations can then be timed accordingly.

REFERENCES CITED

Bailey, J. B., M. P. Hoffmann, L. M. McDonough, and K. N. Olsen. 1988. Field-testing the sex pheromone for *Amorbia cuneana* in avocados. *Calif. Agric.* 42(3):17–18.

Bailey, J. B., K. N. Olsen, L. M. McDonough, N. B. O'Connell, and P. A. Phillips. 1988. Two important worm pests: Monitor program grows as pest threat widens. *Calif. Grower* 12:10, 15–16.

Ebeling, W., and R. J. Pence. 1953. *Avocado pests.* California Agricultural Experiment Station Extension Circular 428.

Fleschner, C. A., D. W. Ricker, and H. G. Johnson. 1957. Parasites of *Amorbia* and the omnivorous looper in avocado orchards. *Calif. Avocado Soc. Yearbk.* 41:107–18.

Table 49.1. Parasites reared from the eggs, larvae, or pupae of *Sabulodes aegrotata* and *Amorbia cuneana* on avocado in southern California, from 1978 through 1980.

Parasite	S. aegrotata			A. cuneana		
	Egg	Larva	Pupa	Egg	Larva	Pupa
HYMENOPTERA						
Braconidae						
Apanteles caberatae Muesebeck		X				
Bracon xanthonotus Ashmead		X				
Meteorus tersus Muesebeck		X				
Microgaster sp.					X	
Zele sp.		X				
Chalcididae						
Brachymeria ovata (Say)			X			X
Eulophidae						
Elachertus proteoteratis Howard					X	
Ichneumonidae						
Casinaria geometrae occidentalis Walkley		X				
Scambus (*Erythroscambus*) hirticauda (Provencher)		X				
Habronyx (*Camposcopus*) sp.						X
Enytus sp.					X	
Scelionidae						
Telenomus sp.	X					
Trichogrammatidae						
Trichogramma platneri Nagarkatti	X			X		
Trichogramma sp.				X		
Pteromalidae						
Dibrachys cavus (Walker)			X			X
DIPTERA						
Tachinidae						
Actia interrupta Curran					X	
Aplomya caesar (Aldrich)		X			X	
Ceromasia auricaudata Townsend					X	
Eurisyropa virilis (Aldrich & Webb)		X			X	
Madremyia saundersii (Williston)		X				
Nemorilla pyste (Walker)					X	
Pseudoperichaeta erecta (Coquillett)		X			X	
Spoggosia tachinomoides (Townsend)		X			X	

McKenzie, H. L. 1935. *Biology and control of avocado insects and mites.* University of California Agricultural Experiment Station Bulletin 592.

Oatman, E. R., and G. R. Platner. 1985. Biological control of two avocado pests. *Calif. Agric.* 39(11–12):21–23.

Oatman, E. R., J. A. McMurtry, M. Waggonner, G. R. Platner, and H. G. Johnson. 1983. Parasitization of *Amorbia cuneana* (Lepidoptera: Tortricidae) and *Sabulodes aegrotata* (Lepidoptera: Geometridae) on avocado in southern California. *J. Econ. Entomol.* 76:52–53.

50 / APPLE ERMINE MOTH

T. R. UNRUH

---

apple ermine moth
Yponomeuta malinellus Zeller
Lepidoptera: Yponomeutidae

INTRODUCTION

Biology and Pest Status

The apple ermine moth, *Yponomeuta malinellus* Zeller, is distributed throughout the Palearctic, where it is a univoltine, monophagous defoliator of apple. A female lays eggs in a mass—about 50 eggs per mass—on 1- to 3-year-old branches from mid- to late summer. Larvae hatch in about 3 weeks, but remain underneath the egg mass through winter. Throughout their immature stages, ermine moths exist in colonies composed mostly of siblings from the egg mass. When buds burst in early spring, the first-instar larvae leave the egg case and mine a single leaf; they become external defoliators in their second through fifth instars, producing a loose silk tent around attacked leaves. They pupate in early summer; adult eclosion follows in 1 to 2 weeks.

The apple ermine moth was once a significant pest of apple in the Palearctic, but it is now uncommon in chemically managed orchards. In Europe unsprayed orchards occasionally suffer damage, but populations there appear to be regulated by a rich parasitoid complex (Junikkala 1960). However, the moth is a continued threat to ornamental and residential apple and crab apple, and it may become a significant problem in commercial orchards in the future if pesticide use for apple pest management is significantly reduced.

Historical Notes

Yponomeuta malinellus was first discovered on Vancouver Island, British Columbia, in 1981 and 1982 on nursery stock. In 1985 large infestations were discovered in the lower Frazer Valley, British Columbia, and in western Whatcom County, Washington (anonymous 1985).

West of the Cascade Mountains in Washington, the moth spread south, reaching northwestern Oregon in the summer of 1991. By 1991 at least nine Washington counties east of the Cascades were infested (Unruh, Congdon, and LaGasa 1993), and by 1992 virtually all Washington counties were infested (Washington State Department of Agriculture, unpubl. data).

RESULTS

Objective A: The Identification and Introduction of Beneficial Organisms

Beginning in 1988 the U.S. Department of Agriculture (USDA) imported parasitoids from Europe and the Far East, releasing them at multiple sites in northwestern Washington. All the specimens had been collected as pupal clusters and shipped to the USDA-ARS Beneficial Insect Research Laboratory in Newark, Delaware. There the parasitoids were allowed to emerge, freed of hyperparasitoids, and sent by next-day parcel service to Washington for release. Approximately 33,000 (in 1988) and 10,500 (in 1989) individuals of the polyembryonic, egg-larval parasitoid *Ageniaspis fuscicollis* Dalman (Encyrtidae) from France were imported and released. Additionally, 100 (in 1989) and 6,700 (in 1990) *A. fuscicollis* from Korea and 28,500 (in 1990) and 5,200 (in 1991) from China were released in the same areas. Beginning in 1989, researchers introduced three other species from France in limited numbers, including the tachinid *Eurysthaea scutellaris* Robineau-Desvoidy, which attacks third- and fourth-instar larvae, and the two ichneumonids *Diadegma armillata* Gravenhorst and *Herpestomus brunnicornis* Gravenhorst, which attack mid-

The following USDA-ARS scientists supplied insects and information that made this project possible: F. Hérard, K. Chen, and K. Hopper at the European Parasite Laboratory in Behoust, France (now the European Biological Control Laboratory in Montpellier, France); and R. Pemberton at the Asian Parasite Laboratory in Seoul, Korea. The quarantine specialists with USDA-ARS in Newark, Delaware—R. Fuester, L. Ertle, and K. Swan—received the foreign shipments, ensured the absence of hyperparasitoids, and sent the parasitoids to Washington for release. In Washington, E. LaGasa and J. Wraspir (Washington State Department of Agriculture) and B. Congdon (Seattle Pacific University) helped with releases.

dle instars and young pupae, respectively. Each species was released in varying amounts in 1989, 1990, and 1991, as follows: approximately 850 then 5,090 then 2,500 *Eurysthaea;* 370 then 800 then 1,135 *Diadegma;* and 1,090 then 475 then 0 *Herpestomus.* In 1989, 150 additional *Herpestomus* from Korea were released.

Ageniaspis has been recovered each summer from 1989 to 1993, and seems permanently established. Its parasitism at all release sites was less than 0.5 pecent in 1989 and 1990 (Unruh, Congdon, and LaGasa 1993), but increased to approximately 3.5, 10, and 24 percent in 1991, 1992, and 1993. As of 1993, the three larval or pupal parasitoids had not been recovered.

RECOMMENDATIONS

After five generations of this univoltine pest had passed, the only parasite recovered one generation after release was *Ageniaspis fuscicollis.* Parasitism by *Ageniaspis* may continue to increase over time and provide a significant level of control. If this species alone does not cause moth populations to drop to acceptable levels, then other parasitoids from the European complex (Dijkerman 1987) should be introduced or reintroduced.

REFERENCES CITED

Anonymous. 1985. *Apple ermine moth new to the United States.* USDA-APHIS PPQ Plant Pest Updates 1–2. Washington, D.C.: U.S. Department of Agriculture.

Dijkerman, H. J. 1987. Parasitoid complexes and patterns of parasitization in the genus *Yponomeuta* Latreille (Lepidoptera: Yponomeutidae). *J. Appl. Entomol.* 104:390–402.

Junikkala, E. 1960. Life history and insect enemies of *Hyponomeuta malinellus* Zell. (Lepidoptera: Hyponomeutidae) in Finland. *Ann. Zool. Soc. "Vanamo"* 21:1–44.

Unruh, T. R., B. D. Congdon, and E. LaGasa. 1993. *Yponomeuta malinellus* Zeller (Lepidoptera: Yponomeutidae), a new immigrant pest of apples in the Northwest: Phenology and distribution expansion, with notes on efficacy of natural enemies. *Pan-Pacific Entomol.* 69(1):57–70.

51 / LEPIDOPTERAN COMPLEX ON TOMATOES

E. R. OATMAN

> ## lepidopteran complex
> ### Lepidoptera: Gelechiidae, Noctuidae, Sphingidae

INTRODUCTION

Biology and Pest Status

The most economically important lepidopterous pests of tomatoes in California are the tomato fruitworm, *Helicoverpa zea* (Boddie); the tomato pinworm, *Keiferia lycopersicella* (Walsingham); the beet armyworm, *Spodoptera exigua* (Hübner); the cabbage looper, *Trichoplusia ni* (Hübner); and the tobacco hornworm, *Manduca sexta* (L.) (Michelbacher, Middlekauf, and Akeson 1948; Oatman and Platner 1971). Except for the cabbage looper, all are primary pests feeding in or on the fruit and must be controlled to meet the marketing standards for tomatoes.

Growers make extensive use of broad-spectrum insecticides to control these pests. Sometimes, however, that causes outbreaks of secondary pests, such as the vegetable leafminer, *Liriomyza sativae* Blanchard (Oatman and Kennedy 1976; Johnson, Oatman, and Wyman 1980). Using selective pesticides that would not disrupt natural enemies could result in a more integrated approach to the management of this pest complex.

Historical Notes

To develop an integrated pest management (IPM) program, researchers have assessed the parasites of the major lepidopterous pests of tomatoes on summer and fall plantings that were grown without insecticides; see table 51.1 (Oatman 1970; Oatman, Wyman, and Platner 1979; Oatman, Platner, et al. 1983).

Southern California's fresh-market tomato culture involves spring, summer, and fall plantings. Each has a slightly different insect pest complex that includes one or more primary—that is, fruit-infesting—pests, though some pests occur on more than one planting. The complex on summer plantings consists mainly of lepidopterous pests: the key primary species is the tomato fruitworm, but the cabbage looper and tobacco hornworm, which feed mainly on foliage, are also often present.

Researchers discovered that these three pests, common on processing tomatoes, could be controlled by mass releases of an egg parasite, *Trichogramma pretiosum* Riley (Oatman and Platner 1971, 1978).

Between 1978 and 1979, researchers conducted studies to determine if the pests could be similarly controlled on summer plantings of fresh-market tomatoes. Because fresh-market tomatoes are such a valuable crop in southern California, applications of Dipel (*Bacillus thuringiensis* Berliner var. *kurstaki*) were combined with releases of *T. pretiosum* to help ensure economically acceptable control.

RESULTS

Objective A: The Identification and Introduction of Beneficial Organisms

In fall plantings of fresh-market tomatoes, the tomato pinworm and the beet armyworm are the key primary pests. Both are less susceptible to control by *Bacillus thuringiensis* var. *kurstaki*. Although *T. pretiosum* does parasitize the eggs of both pests, its effect on the egg masses of the beet armyworm is slight (Oatman and Platner 1978). In 1984 researchers made a preliminary search in northern Queensland, Australia, where the beet armyworm is indigenous (it is known there as the lesser armyworm), and found a *Telenomus* species that commonly parasitized the egg masses of the lawn armyworm, *Spodoptera mauritia* Guenée. This parasite was imported into southern California and released against the beet armyworm in a fall tomato planting; however, it was not recovered in 1985 when egg masses were surveyed on both summer and fall tomato plantings.

Objective C: The Conservation and Augmentation of Beneficial Organisms

Results of the 2-year study on *T. pretiosum* showed that an integrated pest management program of weekly

Table 51.1. Parasites reared from lepidopterous pests (tomato fruitworm, cabbage looper, beet armyworm, and tobacco hornworm) collected on tomato in southern California between 1965 and 1979.

Parasite	Tomato fruitworm			Cabbage looper			Beet armyworm			Tobacco hornworm		
	Egg	Larva	Pupa	Egg	Larva	Pupa	Egg	Larva	Pupa	Egg	Larva	Pupa
HYMENOPTERA												
Braconidae												
Apanteles laeviceps Ashmead					X							
Bracon platynotae Cushman		X										
Chelonus insularis (Cresson)		X						X				
Meteorus leviventris (Wesmael)								X				
Microplitis brassicae Muesebeck					X							
Microplitis plutellae Muesebeck					X							
Microplitis sp. near brassicae								X				
Encyrtidae												
Copidosoma truncatellum (Dalman)					X							
Ichneumonidae												
Campoletis sonorensis (Cameron)								X				
Campoletis flavicincta (Ashmead)		X			X			X				
Campoletis sp.		X			X			X				
Campoplex sp.		X										
Hyposoter exiguae Viereck		X			X			X			X	
Meloboris sp.								X				
Nepiera benevola Gahan								X				
Nepiera fuscifemora Graf		X			X			X				
Pristomerus pacificus pacificus Cresson								X				
Pristomerus spinator (F.)								X				
Pteromalidae												
Pediobius sp. nr. sexdentatus						X						
Trichogrammatidae												
Trichogramma pretiosum Riley	X			X			X			X		
DIPTERA												
Tachinidae												
Lespesia archippivora (Riley)								X				
Voria ruralis (Fallen)					X							
Siphona plusiae Coquillett		X			X							
Siphona sp.					X							

Dipel applications plus twice-weekly releases of the egg parasite could control lepidopterous pests in summer plantings of fresh-market tomatoes nearly as effectively as the commercial program, which consists of weekly applications of methomyl (Oatman, Wyman, et al. 1983). In 1978 the mean percentage of fruit injured by lepidopterous pests was 0.7, 1.6, and 7.1 percent in the commercial, IPM, and untreated control plots, respectively; in 1979 it was 0.3, 0.9, and 5.3 percent. In both years these differences were significant at the 95 percent level.

Although it is more effective at controlling the tomato fruitworm, the commercial program adversely affects predation and egg parasitization while substantially increasing the number of vegetable leafminers. When leafminer populations are high, growers may need to apply additional insecticides to control them. The IPM program had no such adverse effects, thus offering an alternative for growers interested in conserving natural enemies and in reducing pest-upset problems and environmental contamination.

RECOMMENDATIONS

1. Research is needed to find a more effective strain of *B. thuringiensis*, especially for control of the beet armyworm.

2. As the beet armyworm is indigenous to northern Queensland, more intensive searches should be made there for an effective egg parasite.

REFERENCES CITED

Johnson, M. W., E. R. Oatman, and J. A. Wyman. 1980. Effects of insecticides on populations of the vegetable leafminer and associated parasites on summer pole tomatoes. *J. Econ. Entomol.* 73:61–66.

Michelbacher, A. E., W. W. Middlekauf, and N. B. Akeson. 1948. *Caterpillars destructive to tomato.* California Agricultual Experiment Station Bulletin 707.

Oatman, E. R. 1970. Ecological studies of the tomato pinworm on tomato in southern California. *J. Econ. Entomol.* 63:1531–34.

Oatman, E. R., and G. G. Kennedy. 1976. Methomyl-induced outbreak of *Liriomyza sativae* on tomato. *J. Econ. Entomol.* 69:667–68.

Oatman, E. R., and G. R. Platner. 1971. Biological control of the tomato fruitworm, cabbage looper, and hornworms on processing tomatoes in southern California, using mass releases of *Trichogramma pretiosum. J. Econ. Entomol.* 64:501–6.

———. 1978. Effect of mass releases of *Trichogramma pretiosum* against lepidopterous pests in processing tomatoes in southern California, with notes on host egg population trends. *J. Econ. Entomol.* 71:896–900.

Oatman, E. R., J. A. Wyman, and G. R. Platner. 1979. Seasonal occurrence and parasitization of the tomato pinworm on fresh-market tomatoes in southern California. *Environ. Entomol.* 8:661–64.

Oatman, E. R., J. A. Wyman, R. A. Van Steenwyk, and M. W. Johnson. 1983. Integrated control of the tomato fruitworm (Lepidoptera: Noctuidae) and other lepidopterus pests on fresh-market tomatoes in southern California. *J. Econ. Entomol.* 76:1363–69.

Oatman, E. R., G. R. Platner, J. A. Wyman, R. A. Van Steenwyk, M. W. Johnson, and H. W. Browning. 1983. Parasitization of lepidopterous pests on fresh-market tomatoes in southern California. *J. Econ. Entomol.* 76:452–55.

52 / Mango Shoot Caterpillar

D. M. Nafus

mango shoot caterpillar
Penicillaria jocosatrix (Guenée)
Lepidoptera: Noctuidae

INTRODUCTION

Biology and Pest Status

The mango shoot caterpillar, *Penicillaria jocosatrix* (Guenée), can be found from China south into Australia, and from Hawaii westward into India. D. Hill (1983) listed it as a minor pest of mango that feeds on the leaves; I. Schreiner (1987) reported that it consumes flowers as well. On Guam, I. Schreiner found that the caterpillar was a serious pest, completely consuming half of the mango's flower stalks and severely damaging another quarter. Flowering trees treated with insecticides produced twice as many fruit as untreated trees. At that time, mango was being imported into Guam, where it cost about $3 a pound. There was a severe problem with fruit being smuggled in from the Philippines and Hawaii; this smuggled fruit often contained fruit flies, particularly *Dacus dorsalis* Hendel, a species that had been eradicated on Guam in the early 1960s and that is a threat to several of the island's main crops.

Initially studies concentrated on the biology and impact of the caterpillar on mango. The larvae were found to feed on the flowers, buds, and fruit, as well as on the young leaves. When populations were high, once the larvae had consumed all the leaf tissue, they ate the epidermis and other parts of new shoots, which often killed the shoot. In comparing trees treated with carbaryl with untreated ones, researchers found that untreated trees had approximately half the amount of shoot growth of treated trees—an average growth of 11 versus 20 centimeters per year. Untreated trees also produced less total mature leaf area than the treated trees—3.3 versus 5.9 square meters per 25 shoots (Schreiner and Nafus 1991). Trees that were not damaged by caterpillar feeding flushed synchronously twice a year, whereas damaged trees flushed almost twice as often, continuously producing small amounts of new leaves. Treated trees produced flowers on 35 percent of their branches, untreated trees on 13 percent. Also, trees with a leaf area of less than 4.5 square meters per 25 shoots did not produce flowers, which suggests there is a relationship between leaf area and flowering.

On Guam the life cycle of the mango shoot caterpillar is relatively short. Eggs hatch in 2 to 3 days, and there are five larval instars. In the laboratory at 28°C, the larval period lasts about 9.5 days, the first four instars taking about one day each. The fifth instars feed for 2 to 3 days and then leave to pupate in moss, among epiphyte roots or other debris on the tree, or in the litter or cracks in the soil under the tree. They incorporate bits of debris into the cocoon. About 2 to 3 days elapse between cessation of feeding and the tanning of the pupal case. The pupal stage lasts 10 days, and, in the laboratory, adults live for 2 weeks.

Larvae develop fastest on young, growing leaves. When they are fed leaves collected 2 to 4 days after budbreak (when leaves are 3 to 5 cm long), the larval period takes 9.3 days. On 6- to 9-day-old leaves, which are 12 to 15 centimeters long, development takes an additional 0.5 day. On flowers the larval period is 10.4 days, but pupal weights are significantly higher (0.17 grams) than they are when the larvae are reared on leaves (0.13 grams). If the larvae are fed nearly mature leaves, most die in the first instar, and those that survive are slow to develop. Mature leaves can be recognized by a shift in color from reddish brown to light green and by their toughness. The caterpillars do not feed on mature leaves (Nafus, Schreiner, and Dumaliang 1991).

The mango shoot caterpillar had few natural enemies on Guam (Nafus 1991a). *Trichogrammatoidea nana* (Zehntner), *T. guamiensis* Nagaraja, and *Trichogramma chilonis* Ishii collectively parasitize less than 12 percent of the eggs (Nafus 1991a). No larval parasitoids, and only a single species of pupal parasitoid—*Brachymeria lasus* (Walker), which is rare—have been reared. Potter wasps of the genus *Delta* (Vespidae) prey on the caterpillars, but their impact is unknown.

Historical Notes

In the early 1900s, mangoes on Guam were nearly free of pests (Fullaway 1912). Sometime between 1911 and 1936, the mango shoot caterpillar entered Guam. However, little was known about its biology or impact, and the insect was largely ignored. In the 1970s and 1980s, complaints about poor fruit production in both home and orchard trees led researchers to begin studying the impact of the caterpillar. After determining that it was a major pest of mango, researchers initiated a biological control program in 1986. Under the W-84 project, studies began on the caterpillar's natural enemies, its populations, and various aspects of its biology.

RESULTS

Objective A: The Identification and Introduction of Beneficial Organisms

In 1986 researchers released the egg parasitoid *Trichogramma platneri* Nagarkatti, obtained from California. It has not been recovered. Between late 1986 and February 1987, three parasitoids from India were released: *Aleiodes* species, *Blepharella lateralis* Macquart (a tachinid fly), and *Euplectrus* species nr. *parvulus* Ferriere (Nafus and Schreiner 1989).

Aleiodes species, a solitary internal parasitoid that attacks the first three larval instars, was recovered in small numbers through July 1987. No recoveries have been made since, indicating only temporary establishment. At its population peak, *Aleiodes* parasitized a maximum of 7.6 percent of second-instar caterpillars.

Both *B. lateralis* and *Euplectrus* species became established. *B. lateralis* is an internal parasitoid that emerges from the pupae and, occasionally, the fifth-instar larvae; it lays microtype eggs on the new shoots and flowers. However, because it was not possible to rear the fly, only 43 adults were released. Nine months later a few flies were reared from pupae, and 1 month later parasitization had increased to 6.7 percent.

Euplectrus species is a gregarious ectoparasitoid that prefers to lay its eggs on the first three instars but that will oviposit on all instars. The wasp stings the larvae, arresting the caterpillar's development. The eggs then develop rapidly, wasps emerging in only 8 to 10 days. Pupation takes place under the collapsed larval skin of the caterpillar. Within 7 months of its release, *Euplectrus* species had spread from the release areas to all areas surveyed.

By July 1987 the three species of larval parasitoids had parasitized 16.2 percent of the caterpillars. By December mortality rose to 39.8 percent, most of which was due to *Euplectrus*. Since then, parasitization has ranged up to 99 percent.

Objective D: Impact

Before July 1987, caterpillar populations on untreated trees ranged from 0.3 to 1.8 per shoot (an average of 0.8 caterpillars per shoot). After that, parasitization increased and population levels fell, ranging between 0 and 0.9 caterpillars per shoot (an average of 0.2 caterpillars per shoot). Parasitization rates were slightly higher in the dry season—typically over 50 percent. *Euplectrus* species parasitized an average of 68 percent of the caterpillars in the dry season but only 20 percent in the wet season (Nafus 1991b). *B. lateralis,* on the other hand, parasitized about 2 percent of the caterpillars in the dry season and about 22 percent in the wet season. Together the parasites provided complementary control.

Researchers observed extensive damage to the mango leaves before July 1987, but since then damage has decreased, and the trees produce new leaves less frequently. Between 1984 and 1985, mango shoot caterpillars ate 55 percent of the leaf tissue and consumed 90 percent or more of 40 percent of the leaves (Schreiner and Nafus 1991). By 1989 leaf damage had declined to 17 percent (Nafus 1991b). The fruit yields of trees at residences increased significantly. In 1986 monitored trees averaged 0.11 fruits per branch; in 1987 they averaged only 0.03 fruits per branch. However, between 1988 and 1990 yields ranged from 1.7 to 3.3 fruits per branch, a more than fortyfold increase (Nafus 1991b).

RECOMMENDATIONS

This multiple parasitiod introduction provided good results partly because the populations of the two parasitoids *Euplectrus* species and *B. lateralis* fluctuate with different seasons: *Euplectrus* species increased in the dry season and *B. lateralis* in the wet season. Their complementary action provided a fairly stable level of parasitization throughout the year.

REFERENCES CITED

Fullaway, D. T. 1912. *Entomological notes.* Guam Agricultural Experiment Station Report 1911.

Hill, D. 1983. *Agricultural insect pests of the tropics and their control.* Cambridge: Cambridge University Press.

Nafus, D. 1991a. Biological control of the mango shoot moth on Guam. In *Proceedings of the 1989 ADAP Pest Management Conference, Honolulu, Hawaii,* eds. M. W. Johnson, D. E. Ullman, and A. Vargo, 146–49. Manoa: University of Hawaii.

———. 1991b. Biological control of *Penicillaria jocosatrix* (Lepidoptera: Noctuidae) on mango on Guam with notes on the biology of its parasitoids. *Environ. Entomol.* 20:1725–31.

Nafus, D., and I. Schreiner. 1989. Biological control activities in the Mariana Islands from 1911 to 1988. *Micronesica* 22:65–106.

Nafus, D., I. Schreiner, and N. Dumaliang. 1991. Survival and development of mango shoot caterpillar (Lepidoptera: Noctuidae) in relation to leaf age, host, and distribution on host in tropical Anacardiaceae. *Environ. Entomol.* 20:1619–26.

Schreiner, I. 1987. Mango shoot caterpillar control on mango flowers, 1985. *Insecticide and Acaricide Tests* 12:94.

Schreiner, I. H., and D. M. Nafus. 1991. Defoliation of mango trees by the mango shoot caterpillar (Lepidoptera: Noctuidae) and its effect on foliage regrowth and flowering. *Environ. Entomol.* 20:1556–61.

53 / NAVEL ORANGEWORM

D. W. MEALS AND L. E. CALTAGIRONE

> **navel orangeworm**
> *Amyelois transitella* (Walker)
> Lepidoptera: Pyralidae

INTRODUCTION

Biology and Pest Status

The navel orangeworm, *Amyelois transitella* (Walker), is the primary insect pest of almonds and pistachios in California (Rice 1978; UC/IPM 1985), a principal pest of walnuts (UC/IPM 1987), and may soon be an important pest of figs (H. Shorey person. commun.). The moth develops in a large number of host plants, including fruits and nuts such as almond, apple, apricot, fig, loquat, orange, pear, pistachio, pomegranate, and walnut, as well as in noneconomic plants such as *Acacia farnesiana* (L.); the carob tree, *Ceratonia siliqua* L.; the false ebony, *Pithecellobium flexicaule* (Bentham); and the soap tree, *Sapindus saponaria* (L.).

In almonds, the navel orangeworm overwinters as larvae in the old nuts left on the trees. The following spring, emerging adults oviposit on these same nuts, where the first generation develops. Although flights can be identified, because the pest's development is heterogeneous, adults from each generation emerge over a several-week period; eggs are laid continuously from spring to fall. Adults emerging in midsummer oviposit on the new nut crop just as the hulls begin to split and dry. The larvae developing from the eggs of the second- and third-generation moths cause economic damage.

Two factors make it difficult to use chemical control: (1) neither spring- nor summer-generation moths emerge all at once, but rather over the course of several weeks; and (2) because the larvae and pupae develop inside the fruit, only the egg and adult stages are exposed. After early attempts at using insecticides were unsuccessful, the navel orangeworm became a candidate for biological control.

Historical Notes

The navel orangeworm seems to be native to the Neotropics, but it is extending its distribution to the southern Nearctic. It is widespread in Mexico, and has been found as far south as Peru, central Argentina, and Uruguay.

In the United States, this pest is distributed across the south from California to the east coast, and as far north as North Carolina. However, it has become an economic problem only in California, where, following its discovery in the southern part of the state in 1942 (Wade 1961), it gradually extended its range throughout the Central Valley. The first damaging infestations in almond and walnut crops were encountered in the early 1950s (Bacon and Wade 1954; Michelbacher and Davis 1961). Until 1967 only 2 percent of Nonpareil variety almonds were rejected; 10 years later, the damage had risen to 6.1 percent; the following year, in 1978, the figure was 10.2 percent (Almond Board of California 1982). Populations of *A. transitella* in California have not yet stabilized; in fact, they are still increasing.

RESULTS

Objective A: The Identification and Introduction of Beneficial Organisms

Researchers concentrated their initial biological control efforts in three areas: searching the pest's native habitat for biotic agents; introducing natural enemies of the Old World homologue of the navel orangeworm, the carob moth *Ectomyelois ceratoniae* Zeller; and surveying the potential effectiveness of indigenous natural enemies.

An extensive exploration in Mexico—where the navel orangeworm is of no economic importance—resulted in the importation of several natural enemies (see table 53.1). One of these, *Copidosomopsis plethorica* (Caltagirone) (Hymenoptera: Encyrtidae), was mass-reared by the Division of Biological Control of the University of California, Berkeley. With the help of Cooperative Extension Service Farm Advisors, approximately 240 million adults were released in the Central Valley during the 1970s. Establishment occurred in both the Sacramento and San Joaquin valleys.

UC Riverside biological control specialists and South American scientists later explored Uruguay and central Argentina (in the 30° to 35° S latitude); that search resulted in the importation of *Goniozus legneri* Gordh (Hymenoptera: Bethylidae). It, too, became established in parts of California (Legner and Silveira-Guido 1983).

Of the introductions from the Old World, the Israeli *Phanerotoma flavitestacea* Fischer (Hymenoptera: Braconidae) was successfully mass-reared and released in the Central Valley. It became established in several almond-growing areas, parasitizing up to 25 percent of the larvae. Although the wasp overwintered successfully and persisted for several years, it has not been recovered since 1976.

Objective C: The Conservation and Augmentation of Beneficial Organisms

Since the navel orangeworm's appearance in California, numerous indigenous parasitoids and predators have

begun to use it as a host (table 53.1). Of these adventitious insects, it is the parasitoids that have been the most studied.

Researchers have found that in some years *Trichogramma californicum* Nagaraja and Nagarkatti has parasitized 25 to 35 percent of the pest's eggs in late spring and early summer. When they made augmentative releases during these months, the parasitization increased, but there was no significant decrease in crop damage. With the exception of *C. plethorica* and *G. legneri*, the other parasitoids in table 53.1 occur very sporadically and in insignificant numbers.

Objective D: Impact

Copidosomopsis plethorica is an egg-larval polyembryonic parasitoid; some 800 adults develop in each navel orangeworm larva. The parasitoid is well established in all of California's almond-growing areas. In orchards in which pesticides are not used during the growing sea-

Table 53.1. Natural enemies of *Amyelois transitella*, both introduced and native, found in California between 1950 and 1984.

Family	Natural enemy	Native or introduced (origin)	Stage of host attacked
HYMENOPTERA			
Trichogrammatidae	*Trichogramma californicum* Nagaraja and Nagarkatti	N	egg
	Trichogrammatoidea annulata De Santis	I (Argentina)	egg
Encyrtidae	*Copidosomopsis plethorica* (Caltagirone)	I (Mexico)	egg
Braconidae	*Phanerotoma flavitestacea* Fischer	I (Israel)	egg
	Phanerotoma inopinata Caltagirone	N	egg
	Bracon hebetor Say	N	larva
Bethylidae	*Goniozus breviceps* (Krombein)	N	larva
	Goniozus legneri Gordh	I (Uruguay)	larva
	Goniozus emigratus (Rohwer)	I (Texas)	larva
Eulophidae	*Paraolinx typica* (Howard)	N	larva
Ichneumonidae	*Mesostenus gracilis* Cresson	N	larva
	Liotryphon nucicola (Cushman)	N	larva
	Scambus sp.	N	larva
	Venturia canescens (Gravenhorst)	N	larva
	Coccygomimus sp.	N	larva
	Diadegma sp.	I (Australia)	larva
Chalcididae	*Spilochalcis leptis* Burks	N	pupa
DIPTERA			
Tachinidae	*Erynnia* sp.	N	?
COLEOPTERA			
Cleridae	*Cymatodera ovipennis* (Le Conte)	N	larva
	Phyllobaenus sp.	N	egg, small larva
THYSANOPTERA			
Phlaeothripidae	*Leptothrips mali* (Fitch)	N	egg, small larva

son—that is, from April to September—its rate of parasitism of orangeworm larvae in the late summer and fall can reach 75 percent. However, after in-season applications of broad-spectrum insecticides, the parasitoid can take as long as 2 to 3 years to return to prior population levels.

C. plethorica overwinters in immature stages in navel orangeworm larvae. There seems to be a lack of synchrony between the parasitoid and its host, for the parasitoid does not emerge in the spring until several weeks after the pest's first oviposition peak. As a result, the moth's summer population can reach levels that cause economic damage to the current year's crop. Field experiments that sought to compensate for this lack of synchrony by augmentative releases during the early spring months have been inconclusive.

Goniozus legneri is a larval gregarious ectoparasitoid. The adult female's shape enables her to crawl into fruit damaged by the navel orangeworm. Gravid females seek medium-size to fully developed larvae and then sting them, inducing an irreversible paralysis. Eventually the female lays up to 15 eggs on the paralyzed larva. The larvae that hatch from these eggs devour the host larva very quickly, reducing it to an empty head capsule and a shriveled integument. The parasitoid larvae pupate next to the host remains, inside the nut cavity.

Adult females exhibit maternal care of their progeny, not ovipositing again until the present-brood larvae are nearly mature; this reduces the species' fecundity. In his study of the bethylid *Prosierola bicarinata* (Brues), R. L. Doutt (1973) suggested that there must be survival advantages to this behavior, the most likely being that it protects the brood from attack by hyperparasites.

Researchers have observed parasitism levels of 20 to 30 percent, a somewhat conservative estimate, since adult females commonly paralyze more host larvae than they actually use for producing progeny. *G. legneri* is established in various almond-growing areas of the state, particularly in the San Joaquin Valley.

Copidosomopsis plethorica and *G. legneri* appear to complement each other, causing greater mortality of the navel orangeworm together than either does alone (Legner et al. 1982). However, though both parasitoids are well established in the Central Valley, they do not exert enough control to bring damage levels below the economically significant threshold.

RECOMMENDATIONS

The current recommended control measures rely on two cultural practices to reduce the navel orangeworm crop damage: (1) orchard sanitation during the winter to remove nuts that remain in the trees after harvest, and (2) early harvesting, to prevent the third-flight moths from ovipositing. Sanitation effectively reduces the overwintering population, if practiced on an areawide basis, and eliminates egg-laying sites for the first brood.

Growers who do not remove postharvest nuts are usually advised to use one or two applications of insecticide. As an alternative, we suggest early harvesting plus eliminating the use of in-season pesticides against the navel orangeworm. The latter practice allows natural enemies, primarily *C. plethorica* and *G. legneri*, to have their optimal effect.

REFERENCES CITED

Almond Board of California. 1982. *Statistics.* Sacramento: Almond Board of California.

Bacon, O. G., and W. H. Wade. 1954. Davis men discuss problem of navel orangeworms' ravages. *Almond Facts* 19:7,9.

Doutt, R. L. 1973. Maternal care of immature progeny by parasitoids. *Ann. Entomol. Soc. Am.* 66:486–87.

Legner, E. F., and A. Silveira-Guido. 1983. Establishment of *Goniozus emigratus* and *Goniozus legneri* (Hymenoptera: Bethylidae) on navel orangeworm, *Amyelois transitella* (Lepidoptera: Phycitidae), in California, and their biological control potential. *Entomophaga* 28:97–106.

Legner, E. F., G. Gordh, A. Silveira-Guido, and M. E. Badgley. 1982. New wasp may help control navel orangeworm. *Calif. Agric.* 36(5–6):4–5.

Michelbacher, A. E., and C. S. Davis. 1961. The navel orangeworm in northern California. *J. Econ. Entomol.* 54:559–62.

Rice, R. E. 1978. Navel orangeworm: A pest of pistachio nuts in California. *J. Econ. Entomol.* 71:822–24.

UC/IPM. 1985. *Integrated pest management for almonds.* Oakland: University of California Division of Agriculture and Natural Resources, Publication 3308.

———. 1987. *Integrated pest management for walnuts,* 2nd ed. Oakland: University of California Division of Agriculture and Natural Resources, Publication 3270.

Wade, W. H. 1961. Biology of the navel orangeworm, *Paramyelois transitella* (Walker), on almonds and walnuts in northern California. *Hilgardia* 31:129–71.

54 / PINK BOLLWORM

S. E. NARANJO, T. J. HENNEBERRY, AND C. G. JACKSON

> **pink bollworm**
> *Pectinophora gossypiella* (Saunders)
> Lepidoptera: Gelechiidae

INTRODUCTION

Biology and Pest Status

The pink bollworm, *Pectinophora gossypiella* (Saunders), is the key pest of cotton in Arizona and southern California, and a constant threat to the uninfested cotton-growing areas of California's San Joaquin Valley. The feeding of pink bollworm larvae reduces the production of lint and seed, and stains and destroys the lint. Lint and seed quality may also be reduced as a result of infections of boll rot organisms, including *Aspergillus flavus* Link. From 1966 to 1980 annual losses in California's Imperial Valley have averaged 26 percent of crop value. Since 1980 cotton production in the Imperial Valley has dropped from over 100,000 acres to less than 15,000 acres, mainly because of the pink bollworm and associated problems.

Historical Notes

Although the pink bollworm is found worldwide, current studies suggest that it is native to northwestern Australia and to Indonesia (G. Gordh unpub. data). In the United States the pest was first reported in Texas in 1917 (Spears 1968). It spread rapidly to adjoining states: in 1920 infestations were reported in New Mexico, and in 1927 in eastern Arizona. Subsequent eradication programs in Arizona helped keep the pink bollworm from becoming a serious economic threat until 1958. However, when those efforts were relaxed, there was a rapid resurgence of the pest throughout Arizona in the early 1960s. By 1965 the pink bollworm had spread to southern California (Spears 1968).

G. Gordh (unpub. data) recently surveyed the worldwide biological control efforts against the pink bollworm. The early biological control efforts in Texas (1932 to 1955) have been summarized by L. W. Noble (1969); C. G. Jackson (1980) reported on the efforts in western cotton-growing areas in the late 1960s and the 1970s, which involved importing and releasing various hymenopterous parasitoids as well as studying the indigenous arthropod predators. These and other efforts in the western United States by scientists associated with the W-84 project are described in the following sections.

RESULTS

Objective A: The Identification and Introduction of Beneficial Organisms

Soon after the pink bollworm was discovered in southern California, researchers launched efforts to import and release biological control agents (Bartlett and González 1970). After searching for exotic parasites in India, eastern Africa, southern Europe, Australia, and Hawaii, E. F. Legner and R. A. Medved (1979) released 14 species in California between 1969 and 1976 and in Arizona between 1977 and 1978. Although the rates of host parasitism were consistently low, certain species showed some promise, including *Chelonus blackburni* Cameron, a *Chelonus* species nr. *curvimaculatus* Cameron from Ethiopia, a *Chelonus* species nr. *curvimaculatus* from Australia, and *Pristomerus hawaiiensis* Ashmead. Eight of these species reproduced during release years; however, none overwintered successfully, despite evidence that several were capable of overwintering in Arizona and southern California (Fye and Jackson 1973; Legner 1979). Researchers cited parasitoid and host dispersal, limited genetic heterogeneity of parasitoid populations, and parasitoid-host asynchrony between seasons as reasons for the generally poor results (Legner and Medved 1979). In large-scale field releases in Arizona from 1971 to 1973, *Bracon kirkpatricki* (Wilkinson) had only a limited impact on pink bollworm early in the season, and *C. blackburni* was generally ineffective in releases made later in the season (Bryan et al. 1973a,b, 1976). Researchers released *Goniozus pakmanus* Gordh in small numbers in Arizona and California in 1984 and 1985, respectively, but follow-up studies were not conducted to determine establishment (Gordh and Medved 1986).

Releases of an Australian egg parasitoid, *Trichogrammatoidea bactrae* Nagaraja, were begun in 1986; work on this species continues (G. Gordh person. commun.)

Relatively little is known about the pink bollworm's indigenous natural enemies in the southwestern United States. D. N. Ferro and R. E. Rice (1970) reported two ichneumonid parasitoids—*Gambrus ultimus* (Cresson) and an unidentified species in the subfamily Pimplinae—and one Pteromalidae, *Dibrachys cavus* (Walker), emerging from diapause larvae. Another native parasitoid, *Bracon platynotae* (Cushman), occurs in the Southwest, but its impact in desert cotton has not been examined (Jackson and Patana 1980). Most of the common predators found in desert cotton fields are capable of feeding on one or more stages of the pink bollworm, and the limited field studies that have been done so far suggest that their predation of artificially placed pink bollworm eggs can be significant (Henneberry and Clayton 1982). Based on its performance in laboratory, greenhouse, and field cage studies, *Chrysoperla* (=*Chrysopa*) *carnea* (Stephens) also has potential as a predator of bollworm eggs (Orphanides, González, and Bartlett 1971; Irwin, Gill, and González 1974; Henneberry and Clayton 1985). *Geocoris pallens* Stål, *G. punctipes* (Say), *Nabis americoferus* (Carayon), and *Orius tristicolor* (White) also show promise. However, key predator species need to be identified and their role in regulating pink bollworm populations needs to be assessed under realistic conditions.

Several entomopathogens are known to affect pink bollworm, including (1) *Bacillus thuringiensis* Berliner, (2) a cytoplasmic polyhedrosis virus, and (3) *Autographa californica* (Speyer), a nuclear polyhedrosis virus (Bell and Henneberry 1980). All three appeared to be effective in the laboratory, but none had a significant impact on pink bollworm in the field (Graves and Watson 1970; Bell and Henneberry 1980).

Objective B: Ecological and Physiological Studies

Researchers have performed life-history studies on several natural enemies in order to predict their ecological adaptability, to aid in developing mass-rearing methods, and to measure attack efficiencies. Temperature-dependent studies of development, survival, and fecundity have been done for *B. kirkpatricki* (Bryan et al. 1971; Engroff and Watson 1975), *C. blackburni* (Jackson, Delph, and Neeman 1978), *B. platynotae* (Jackson and Patana 1980), and *Bracon greeni* Ashmead, *B. hebetor* Say, and *B. brevicornis* Wesmael (Jackson and Butler 1984). These species' short development times, high rates of fecundity, and good survival under high temperatures suggest that they may have potential for biological control in desert cotton.

Detailed biological studies have also been conducted for *Goniozus aethiops* Evans (Gordh and Evans 1976), *G. legneri* Gordh (Gordh, Woolley, and Medved 1983), and *G. pakmanus* (Gordh and Medved 1986). These authors concluded that *G. legneri* and *G. pakmanus* are promising biological control candidates. Recent studies have also shown that pink bollworm is a highly suitable host for *Trichogrammatoidea bactrae* Nagaraja, which indicates that inundative releases may be effective for control (Hutchison, Moratorio, and Martin 1990).

Researchers have also done laboratory studies on some life-history parameters of the pink bollworm's insect predators. *Chrysoperla carnea*, *Collops vittatus* (Say), and *Hippodamia convergens* Guerin-Meneville all completed larval development on pink bollworm eggs, and their developmental times compared favorably with those on other prey species (Henneberry and Clayton 1985). The reduviids, *Sinea confusa* Caudell and *Zelus renardii* Kolenati, did not display a preference for pink bollworm eggs and fourth-stage larvae, but they did complete development in the fourth nymphal stage on larvae (Fye 1979). In the same study, *Nabis alternatus* Parshley readily attacked pink bollworm eggs and completed nymphal development on a mixture of eggs and fourth-stage larvae.

Little attention has been given to natural enemy behavior, particularly host- or prey-finding behavior. A. A. Chiri and E. F. Legner (1982) isolated a kairomone from pink bollworm moth scales that elicited locomotory arrestment, antennation, ovipositor probing, and inverse klinokinesis in female *C.* species nr. *curvimaculatus* from Ethiopia. However, field application of extracts from other moth species to cotton appeared to distract rather than enhance parasitoid attack on pink bollworm eggs (Chiri and Legner 1983). R. E. Fye (1979) studied the searching behavior of several predator species in the laboratory and greenhouse, and M. E. Irwin, R. W. Gill, and D. González (1974) showed that a number of predators were capable of finding pink bollworm eggs hidden under cotton boll calyces.

Objective C: The Conservation and Augmentation of Beneficial Organisms

To date, cotton growers' heavy use of insecticides has severely limited the establishment of exotic parasitoids and the utility of naturally occurring parasitoids and predators. However, compared with fields treated only with insecticides, fields treated with two commercial formulations of gossyplure starting around flowering time required significantly fewer insecticide treatments through the season and had higher populations of various predator species, including *Orius* and *Geocoris* (Beasley and Henneberry 1985). When permethrin was added to the sticker to enhance gossyplure's effectiveness, some predator species were reduced slightly, but an independent study found that it had little impact on total predator populations (Butler and Las 1983). Combining parasitoid releases with gossyplure reduced pink bollworm infestation more than did the application of pheromone alone (Legner and Medved 1981).

To a limited degree, researchers have explored the use of entomopathogens in inundative releases. *Bacillus thuringiensis* can cause significant mortality in pink bollworm in the laboratory; however, limited field tests in Arizona demonstrated a poor level of control (Graves and Watson 1970). Researchers improved the efficacy of the *A. californica* nuclear polyhedrosis virus in the laboratory by formulating a bait consisting of the virus and a feeding stimulant (Bell and Kanavel 1975). Field applications of this bait reduced larval populations by 69 percent and boll damage by 62 percent (Bell and Kanavel 1977). Although partial control could be attributed to the effect of the feeding stimulant alone, and the cost of the virus dosages was considered economically impractical, the nuclear polyhedrosis virus may be useful in an overall integrated management program as an early-season insecticide substitute that would help preserve natural enemies.

RECOMMENDATIONS

1. Continue exploring and importing exotic natural enemy species, paying particular attention to northwestern Australia, Indonesia, and surrounding areas.

2. Conduct studies on the biological, ecological, and behavioral parameters that are important for allowing the establishment of candidate natural enemies in southwestern cotton-growing areas.

3. Identify key indigenous natural enemies, particularly arthropod predators, of the pink bollworm under field-realistic conditions, and quantify their roles in regulating populations.

4. Continue studies to integrate biological control with other management tactics—including pheromones, host-plant resistance, and cultural methods—that will reduce reliance on insecticides and minimize the effects on natural enemies when insecticides are necessary.

REFERENCES CITED

Bartlett, B. R., and D. González. 1970. A progress report: Biological control of pink bollworm in cotton. *Calif. Agric.* 24(1):12–14.

Beasley, C. A., and T. J. Henneberry. 1985. *Pink bollworm–gossyplure studies in the Palo Verde Valley, California.* USDA-ARS-39. Washington, D.C.: U.S. Department of Agriculture.

Bell, M. R., and T. J. Henneberry. 1980. Entomopathogens for pink bollworm control. In *Pink bollworm control in the western United States,* ed. H. M. Graham, 76–81. Science and Education Administration, ARM, ARS-W-16. Washington, D.C.: U.S. Department of Agriculture.

Bell, M. R., and R. F. Kanavel. 1975. Potential of bait formulations to increase effectiveness of nuclear polyhedrosis virus against the pink bollworm. *J. Econ. Entomol.* 68:389–91.

———. 1977. Field tests of a nuclear polyhedrosis virus in a bait formulation for control of pink bollworm and *Heliothis* spp. on cotton in Arizona. *J. Econ. Entomol.* 70:625–29.

Bryan, D. E., R. E. Fye, C. G. Jackson, and R. Patana. 1973a. *Releases of* Bracon kirkpatricki *(Wilkinson) and* Chelonus blackburni *Cameron for pink bollworm control in Arizona.* USDA-ARS Production Research Report 150. Washington, D.C.: U.S. Department of Agriculture.

———. 1973b. *Releases of parasites for suppression of pink bollworm in Arizona.* USDA-ARS-W-7. Washington, D.C.: U.S. Department of Agriculture.

———. 1976. *Nonchemical control of pink bollworms.* USDA-ARS-W-39. Washington, D.C.: U.S. Department of Agriculture.

Bryan, D. E., C. G. Jackson, R. Patana, and E. G. Neemann. 1971. Field cage and laboratory studies with *Bracon kirkpatricki,* a parasite of the pink bollworm. *J. Econ. Entomol.* 64:1236–41.

Butler, G. D. Jr., and A. S. Las. 1983. Predaceous insects: Effect of adding permethrin to the sticker used in gossyplure applications. *J. Econ. Entomol.* 76:1448–51.

Chiri, A. A., and E. F. Legner. 1982. Host-searching kairomones alter behavior of *Chelonus* sp. nr. *curvimaculatus,* a hymenopterous parasite of the pink bollworm, *Pectinophora gossypiella* (Saunders). *Environ. Entomol.* 11:453–55.

———. 1983. Field applications of host-searching kairomones to enhance parasitization of the pink bollworm (Lepidoptera: Gelechiidae). *J. Econ. Entomol.* 76:254–55.

Engroff, B. W., and T. F. Watson. 1975. Influence of temperature on adult biology and population growth of *Bracon kirkpatricki.* *Ann. Entomol. Soc. Am.* 68:1121–25.

Ferro, D. N., and R. E. Rice. 1970. Parasites of pink bollworm in southern California. *Ann. Entomol. Soc. Am.* 63:1783–84.

Fye, R. E. 1979. Cotton insect populations: Development and impact of predators and other mortality factors. USDA-ARS Techical Bulletin 1592. Washington, D.C.: U.S. Department of Agriculture.

Fye, R. E., and C. G. Jackson. 1973. Overwintering of *Chelonus blackburni* in Arizona. *J. Econ. Entomol.* 66:807–8.

Gordh, G., and H. E. Evans. 1976. A new species of *Goniozus* imported into California from Ethiopia for the biological control of pink bollworm and some notes on taxonomic status of *Parasierola* and *Goniozus.* *Proc. Entomol. Soc. Wash.* 78:479–89.

Gordh, G., and R. E. Medved. 1986. Biological notes on *Goniozus pakmanus* Gordh (Hymenoptera: Bethylidae), a parasite of pink bollworm, *Pectinophora gossypiella* (Lepidoptera: Gelechiidae). *J. Kansas Entomol. Soc.* 59:723–34.

Gordh, G., J. B. Woolley, and R. A. Medved. 1983. Biological studies on *Goniozus legneri* Gordh (Hymenoptera: Bethylidae) a primary external parasite of the navel orangeworm, *Amyelois transitella*, and pink bollworm, *Pectinophora gossypiella* (Lepidoptera: Pyralidae, Gelechiidae). *Contributions Am. Entomol. Institute* 20:433–68.

Graves, G. N., and T. F. Watson. 1970. Effect of *Bacillus thuringiensis* on the pink bollworm. *J. Econ. Entomol.* 63:1828–30.

Henneberry, T. J., and T. E. Clayton. 1982. Pink bollworm: Seasonal oviposition, egg predation, and square and boll infestations in relation to cotton plant development. *Environ. Entomol.* 11:663–66.

———. 1985. Consumption of pink bollworm (Lepidoptera: Gelechiidae) and tobacco budworm (Lepidoptera: Noctuidae) eggs by some predators commonly found in cotton fields. *Environ. Entomol.* 14:416–19.

Hutchison, W. D., M. Moratorio, and J. M. Martin. 1990. Morphology and biology of *Trichogrammatoidea bactrae* (Hymenoptera: Trichogrammatidae), imported from Australia as a parasitoid of pink bollworm (Lepidoptera: Gelechiidae) eggs. *Ann. Entomol. Soc. Am.* 83:46–54.

Irwin, M. E., R. W. Gill, and D. González. 1974. Field-cage studies of native egg predators of the pink bollworm in southern California cotton. *J. Econ. Entomol.* 67:193–96.

Jackson, C. G. 1980. Entomophagous insects attacking *Pectinophora gossypiella*. In *Pink bollworm control in the western United States,* ed. H. M. Graham, 71–75. Science and Education Administration, ARM, ARS-W-16. Washington, D.C.: U.S. Department of Agriculture.

Jackson, C. G., and G. D. Butler, Jr. 1984. Development time of three species of *Bracon* (Hymenoptera: Braconidae) on pink bollworm (Lepidoptera: Gelechiidae) in relation to temperature. *Ann. Entomol. Soc. Am.* 77:539–42.

Jackson, C. G., and R. Patana. 1980. Effect of temperature on the development and survival of the immature stages of *Bracon platynotae*, a native parasite of *Pectinophora gossypiella* (Lepidoptera: Gelechiidae). *Southwest. Entomol.* 5:65–68.

Jackson, C. G., J. S. Delph, and E. G. Neeman. 1978. Development, longevity, and fecundity of *Chelonus blackburni* (Hymenoptera: Braconidae) as a parasite of *Pectinophora gossypiella*. *Entomophaga* 23:35–42.

Legner, E. F. 1979. Emergence patterns and dispersal in *Chelonus* spp. near *curvimaculatus* and *Pristomerus hawaiiensis*, parasitic on *Pectinophora gossypiella*. *Ann. Entomol. Soc. Am.* 72:681–86.

Legner, E. F., and R. A. Medved. 1979. Influence of parasitic Hymenoptera on the regulation of pink bollworm, *Pectinophora gossypiella*, on cotton in the lower Colorado desert. *Environ. Entomol.* 8:922–30.

———. 1981. Pink bollworm, *Pectinophora gossypiella* (Diptera: Gelechiidae), suppression with gossyplure, a pyrethroid, and parasite releases. *Can. Entomol.* 113:355–57.

Noble, L. W. 1969. *Fifty years of research on the pink bollworm in the United States.* USDA-ARS Handbook 357. Washington, D.C.: U.S. Department of Agriculture.

Orphanides, G. M., D. González, and B. R. Bartlett. 1971. Identification and evaluation of pink bollworm predators in southern California. *J. Econ. Entomol.* 64:421–24.

Spears, J. H. 1968. The westward movement of the pink bollworm. *Bull. Entomol. Soc. Am.* 14:118–19.

55 / POINCIANA LOOPER

D. M. NAFUS AND J. R. NECHOLS

> **poinciana looper**
> *Pericyma cruegeri* (Butler)
> Lepidoptera: Noctuidae

INTRODUCTION

Biology and Pest Status

Royal poinciana, *Delonix regia*, is a widely planted tropical ornamental tree, noted for both its attractive shape and floral display. During certain seasons the tree flowers in synchrony and is completely covered with brilliant red-and-yellow flowers.

The noctuid caterpillar *Pericyma cruegeri* (Butler) feeds on the foliage of poinciana and, in certain localities, causes extensive damage. On Guam, R. Muniappan (1973) found that these caterpillars killed small branches by defoliating them, reduced the vigor of affected trees, caused increased infestation by bark borers and fungal diseases, and disrupted or prevented flowering. Of primary importance from an esthetic viewpoint, the caterpillar caused a shift from synchronous flowering to erratic flowering by single branches or no flower production at all. In addition to royal poinciana, the looper also attacks *Desmanthus virgatus* and yellow poinciana, *Peltophorum pterocarpum*.

On Guam the looper is more abundant in the wetter parts of the year and scarce during drier periods. It lays its yellowish or bluish green eggs singly on the leaflets; the eggs hatch within 2 to 3 days. There are five larval instars. The mean number of days for development are, for the first instar, 5.1; for the second, 3.7; for the third, 5.1; for the fourth, 4.4; for the fifth, 9.9; and for the pupa, 10.2. The looper takes an average of 40 days to develop from egg to adult (Muniappan 1973).

Historical Notes

The poinciana looper is native to tropical and subtropical areas of Southeast Asia; in recent years it has spread to other areas of the Pacific including Guam, Palau, Papua New Guinea, and Hawaii. The poinciana looper entered Guam in 1971 and quickly became a serious pest. By 1972 it was widespread and had nearly defoliated most of the poincianas (Muniappan 1973). It also defoliated poinciana on Palau and Papua New Guinea, but it has not been a problem in Hawaii.

RESULTS

Objective A: The Identification and Introduction of Beneficial Organisms

W-84 project participants began work on this pest in 1978. Earlier, R. Muniappan (1973) had surveyed the natural enemy complex attacking the poinciana looper on Guam and identified the tachinid *Exorista civiloides* (Bar.) and the mantid *Hierodula patallifera* (Serville). L. Stevens (1978, 1979) also observed predatory wasps of *Delta* species. However, none of these natural enemies had an appreciable effect on looper populations (Stevens 1978, 1979). *Brachymeria lasus* (Walker), a chalcid pupal parasite released by Guam's Department of Agriculture in 1973 (Nafus and Schreiner 1989), became established but had little effect on looper populations (Stevens 1979; Nechols 1981) because of its low numbers and because it is known to attack a wide variety of nontarget hosts.

Near the end of the wet season in 1981, J. R. Nechols (1981) recorded defoliation levels of up to 80 percent; looper populations reached 15 caterpillars per branch. I. Schreiner (unpub. data) found low populations in 1987—never exceeding four caterpillars per branch—after an unusual dry season in which the trees did not shed their leaves. That year, researchers noted a substantial recovery of the trees, and approximately 60 percent of the branches flowered in June 1988. However, in 1988 the pest resurged and defoliated most of the trees at least twice. In November of that year, looper populations peaked at 12 caterpillars per branch.

Changes in looper populations appear to be associated with weather conditions rather than with natural enemies. Populations are high in the wet season—from August to February—and drop to very low levels in the dry season (Nechols 1981; Schreiner unpub. data).

Objective C: The Conservation and Augmentation of Beneficial Organisms

L. Stevens (1979), in studies on the impact of various systemic insecticides and *Bacillus thuringiensis* on the looper and its natural enemies on royal poincianas, found that *B. thuringiensis* was effective but too expensive for general use. None of the systemic insecticides tested was effective.

RECOMMENDATIONS

Effective biological control of the poinciana looper has been hampered by searching for natural enemies in inappropriate locations. For example, early searches were concentrated in Papua New Guinea, where the poinciana looper is also an introduced pest and causes periodic defoliation. The introduction of *Brachymeria lasus* to Guam from Papua New Guinea appears to have been unsuccessful in part because that species is too general a feeder. Searches for natural enemies should concentrate on Southeast Asia, where the poinciana looper is thought to have originated. The looper is present but rare in both Malaysia and Thailand; both areas have climates similar to that of Guam.

REFERENCES CITED

Muniappan, R. 1973. Biology of the poinciana looper, *Pericyma cruegeri* (Butler), on Guam. *Micronesica* 10:273–78.

Nafus, D., and I. Schreiner. 1989. Biological control activities in the Mariana Islands from 1911 to 1988. *Micronesica* 22:65–106.

Nechols, J. R. 1981. *Entomology: Biological control*, 16–20. Annual Report of the Guam Experiment Station .

Stevens, L. 1978. *Entomology: Biological control*, 18–20. Annual Report of the Guam Experiment Station.

———. 1979. *Entomology: Biological control*, 15–19. Annual Report of the Guam Experiment Station.

56 / Potato Tuberworm and Tomato Pinworm

E. R. Oatman

potato tuberworm	tomato pinworm
Phthorimaea operculella (Zeller)	*Keiferia lycopersicella* (Walsingham)
Lepidoptera: Gelechiidae	Lepidoptera: Gelechiidae

INTRODUCTION

Biology and Pest Status

The potato tuberworm, *Phthorimaea operculella* (Zeller), is a worldwide pest of potato, tomato, tobacco, and other Solanaceae. It is especially serious in dry, subtropical climates and during dry seasons in areas that normally have adequate rainfall (Lloyd 1972). The potato tuberworm occurs in all of the major agricultural areas of central and southern California, but not in the extreme north of the state (Bacon 1960). In southern California, where two irrigated crops of potatoes are grown annually, the pest is active throughout the year (Oatman and Platner 1974a).

The cultivated potato, *Solanum tuberosum* L., is the potato tuberworm's most common host. The pest mines the foliage, stems, and tubers. Most of the economic damage is done to exposed tubers; if growers do not use preventive measures, such as deep planting and ridging, up to 100 percent of the tubers may be infested (Shelton and Wyman 1979). Where water is plentiful, another good preventive measure is irrigating frequently, to prevent the soil from drying out and cracking (Oatman unpub. observation).

In the United States the tomato pinworm, *Keiferia lycopersicella* (Walsingham), is a serious economic pest of both processing and fresh-market tomatoes—especially in southern California (Oatman 1970), Florida (Wolfenbarger and Poe 1973), and Texas (Harding 1971). By my observation, it is also a serious pest of tomatoes in Hawaii and commonly infests commercial eggplant in southern California.

Historical Notes

Because of the importance of potatoes and tomatoes to California's agricultural economy and the severity of potato tuberworm and tomato pinworm infestations, in 1963 I and my colleagues in southern California began studying the parasite complexes associated with both

pests on their respective primary hosts (Oatman 1970; Oatman and Platner 1974a; Oatman, Wyman, and Platner 1979). During the course of those studies, I explored the potato- and tomato-growing regions of Mexico and of Central and South America, to delineate the areas of origin of the two pests and to collect exotic parasites of each.

From 1963 through 1965, the Commonwealth Institute of Biological Control cooperated with the Department (later Division) of Biological Control of the University of California, Riverside, by shipping parasites of the potato tuberworm to the Department's quarantine facility. Commonwealth Institute field agents collected parasitized material from potato foliage in Argentina and Uruguay during the summers of 1962–63 and 1963–64, and in Peru during the summer of 1964–65 (Lloyd 1972). Commonwealth Institute facilities in South Africa and India provided additional potato tuberworm parasites.

RESULTS

Objective A: The Identification and Introduction of Beneficial Organisms

I explored Mexico, Costa Rica, Colombia, and Guatemala to collect parasites of both the potato tuberworm and the tomato pinworm, and I made trips to Hawaii and to the lower Rio Grande Valley in Texas to collect pinworm parasites. The parasitized material (all from unsprayed plants) was shipped by air to the Department of Biological Control, where it was processed through the quarantine facility and where selected parasites were mass-produced and released. Additional tomato pinworm parasites were collected in 1977 by R. Burkhart of the Division of Plant Industry in Hawaii's Department of Agriculture, while exploring in Arizona, California, Texas, Mexico, Guatemala, Honduras, and El Salvador. Tables 56.1 and 56.2 show which parasites were reared from each pest. Parasites of

Table 56.1. Parasites reared from the potato tuberworm, *Phthorimaea operculella*, on potato.

Parasite	Collection site
Braconidae	
Agathis gibbosa (Say)	California
Agathis tandilensis Blanchard[*]	Argentina
Agathis unicolor (Schrottky)[†]	Argentina
Apanteles dignus Muesebeck	California
Apanteles scutellaris Muesebeck	California, Mexico
Apanteles subandinus Blanchard[†]	Argentina, Peru
Apanteles spp.[†‡]	Peru
Bracon cuyanus Blanchard[*]	Argentina
Bracon gelechiae Ashmead	California
Chelonus curvimaculatus Cameron[†]	South Africa
Chelonus kellieae Marsh	Costa Rica
Chelonus phthorimaeae Gahan	California, Mexico
Microplitis minuralis Muesebeck[†]	Uruguay
Orgilus jennieae Marsh	Costa Rica
Orgilus lateralis (Cresson)	California
Orgilus lepidus Muesebeck[†]	California, Argentina
Orgilus parcus Turner[†]	South Africa
Orgilus spp.	California, Costa Rica
Rogas spp.[*]	South America
Ichneumonidae	
Campoplex haywardi Blanchard[†]	Argentina, Uruguay
Campoplex phthorimaeae (Cushman)	California, Mexico
Diadegma blackburni (Cameron)	California, Hawaii
Diadegma compressus (Cresson)	California
Diadegma stellenboschense (Cameron)[†]	South Africa
Diadegma sp.	California
Nepiera fuscifemora (Graf)	California
Nythobia spp.[†‡]	Cyprus, India
Pristomerus spinator (F.)	California
Temelucha spp.[†]	California, Argentina, Costa Rica
Trathala sp.	Costa Rica
Encyrtidae	
Apsilophrys oeceticola (DeSantis)[*]	Peru
Copidosoma koehleri (Blanchard)[†]	Argentina, Brazil, Chile, Peru
Eulophidae	
Elachertus sp.	Costa Rica
Hyssopus sp.	Costa Rica
Rhetisympiesis phthorimaeae Blanchard[†]	Argentina, Peru
Sympiesis stigmatipennis Girault	California

[*]Reported as parasites of *Phthorimaea operculella* in South America by the CIBC. None of these was collected and shipped to California.
[†]Received in quarantine between 1963 and 1967 by the Department of Biological Control, UC Riverside, from the Commonwealth Institute of Biological Control (CIBC).
[‡]Two species were present.

Table 56.2. Parasites reared from the tomato pinworm, *Keiferia lycopersicella*, on tomato.

Parasite	Collection site
Braconidae	
Agathis sp.	Texas
Apanteles dignus Muesebeck	Arizona, California, Hawaii, Texas, El Salvador, Guatemala, Honduras, Mexico
Apanteles scutellaris Muesebeck	California, Hawaii, Texas, Mexico
Bracon gelechiae Ashmead	California
Bracon spp.	Texas, Guatemala, Mexico
Chelonus blackburni Cameron	Hawaii
Chelonus phthorinmaeae Gahan	Arizona, California, Texas, Guatemala, Mexico
Orgilus spp.	California, Texas, Guatemala, Mexico
Parahormius pallidipes (Ashmead)	Arizona, California, Texas, Guatemala, Mexico
Ichneumonidae	
Campoplex phthorimaeae (Cushman)	California
Campoplex n. sp.[*]	Texas, Mexico
Pristomerus hawaiiensis Perkins	Hawaii
Pristomerus spinator (F.)	Hawaii, Texas, Guatemala, Mexico
Trathala flavoorbitalis Cameron	Hawaii
Eulophidae	
Elasmus nigripes Howard[*]	Texas, Guatemala, Honduras, Mexico
Sympiesis stigmatipennis Girault	California
Pteromalidae	
Zatropis sp. (nr. *tortricidis* Crawford)[*]	Guatemala
Bethylidae	
Goniozus sp. (nr. *platynotae* [Ashmead])	California, Guatemala
Trichogrammatidae	
Trichogramma pretiosum Riley	California

[*]Species collected by R. Burkhart that differed from those collected by E. R. Oatman.

the potato tuberworm received from the Commonwealth Institute are also shown in table 56.1.

Except for the braconid *Microplitis minuralis* Muesebeck, the eulophid *Rhetisympiesis phthorimaeae* Blanchard, and two species of *Apanteles* (Braconidae) from Peru, the laboratory obtained sufficient numbers of each parasite to establish insectary cultures. All of the cultured species were released in southern California against the potato tuberworm (Oatman and Platner 1974b). At the same time, cultures of the South American parasites were shipped to the Indian station, the Commonwealth Institute of Biological Control in Bangalore, India, for propagating and shipping to the commonwealth countries participating in the Commonwealth Institute's potato tuberworm control project (Lloyd 1972). Additional species that were reported as parasites of the potato tuberworm in South America (see table 56.1) were not collected.

Of the 10 hymenopterous parasites of the potato tuberworm received from the Commonwealth Institute and later released in southern California, only the encyrtid *Copidosoma koehleri* (Blanchard) and the braconid *Orgilus lepidus* Muesebeck became established (Oatman and Platner 1974b). However, *C. koehleri* was not recovered in the 1968–69 surveys and thus may have failed to survive.

It is interesting to note that the potato tuberworm parasites collected about 100 miles north of Mexico City and those collected in southern California were the same species. Also, of the seven parasites collected near Cartago in Costa Rica, only one ichneumonid and two eulophids were in genera different from those collected in California (see table 56.1). In fact, the North and South American potato tuberworm parasite complexes are remarkably similar, which suggests the existence of ecological homologues on the two continents. Studies have been published on the biologies of most of these (Leon and Oatman 1968; Oatman, Platner, and Greany 1969; Djamin 1970; Cardona and Oatman 1971, 1975; Odebiyi and Oatman 1972, 1977; Oatman and Platner 1974a; Flanders and Oatman 1982; Teran 1983; Powers and Oatman 1985).

Researchers have recorded 21 species of primary parasites of the tomato pinworm (see table 56.2): 12 from Texas, 10 from southern California, 9 each from Guatemala and Mexico, and 6 from Hawaii. Of the 10 recorded from southern California, 6 are common (Oatman 1970; Oatman, Wyman, and Platner 1979). The braconid *Apanteles dignus* Muesebeck—whose biology has been published (Cardona and Oatman 1971)—is the predominant parasite, having been found in all of the areas surveyed. Except for the species from Hawaii, all of the parasites listed in table 56.2 occur naturally in the area where they were collected. Based on foreign

exploration trips and personal observation, the tomato pinworm does not occur in Costa Rica, Argentina, or Colombia; available evidence strongly indicates that it originated in what is now Mexico and Guatemala. The large numbers of primary parasites associated with this pest there support this assumption. In fact, the largest number of different primary parasites collected in one area—seven—were found by R. Burkhart in 1977 near Sanarate, Guatemala. The principal host, tomato (*Lycopersicon esculentum* Miller), is likewise thought to have originated somewhere in Central America.

Neither the potato tuberworm nor the tomato pinworm were collected during trips to Colombia in 1973 (Oatman and Platner 1989). The only gelechiid collected from both commercial potato and tomato there was *Scrobipalpula absoluta* (Meyrick), which mines the leaves of potato but does not infest the tubers. In tomato, *S. absoluta* both mines the leaves and infests the fruit. It is a serious economic pest in that it infests both green and ripe tomatoes, injuring up to 100 percent of the fruit in spite of repeated insecticide applications.

RECOMMENDATIONS

1. The large parasite complex associated with the potato tuberworm in southern California, and the occurrence of the same parasite species in south-central Mexico, suggest that the pest has been here as long as it has been in South America. Those interested in collecting parasites of the potato tuberworm would do well to concentrate their search in the drier, subtropical areas of North America.

2. The tomato pinworm's absence from Costa Rica, Colombia, and Venezuela, and the large complex of parasites associated with the pest in southern California, Mexico, and Guatemala, suggest that searches for more effective parasites should also be concentrated in the drier, subtropical areas of North America.

3. The 25 primary parasites that were reared from *S. absoluta* on tomato and potato and from *Scrobipalpula isoclora* Povolny on *Solanum saponaceum* Dun., a native solanaceous weed, near Palmira, Colombia (Oatman and Platner 1989), should be considered for importation and release against both the potato tuberworm and the tomato pinworm. All four gelechiid species are closely related and feed on Solanaceae; three are very serious pests.

REFERENCES CITED

Bacon, O. G. 1960. Control of the potato tuberworm in potatoes. *J. Econ. Entomol.* 53:853–71.

Cardona, C., and E. R. Oatman. 1971. Biology of *Apanteles dignus*

(Hymenoptera: Braconidae), a primary parasite of the tomato pinworm. *Ann. Entomol. Soc. Am.* 64:996–1007.

———. 1975. Biology and physical ecology of *Apanteles subandinus* Blanchard (Hymenoptera: Braconidae), with notes on temperature responses of *Apanteles scutellaris* Muesebeck and its host, the potato tuberworm. *Hilgardia* 43:1–51.

Djamin, A. 1970. Biologies of the tomato pinworm, *Keiferia lycopersicella* (Walsingham) (Lepidoptera: Gelechiidae), and its parasite, *Apanteles scutellaris* Muesebeck (Hymenoptera: Braconidae), with special reference to the influence of temperature on population increase. Ph.D. diss., University of California, Berkeley.

Flanders, R. V., and E. R. Oatman. 1982. Laboratory studies on the biology of *Orgilus jennieae* (Hymenoptera: Braconidae), a parasitoid of the potato tuberworm, *Phthorimaea operculella* (Lepidoptera: Gelechiidae). *Hilgardia* 50:1–33.

Harding, J. A. 1971. Field comparisons of insecticidal sprays for control of four tomato insects in south Texas. *J. Econ. Entomol.* 64:1302–4.

Leon, J. K. L., and E. R. Oatman. 1968. Biology of *Campoplex haywardi* (Hymenoptera: Ichneumonidae), a primary parasite of the potato tuberworm. *Ann. Entomol. Soc. Am.* 61:26–36.

Lloyd, D. C. 1972. Some South American parasites of the potato tuber moth, *Phthorimaea operculella* (Zeller), and remarks on those in other continents. *CIB Commonw. Agric. Bur. Tech. Bull.* 15:35–49.

Oatman, E. R. 1970. Ecological studies of the tomato pinworm on tomato in southern California. *J. Econ. Entomol.* 63:1531–34.

Oatman, E. R., and G. R. Platner. 1974a. Biology of *Temelucha* sp., *Platensis* group (Hymenoptera: Ichneumonidae), a primary parasite of the potato tuberworm. *Ann. Entomol. Soc. Am.* 67: 275–80.

———. 1974b. Parasitization of the potato tuberworm in southern California. *Environ. Entomol.* 3:262–64.

———. 1989. Parasites of the potato tuberworm, tomato pinworm, and other closely related gelechiids. *Proc. Hawaii. Entomol. Soc.* 29:23–30.

Oatman, E. R., G. R. Platner, and P. D. Greany. 1969. Biology of *Orgilus lepidus* (Hymenoptera: Braconidae), a primary parasite of the potato tuberworm. *Ann. Entomol. Soc. Am.* 62:1407–14.

Oatman, E. R., J. A. Wyman, and G. R. Platner. 1979. Seasonal occurrence and parasitization of the tomato pinworm on fresh-market tomatoes in southern California. *Environ. Entomol.* 8:661–64.

Odebiyi, J. A., and E. R. Oatman. 1972. Biology of *Agathis gibbosa* (Hymenoptera: Braconidae), a primary parasite of the potato tuberworm. *Ann. Entomol. Soc. Am.* 64:1104–14.

———. 1977. Biology of *Agathis unicolor* (Schrottky) and *Agathis gibbosa* (Say) (Hymenoptera: Braconidae), primary parasites of the potato tuberworm. *Hilgardia* 45:123–51.

Powers, N. R., and E. R. Oatman. 1985. Biology and temperature responses of *Chelonus kellieae* and *Chelonus phthorimaea* (Hymenoptera: Braconidae) and their host, the potato tuberworm, *Phthorimaea operculella* (Lepidoptera: Gelechiidae). *Hilgardia* 52:1–32.

Shelton, A. M., and J. A. Wyman. 1979. Potato tuberworm damage to potatoes under different irrigation and cultural practices. *J. Econ. Entomol.* 72:261–64.

Teran, J. B. 1983. Comparative biologies of *Parahormius pallididipes* (Ashmead) (Hymenoptera: Braconidae) and *Sympiesis stigmatipennis* Girault (Hymenoptera: Eulophidae), ectoparasites of the tomato pinworm, *Keiferia lycopersicella* (Walsingham) (Lepidoptera: Gelechiidae). Ph.D. diss., University of California, Riverside.

Wolfenbarger, D. O., and S. L. Poe. 1973. Tomato pinworm control. *Proc. Fla. St. Hortic. Soc.* 86:139–43.

57 / TOBACCO BUDWORM

T. J. HENNEBERRY AND T. F. WATSON

tobacco budworm
Heliothis virescens (Fabricius)
Lepidoptera: Noctuidae

INTRODUCTION

Biology and Pest Status

The widespread use of insecticides against the pink boll-worm, *Pectinophora gossypiella* (Saunders)—which reduces the population of natural enemies (Watson 1980)—has made the tobacco budworm, *Heliothis virescens* (Fabricius), an increasingly important pest in the southwestern desert cotton-growing areas of Arizona and southern California. The problem has been aggravated by the budworm's increasing tolerance of insecticides (Crowder, Tollefson, and Watson 1979; Watson, Crowder, and Kelly 1986; Watson and Kelly 1991). Its potential for developing resistance to the pyrethroid insecticides makes a broad-based management scheme even more imperative (Watson and Kelly 1991). In recent years, the tobacco budworm together with the bollworm *Helicoverpa* (=*Heliothis*) *zea* (Boddie) have caused the highest control costs and yield losses across the Cotton Belt (Frisbie and Walker 1981).

Egg-larval economic thresholds and chemical control remain the principal approaches to avoiding crop losses. Sterility methods, pheromones, and release of interspecific sterile hybrids have been considered as potential control technologies. Such strategies would facilitate biological control components in IPM systems.

Historical Notes

The cotton ecosystem contains so many beneficial species that it is difficult to quantify the impact of any one species or group of species. However, the first step toward understanding interactions within and between natural enemy species, the host species, and the host plants, is to gather information on the identification, seasonal abundance, and biologies of the beneficial species. The W-84 project has emphasized these objectives; it has also evaluated potential exotic or indigenous natural enemies and developed cultural, microbial, and behavioral control methods to conserve natural enemies.

RESULTS

Objective B: Ecological and Physiological Studies

Flight-trap collections of adult parasitoids in untreated cotton plots near Phoenix, Arizona, have revealed 16 species of tachinid flies (Werner and Butler 1979). A high percentage were *Eucelatoria* species, which parasitize the larvae of the bollworm–tobacco budworm complex as well as other economically important lepidopterous pests. Eight of the parasitoid species that were collected in lesser numbers have also been reported on bollworm–tobacco budworm larval hosts.

In other studies, adults of 18 tachinid species, 6 or 7 braconid species, 6 to 8 ichneumonid species, and 2 chalcid species were commonly collected in cotton, alfalfa, and corn (Butler et al. 1982). (Among the frequently collected tachinid species were *Eucelatoria* species, *Lespesia archippivora* Riley, *Ceratomyiella bicincta* Rainhard, *Periscepsia laevigata* (Wulp), and *Micromintho melania* Townsend.) A number of the tachinid and hymenopterous species collected in these studies are known parasitoids of lepidopterous insect pest species. These results suggest that an active parasitoid complex exists in Arizona's major cultivated crop ecosystems.

Equally important may be the closely related species found on noncrop hosts. At times these parasitoids may provide continuity of the beneficial species in a crop ecosystem. Also playing a major role are various plant species that serve as hosts of *Heliothis virescens*, *Helicoverpa zea*, and *Helicoverpa phloxiphaga* (Grote and Robinson), including redstem filaree, *Erodium cicutarium* (Linnaeus); yellow bird-of-paradise, *Caesalpinnia gillesii* Wall; daisy fleabane, *Erigeron divergens* Torr. and Gray; Coulter's globemallow, *Sphaeralcea coulteri* (Watson); and cultivated garbanzo bean, *Cicer arietinum* Linneaus. A 2-year study of *Helicoverpa* larvae collected from these hosts yielded seven species of hymenoptera and four dipterous parasitoids (Rathman and Watson 1985).

Researchers have published numerous reports documenting the numbers and species of bollworm–tobacco budworm insect predators in cotton fields. However, adequate research has not yet been done to quantify their impact. Preliminary studies showed that a predator complex consisting of *Geocoris* species, *Orius tristicolor* (White), *Nabis* species, *Hippodamia convergens* Guerin-Meneville, spiders, *Chrysoperla* (=*Chrysopa*) *carnea* (Stephens), and *Collops vittatus* (Say) was generally abundant in cotton from May through July (Henneberry and Beasley 1986). The populations decreased dramatically in August and increased slightly in September. In Arizona—where reduviids are important predators—R. J. Rakickas and T. F. Watson (1974) found *Zelus renardii* Kolenati to be primarily a late-summer predator, whereas *Z. socius* Uhler was prevalent throughout the summer, although at lower numbers.

Under laboratory conditions *H. convergens* adults and larvae and *C. carnea* larvae consumed 62, 34, and 28 tobacco budworm eggs per day, respectively, indicating that these predators could have a significant potential impact on tobacco budworm populations (Henneberry and Clayton 1985). R. K. Lawrence and T. F. Watson (1979) conducted laboratory predator-prey studies between *Geocoris punctipes* (Say) and the tobacco budworm to determine the impact of each predator instar on the tobacco budworm's eggs and various larval instars. They found that a *G. punctipes* nymph consumed an average of 151 eggs or 77 first-instar larvae before completing its immature development.

Studies on the development rates of the various life stages of *Collops vittatus* (Say), which is abundant in Arizona cotton fields, indicate that it is a warm-weather insect (Butler and Wardecker 1973). Temperature-dependent development rate tables were also developed for *Geocoris pallens* Stål, *G. punctipes* (Say), *Chrysoperla carnea*, *Hippodamia convergens*, *Zelus renardii*, *Z. socius*, *Orius tristicolor*, and *Sinea confusa* Caudell (Butler and Henneberry 1976). A similar study by A.-S. A. Ali and T. F. Watson (1978) showed the importance of *Z. renardii* in late summer in Arizona, since the highest nymph survival rates and the greatest adult longevity occurred at 25°C.

Objective C: The Conservation and Augmentation of Beneficial Organisms

As mentioned previously, *Collops vittatus* and *Chrysoperla carnea* are important predators in western cotton ecosystems. The discovery that caryophyllene alcohol and caryophyllene attract *C. vittatus* and *C. carnea* into cotton fields has suggested a potential augmentation strategy (Flint, Salter, and Waters 1979; Flint, Merkle, and Sledge 1981). These compounds might be used to improve predator-prey interactions, by attracting those two important predators to, and retaining them in, cotton fields.

For many years researchers have recognized the potential value of microbial control—which affects the pest species directly and conserves natural enemies—as a component in pest management systems. Under laboratory conditions, the tobacco budworm and the bollworm proved susceptible to the nuclear polyhedrosis virus, which was isolated from the alfalfa looper, *Autographa californica* (Speyer) (Vail et al. 1978). To improve the effectiveness of the virus under field conditions, researchers developed a feeding attractant, which has been produced commercially under the trade name Coax (Bell and Kanavel 1978). When the feeding attractant was combined with either or both the alfalfa looper virus and *Bacillus thuringiensis* Berliner, it improved tobacco budworm control, resulted in less damage to cotton fruiting forms, and had no adverse effect on natural enemies. In some cases, cotton yields were increased (Bell and Romine 1980).

A.-S. A. Ali and T. F. Watson (1982a) studied how different rates of *B. thuringiensis* combined with varying densities of *Geocoris punctipes* affected tobacco budworm control on greenhouse and field-caged cotton. Their results indicated the potential for good control with a *B. thuringiensis* rate as low at 282 grams per hectare and one *G. punctipes* nymph for every four plants. On the other hand, laboratory studies have indicated that under field conditions *B. thuringiensis* applications would probably adversely affect immature stages of the parasitoid *Hyposoter exiguae* (Viereck) by causing premature host death (Thoms and Watson 1986).

The nuclear polyhedrosis virus *Baculovirus heliothis* has proved extremely effective against tobacco budworm under laboratory conditions, but less effective and more erratic under field conditions (Watson unpub. data). M. F. Potter and T. F. Watson (1983a) found the timing of the applications to be extremely important, since tobacco budworm larvae are capable of ingesting a lethal inoculation of the virus during eclosion from treated eggs. When applications were delayed, the addition of the feeding stimulant Coax improved efficacy. Adding the feeding stimulant significantly extended the activity of virus residues on cotton terminals that were bioassayed with the tobacco budworm in the laboratory, but the addition did not increase mortality when larvae were allowed to feed on treated plants in the field (Potter and Watson 1973b, 1984).

A.-S. A. Ali and T. F. Watson (1982b) found that neonate tobacco budworms emerging from eggs sprayed with *B. thuringiensis* had a high mortality rate. They also found that tobacco budworm larvae could recover from short-term feeding on *B. thuringiensis*–treated foliage (Ali and Watson 1982c); however, adding a feeding stimulant increased the larvae's feeding and thus increased the *B. thuringiensis*–induced mortality (Ali 1981).

These results suggest that microbial and behavioral control have potential in integrated pest management programs in cotton by reducing insecticide use and hence the impact on other biological agents.

Objective D: Impact

A 2-year study in the Imperial Valley of California evaluated the effect of parasitoids and the nuclear polyhedrosis virus on tobacco budworm and bollworm larval mortality in cotton, alfalfa, and lettuce (Henneberry et al. 1991). The results showed that the highest numbers of tobacco budworm and bollworm larvae were in alfalfa, with populations peaking in August. The highest numbers of larvae occurred in cotton in July and August, and in lettuce in November. The parasitoids caused larval mortality rates ranging from 0 to 33.3 percent, 0 to 3.8 percent, and 2.3 to 4.3 percent in alfalfa, cotton, and lettuce, respectively. The rates from viral infections were 2.1 in alfalfa, 1.6 in cotton, and 0 in lettuce. In this study the researchers identified nine parasitoid species—three other species were not identified—from tobacco budworm and bollworm larvae. The most abundant parsitoids were the tachinid *Lespesia archippivora* (Riley) and the ichneumonid *Hyposoter exiguae*.

RECOMMENDATIONS

1. Continue research to identify the natural enemies of tobacco budworm and bollworm species in cotton ecosystems, and initiate studies to quantify their role in regulating population.

2. Continue efforts to conserve and augment biological control in cotton ecosystems using microbial insecticides and behavioral control methods.

3. Initiate research to determine the impact of crop systems sequencing and wild hosts on natural enemies and tobacco budworm population dynamics.

4. Expand the study of kairomones and their potential for manipulating entomophagous species.

5. Determine the effect of tobacco budworm dispersal and migration on crop infestations.

REFERENCES CITED

Ali, Abdul-Sattar A. 1981. Efficacy of *Bacillus thuringiensis* Berliner and its compatibility with the predator *Geocoris punctipes* (Say) for control of *Heliothis virescens* (F.), on cotton in Arizona. Ph.D. diss., University of Arizona, Tucson.

Ali, Abdul-Sattar A., and T. F. Watson. 1978. Effect of temperature on development and survival of *Zelus renardii*. *Environ. Entomol.* 7:889–90.

———. 1982a. Efficacy of Dipel and *Geocoris punctipes* (Hemiptera: Lygaeidae) against the tobacco budworm (Lepidoptera: Noctuidae) on cotton. *J. Econ. Entomol.* 75:1002–4.

———. 1982b. Effects of *Bacillus thuringiensis* var. *kurstaki* on tobacco budworm (Lepidoptera: Noctuidae) adult and egg stages. *J. Econ. Entomol.* 75:596–98.

———. 1982c. Survival of tobacco budworm (Lepidoptera: Noctuidae) larvae after short-term feeding periods on cotton treated with *Bacillus thuringiensis*. *J. Econ. Entomol.* 75:630–32.

Bell, M. R., and R. F. Kanavel. 1978. Tobacco budworm: Development of a spray adjuvant to increase effectiveness of a nuclear polyhedrosis virus. *J. Econ. Entomol.* 71:350–52.

Bell, M. R., and C. L. Romine. 1980. Tobacco budworm: Field evaluation of microbial control in cotton using *Bacillus thuringiensis* and a nuclear polyhedrosis virus with a feeding adjuvant. *J. Econ. Entomol.* 73:427–30.

Butler, G. D., Jr., and T. J. Henneberry. 1976. *Temperature-dependent development rate tables for insects associated with cotton in the Southwest*. USDA-ARS-W-38. Washington, D.C.: U.S. Department of Agriculture.

Butler, G. D., Jr., and J. D. Lopez. 1980. *Trichogramma pretiosum*: Development in two hosts in relation to constant and fluctuating temperatures. *Ann. Entomol. Soc. Am.* 73:671–73.

Butler, G. D., Jr., and A. L. Wardecker. 1973. *Collops vittatus* (Coleoptera: Malachiidae): Development at constant temperatures. *Ann. Entomol. Soc. Am.* 66:1168–70.

Butler, G. D., Jr., T. J. Henneberry, F. G. Werner, and J. M. Gillespie. 1982. *Seasonal distribution, hosts, and identification of parasites of cotton insects*. USDA-ARM-ARS-W-27. Washington, D.C.: U.S. Department of Agriculture.

Crowder, L. A., M. S. Tollefson, and T. F. Watson. 1979. Dosage-mortality studies of synthetic pyrethroids and methyl parathion on the tobacco budworm in central Arizona. *J. Econ. Entomol.* 72:1–3.

Flint, H. M., J. R. Merkle, and M. Sledge. 1981. Attraction of male *Collops vittatus* in the field by caryophyllene alcohol. *Environ. Entomol.* 10:301–3.

Flint, H. M., S. S. Salter, and S. Waters. 1979. Caryophyllene: An attractant for the green lacewing. *Environ. Entomol.* 8:1123–25.

Frisbie, R. E., and J. K. Walker. 1981. Pest-management systems for cotton insects. In *Handbook of pest management for agriculture*, ed. D. Pimental, 187–202. Boca Raton, Fla.: CRC Press.

Henneberry, T. J., and C. A. Beasley. 1986. Beneficial arthropod species in southwestern desert cotton fields. In *Proceedings of the 1986 Beltwide Cotton Production Resources Conference, Las Vegas, Nevada*, eds. T. Cotton Nelson and J. M. Brown, 196–98. Memphis, Tenn.: National Cotton Council of America.

Henneberry, T. J., and T. E. Clayton. 1985. Consumption of pink bollworm (Lepidoptera: Gelechiidae) and tobacco budworm (Lepidoptera: Noctuidae) eggs by some predators commonly found in cotton fields. *Environ. Entomol.* 14:416–19.

Henneberry, T. J., P. V. Vail, A. C. Pearson, and V. Sevacherian. 1991. Biological control agents of noctuid larvae (Lepidoptera: Noctuidae) in the Imperial Valley of California. *Southwest. Entomol.* 16:81–89.

Lawrence, R. K., and T. F. Watson. 1979. Predator-prey relationship of *Geocoris punctipes* and *Heliothis virescens*. *Environ. Entomol.* 8:245–48.

Potter, M. F., and T. F. Watson. 1983a. Timing of nuclear polyhedrosis virus–bait spray combinations for control of egg and larval stages of tobacco budworm (Lepidoptera: Noctuidae). *J. Econ. Entomol.* 76:446–48.

———. 1983b. Laboratory and greenhouse performance of *Baculovirus heliothis*, combined with feeding stimulants for control of neonate tobacco budworm. *Protection Ecology* 5:161–65.

———. 1984. Field persistence of Elcar (*Baculovirus heliothis*) applied in a bait formulation for control of tobacco budworm in Arizona cotton. *J. Agric. Entomol.* 1:78–81.

Rakickas, R. J., and T. F. Watson. 1974. Population trends of *Lygus* spp. and selected predators in strip-cut alfalfa. *Environ. Entomol.* 3:781–84.

Rathman, R. J., and T. F. Watson. 1985. A survey of early-season host plants and parasites of *Heliothis* spp. in Arizona. *J. Agric. Entomol.* 2:388–94.

Thoms, E. M., and T. F. Watson. 1986. Effect of Dipel (*Bacillus thuringiensis*) on the survival of immature and adult *Hyposoter exiguae* (Hymenoptera: Ichneumonidae). *J. Invert. Pathol.* 47:178–83.

Vail, P. V., D. L. Jay, F. D. Stewart, A. J. Martinez, and H. T. Dulmage. 1978. Comparative susceptibility of *Heliothis virescens* and *H. zea* to the nuclear polyhedrosis virus isolated from *Autographa californica. J. Econ. Entomol.* 71:293–96.

Watson, T. F. 1980. Methods for reducing winter survival of the pink bollworm. In *Pink bollworm control in the western United States,* ed. H. M. Graham, 24–34. USDA Science and Education Administration, ARM, ARS-W-16. Washington, D.C.: U.S. Department of Agriculture.

Watson, T. F., and S. E. Kelly. 1991. Inheritance of resistance to permethrin by the tobacco budworm, *Heliothis virescens* (F.): Implications for resistance management. *Southwest. Entomol.* Suppl. 15:135–41.

Watson, T. F., L. A. Crowder, and S. Kelly. 1986. Pyrethroid and methyl parathion susceptibility of the tobacco budworm in Arizona. *Southwest. Entomol.* 11:281–86.

Werner, F. G., and G. D. Butler, Jr. 1979. Tachinid flies collected in a Phoenix, Arizona, cotton field. *Southwest. Entomol.* 4:282–84.

58 / WESTERN TENTIFORM LEAFMINER

B. A. BARRETT, S. C. HOYT, AND J. F. BRUNNER

> **western tentiform leafminer**
> *Phyllonorycter elmaella* Doganlar & Mutuura
> **Lepidoptera: Gracillariidae**

INTRODUCTION

Biology and Pest Status

Phyllonorycter elmaella Doganlar & Mutuura, the western tentiform leafminer, is one of three major *Phyllonorycter* species attacking apple in North America. *Phyllonorycter blancardella* (F.), the spotted tentiform leafminer, and *P. crataegella* (Clemens), the apple blotch leafminer, occur in eastern North America; *P. elmaella* occurs in western North America. The biology, life history, and morphology of all three species—as reported by Pottinger and LeRoux (1971), by Beckham, Hough, and Hill (1950), and by Gibb (1983), respectively—are very similar.

The preferred host for the western tentiform leafminer is apple, but cherry and prune may also be heavily infested. During the past several years infestations of the western tentiform leafminer have become more frequent and severe in many fruit-growing areas of western North America, including Washington, Oregon, California, Utah, and British Columbia. However, some evidence suggests that the leafminer species in California and British Columbia may not be *P. elmaella* but rather a closely related, previously undescribed species (Cossentine and Jensen 1992; Varela and Welter 1992).

The western tentiform leafminer lays its eggs singly, on the leaves' undersurface. Depending on temperature, the eggs hatch in 6 to 16 days. There are five larval instars. The first three, referred to as sap-feeders, are flattened, wedge shaped, legless, and white, with a pointed head. They separate the lower layer of the leaf surface from the tissue above, creating a mine that can only be seen from the underside of the leaf. The fourth and fifth instars, referred to as tissue-feeders, are cylindrical and have legs. The fourth instar is white, the fifth yellowish. Full-grown larvae are about 4 millimeters in length and consume the tissue between the upper and lower leaf surfaces, producing tentlike mines with spots that are visible on the upper leaf surface. Total larval development requires about 24 days. A complete gener-

ation from egg to adult requires 35 to 55 days in the first and second generations, and longer in the third or partial fourth generations.

The leafminer causes damage indirectly, by destroying leaf area with its characteristic tentiform mines. Mine densities of 10 or more per leaf can result in a reduction of fruit size, lowered soluble solids, and failure of the fruit to mature properly. Both the number of mines per leaf and the time of year relative to harvest when a high number of mines develop are critical. Apples, for example, tolerate considerable leaf mining with little or no effect on the fruit, particularly when the majority of the mines develop in September.

In 1985, 1986, and 1987 there were three complete western tentiform leafminer generations in central Washington, with portions of a fourth generation evident in the sap-feeding and tissue-feeding stages (Barrett 1988). Phenological data showed that third- and fourth-generation sap-feeders overlapped extensively. In eastern North America a similar phenomenon occurs among some populations of the spotted tentiform leafminer; researchers have determined that it is the result of a summer diapause (Laing, Heraty, and Corrigan 1986). Whether portions of western tentiform leafminer sap-feeder populations have a similar delayed development during the last summer generation is still in question.

Western tentiform leafminers overwinter as pupae in the mines on fallen leaves. When freezing temperatures occurred in late October 1987 in central Washington, the remaining population consisted of 60 percent sap-feeders and 40 percent tissue-feeders. Adults had continued to emerge until the freeze, meaning that the population was essentially devoid of pupae. By the first of December, the only healthy (living) stage remaining was that of the tissue-feeders. The low levels of leafminer pupae typically found in Washington apple orchards in the late fall suggest that this species suffers high overwintering mortality. Survival may therefore be a random or chance event, dependent on the composition of the leafminer population at the time when environmental conditions no longer support development. It is this characteristic that makes it very difficult to forecast leafminer population levels from year to year and from orchard to orchard in the Northwest. In contrast, in Quebec, Canada, R. P. Pottinger and E. J. LeRoux (1971) found that the overwintering generation of the spotted tentiform leafminer began to pupate by October, with 60 percent already in the pupal stage by late November.

Historical Notes

A tentiform leafminer had been reported on apples in the northwestern United States for many years, but before 1980 it was considered a minor pest. However, an outbreak that began in 1980 in one orchard on the lower Columbia River later spread to virtually all the apple-growing areas of central Washington. The species responsible for the outbreak was identified as *Phyllonorycter elmaella* Doganlar & Mutuura, and was referred to as the western tentiform leafminer.

Researchers believe that this outbreak in Washington state was caused by the introduction or selection of a strain resistant to several pesticides; the species is now a serious problem in many parts of California, Oregon, Washington, British Columbia, and Utah.

RESULTS

Objective A: The Identification and Introduction of Beneficial Organisms

Their combined habits of parasitization and adult host-feeding make hymenopterous parasitoids the principal natural enemies of *Phyllonorycter* species (Askew and Shaw 1974, 1979). Hymenopterous parasitoids have been reported to control, to varying degrees, *Phyllonorycter* populations throughout the world.

Pnigalio flavipes (Ashmead) was found to be the major parasitoid species in all of Washington's western tentiform leafminer generations (Barrett 1988), constituting about 85 percent of all parasitoid species reared from the sampled mines. *Sympiesis marylandensis* Girault was a distant second, followed by *Zagrammosoma americanum* Girault. In all, seven parasitoid species have been reared from western tentiform leafminer mines in Washington. Similarly, in Utah, researchers found six species of parasitoids attacking leafminers (Barrett and Jorgensen 1986).

Several other parasitoid species that are important in the biological control of *Phyllonorycter* species have been found in other areas of the world (Beckham, Hough, and Hill 1950; Askew and Shaw 1979; Wang and Laing 1989). So far no attempt has been made to introduce exotic parasitoids into Washington orchards, but such an introduction could supplement the control provided by *P. flavipes*.

Objective B: Ecological and Physiological Studies

All the immature stages of *P. flavipes* are found within the leaf mine. The parasitoid's eggs are white, elongate, and cylindrical, averaging 0.38 millimeters in length. The larva is maggotlike and, depending on the instar, 0.3 to 1.5 millimeters long (Barrett unpub. data). Larvae are slightly opaque, though a brown to black gut may be visible in later instars. Pupae are glossy black and naked, not found in a silken cocoon; adults escape from the mine by chewing a small circular hole through the leaf.

Studies have found that throughout most of each season, all stages of *P. flavipes* were present concurrently in orchards (Barrett 1988). This is probably a result of its rapid development rate, which, assuming suitable host stages, would allow it to complete almost two generations in the time required to produce one leafminer generation. In general, *P. flavipes* appears to have at least the same number of generations as its leafminer host. *Sympiesis marylandensis*, a minor member of the leafminer parasitoid complex in Washington, is reported to have three to four generations per year on spotted tentiform leafminers and apple blotch leafminers infesting orchards in the eastern United States (Gambino and Sullivan 1982).

In 1987 adult female *P. flavipes* were active and laying eggs in the orchards in November. Similarly, R. P. Pottinger and E. J. LeRoux (1971) observed *S. marylandensis* parasitizing the spotted tentiform leafminer in Quebec orchards in late October. By the first of December 1987, approximately 70 percent of the *P. flavipes* sampled were still in the larval stage; the other 30 percent had pupated. The overwintering stage of *P. flavipes* is unknown. However, *S. marylandensis* is reported to overwinter in the larval or pupal stage (Pottinger and LeRoux 1971). The ability to overwinter in more than one stage may explain why parasitoids of the spotted tentiform leafminer and of *Phyllonorycter ringoniella* Matsumura are reported to suffer less overwintering mortality than their host (Pottinger and LeRoux 1971; Yamada, Sekita, and Oyama 1986). It is possible that *P. flavipes* is not restricted to one overwintering stage, unlike its host.

The average level of parasitoid-induced mortality of the western tentiform leafminer usually decreases in successive generations. In a 3-year study, the decrease in mortality from first to second generation ranged from 3.2 to 33.4 percent (Barrett 1988). Second- to third-generation decreases ranged from 7 to 21 percent. In contrast, second- to third-generation mortality levels increased in one orchard during all 3 years. The intraseasonal decline in parasitoid-induced mortality was reversed in 1987, as fourth-generation levels increased 6.6 to 24 percent over third-generation levels. Typically, the highest mortality levels recorded were during the first generation, when leafminer densities were at their lowest. Conversely, the lowest level of mortality was found in the third generation, when leafminer densities were near their seasonal high.

P. flavipes oviposited on leafminer tissue-feeders five times more frequently than on sap-feeders. However, sap-feeders were the preferred stage for host-feeding: adult females fed on the body liquids of paralyzed hosts (Askew 1971) 12 times more frequently than they did on tissue-feeders. Sap-feeders killed by host-feeding have a characteristic appearance: the body is completely or partly void of its liquid contents, and it is usually attached to the inner mine surface at a localized dark spot.

The conventional method for determining density-dependent relationships is to plot average mortality—usually as percentages of k-values—against the average population density per corresponding generation; a positive correlation indicates a density-dependent relationship (Hassell 1985). When using this approach to analyze the interaction of the western tentiform leafminer and *P. flavipes,* Barrett (1988) found a slight negative correlation, indicating an inverse density-dependent relationship. This does not necessarily mean that *P. flavipes* populations are not responding positively to leafminer populations. When Barrett looked at the data on a relative basis—that is, at the number of parasitoid-attacked mines per leaf and healthy leafminer mines per leaf over the season—he found that *P. flavipes* responded positively to the leafminers: as the number of healthy mines increased over the season, so did the number of parasitoid-attacked mines. But because of some limitation—in either reproductive capability or time constraints on searching or handling—*P. flavipes* activity (parasitism) was not as great as the leafminer's rate of increase. Consequently, the percentage of parasitism decreased throughout the season. However, the activity of *P. flavipes* during the season may be sufficient to limit the pest's population growth rate and prevent it from reaching damaging levels. In all orchard blocks examined during the 3-year study mentioned earlier, leafminer densities never exceeded the proposed treatment thresholds of one, two, and five mines per leaf for first, second, and third generations, respectively (Beers et al. 1993).

Objective C: The Conservation and Augmentation of Beneficial Organisms

Currently, oxamyl is the only chemical recommended for leafminer control in Washington. Researchers recommend that this product be used conservatively, for the following reasons: (1) to limit possible detrimental effects on *P. flavipes*, (2) to reduce selection for leafminer resistance, and (3) to avoid disrupting the biological control afforded by predatory mites, to whom oxamyl is moderately toxic. However, azinphosmethyl, which is applied for codling moth control, did not appear to have a serious effect on the interaction between *P. flavipes* and its leafminer host. So far, there is no information on the selectivity of other insecticides with regard to *P. flavipes*. Researchers have observed, however, that the western tentiform leafminer is not a problem in organic orchards. This could be due in part to improved biological control where broad-spectrum, synthetic pesticides are not used. Because leafminer adults and *P. flavipes* emerge at the same time, parasitoid adults have to survive for 2 to 3 weeks before their preferred host stages—namely, tissue-feeders—are available for oviposition. This delay could subject *P. flavipes* adults to an excessive exposure to residual insecticides, reducing the likelihood of their survival. More information on the toxicity of pesticides is critical to parasitoid conservation efforts.

Objective D: Impact

Researchers have developed an integrated pest management program for the western tentiform leafminer that takes into account its overwinter survival, parasitism by *Pnigalio flavipes*, the within-season development of leafminer generations, the tolerance of apple to leafminer infestation, and as-needed applications of oxamyl. In 1988 the program was monitored in 20 orchards; 15 required no chemicals to control the leafminer. This program, however, requires extensive and timely monitoring of orchards for mine density and the percentage of parasitism, and it has not yet been widely accepted by growers or consultants. A simpler, less time-consuming method of sampling tentiform leafminers has recently been developed (Jones 1991). We believe the potential for the biological control of this pest to be promising, particularly if all or part of the planned research objectives described under "Recommendations" are completed.

RECOMMENDATIONS

1. Introduce and evaluate additional species of parasitoids. Introductions currently planned include *Holcothorax testaceipes* Ratzeburg, a species that attacks eggs and develops polyembryonically within larvae, and *Pholetesor ornigis* (Weed), a larval endoparasitoid.

2. Evaluate the toxicity of commonly used agricultural chemicals to the parasitoids. Determine whether selective chemicals, dosages, or timings can be used to control other pests and can be integrated with the biological control of the leafminer.

3. Determine whether mass trapping of males can be used in conjunction with biological control to regulate the pest's populations. Since relatively small numbers of leafminers survive the winter in Washington, this tactic may help reduce the mating of early-generation females there.

REFERENCES CITED

Askew, R. R. 1971. *Parasitic insects.* New York: American Elsevier.

Askew, R. R., and M. R. Shaw. 1974. An account of the Chalcidoidea (Hymenoptera) parasitizing leaf-mining insects of deciduous trees in Britain. *Biol. J. Linn. Soc.* 6:289–335.

———. 1979. Mortality factors affecting the leafmining stages of *Phyllonorycter* (Lepidoptera: Gracillariidae) on oak and birch: I. Analysis of the mortality factors. II. Biology of the parasite species. *Zool. J. Linn. Soc.* 67:31–64.

Barrett, B. A. 1988. The population dynamics of *Pnigalio flavipes* (Hymenoptera: Eulophidae), the major parasitoid of *Phyllonorycter elmaella* (Lepidoptera: Gracillariidae) in central Washington apple orchards. Ph.D. diss., Washington State University, Pullman.

Barrett, B. A., and C. D. Jorgensen. 1986. Parasitoids of the western tentiform leafminer, *Phyllonorycter elmaella* (Lepidoptera: Gracillariidae), in Utah apple orchards. *Environ. Entomol.* 15:635–41.

Beckham, C. M., W. S. Hough, and C. H. Hill. 1950. Biology and control of the spotted tentiform leafminer on apple trees. *Va. Agric. Exp.Sta. Bull.* 114:3–12.

Beers, E. H., J. F. Brunner, M. J. Willett, and G. M. Warner. 1993. *Orchard pest management: A resource book for the Pacific Northwest.* Yakima, Wash.: GoodFruit Grower.

Cossentine, J. E., and L. B. Jensen. 1992. Establishment of a *Phyllonorycter* sp. (Lepidoptera: Gracillariidae) and its parasitoid, *Pnigalio flavipes* (Hymenoptera: Eulophidae), in fruit orchards in the Okanagan and Similkameen valleys of British Columbia. *J. Entomol. Soc. British Columbia* 89:18–24.

J. Gambino, P., and D. J. Sullivan. 1982. Phenology of emergence of the spotted tentiform leafminer, *Phyllonorycter crataegella* (Lepidoptera: Gracillariidae), and its parasitoids in New York. *J. N.Y. Entomol. Soc.* 90:229–36.

Gibb, T. J. 1983. Biology of the western tentiform leafminer, *Phyllonorycter elmaella*, in Utah. Master's thesis, Brigham Young University, Provo, Utah.

Hassell, M. P. 1985. Insect natural enemies as regulating factors. *J. Anim. Ecol.* 54:323–34.

Hoyt, S. C. 1983. Biology and control of the western tentiform leafminer. *Proc. Wash. St. Hort. Assoc.* 79:115–18.

Jones, V. P. 1991. Binomial sampling plans for tentiform leafminer (Lepidoptera: Gracillariidae) on apple in Utah. *J. Econ. Entomol.* 84:484–88.

Laing, J. E., J. M. Heraty, and J. E. Corrigan. 1986. Summer diapause in *Phyllonorycter blancardella* (Fabr.) (Lepidoptera: Gracillariidae) in Ontario. *Can. Entomol.* 118:17–28.

Pottinger, R. P., and E. J. LeRoux. 1971. The biology and dynamics of *Lithocolletis blancardella* (Lepidoptera: Gracillariidae) on apple in Quebec. Memoirs of the Entomological Society of Canada 77. Ottawa: Entomological Society of Canada.

Varela, L. G., and S. C. Welter. 1992. Parasitoids of the leafminer, *Phyllonorycter* nr. *elmaella* (Lepidoptera: Gracillariidae), on apple in California: Abundance, impact on leafminer, and insecticide-induced mortality. *Biological Control* 2: 124–30.

Wang, T., and J. E. Laing. 1989. Diapause termination and morphogenesis of *Holcothorax testaceipes* Ratzeburg (Hymenoptera: Encyrtidae), an introduced parasitoid of the spotted tentiform leafminer, *Phyllonorycter blancardella* (F.) (Lepidoptera: Gracillariidae). *Can. Entomol.* 121:65–74.

Yamada, M., N. Sekita, and N. Oyama. 1986. *Studies on the biology and the population dynamics of the apple leaf miner, Phyllonorycter ringoniella Matsumura (Lepidoptera: Gracillariidae).* Aomori Apple Experiment Station Bulletin 23.

59 / SERPENTINE LEAFMINERS

M. W. JOHNSON, E. R. OATMAN, D. M. NAFUS, AND J. W. BEARDSLEY

> **serpentine leafminers**
> *Liriomyza* spp.
> Diptera: Agromyzidae

INTRODUCTION

Biology and Pest Status

Several members of the genus *Liriomyza* are pests throughout North America and the Pacific Basin (Parrella 1982; Waterhouse and Norris 1987). W-84 project research has targeted four species: *Liriomyza sativae* Blanchard, *Liriomyza trifolii* (Burgess), *Liriomyza huidobrensis* (Blanchard), and *Liriomyza brassicae* (Riley). Following its accidental introduction into many countries, *L. trifolii* gained worldwide prominence in the 1980s because of its wide plant-host range and ability to rapidly develop resistance to pesticides (Parrella 1987).

The biological control of agromyzid leafminers can be disrupted in two ways. First, leafminers may be introduced into new locations—such as Guam and Hawaii—without their natural enemy complexes. More commonly, the use of pesticides inhibits or eliminates established leafminer parasitoids, thus promoting increases in *Liriomyza* densities that are classified as either pest resurgences (Johnson 1987) or secondary pest upsets (Oatman and Kennedy 1976; Johnson, Oatman, and Wyman 1980a). Also, the degree of biological control can vary, since a host crop influences a given leafminer species' parasitoid complex (Johnson and Hara 1987).

Historical Notes

W-84 projects in California, Guam, and Hawaii had common research objectives, one of the foremost of which—because *Liriomyza* species are polyphagous— was properly identifying these pests and their associated parasitoids (Johnson and Hara 1987). Other objectives included evaluating both introduced and native parasitoids and quantifying the impact of field applications of pesticides on natural enemies.

In California. The first W-84-coordinated activities on *Liriomyza* species began in 1975, when southern California tomato growers experienced problems with *L.* *sativae* (Oatman and Kennedy 1976). Most of this W-84 work was completed before *L. trifolii* also became a tomato pest in the 1980s. W-84 researchers developed simple sampling methodologies for integrated pest management programs.

In Guam. Although *L. sativae, L. brassicae,* and *L. trifolii* are all present on Guam, only the latter causes economic damage. After being introduced around 1978, *L. trifolii* became a problem on beans, causing premature leaf drop and yield losses. Insecticides often failed to control the leafminer, and pest densities sometimes became so high that the leaves were completely mined. On cucurbits, leafminers were present but not a problem until 1985, when growers increased their use of insecticide to control *Thrips palmi* Karny. W-84 researchers introduced exotic natural enemies to Guam, evaluated the effectiveness of *Ganaspidium utilis* Beardsley, and quantified the economic impact of *L. trifolii* on beans. Because leafminer problems on this crop were so serious, managing *L. trifolii* on beans was their priority.

In Hawaii. The species targeted for control in this state included *L. sativae, L. trifolii,* and *L. huidobrensis*. W-84 researchers described several newly discovered parasitoid species that were introduced for leafminer control; evaluated, conserved, and augmented introduced parasitoids on vegetable crops that were receiving pesticide treatments for primary pests such as aphids and thrips; determined the susceptibilities of *Liriomyza* species and associated parasitoids to pyrethroid insecticides; and analyzed leafminer and parasitoid distributions on watermelon.

RESULTS

Objective A: The Identification and Introduction of Beneficial Organisms

In California. Among the six major parasitoids reported attacking *L. sativae* on tomato, the predominant were

Diglyphus begini (Ashmead), *Chrysonotomyia punctiventris* (Crawford), and *Chrysocharis oscinidis* Ashmead (=*parksi* Crawford) (Johnson, Oatman, and Wyman 1980a,b).

In Guam. Researchers recorded several natural enemies, including the eucoilids *Gronotoma micromorpha* (Perkins) and *Disorygma pacifica* (Yoshimoto), and the eulophids *Hemiptarsenus semialbiclavus* Girault and *Chrysonotomyia formosa* (Westwood) (Schreiner, Nafus, and Bjork 1986). On beans the eucoilids were sparse, parasitizing about 2 percent of leafminers; *H. semialbiclavus* and *C. formosa* were the dominant parasitoids. On cole crops and cucurbits, *G. micromorpha* and *D. pacifica* were the dominant parasitoids, closely followed by *H. semialbiclavus* and *C. formosa*. On tomato *H. semialbiclavus* was the most abundant parasitoid; however, the eucoilids were also important.

In attempts to introduce additional biological control agents, researchers released *G. utilis* and *D. begini*. Groups of 200 *G. utilis* individuals were released at several locations over several months, and the species became established. With *D. begini*, three releases of low numbers were made at a time when leafminers were heavily parasitized by local parasitoids. Continuous monitoring over several years indicated that *D. begini* did not become established.

In Hawaii. Researchers conducted surveys of leafminer and associated parasitoid species in beans, cucumbers, onions, tomatoes, and watermelon. Where pesticides were not used, *L. sativae* and *L. trifolii* were usually the first- and second-most-common leafminer species, respectively, on beans (Mothershead 1978; Johnson and Mau 1986), cucumber (Herr 1987), tomatoes (Herr 1987), and watermelon (Johnson 1987). On tomatoes and watermelon, the relative importance of *L. trifolii* increased when pesticides were used frequently (Johnson 1987). Which species infested onions varied with the onion cultivar—bulb versus green onion—and location. On Maui *L. huidobrensis* and *L. trifolii* infested bulb onions at elevations of 500 to 1,000 meters (Johnson and Mau 1986). On Oahu only *L. sativae* commonly infested green onions and shallots grown near sea level (Herr 1987).

From leafminers infesting beans, onions, tomato, and watermelon, researchers reared 8, 7, 9, and 10 parasitoid species, respectively. They found that crop, location, and season all played a role in determining which parasitoid attacked leafminers. Only four parasitoid species were reared from leafminers attacking bulb onions, compared with eight species from leafminers on adjacent beans (Johnson and Mau 1986). On beans grown on Oahu, *Closterocerus utahensis* Crawford was the predominant parasitoid (Herr 1987), whereas on Maui *D. begini* was the most common on beans grown at an elevation of about 670 meters (Johnson and Mau 1986).

In systematic studies of leafminer parasitoids, researchers focused on the Eucoilidae (Cynipoidea), an important group whose taxonomy is poorly understood. They found that the eucoilid purposely introduced into Hawaii from Texas in 1976 under the name *Cothonaspis* n. sp. belonged in the genus *Ganaspidium* Weld (Beardsley 1986). Initially, the species was misidentified as *Ganaspidium hunteri* (Crawford), previously known only from the holotype; however, further study showed that although closely related, the species established in Hawaii was undescribed. It was named *Ganaspidium utilis* (Beardsley 1988). This parasitoid is now established throughout the Hawaiian islands and on Guam; it has also been sent to Tonga. Other eucoilid parasitoids that parasitize *Liriomyza* species in Hawaii and Guam are *Disorygma pacifica* (Yoshimoto) and *Gronotoma micromorpha* (Perkins) (Beardsley 1988, unpub. data). Both species are believed to have been accidentally introduced into Hawaii and Guam from the continental United States. *Disorygma pacifica* occurs in Texas and Mexico, *G. micromorpha* in Florida (Beardsley 1988).

Objective B: Ecological and Physiological Studies

In California. Researchers found that the percentage of parasitization of *L. sativae* by *C. oscinidis* was correlated with the densities of leafminer pupae (Johnson, Oatman, and Wyman 1980c).

In Hawaii. Host-searching and ovipositional behavior were analyzed for the larval-pupal parasitoids *G. utilis* and *C. oscinidis,* using *L. sativae* as a host (Petcharat 1987; Petcharat and Johnson 1988). They found that the parasitoids used the serpentine mines as trails to locate their hosts. Under laboratory conditions, *G. utilis* was able to discriminate between previously investigated mines and newly encountered mines, but *C. oscinidis* was not. Also, host-handling times were significantly shorter for *G. utilis* than for *C. oscinidis*. A review of the literature on *Liriomyza* parasitoids suggests that host-crop habitat influences the relative abundance of parasitoid species attacking *Liriomyza* species (Johnson and Hara 1987).

On Oahu, *C. punctiventris* and *Halticoptera circulus* (Walker) were the first- and second-most-common species, respectively, on spring plantings of watermelon; *G. utilis* and *C. punctiventris* were the most common on summer plantings (Johnson 1987).

Objective C: The Conservation and Augmentation of Beneficial Organisms

In California. Various rates—those less than or equal to 1 kilogram of active ingredient per hectare—of the broad-spectrum insecticide methomyl caused increases in *L. sativae* densities (Oatman and Kennedy 1976; Johnson, Oatman, and Wyman 1980a,b) by reducing the populations of *D. begini*, *C. punctiventris*, and *C. oscinidis* (Johnson, Oatman, and Wyman 1980a,b). In contrast, applications of Dipel (*Bacillus thuringiensis* Berliner var. *kurstaki*) for lepidopterous pest control did not adversely affect the leafminer or its parasitoids (Johnson, Oatman, and Wyman 1980a,b). A combination of Dipel and a reduced rate of methomyl (0.25 kilograms of active ingredient per hectare), however, did increase *L. sativae* densities compared with those in an untreated check (Johnson, Oatman, and Wyman 1980b).

Researchers also developed a simple sampling technique, placing styrofoam trays (22.9 × 27.9 × 1.6 cm) beneath tomato plants to trap the prepupal *L. sativae* larvae falling from foliage (Johnson et al. 1980). They found that the densities of medium and large live *L. sativae* larvae were significantly correlated with the mean numbers of pupae collected in the trays.

In Guam. Researchers observed that insecticides had different effects on the parasitoids: fenvalerate was the least toxic; permethrin (Ambush) and naled were intermediate; and carbaryl, ethion, diazinon, dimethoate, and malathion were highly toxic.

Insecticide trials on yardlong beans showed that as *L. trifolii* mean densities increased from 5 to 70 mines per leaf, the yield dropped from 300 to 75 kilograms of beans per 100-meter row (Schreiner, Nafus, and Bjork 1986). In fact, at densities of more than 40 mines per leaf, maturing leafminers destroyed virtually all of the photosynthetic tissue, and yield loss did not increase with further increases in leafminer densities. Yield losses were nonlinear, best fitting the following quadratic equation

$$Y = 269.53 - 7.878M + 0.77M2$$

where *Y* equals yield and *M* equals the mean number of *L. trifolii* mines per leaflet (seasonal average).

In Hawaii. Researchers found a way to conserve natural enemies in commercial watermelon plantings by modifying pesticide application methods and reducing their frequency during crop cycles (Johnson 1987). Malathion treatments applied directly to the crop to control the melon fly, *Dacus cucurbitae* Coquillett, were discontinued and replaced with malathion bait sprays applied to adjacent corn borders.

The decision whether pesticide treatments were necessary to suppress pests such as leafminers, aphids, thrips, and spider mites was based on pest densities with respect to the established nominal density treatment levels. Researchers monitored plantings weekly to determine whether treatments were necessary. In sur-

veys conducted in 43 watermelon plantings on Oahu and Molokai, researchers found that *L. sativae* and *L. trifolii* were present in crops on more than 96 percent of the survey dates, but that density treatment levels were surpassed on less than 9 percent of the dates (Johnson et al. 1989). On Oahu alone, using these guidelines reduced pesticide use for *Liriomyza* control by more than 90 percent.

In bulb and green onion crops, researchers attempted to augment the control of leafminers by intercropping with beans; the aim was to increase parasitoid species diversity and densities. For bulb onions, these attempts were unsuccessful—only four parasitoid species attacked *L. huidobrensis* and *L. trifolii*, compared with eight species parasitizing *L. sativae* and *L. trifolii* in beans (Johnson and Mau 1986). On green onions, whose dominant leafminer was *L. sativae*, intercropping increased the number of parasitoid species from six to seven (Herr 1987). However, the increase did not enhance the biological control of *L. sativae*.

Studies were done on *D. begini*, *C. punctiventris*, *G. utilis*, *C. oscinidis*, and *H. circulus* tolerances to permethrin and fenvalerate residues (Mason and Johnson 1988). *D. begini* showed significantly more tolerance of pyrethroids than did *C. punctiventris*. *H. circulus* and *G. utilis* exhibited the least tolerance to permethrin and fenvalerate, respectively. LC_{50}'s for all the parasitoids assayed were higher with fenvalerate than with permethrin. Compared with *L. sativae* and *L. trifolii*, parasitoids had medium to high tolerances. *D. begini* had significantly higher LC_{50}'s for both pyrethroids than did the *Liriomyza* species.

Researchers also conducted field studies for *L. sativae* and *L. trifolii* larvae and associated parasitoids, in order to evaluate the accuracy of random versus stratified sampling of watermelon foliage with respect to leaf size and distance from the plant base (Lynch and Johnson 1987). Before full leaf-canopy establishment, researchers recorded significantly greater densities of *Liriomyza* larvae per leaf at the plant base compared with the distal end of the vine. Using leaf-size stratification significantly increased the precision of the sample mean estimates for both *Liriomyza* larvae and their natural enemies. Stratification with respect to distance from the plant base—before full leaf-canopy establishment—resulted in a greater precision of *Liriomyza* density estimates.

Objective D: Impact

In Guam. After becoming established, *G. utilis* proved to be the dominant parasitoid on beans. It parasitized up to 78 percent of leafminers, causing so dramatic a decrease in pest densities that leafminers are no longer a problem in unsprayed bean plantings.

RECOMMENDATIONS

1. Natural enemy introductions should continue in areas where complete biological control has not been achieved on all crops.

2. Further studies should be made on the crop-habitat preferences of the major leafminer parasitoids, to maximize the effectiveness of biological control introduction and augmentation programs.

3. The impact of *Liriomyza*-induced damage on crop yields should be determined, to provide guidelines for setting density treatment levels for use in integrated pest management monitoring programs as well as to quantify what levels of control are necessary for complete biological control.

4. Sampling programs suitable for growers and consultants should be developed, to implement established density treatment thresholds.

5. On crops in which leafminers occur together with several other primary pest species, and in which natural enemies are disrupted by pesticide use, the impact of commonly applied pesticides on predominant parasitoids needs to be studied.

6. Efforts should be made to identify selective pesticides and natural enemy populations that exhibit resistance to common pesticides.

7. In some situations, major parasitoids (for example, *G. utilis*) should undergo laboratory selection to enhance their pesticide resistance.

REFERENCES CITED

Beardsley, J. W. 1986. Taxonomic notes on the genus *Ganaspidium* Weld (Hymenoptera: Cynipoidea: Eucoilidae). *Proc. Hawaii. Entomol. Soc.* 26:35–39.

———. 1988. Eucoilid parasites of agromyzid leafminers in Hawaii (Hymenoptera: Cynipoidea). *Proc. Hawaii. Entomol. Soc.* 28:33–47.

Herr, J. C. 1987. Influence of intercropping on the biological control of *Liriomyza* leafminers in green onions. Master's thesis, University of Hawaii at Manoa, Honolulu.

Johnson, M. W. 1987. Parasitization of *Liriomyza* spp. (Diptera: Agromyzidae) infesting commercial watermelon plantings in Hawaii. *J. Econ. Entomol.* 80:56–61.

Johnson, M. W., and A. H. Hara. 1987. Influence of host crop on parasitoids (Hymenoptera) of *Liriomyza* spp. (Diptera: Agromyzidae). *Environ. Entomol.* 16:339–44.

Johnson, M. W., and R. F. L. Mau. 1986. Effects of intercropping beans and onions on populations of *Liriomyza* spp. and asso-

ciated parasitic Hymenoptera. *Proc. Hawaii. Entomol. Soc.* 27:95–103.

Johnson, M. W., E. R. Oatman, and J. A. Wyman. 1980a. Effects of insecticides on populations of the vegetable leafminer and associated parasites on summer pole tomatoes. *J. Econ. Entomol.* 73:61–66.

———. 1980b. Effects of insecticides on populations of the vegetable leafminer and associated parasites on fall pole tomatoes. *J. Econ. Entomol.* 73:67–71.

———. 1980c. Natural control of *Liriomyza sativae* on pole tomatoes in southern California. *Entomophaga* 25:193–98.

Johnson, M. W., R. F. L. Mau, A. P. Martinez, and S. Fukuda. 1989. Foliar pests of watermelon in Hawaii. *Trop. Pest Manage.* 35:90–96.

Johnson, M. W., E. R. Oatman, J. A. Wyman, and R. A. Van Steenwyk. 1980. A technique for monitoring *Liriomyza sativae* in fresh-market tomatoes. *J. Econ. Entomol.* 73:552–55.

Lynch, J. A., and M. W. Johnson. 1987. Stratified sampling of *Liriomyza* spp. (Diptera: Agromyzidae) and associated hymenopterous parasites on watermelon. *J. Econ. Entomol.* 80:1254–61.

Mason, G. A., and M. W. Johnson. 1988. Tolerance to permethrin and fenvalerate in hymenopterous parasitoids associated with *Liriomyza* spp. (Diptera: Agromyzidae). *J. Econ. Entomol.* 81:123–26.

Mothershead, P. D. 1978. An evaluation of the effectiveness of established and recently introduced hymenopterous parasites of *Liriomyza* on Oahu, Hawaii. Master's thesis, University of Hawaii at Manoa, Honolulu.

Oatman, E. R., and G. G. Kennedy. 1976. Methomyl-induced outbreak of *Liriomyza sativae* on tomato. *J. Econ. Entomol.* 69:667–68.

Parrella, M. P. 1982. A review of the history and taxonomy of economically important serpentine leafminers (*Liriomyza* spp.) in California (Diptera: Agromyzidae). *Pan-Pacific Entomol.* 58:302–8.

———. 1987. Biology of *Liriomyza*. *Annu. Rev. Entomol.* 32:201–24.

Petcharat, J. 1987. Biology and searching behavior of *Ganaspidium hunteri* (Crawford) (Hymenoptera: Eucoilidae) and *Chrysocharis parksi* Crawford (Hymenoptera: Eulophidae), parasites of the vegetable leafminer, *Liriomyza sativae* Blanchard (Diptera: Agromyzidae). Ph.D. diss., University of Hawaii at Manoa, Honolulu.

Petcharat, J., and M. W. Johnson. 1988. Biology of the leafminer parasitoid *Ganaspidium utilis* Beardsley (Hymenoptera: Eucoilidae). *Ann. Entomol. Soc. Am.* 81:477–80.

Schreiner, I., D. Nafus, and C. Bjork. 1986. Control of *Liriomyza trifolii* (Burgess) (Diptera: Agromyzidae) on yard-long (*Vigna unguiculata*) and pole beans (*Phaseolus vulgaris*) on Guam: Effect on yield loss and parasite numbers. *Trop. Pest Manage.* 32:333–37.

Waterhouse, D. F., and K. R. Norris. 1987. *Biological control: Pacific prospects*. Melbourne: Inkata Press.

60 / APPLE MAGGOT

M. T. ALINIAZEE, A. B. MOHAMMAD, AND V. P. JONES

---

apple maggot
Rhagoletis pomonella (Walsh)
Diptera: Tephritidae

INTRODUCTION

Biology and Pest Status

The apple maggot, *Rhagoletis pomonella* (Walsh), is a major pest throughout the northeastern apple-growing regions of the United States as well as in Canada. In the western United States it is a recent introduction.

M. T. AliNiazee and R. L. Westcott (1987) studied the pest's seasonal emergence, flight activity, oviposition, and larval development over a period of 6 years in apple and in hawthorn (*Crataegus* species) hosts in the Willamette Valley and southern Oregon. Their results showed that in Oregon apple maggots usually complete one generation per year; occasionally, a partial second generation may occur. Apple maggot populations in apple hosts emerged in late June and peaked slightly earlier than did those in hawthorn. Peak flight activity generally occurred between mid-August and early September, and oviposition and hatching of larvae closely followed adult emergence patterns.

In studies in Oregon and Washington, K. T. Tracewski et al. (1987) also found that apple maggots showed activity in apples earlier than in hawthorn. However, where the host plants were contiguous, they detected activity in both hosts at about the same time. Adult activity peaked 3 to 4 weeks earlier—in late July to late August—on apples and on *Crataegus douglasii* Lindley than it did on *C. monogyna* Jacquin. On the latter, peak activity occurred in early to late September.

In Utah, studies on the phenology of apple maggots on sweet and tart cherries and hawthorn (*C. douglasii*) between 1985 and 1988 (Jones et al. 1989) indicated that apple maggot populations in different parts of the state emerged up to 4 weeks apart, although the emergence curves of apple maggots associated with these three hosts were similar on a physiological time scale. V. P. Jones et al. (1989) concluded that the phenology models of apple maggot developed in one region were not relevant to other areas, mainly because of differences

in when the first flies emerged. In Colorado, apple maggot populations are primarily confined to a native species of hawthorn, *C. rivularis* Nutt., and it appears that the pest has yet to attack apples there (Kroening, Kondratieff, and Nelson 1989).

Historical Notes

The apple maggot, a native North American insect, became a key pest of apples (*Malus domesticus* Borkh.) in the eastern United States a century ago when it shifted from its native hosts, the *Crataegus* species (Dean and Chapman 1973). It was first recorded in the western United States in 1979 in Portland, Oregon (AliNiazee and Penrose 1981); since then, it has been recorded in six western states: Oregon, Washington, Idaho, California, Utah, and Colorado (AliNiazee and Brunner 1986). In these states the apple maggot attacks apples, cherries, and two species of hawthorn—one a native species, *Crataegus douglasii*, and the other an introduced ornamental species, *C. monogyna*. Researchers have recently conducted studies on the phenology and monitoring of apple maggots in the western United States on these three hosts (AliNiazee and Westcott 1987; Tracewski et al. 1987; Jones et al. 1989; Kroening, Kondratieff, and Nelson 1989).

RESULTS

Objective A: The Identification and Introduction of Beneficial Organisms

In the eastern United States, larval-pupal parasitoids are important in regulating apple maggot populations (Monteith 1971; Dean and Chapman 1973; Cameron and Morrison 1977). Researchers have reported several species of opiine braconids from these areas, including *Opius canaliculatus* Gahan and *Diachasma alloeum* (Muesebeck) in Quebec (Rivard 1967); *Biosteres melleus* (Gahan), *O. lectus* Gahan, and *O. alloeus* (Muesebeck) in

221

New York (Dean and Chapman 1973); and *B. melleus*, *O. lectus*, *D. alloeum*, *D. ferrugineum* (Gahan), and *O. downesi* Gahan in Connecticut (Maier 1981). In hawthorn the opiines rather significantly reduced apple maggot populations. However, their impact on pest populations in apple was only marginal, even though the synchrony of the parasitoid's and pest's life cycles was excellent (Dean and Chapman 1973). This is possibly because the ovipositors of the opiines—*O. lectus* and *O. downesi*—are fairly short and may be unable to reach the larvae in the larger apple fruit. Another explanation may be found in the cannibalistic nature of the opiines during larval development (Dean and Chapman 1973).

In Oregon, M. T. AliNiazee (1985) reported that two species of opiines, *O. lectoides* Gahan and *O. downesi*, both of which commonly parasitize a closely related species, the snowberry maggot (*R. zephyria* Snow), were associated with apple maggots, mainly on hawthorn hosts. These two species parasitized as much as 60 percent of apple maggot pupae on native hawthorn (*C. douglasii*), but less than 2 percent of pupae on apples. Nearly 91 percent of the parasitoids were *O. downesi*. Further laboratory studies (AliNiazee unpub. data) on emergence patterns indicated an excellent synchrony between apple maggots and *O. downesi*. Since this parasitoid oviposits in the mature larvae before they have an opportunity to exit from the fruit, a 40- to 50-day difference between the emergence of adult flies and the adult parasitoid is highly appropriate.

In southwestern Washington four species of parasitoids have been recorded from apple maggot pupae (Gut and Brunner 1987), including *O. downesi* as well as *O. lectoides*, from larvae feeding in the fruit of the ornamental hawthorn, *C. monogyna*. *Opius downesi* was the dominant species in 8 of 12 sites sampled, and its rate of parasitization ranged from 10 percent to—at one site—a high of 50 percent. The researchers also found *Biosteres melleus* and an unidentified species of chalcidoid attacking apple maggot larvae on the native hawthorn, *C. douglasii*.

In Utah, apple maggot pupae from cherries and hawthorn are attacked by three species of parasitoids (F. J. Messina person. commun.). Around 10 to 17 percent of pupae from hawthorn and 13 percent from cherries were parasitized by *O. downesi* and *Opius* sp. nr. *lectoides* Gahan as well as by a diapriid, *Coptera* sp. nr. *occidentalis* Muesebeck, a pupal parasitoid. More than 90 percent of the parasitoid adults were *O. downesi*. Adults of the *Coptera* species emerged after two overwintering periods.

Most of the biological control work on this pest has emphasized documenting and identifying natural enemies and assessing their impact under undisturbed natural conditions. No attempts have yet been made to conserve and augment the apple maggot parasitoids on a large scale, in part because biological control alone is not sufficient to keep pest populations below levels that cause economic damage. Our relatively scant knowledge of the biology of these natural enemies has likewise contributed to the lack of effort to manipulate their populations. Also, the effectiveness of these parasitoids in commercial orchards is reduced by insecticide use, which is a major problem.

Several generalist predators, including carabids, ants, spiders, and birds, have been found feeding on the various life stages of apple maggots (Dean and Chapman 1973). The mature maggots leave the host fruit just before pupating, and the newly emerged adults appear highly vulnerable to predation. However, predators do not keep pest populations below levels that cause economic damage. Thus, research on the biological control of apple maggots will continue to focus on parasites.

Because the apple maggot is of North American origin, researchers have not explored for natural enemies in foreign lands. However, in 1954 two subspecies of Opiinae, *Opius longicaudatus compensans* (Silvestri) and *O. longicaudatus taiensis* Fullaway, were imported from Hawaii and released against apple maggots in West Virginia; both failed to become established (Clausen 1978). Researchers are currently working on releasing eastern opiine species in the western United States. Parasitoids of the exotic European cherry fruit fly, *Rhagoletis cerasi* L., and of the walnut huskfly, *R. completa* Cresson, may also be promising biocontrol agents.

RECOMMENDATIONS

1. Using natural enemies to control apple maggots in commercial orchards is difficult because pesticide applications cause repeated disruption, and only very clean fruit can be marketed. Further research should be conducted on the possibility of a mass-release program using laboratory-reared parasitoids, and on better pesticide management.

2. Attempts should be made to import, colonize, and release the natural enemies of this pest from the eastern United States into western areas where it has recently become established.

REFERENCES CITED

AliNiazee, M. T. 1985. Opiine parasitoids (Hymenoptera: Braconidae) of *Rhagoletis pomonella* and *R. zephyria* (Diptera: Tephritidae) in the Willamette Valley, Oregon. *Can. Entomol.* 117:163–66.

AliNiazee, M. T., and J. F. Brunner. 1986. Apple maggot in the western United States: A review of its establishment and current approaches to management. *J. Entomol. Soc. Brit. Columbia*. 83:49–53.

AliNiazee, M. T., and R. L. Penrose. 1981. Apple maggot in Oregon: A possible threat to the Northwest apple industry. *Bull. Entomol. Soc. Am*. 27:245–46.

AliNiazee, M. T., and R. L. Westcott. 1987. Flight period and seasonal development of the apple maggot, *Rhagoletis pomonella* (Diptera: Tephritidae). *Ann. Entomol. Soc. Am*. 83:1143–48.

Cameron, P. J., and F. O. Morrison. 1977. Analysis of mortality in the apple maggot, *Rhagoletis pomonella* (Diptera: Tephritidae), in Quebec. *Can. Entomol*. 109:769–88.

Clausen, C. P. 1978. *Introduced parasites and predators of arthropod pests and weeds: A world review*. USDA-ARS Handbook 480. Washington, D.C.: U.S. Department of Agriculture.

Dean, R. W., and P. J. Chapman. 1973. Bionomics of the apple maggot in eastern New York. *Search Agriculture* 3:1–64.

Gut, L. J., and J. F. Brunner. 1987. Parasitoids of the apple maggot, *Rhagoletis pomonella* (Walsh), in southwestern Washington. In *Research Report from the 61st Annual Western Orchard Pest and Disease Management Conference, 14–16 January 1987, Portland, Oregon,* 26.

Jones, V. P., D. W. Davis, S. L. Smith, and D. B. Allred. 1989. Phenology of apple maggot (Diptera: Tephritidae) associated with cherry and hawthorn in Utah. *J. Econ. Entomol*. 82:788–92.

Kroening, M. K., B. C. Kondratieff, and E. E. Nelson. 1989. Host status of apple maggot (Diptera: Tephritidae) in Colorado. *J. Econ. Entomol*. 82:886–90.

Maier, C. T. 1981. Parasitoids emerging from puparia of *Rhagoletis pomonella* (Diptera: Tephritidae) infesting hawthorn and apple in Connecticut. *Can. Entomol*. 113:867–70.

Monteith, L. G. 1971. The status of parasites of the apple maggot, *Rhagoletis pomonella* (Diptera: Tephritidae), in Ontario. *Can. Entomol*. 103:507–12.

Rivard, I. 1967. *Opius lectus* and *O. alloeus*, larval parasites of the apple maggot, *Rhagoletis pomonella* (Diptera: Trypetidae), in Quebec. *Can. Entomol*. 99:895–96.

Tracewski, K. T., J. F. Brunner, S. C. Hoyt, and S. R. Dewey. 1987. Occurrence of *Rhagoletis pomonella* (Walsh) in hawthorn, *Crataegus*, of the Pacific Northwest. *Melanderia* 45: 20–25.

61 / Walnut Husk Fly

K. S. HAGEN, R. L. TASSAN, M. FONG, AND M. T. ALINIAZEE

> **walnut husk fly**
> *Rhagoletis completa* Cresson
> Diptera: Tephritidae

INTRODUCTION

Biology and Pest Status

The walnut husk fly, *Rhagoletis completa* Cresson, is a serious pest of walnuts (*Juglans* species) in Arizona, California, Oregon, Utah, and Washington, where it is usually controlled by insecticides (AliNiazee and Fisher 1985). In addition to attacking most varieties of Persian walnuts, the walnut husk fly is also an occasional pest of peaches and nectarines. At one time it only attacked peaches growing near walnuts (Boyce 1934), but it now attacks peaches and nectarines that are somewhat remote from walnut trees, particularly in urban areas. Because it is an introduced insect, the walnut husk fly became a candidate for classical biological control. Researchers first attempted to introduce natural enemies in southern California in the early 1930s. New research on biological control began in 1966 and has continued sporadically since then.

In his extensive biology of the walnut husk fly, A. M. Boyce (1934) found that it attacked all *Juglans* species, but that damage varied with the walnut husk's hardness and the time in the season when adults emerged. The biology of *R. completa* is quite similar to other *Rhagoletis* species (Boller and Prokopy 1976): it is univoltine and overwinters in puparia in the soil, but occasionally a small proportion of any given population may develop over 2, 3, or even 4 years. About 5 percent of the puparia formed in the fall will yield flies, and the rest will enter diapause. In the laboratory, new diapausing puparia held at 5°C for at least 6 weeks and then transferred to 22°C, produced adult flies after 6 more weeks, making it possible to produce adult flies throughout the year. A detailed study has suggested that continuous—that is, nondiapause—emergence of this insect is possible, and that this species possesses the greatest flexibility in diapause development and postdiapause growth requirements of any *Rhagoletis* species (AliNiazee 1988).

In the field, adults begin to emerge in July and August, the exact timing depending on spring and summer temperatures. To produce eggs, adult females require a complex diet, including vitamins and amino acids (Tsiropoulos 1978; Tsiropoulos and Hagen 1987). A. M. Boyce (1934) suspected that adults probably fed mostly on aphid honeydew. In laboratory studies, K. S. Hagen (unpub. data) obtained 280 eggs per female over 80 days from flies fed *Chromaphis juglandicola* (Kaltenbach) honeydew and compared them with 405 eggs per female from flies fed sucrose plus yeast hydrolysate solution. Because honeydews rarely contain all the essential amino acids, and the walnut husk fly requires all ten to produce eggs, researchers suspect that bacterial symbionts carried by adults probably synthesize the missing amino acids (Tsiropoulos 1976). Adults mate during the day, 6 to 14 days after emergence, and females begin ovipositing 10 to 20 days after emergence.

The females deposit 10 to 20 eggs in small cavities formed by the ovipositor just beneath the surface of the green husk. Then the female smears an oviposition-deterring pheromone over the area, which accounts for the fact that nuts usually have only a single oviposition puncture (Cirio 1972). The eggs hatch in 4 to 5 days.

The larval stages, consisting of three instars, are spent within the husk tissue where active feeding takes place, and last for 25 to 38 days. The mature larvae may emerge from the husks while the nuts are still in the trees, or they may emerge after the nuts fall. They then burrow about 25 to 50 millimeters into the soil and form puparia, enter diapause, and remain dormant until the next summer, when they transform into adult flies.

Historical Notes

Rhagoletis completa is one of five Nearctic tephritid species—including *R. juglandis* Cresson—in the *Suavis* species group, all of which attack the husks of native *Juglans* species. The original distribution of *R. completa* is the south-central United States (Bush 1966). In southern California it was first found near Riverside in 1926. In 1954 it appeared in northern California, in Sonoma

County, and by 1957 it had spread to the Santa Clara Valley (Michelbacher and Ortega 1958). A few years later it was found in all the areas of California where commercial walnuts are grown. It was first noticed in Oregon in the early 1960s, and has since spread to Washington, British Columbia, Arizona, and Utah.

RESULTS

Objective A: The Identification and Introduction of Beneficial Organisms

A. M. Boyce (1934) commented how remarkably free the walnut husk fly was from important natural enemies in southern California. A few general predators, such as spiders, anthocorids, chrysopid larvae, and ants were observed occasionally preying on various stages of the pests, and Boyce reported that in Kansas two parasitoids, *Spalangia rugosicollis* Ashmead (now considered to be a synonym of *S. nigra* Latreille) and a *Galesus* species (now placed in the genus *Coptera* Say), had emerged from walnut husk fly puparia. The latter was placed taxonomically near *C. atricornis* (Ashmead), but it may have been *C. evansi* Muesebeck (Muesebeck 1980), which has been reared from the puparia of *Rhagoletis* species in several western states. Between 1929 and 1933, A. M. Boyce and H. S. Smith reared thousands of flies from field-collected fly larvae and puparia in southern California and observed no parasitoids, nor were any found attacking larvae in Kansas. In the early 1960s K. S. Hagen reared thousands of flies from puparia collected in the Manteca-Modesto area of California's Central Valley; he, too, observed no parasitoids. However, in 1968 G. Buckingham and R. Tassan reared *Coptera occidentalis* Muesebeck from puparia from Napa County, California. M. T. AliNiazee also recorded an unidentified species of *Coptera* in Oregon.

The first intentional introductions of natural enemies against *R. completa* began in 1931 and 1932, when H. S. Smith released *Opius humilis* Silvestri and *Biosteres tryoni* (Cameron) in southern California. These solitary larval-pupal parasitoids, which were reared successfully on the walnut husk fly in the laboratory, were obtained from Hawaii, where they attack the Mediterranean fruit fly. One year later, in 1932, there was a recovery of *O. humilis* in the release area (Boyce 1934), but there has been none since. During the fall of 1984, researchers received 683 *B. tryoni* males and 713 females from Hawaii, and released them against the walnut husk fly in an untreated commercial walnut orchard in Solano County, California. A recovery was made the same year, but despite extensive sampling no further recoveries were made in later years. Both *O. humilis* and *B. tryoni* are multivoltine, and thus may not be able to enter diapause during winters in temperate climates.

During 1966 a search for natural enemies was made in Arizona and New Mexico. Only *Rhagoletis juglandis* Cresson was collected in Arizona from native walnuts, and *R. completa* was found only in southeastern New Mexico near Roswell, where it occurred along with *R. juglandis*.

In Arizona researchers collected many *R. juglandis* puparia from beneath native *Juglans rupestris* trees in the Chiricahua Mountains. These puparia were sent to the University of California quarantine facility at Albany, where they produced two parasitoid species, *Biosteres juglandis* (Muesebeck) and a *Pseudeucoila* species. Both parasitoids oviposited in *R. completa* larvae, and adults emerged a year later from the walnut husk fly puparia. Between 1967 and 1970, 950 *B. juglandis* individuals were released in San Benito County, California, in untreated commercial walnut orchards and around black walnuts (*Juglans hindsi*) infested with the walnut husk fly; in 1968, 18 *Pseudeucoila* were released in the same area.

During the summer of 1977 over 5,000 more *R. juglandis* puparia were collected in Arizona, again mainly from the Chiricahua Mountains, and sent to the quarantine facility at Albany. In 1978 these yielded 25 *B. juglandis*, 7 *Pseudeucoila* species, and 335 *Coptera evansi*. Thus, about 7 percent of the field-collected puparia yielded parasitoids. These species were later reared in the insectary and then released. However, neither *B. juglandis* nor the *Pseudeucoila* species was recovered at any of the release sites, and these species were not further released.

Since 1977 we have devoted most of our attention to culturing and releasing the two species of *Coptera* and a new husk fly parasitoid from Texas, *Biosteres sublaevis* Wharton. Hundreds of *C. occidentalis* and *C. evansi* individuals were released against the walnut husk fly in both commercial walnut orchards and isolated black walnut trees in Alameda, Contra Costa, Humboldt, Napa, San Luis Obispo, Siskiyou, Solano, and Stanislaus counties in California. Furthermore, *C. occidentalis* was sent to Greece, where it successfully attacked puparia of the olive fly and the Mediterranean fruit fly in the laboratory.

In California, *C. occidentalis* became successfully established against the walnut husk fly at Albany and Paso Robles, and against the western cherry fruit fly, *R. indifferens* Curran, on the slopes of Mt. Shasta. In Albany, *C. evansi* became successfully established against the walnut husk fly on black walnuts. Both *Coptera* species are still being cultured and released at new sites, even though C. F. W. Muesebeck (1980) has recorded both species in other western localities on other tephritids.

While studying native populations of *R. completa* in the Davis Mountains of Texas, S. H. Berlocher of the University of Illinois (person. commun.) reared a braconid, *Biosteres sublaevis* Wharton (Wharton and Marsh 1978), from walnut husk fly puparia he had collected. Berlocher sent puparia collected from *Juglans microcarpa* fruit from the same locality to E. F. Legner at Riverside, who forwarded 600 puparia to the Albany laboratory in 1975. From this material about 50 *B. sublaevis* emerged. In 1978, 3,642 puparia were sent to Albany from Texas, and in 1979, 1,196 *B. sublaevis* emerged, indicating that about 33 percent of the puparia were parasitized. During a 4-year survey (conducted in 1978 and then again between 1980 and 1983) of natural parasitism of the walnut husk fly infesting *Juglans microcarpa* in western Texas and southeastern New Mexico, E. F. Legner and R. D. Goeden (1987) recorded parasitism by *B. sublaevis* ranging from 0 to 88 percent, and by *Trybliographa* species (Eucoilidae) from 0 to 12.9 percent.

In the laboratory we reared *B. sublaevis* on husk fly larvae infesting Persian and black walnuts. This solitary larval-pupal endoparasitoid is univoltine. In 1979 we began to release it, liberating 784 individuals—357 females and 427 males—in a commercial unsprayed orchard in Solano County, California, and 24 individuals—13 females and 11 males—in two black walnut trees at our research station at Albany, California. Further releases of 521 males and 454 females were made in the same Solano County orchard between 1980 and 1983. Also, between 1983 and 1986 we released several hundred *B. sublaevis* in black walnuts intermixed with Persian walnut trees near Paso Robles, California, and more recently, in black walnuts in Contra Costa and Santa Clara counties. So far, *B. sublaevis* has only been recovered at one release site—the two black walnut trees located at the Albany research station, where it has been recovered every year since 1981.

Thus, *Coptera occidentalis*, *C. evansi*, and *Biosteres sublaevis* are the only parasitoids to have become established against the walnut husk fly in California; *C. occidentalis* also became established against *R. indifferens* in Siskiyou County, California. Although *Biosteres juglandis*, *B. tryoni*, and *B. humilis* were reared on the walnut husk fly in the laboratory, they were never recovered. *Biosteres arisanus* Sonon and *B. longicaudatus* Ashmead, obtained from Hawaii where they parasitize many tropical fruit flies, did not attack walnut husk fly eggs or larvae in walnuts; it appeared that these wasps did not recognize walnuts as a tephritid host.

Objective B: Ecological and Physiological Studies

Researchers have determined the nutrients that husk fly adults require for reproduction and have developed effective diets (Tsiropoulos 1974, 1976, 1978, 1981; Tsiropoulos and Hagen 1979, 1987). A. I. Ciociola (1982) has made walnut husk fly larval nutrition studies and developed an artificial diet that allowed up to 50 percent of the larvae to pupate.

Objective C: The Conservation and Augmentation of Beneficial Organisms

To determine whether inundative releases of *C. occidentalis* could be used to control the walnut husk fly, 600 adults were released at the rate of 15 per square meter over 45 square meters of soil where 10,000 husk fly puparia had been buried just beneath the surface. The results were disappointing: when they were excavated 6 weeks later, fewer than 5 percent of the puparia were parasitized. We suspect that because of a lack of walnut husks, and the presence of manure in the soil used to bury the puparia, the synomones (chemical attractants) needed to stimulate the wasps to search were missing: hence the lower-than-expected parasitism rate.

Objective D: Impact

Out of 368 *R. completa* puparia collected monthly between January 15 and July 13, 1990, from soil beneath the two black walnut trees at the Albany research station, 10.5 percent were parasitized by the *Coptera* species and 40.2 percent by *B. sublaevis*. Thus, about 50 percent of the puparia were parasitized; however, virtually all the nuts on both trees were infested by the walnut husk fly.

RECOMMENDATIONS

1. Release *Coptera evansi* and *C. occidentalis* against the walnut husk fly where those parasitoids do not exist.

2. Release *Biosteres sublaevis* in infested black walnut trees near commercial walnut orchards.

3. Continue to evaluate the impact of the three established parasitoids.

4. Improve the artificial diet for rearing walnut husk fly larvae.

5. Determine the role of semiochemicals in the searching behavior of adult *Coptera* species.

REFERENCES CITED

AliNiazee, M. T. 1988. Diapause modalities in some species of *Rhagoletis* flies. In *Ecology and management of economically important fruit flies*, ed. M. T. AliNiazee, 13–25. Oregon Agricultural Experiment Station Special Report 830.

AliNiazee, M. T., and G. C. Fisher. 1985. *Controlling walnut husk fly in Oregon.* Oregon State University Extension Publication FS168.

Boller, E. F., and R. J. Prokopy. 1976. Bionomics and management of *Rhagoletis. Annu. Rev. Entomol.* 21:223–46.

Boyce, A. M. 1934. Bionomics of the walnut husk fly, *Rhagoletis completa. Hilgardia* 8:363–579.

Bush, G. L. 1966. The taxonomy, cytology, and evolution of the genus *Rhagoletis* in North America (Diptera: Tephritidae). *Bull. Mus. Comparative Zool.* 134:431–562.

Ciociola, A. I. 1982. Larval nutrition of walnut husk fly, *Rhagoletis completa* Cresson (Diptera: Tephritidae). Ph.D. diss., University of California, Berkeley.

Cirio, U. 1972. Observazioni sul comportamento de ovideposizione della *Rhagoletis completa* in laboratorio. In *Proceedings of the 9th Congress of the Italian Entomological Society,* 99–117.

Legner, E. F., and R. D. Goeden. 1987. Larval parasitism of *Rhagoletis completa* (Diptera: Tephritidae) on *Juglans microcarpa* (Juglandaceae) in western Texas and southeastern New Mexico. *Proc. Entomol. Soc. Wash.* 89:739–43.

Michelbacher, A. E., and J. C. Ortega. 1958. *A technical study of insects and related pests attacking walnuts.* California Agricultural Experiment Station Bulletin 764.

Muesebeck, C. F. W. 1980. *The Nearctic parasitic wasps of the genera* Psilus *Panzer and* Coptera *Say (Hymenoptera, Proctotrupoidea, Diapriidae).* USDA Technical Bulletin 1617.

Tsiropoulos, G. J. 1974. Ecophysiology of adult reproduction of walnut husk fly, *Rhagoletis completa.* Ph.D. diss., University of California, Berkeley.

———. 1976. Bacteria associated with walnut husk fly, *Rhagoletis completa. Environ. Entomol.* 5:83–86.

———. 1978. Holidic diets and nutritional requirements for survival and reproduction of adult walnut husk fly. *J. Insect Physiol.* 24:239–42.

———. 1981. Effect of antibiotics incorporated into defined adult diets on survival and reproduction of the walnut husk fly, *Rhagoletis completa* Cress. (Diptera: Trypetidae). *Zeit. angew. Entomol.* 91:100–106.

Tsiropoulos, G. J., and K. S. Hagen. 1979. Ovipositional response of walnut husk fly, *Rhagoletis completa,* to artificial substrates. *Zeit. angew. Entomol.* 88:547–50.

———. 1987. Effect of nutritional deficiencies, produced by antimetabolites, on reproduction of *Rhagoletis completa* Cresson (Diptera: Tephritidae). *Zeit. angew. Entomol.* 103:351–54.

Wharton, R. A., and P. M. Marsh. 1978. New World Opiinae (Hymenoptera: Braconidae) parasitic on Tephritidae (Diptera). *J. Wash. Acad. Sci.* 68:147–67.

PART 3

WEED CASE HISTORIES

62 / BULL THISTLE

S. S. ROSENTHAL AND G. L. PIPER

---

bull thistle
Cirsium vulgare (Savi) Ten.
Asteraceae

INTRODUCTION

Biology and Pest Status

Bull thistle, an herbaceous biennial weed, is found throughout the United States but is most serious in the Northeast (Reed and Hughes 1970). During its first year of growth, a rosette develops from the thistle's thick taproot; the next year the rosette gives rise to flowering shoots up to 2 meters high. Bull thistles reproduce only by seed, and a large plant may produce over 300 seedheads. Although very widespread and a highly competitive weed of fields, pastures, and wasteland, the bull thistle does not survive well on cultivated land.

Historical Notes

Bull thistle is found throughout Europe, Southwest Asia, North Africa, and Siberia, and is naturalized in North America, Central America, and Australia (Davis and Parris 1975). Researchers have studied its natural enemies, as well as those attacking other species of *Cirsium* and *Centaurea* surveyed by the Commonwealth Agricultural Bureaux International Institute of Biological Control in Europe (Zwölfer 1972; Harris and Wilkinson 1984). When researchers surveyed the fauna of this weed in Canada and found no specialized insects attacking the flower heads, P. Harris and A. T. S. Wilkinson (1984) proposed that the tephritid *Urophora stylata* F. might be a promising biological control agent for Canada.

In southern California a survey of phytophagous insects attacking *C. vulgare* (Goeden and Ricker 1986) showed 13 polyphagous and 2 stenophagous insects feeding in its flower heads. In Oregon, *Vanessa cardui* L., as well as a biotype of *Rhinocyllus conicus* Froelich from Italian thistle and slender-flowered thistle (*C. pycno-*

cephalus L. and *C. tenuiflorus* L.), have also been found using bull thistle as an incidental host (Coombs person. commun.).

Beginning in 1973, researchers imported *U. stylata* into British Columbia from Germany and Switzerland; it was recovered the following year (Harris and Wilkinson 1984). The fly increased rapidly at several sites, and by 1978 over 90 percent of the heads at two sites near Cloverdale, British Columbia, were infested. P. Harris and A. T. S. Wilkinson (1984) calculated that if the fly population infested 88 percent of the thistle heads, it would reduce the production of plump seeds by at least 60 percent. Theoretically, the reduction in thistle density should parallel the reduction of fertile seed in the soil seed bank.

RESULTS

Objective A: The Identification and Introduction of Beneficial Organisms

M. Redfern (1968) has studied the biology of *U. stylata* in England. Adults are active in June and July; males first appear in June and mating follows female emergence in July. Throughout July, females lay eggs in small batches in unopened buds. Larvae induce multilocular galls of the achenes and receptacle in the flower heads. There is one generation per year. H. Zwölfer (1972) concluded from available field records, screening tests, host-recognition studies, and the fly's cecidicole habit that *U. stylata* was highly host-specific.

Based on these studies of biology and host specificity, in 1982 the USDA's Agricultural Research Service obtained clearance to release *U. stylata* in the United States, and arranged with P. Harris and A. T. S. Wilkinson to import the fly from British Columbia. That fall, Wilkinson sent infested seedheads to the

We thank E. M. Coombs, S. D. Hight, K. Mann, K. Mowrer, and N. E. Rees for their contributions to this report. We also appreciate the support and assistance of the state and federal collaborators who helped distribute *U. stylata* in their areas.

USDA-ARS Biological Control of Weeds Laboratory at Albany, California. During July and August 1983, 1,900 of the adults that emerged were sent to Washington state and 1,000 to Maryland.

The insects became readily established in Washington (Piper 1985). More *U. stylata* were imported in 1984 and 1985; adults were released again in Washington and Maryland as well as in Colorado. Establishment in Colorado has not yet been confirmed (Mowrer person. commun.). In Maryland, the adults released in 1983 and 1985 did not become established, apparently because their emergence was out of synchrony with their host; however, the 150 adults received in 1984 produced 455 galls in a field cage (Batra and Hight 1987). These increased to 500 galls on caged thistle the following year, and in 1986, researchers made both enclosed and open field releases. The population level has not been determined recently.

In the fall of 1988, approximately 250 infested seedheads were shipped to Oregon from British Columbia. A total of 6 releases of 25 seedheads each were made in four western Oregon counties (Coombs person. commun.). Another 25 seedheads from this shipment were sent from Oregon to the USDA-ARS facility in Bozeman, Montana, where the 50 adults reared from them were released (Mann person. commun.).

RECOMMENDATIONS

Efforts should be made to evaluate the effectiveness of *U. stylata* in the United States.

REFERENCES CITED

Batra, S. W. T., and S. D. Hight. 1987. *Biological control of thistles in Maryland.* Final Report to the Maryland Department of Transportation, Project AW-086-284-046.

Davis, P. H., and B. S. Parris. 1975. Compositae Genus 62. *Cirsium* Miller. In *Flora of Turkey,* ed. P. H. Davis, vol. 5, 370–41. Edinburgh: University Press.

Goeden, R. D., and D. W. Ricker. 1986. Phytophagous insect faunas of two introduced *Cirsium* thistles, *C. ochrocentrum* and *C. vulgare*, in southern California. *Ann. Entomol. Soc. Am.* 79:945–52.

Harris, P., and A. T. S. Wilkinson. 1984. *Cirsium vulgare* (Savi) Ten. Bull thistle (Compositae). In *Biological control programmes against insects and weeds in Canada 1969–1980,* eds. J. S. Kelleher and M. A. Hulme, chap. 33, 147–53. Farnham Royal, Slough, U. K.: Commonwealth Agricultural Bureaux International.

Piper, G. L. 1985. Biological control of weeds in Washington: Status report. In *Proceedings of the 6th International Symposium on Biological Control of Weeds, 19–25 August 1984, Vancouver, British Columbia,* ed. E. S. Delfosse, 817–26. Ottawa: Agriculture Canada.

Redfern, M. 1968. The natural history of spear thistle-heads. *Field Studies* 2(5):669–717.

Reed, C. F., and R. Hughes. 1970. *Common weeds of the United States.* USDA-ARS Handbook 366. Washington, D.C.: U.S. Department of Agriculture.

Zwölfer, H. 1972. Investigations on *Urophora stylata* Fabr., a possible agent for the biological control of *Cirsium vulgare* in Canada. Weed projects for Canada. Commonwealth Institute of Biological Control Progress Report 29.

63 / CANADA THISTLE

G. L. PIPER AND L. A. ANDRES

> **Canada thistle**
> *Cirsium arvense* (L.) Scop.
> Asteraceae

INTRODUCTION

Biology and Pest Status

Canada thistle, a herbaceous perennial indigenous to temperate Europe and Asia, has been recorded in 39 countries (Holm et al. 1979) and ranks as one of the world's worst weeds (Holm et al. 1977). In North America it is widely distributed across the northern half of the United States and southern Canada (Reed and Hughes 1970; Moore 1975), where it infests pastures and rangeland, cropland, residential properties, wasteland, ditch and stream banks, and transportation rights-of-way.

Various investigators have detailed the life history, ecology, and economic importance of Canada thistle (Detmers 1927; Hodgson 1968; Moore 1975). The spiny plant is a pest primarily because of its extensive system of horizontal and deep vertical roots, which make it a strong competitor of crops and other vegetation.

Aerial shoots develop from buds formed along the horizontal roots; vertical roots function as carbohydrate repositories. Fragments of either root type can give rise to new plants. *Cirsium arvense* is functionally dioecious, and male and female plants flower from mid-June to September.

The plumed seeds are dispersed by wind and water; by adhering to vehicles, animals, and humans; or as a contaminant of crop seeds. Seeds can remain viable in the soil for over 20 years (Goss 1925). Several Canada thistle ecotypes have been identified in North America, the forms being distinguished by differences in morphology, phenology, and susceptibility to herbicides (Hodgson 1964).

Historical Notes

Canada thistle was apparently first introduced into Canada and the United States as a contaminant of crop seeds during the mid-17th century (Moore 1975). In many American states it has become the most widespread and serious agricultural weed (Piper 1985; Story, DeSmet-Moens, and Morrill 1985; French and Burrill 1989).

Beginning in 1959, surveys of the phytophagous insects associated with the plant in western Europe and Japan were conducted by the Commonwealth Agricultural Bureaux International Institute of Biological Control (CIBC) in Delémont, Switzerland, at the request of Agriculture Canada. These surveys were augmented, to a lesser extent, by the efforts of the USDA-ARS Biological Control of Weeds Laboratory in Rome, Italy (Zwölfer 1965a; Schroeder 1980). From the diverse European entomofauna associated with Canada thistle, Canadian entomologists selected and introduced the leaf-feeding chrysomelids, *Altica carduorum* Guerin-Meneville and *Lema cyanella* (L.); the leaf- and stem-mining weevil, *Ceutorhynchus litura* (F.); and the gall-forming tephritid *Urophora cardui* (L.). D. P. Peschken (1971, 1984) has summarized the biological control program in Canada. In the United States, research efforts on the biological control of the weed have focused on acquiring and colonizing all but *L. cyanella*.

Researchers have also conducted surveys in Canada (Maw 1976), Idaho (Barr person. commun.), and Montana (Story, DeSmet-Moens, and Morrill 1985) to ascertain whether potentially useful phytophages were already attacking Canada thistle in North America. Although a number of arthropods were found feeding on the weed, their impact proved limited.

We thank E. M. Coombs, J. L. Littlefield, J. P. McCaffrey, and J. M. Story for their contributions to this report. We also acknowledge the numerous state and federal agency personnel and private landowners who have assisted with *C. arvense* natural enemy releases and redistributions in the United States. Appreciation is extended to R. W. Carlson, E. E. Grissell, and P. M. Marsh of the USDA Systematic Entomology Laboratory, for providing parasitoid identifications.

RESULTS

Objective A: The Identification and Introduction of Beneficial Organisms

In 1963 researchers made the first North American release of *Altica carduorum* in Canada, with stock obtained from Switzerland (Julien 1987). In 1964 adult beetles from the Canadian colony were received in quarantine at the USDA-ARS Biological Control of Weeds Laboratory in Albany, California, where researchers conducted supplemental host-specificity studies. Between 1966 and 1972, a total of 6,145 laboratory-reared adults were liberated in California, Colorado, Delaware, Idaho, Indiana, Maryland, Minnesota, Montana, Nevada, New Jersey, Oregon, South Dakota, Washington, and Wisconsin. Beetles from a population originating along the French Atlantic coast and reared at Albany were also released in Maryland, New Jersey, and South Dakota. It is unclear why the beetle failed to become established in North America (Peschken 1977).

H. Zwölfer (1965b) and D. P. Peschken et al. (1970) have studied the life history of *A. carduorum*. In brief, adults emerge in the spring, and females deposit eggs on the undersurface of leaves. Both adults and larvae are leaf skeletonizers. The beetles pupate in the soil, and the adults that emerge feed for a time before once again entering the soil to overwinter.

In 1965 the weevil *Ceutorhynchus litura* was introduced into Canada from Switzerland. In 1971 field-collected adults from Dachau, Germany, were released in California, Colorado, Idaho, Maryland, Montana, New Jersey, Oregon, South Dakota, Washington, and Wyoming, but the weevil became established only in Idaho, Montana, Oregon, and Wyoming (Story, DeSmet-Moens, and Morrill 1985; Julien 1987; Harmon and McCaffrey 1989; Littlefield person. commun.). However, even in those states populations have not reached high numbers or dispersed very far from release sites. Widespread redistribution has not been attempted in any state where *C. litura* is established.

H. Zwölfer and P. Harris (1966) and D. P. Peschken and R. W. Beecher (1973) have provided life history accounts of *C. litura*. The weevil is univoltine, and larvae hatch from eggs laid in the midribs of rosette leaves in the spring. The larvae soon penetrate and develop in the plant crown and, on reaching maturity, exit to pupate in the soil. Adults emerge in late summer or early fall and are active for a short time before entering the soil to overwinter.

In 1974 *Urophora cardui* was released in Canada, where it has become well established in several eastern provinces (Peschken 1984). In the United States, flies reared from galls collected near Pressbaum, Austria, were first released in California in 1977. Since 1978, releases of additional Austrian material and of adults reared from galls obtained near Mulhouse, France, have been made in Colorado, Idaho, Iowa, Maryland, Montana, Nevada, Oregon, Virginia, Washington, and Wyoming; the fly has become established in all states except Colorado, Idaho, and Iowa (Julien 1987; Littlefield and McCaffrey person. commun.). In Oregon (Coombs person. commun.) and Washington, *U. cardui* was extensively redistributed.

Accounts of the biology and ecology of *U. cardui* have been provided by H. Zwölfer (1967); by H. Zwölfer, W. Englert, and W. Pattullo (1970); and by D. P. Peschken, D. B. Finnamore, and A. K. Watson (1982). Adults emerge during May and June, and females oviposit in developing stems. Larvae feed within and overwinter in multilocular galls and pupariate the following spring.

The leaf-feeding beetle *L. cyanella* was released in Canada in 1978 and again in 1983, but it did not become established (Julien 1987). In the United States the insect has not yet been approved for field release because of concerns about its ability to develop on several indigenous *Cirsium* species (Peschken 1984).

D. P. Peschken and G. R. Johnson (1979) have reported on the life history of *L. cyanella*. Overwintered adults emerge, feed, and oviposit on young thistle rosettes in the spring. Larvae consume the foliage and eventually pupate in the soil. There is only one generation annually.

Unfortunately, little has been published on the bionomics of the various species that have been released in the United States during the last 25 years.

Objective D: Impact

In Wyoming researchers investigated the interactions among *U. cardui* and different biotypes of Canada thistle under greenhouse conditions. All of the plant biotypes exposed to the fly were galled, but researchers found differences in infestation rates and gall loads both within a given biotype and among different biotypes (Littlefield 1986).

Observations of *U. cardui* release sites in Oregon, Montana, and Washington suggest that the fly develops more quickly and survives longer when Canada thistle is in partially shaded, moist habitats (Coombs and Story person. commun.); this supports observations made in Europe (Zwölfer, Englert, and Pattullo 1970). In order for the adults to emerge, moisture must infiltrate the gall tissue in the spring, thus starting its decomposition (Lalonde and Shorthouse 1982; Rotheray 1986).

Researchers found that the synchrony of gall induction and plant development as well as the position of the gall within the crown influence the damage done by *U. cardui* (Littlefield 1986). Galls on the primary stem stunted the plant, reduced flower formation, and diminished the carbohydrate reserves in the roots; galls on lateral branches did not have these effects. In general, *U. cardui* galls caused little stress on Canada thistle, as S. F. Forsyth and A. K. Watson have also reported (1985).

In the western United States, researchers have reported predation and parasitism of the immature stages of *U. cardui*. During the fall, winter, and spring, galls are often damaged by birds (Coombs person. commun.; Piper unpub. data), rodents, and grasshoppers (Littlefield person. commun.). In Washington and Oregon, *Campoletis* sp., *Pteromalus* sp., *Torymus* n. sp. nr. *chloromerus* Walker, and *Bracon* sp. have all been identified as parasitoids of the tephritid. However, the incidence of parasitism was less than 1 percent (Piper unpub. data; Coombs person. commun.).

So far, though plant damage does occur, none of the agents released in North America has had any controlling impact on Canada thistle.

RECOMMENDATIONS

1. Further evaluation of *C. litura* and *U. cardui* in the United States should be conducted before attempting to enhance the distributions of these bioagents.

2. Personnel of the Commonwealth Institute of Biological Control or the USDA should conduct surveys in eastern Europe and Asia to discover additional arthropods and plant pathogens suitable for introduction into North America.

REFERENCES CITED

Detmers, F. 1927. *Canada thistle*, Cirsium arvense *Tourn.* Ohio Agricultural Experiment Station Bulletin 414.

Forsyth, S. F., and A. K. Watson. 1985. Stress inflicted by organisms on Canada thistle. In *Proceedings of the 6th International Symposium on Biological Control of Weeds, 19–25 August 1984, Vancouver, British Columbia,* ed. E. S. Delfosse, 425–31. Ottawa: Agriculture Canada.

French, K., and L. C. Burrill. 1989. *Problem thistles of Oregon.* Oregon State University Cooperative Extension Service Circular 1288.

Goss, W. L. 1925. The vitality of buried seeds. *J. Agric. Res.* 29:349–62.

Harmon, B. L., and J. P. McCaffrey. 1989. *Biological control of weeds in Idaho: Bioagent release records.* University of Idaho Agricultural Experiment Station Bulletin 707.

Hodgson, J. M. 1964. Variations in ecotypes of Canada thistle. *Weeds* 12:167–71.

———. 1968. *The nature, ecology, and control of Canada thistle.* USDA Technical Bulletin 1386.

Holm, L. G., D. L. Plucknett, J. V. Pancho, and J. P. Herberger.

1977. *The world's worst weeds: Distribution and biology.* Honolulu: University of Hawaii Press.

———. 1979. *A geographical atlas of world weeds.* New York: Wiley.

Julien, M. H. 1987. *Biological control of weeds: A world catalogue of agents and their target weeds,* 2nd ed. Wallingford, U. K.: Commonwealth Agricultural Bureaux International.

Lalonde, R. G., and J. D. Shorthouse. 1982. Exit strategy of *Urophora cardui* (Diptera: Tephritidae) from its gall on Canada thistle. *Can. Entomol.* 114:873–78.

Littlefield, J. L. 1986. The suitability of various clones of Canada thistle, *Cirsium arvense* (L.) Scop., to host selection and gall induction by *Urophora cardui* (L.) (Diptera: Tephritidae), an introduced biological control agent. Ph.D. diss., University of Wyoming, Laramie.

Maw, M. G. 1976. An annotated list of insects associated with Canada thistle (*Cirsium arvense*) in Canada. *Can. Entomol.* 108:235–44.

Moore, R. J. 1975. The biology of Canadian weeds: 13. *Cirsium arvense* (L.) Scop. *Can. J. Plant Sci.* 55:1033–48.

Peschken, D. P. 1971. *Cirsium arvense* (L.) Scop., Canada thistle (Compositae). In *Biological control programmes against insects and weeds in Canada, 1959–1968,* 79–83. Commonwealth Institute of Biological Control Technical Communication 4.

———. 1977. Biological control of creeping thistle (*Cirsium arvense*): Analysis of releases of *Altica carduorum* (Col.: Chrysomelidae) in Canada. *Entomophaga* 22:425–28.

———. 1984. *Cirsium arvense* (L.) Scop., Canada thistle (Compositae). In *Biological control programmes against insects and weeds in Canada 1969–1980,* eds. J. S. Kelleher and M. A. Hulme, 139–46. Farnham Royal, Slough, U. K.: Commonwealth Agricultural Bureaux International.

Peschken, D. P., and R. W. Beecher. 1973. *Ceutorhynchus litura* (Coleoptera: Curculionidae): Biology and first release for biological control of the weed Canada thistle (*Cirsium arvense*) in Ontario, Canada. *Can. Entomol.* 105:1489–94.

Peschken, D. P., and G. R. Johnson. 1979. Host specificity and suitability of *Lema cyanella* (Coleoptera: Chrysomelidae), a candidate for the biological control of Canada thistle (*Cirsium arvense*). *Can. Entomol.* 111:1059–68.

Peschken, D. P., D. B. Finnamore, and A. K. Watson. 1982. Biocontrol of the weed Canada thistle (*Cirsium arvense*): Releases and development of the gall fly *Urophora cardui* (Diptera: Tephritidae) in Canada. *Can. Entomol.* 114:349–57.

Peschken, D. P., H. A. Friesen, N. V. Tonks, and F. L. Banham. 1970. Releases of *Altica carduorum* (Chrysomelidae: Coleoptera) against the weed Canada thistle (*Cirsium arvense*) in Canada. *Can. Entomol.* 102:264–71.

Piper, G. L. 1985. Biological control of weeds in Washington: Status report. In *Proceedings of the 6th International Symposium on Biological Control of Weeds, 19–25 August 1984, Vancouver, British Columbia,* ed. E. S. Delfosse, 817–26. Ottawa: Agriculture Canada.

Reed, C. F., and R. O. Hughes. 1970. *Selected weeds of the United*

States. USDA-ARS Handbook 366. Washington, D.C.: U.S. Department of Agriculture.

Rotheray, G. E. 1986. Effect of moisture on emergence of *Urophora cardui* (L.) (Diptera: Tephritidae) from its gall on *Cirsium arvense* (L.). *Entomol. Gaz.* 37:41–44.

Schroeder, D. 1980. The biological control of thistles. *Biocont. News Info.* 1:9–26.

Story, J. M. l985. Status of biological weed control in Montana. In *Proceedings of the 6th International Symposium on Biological Control of Weeds, 19–25 August l984, Vancouver, British Columbia,* ed. E. S. Delfosse, 837–42. Ottawa: Agriculture Canada.

Story, J. M., H. DeSmet-Moens, and W. L. Morrill. 1985. Phytophagous insects associated with Canada thistle, *Cirsium arvense* (L.) Scop., in southern Montana. *J. Kansas Entomol. Soc.* 58:472–78.

Zwölfer, H. 1965a. Preliminary list of phytophagous insects attacking wild Cynareae (Compositae) species in Europe. *CIBC Tech. Bull.* 6:81–154.

———. 1965b. Observations on the distribution and ecology of *Altica carduorum* Guer. (Col.: Chrysomelidae). *Commonwealth Institute of Biological Control Technical Bulletin* 5:129–41.

———. 1967. *Observations on* Urophora cardui *L. (Trypetidae).* Commonwealth Institute of Biological Control Weed Projects for Canada Progress Report 19.

Zwölfer, H., and P. Harris. 1966. *Ceutorhynchus litura* (F.) (Col.: Curculionidae), a potential insect for the biological control of thistle, *Cirsium arvense* (L.) Scop., in Canada. *Can. J. Zool.* 44:23–38.

Zwölfer, H., W. Englert, and W. Pattullo. 1970. *Investigations on the biology, population ecology, and distribution of* Urophora cardui *L.* CIBC Weed Projects for Canada Progress Report 27.

64 / DIFFUSE KNAPWEED

G. L. PIPER AND S. S. ROSENTHAL

> **diffuse knapweed**
> *Centaurea diffusa* **Lamarck**
> **Asteraceae**

INTRODUCTION

Biology and Pest Status

Diffuse knapweed is a herbaceous biennial or short-lived perennial native to the Balkans, southern Russia, and Asia Minor (Moore and Frankton 1974). The weed is an aggressive pioneer species that thrives in disturbed semi-arid environments, including overgrazed pasture and rangeland, wastelands, and transportation rights-of-way. It is a strong competitor for water and nutrients. Dense infestations can decrease livestock-carrying capacity and revenues as well as the amount of forage available to wildlife by 88 percent (Harris and Cranston 1979; Maddox 1979). Additionally, when *C. diffusa* infests land, it diminishes the land's recreational uses, aesthetics, and property values.

Diffuse knapweed seeds germinate in the fall or spring, and seedlings develop into rosettes (Roché, Piper, and Talbott 1987). Rosettes 1 year old or older bolt during early May to yield single, upright, branched stems that flower in midsummer and that by mid-August can—depending on the availability of moisture—produce 900 to 18,000 seeds per plant (Watson and Renney 1974). The seeds are disseminated by wind, water, animals, vehicular traffic, and the movement of contaminated hay. It is not known how long seeds are viable in the soil, but they can germinate after at least 5 years (Schirman 1984).

Historical Notes

Diffuse knapweed was first collected in western North America in 1907 from an alfalfa field in Bingen, Washington (Roché and Talbott 1986). It may have been introduced with Turkestan alfalfa from the Caspian Sea region (Harris and Myers 1976) or with hybrid alfalfa seed from Germany (Maddox 1982). The weed currently infests over 30,300 hectares in southwestern Canada

and 1,264,000 hectares in the western United States. The worst infestations are in British Columbia, Idaho, Oregon, and Washington (Lacey 1989). C. J. Talbott (1987) calculated that each year the weed increased its distribution by 17.8 percent.

In 1961 the European Station of the Commonwealth Agricultural Bureaux International Institute of Biological Control at Delémont, Switzerland, on behalf of Agriculture Canada, began surveys in Europe of the insect associates of *C. diffusa* and of spotted knapweed, *C. maculosa* Lamarck, to assess their potential for biological control in North America (Zwölfer 1965; Schroeder 1985). In the late 1970s and early 1980s, further surveys were conducted by personnel of the Commonwealth Institute and the USDA's Agricultural Research Service (ARS) at the Biological Control of Weeds Laboratories in Rome, Italy, and Thessaloniki, Greece (Müller and Schroeder 1989). So far, researchers have introduced eight insects into Canada and the United States against diffuse and spotted knapweed; the introduction of six more is expected by 1995 (Story 1989).

RESULTS

Objective A: The Identification and Introduction of Beneficial Organisms

The first insect released in North America for diffuse knapweed control was the seedhead gall fly, *Urophora affinis* Frauenfeld (Maddox 1982). Researchers released this tephritid, collected in Russia near Krasnodar, in British Columbia in 1970, where it readily became established (Harris 1980a). Using material supplied by the USDA-ARS Biological Control of Weeds Laboratory in Albany, California, the fly was introduced in Montana and Oregon in 1973 (Maddox 1982), in Idaho and

We thank E. M. Coombs, J. P. McCaffrey, and J. M. Story for their contributions to this report. We also acknowledge the many state and federal agency personnel and private landowners who from the early 1970s to the present have assisted with natural enemy releases and redistributions in the western United States.

Washington in 1974 (Piper 1985; Story 1989), in California in 1976 (Maddox 1982), and in Wyoming in 1984 (Littlefield person. commun.). Within 5 to 6 years, fly populations in all of these states were large enough to permit collection for intrastate redistribution. Several researchers have chronicled these intial releases in detail (Maddox 1982; Piper 1985; Story 1985a; Harmon and McCaffrey 1989).

A second gall-forming tephritid, *U. quadrifasciata* (Meigen), was released in 1972 in British Columbia, about 193 kilometers north of the Canada–United States border (Harris 1980a). This fly, also obtained near Krasnodar, readily became established and quickly dispersed to infest stands of diffuse and spotted knapweed. In 1979 *U. quadrifasciata* was discovered to have spread unaided to Washington (Piper 1985); by 1980 it had reached Idaho, Oregon, and Montana (Story 1985b, 1989). It is now widespread in most of the knapweed-infested areas in these states. Federal approval for interstate redistribution was obtained in 1989 (Rees person. commun.), and researchers were able to introduce the fly into Wyoming that same year (Lavigne person. commun.).

Since 1979 various local, state, and federal weed control practitioners and private landowners, following established protocols, have redistributed millions of both fly species (Story 1984; McCaffrey et al. 1988). In Montana and the Pacific Northwest, much of the annual redistribution effort has been undertaken by USDA Animal and Plant Health Inspection Service–Plant Protection and Quarantine (APHIS-PPQ) personnel (Nowierski 1985).

Several other insects that infest the capitula of diffuse knapweed may soon become available for distribution in the United States and Canada. These include the weevils *Bangasternus fausti* (Reitter), *Larinus minutus* Gyllenhal, and *L. obtusus* Gyllenhal, and the tephritid *Chaetorellia acrolophi* White & Marquardt, all of which destroy developing seeds (Dunn and Campobasso 1987; Groppe 1988, 1990; Groppe and Marquardt 1989).

In 1976 researchers obtained the root-boring buprestid *Sphenoptera jugoslavica* Obenberger from Yugoslavia and released it against *C. diffusa* in southern British Columbia (Powell and Myers 1988). This beetle was released in California, Idaho, Oregon, and Washington in 1980 (Piper 1985; Harmon and McCaffrey 1989; Story 1989), and in Montana in 1983 (Story 1989); the material was from Greece, and was provided by the USDA-ARS Biological Control of Weeds facility at Albany, California. Since 1987 state and federal weed control workers have redistributed adults collected in British Columbia, Oregon, and Washington in the western United States (Piper 1989). The insect is now established in Idaho, Oregon, Montana, and Washington, but its fate in California has not been

ascertained. J. P. McCaffrey et al. (1988) have described the collection and release methods; annual redistributions are being made by USDA-ARS, USDA-APHIS-PPQ, and state scientists.

In 1982 *Agapeta zoegana* (L.), a moth that feeds primarily on spotted knapweed rosette roots and occasionally on diffuse knapweed roots (Müller, Schroeder, and Gassmann 1988), was introduced into Canada from Austria, Hungary, and the Soviet Union (Story 1989). In 1984 it was introduced into Montana (Story 1989), in 1987 into Oregon (Coombs person. commun.), in 1987 into Washington, and in 1989 into Idaho (McCaffrey person. commun.). Establishment has been confirmed in all those states, as well as in British Columbia.

Attempts to establish the root-boring moth *Pelochrista medullana* (Staudinger) from Austria, Hungary, and Romania failed in British Columbia in 1982 and in Montana in 1984 (Gassmann, Schroeder, and Müller 1982). Similar efforts with a second root moth, *Pterolonche inspersa* (Staudinger), in Canada, Idaho, and Oregon in 1986, Washington in 1987, and Montana in 1988 also failed (Dunn et al. 1989). These failures are probably related to the small numbers of moths released and the cold 1988–1989 winter.

The root-mining weevil *Cyphocleonus achates* (Fahr.), introduced into British Columbia in 1987 and Montana in 1988, has become established (Story 1989). European researchers reported that its infestation rates are greater in the larger-diameter roots of spotted knapweed than in those of *C. diffusa* (Stinson 1987).

USDA-ARS and Italian scientists in Europe are currently studying the eriophyid mite *Aceria centaureae* (Nalepa), which galls diffuse knapweed rosette and shoot leaves. Through its feeding and gall formation, the mite inhibits plant development and can kill rosettes (Schroeder 1977).

Objective C: The Conservation and Augmentation of Beneficial Organisms

Although laboratory studies showed that honey is an effective diet supplement for *U. affinis* and *U. quadrifasciata* (Vogt 1986), in Washington food-spray field tests, honey and five other carbohydrate or protein supplements failed to enhance fly reproduction, as measured by increased gall production (Norambuena 1988).

Objective D: Impact

Surveys in Washington in 1983 (Piper unpub. data) and Montana in 1984 (Story, Nowierski, and Boggs 1987) on the abundance of *U. affinis* and *U. quadrifasciata* on both knapweeds identified the areas where the flies were absent and where they were subsequently released and established.

Researchers have investigated the biology and behavior of *U. affinis* and *U. quadrifasciata* thoroughly in western North America (Story and Anderson 1978; Roze 1981; Gillespie 1983; Harris 1986). *U. affinis* is a facultative bivoltine tephritid; *U. quadrifasciata* is obligately bivoltine (Johansen and McCaffrey 1989). The seedheads attacked by *U. affinis* exhibit one or more lignified unilocular receptacle galls (Shorthouse 1977a) that function as nutritive sinks, sequestering energy and reducing seed production even in uninfested capitula on the same plant. The nonlignified unilocular galls that *U. quadrifasciata* forms within individual ovaries (Shorthouse 1977b), on the other hand, sequester less energy, and thus reduce seed numbers only in infested capitula (Harris 1980b).

The action of these flies appears to be complementary and has reduced seed crop output by 75 to 95 percent (Harris 1980b; Piper unpub. data). However, though this drop in seeds has retarded the rate at which the weed spreads, it has not appreciably lowered stand density, because sufficient seeds remain.

In Montana (Story and Nowierski 1984) and Idaho (Wheeler 1985), researchers have examined the role of spiders as predators of *Urophora* adults. *Dictyna coloradensis* Chamberlin and *D. major* Gertsch were consistently associated with gall fly populations, but researchers did not find that they significantly affected them (Wheeler 1985). At some sites in Montana, rodents preyed extensively on immature flies within the seedheads during the fall, winter, and spring, but they did not prevent the *Urophora* population from reaching damaging levels (Story and Nowierski 1984).

Researchers have also examined the prospects of integrating the herbicides 2,4-D and picloram and the *Urophora* species into a knapweed management scheme (McCaffrey and Callihan 1988; Story, Boggs, and Good 1988). Properly timed herbicide applications do not interfere with fly activity or survival.

Field observations of *Sphenoptera jugoslavica* release sites in Oregon and Washington indicate that the insect has survived well and spread rapidly. At one locality in Oregon, *S. jugoslavica* had spread 16 kilometers from the initial release site in 8 years and had infested 82 percent of *C. diffusa* roots (Coombs person. commun.). An ongoing cooperative study by state and federal entomologists was begun in 1988 to assess beetle impact on diffuse knapweed in three climatically and ecologically diverse areas of Idaho, Oregon, and Washington.

RECOMMENDATIONS

1. Continue the redistribution of *U. affinis*, *U. quadrifasciata*, and *S. jugoslavica*.

2. Develop techniques to provide sufficient stocks for field releases—namely, mass-rearing—of the root-infesting moths and beetles.

3. Devote more research attention to integrating biocontrol organisms with other diffuse knapweed management techniques.

4. Continue to evaluate and introduce new biological control organisms expeditiously.

REFERENCES CITED

Dunn, P. H., and G. Campobasso. 1987. *A petition for the introduction into quarantine for testing* Bangasternus fausti *(Reitter) (Coleoptera: Curculionidae), a potential biocontrol agent of diffuse knapweed* (Centaurea diffusa *Lam.).* Rome: USDA-ARS Biological Control of Weeds Laboratory.

Dunn, P. H., S. S. Rosenthal, G. Campobasso, and S. M. Tait. 1989. Host specificity of *Pterolonche inspersa* (Lep.: Pterolonchidae) and its potential as a biological control agent for *Centaurea diffusa*, diffuse knapweed, and *C. maculosa*, spotted knapweed. *Entomophaga* 34:435–46.

Gassmann, A., D. Schroeder, and H. Müller. 1982. *Investigations on* Pelochrista medullana *(Stgr.)(Lep.: Tortricidae), a possible biocontrol agent of diffuse and spotted knapweed,* Centaurea diffusa *Lam., and* C. maculosa *Lam. (Compositae), in North America.* Commonwealth Institute of Biological Control Final Report.

Gillespie, R. L. 1983. Bionomics of *Urophora affinis* Frauenfeld and *U. quadrifasciata* Meigen (Diptera: Tephritidae) in northern Idaho. Master's thesis, University of Idaho, Moscow.

Groppe, K. 1988. *Diffuse and spotted knapweed work in Europe in 1988.* Commonwealth Institute of Biological Control Report.

———. 1990. Larinus minutus *Gyll. (Coleoptera: Curculionidae), a suitable candidate for the biological control of diffuse and spotted knapweed in North America.* Commonwealth Institute of Biological Control Report.

Groppe, K., and K. Marquardt. 1989. Chaetorellia acrolophi *White & Marquardt (Diptera: Tephritidae), a suitable candidate for the biological control of diffuse and spotted knapweed in North America.* Commonwealth Institute of Biological Control Report.

Harmon, B. L., and J. P. McCaffrey. 1989. *Biological control of weeds in Idaho: Bioagent release records.* University of Idaho Agricultural Experiment Station Bulletin 707.

Harris, P. 1980a. Establishment of *Urophora affinis* Frfld. and *U. quadrifasciata* (Meig.) (Diptera: Tephritidae) in Canada for the biological control of diffuse and spotted knapweed. *Zeit. angew. Entomol.* 89:504–14.

———. 1980b. Effects of *Urophora affinis* Frfld. and *U. quadrifasciata* (Meig.) (Diptera: Tephritidae) on *Centaurea diffusa* Lam. and *C. maculosa* Lam. (Compositae). *Zeit. angew. Entomol.* 90:190–201.

———. 1986. Biological control of knapweed with *Urophora*

quadrifasciata Mg. *Canadex* 641.613:2.

Harris, P., and R. Cranston. l979. An economic evaluation of control methods for diffuse and spotted knapweed in western Canada. *Can. J. Plant Sci.* 59:375–82.

Harris, P., and J. Myers. 1976. *Centaurea diffusa* Lam. and *C. maculosa* Lam. *s. lat.*, diffuse and spotted knapweed (Compositae). In *Biological control programmes against insects and weeds in Canada l969–l980,* eds. J. S. Kelleher and M. A. Hulme, 127–37. Farnham Royal, Slough, U.K.: Commonwealth Agricultural Bureaux International.

Johansen, E. W., and J. P. McCaffrey. 1989. Preliminary observations on the phenology and resource utilization of *Urophora* spp. (Diptera: Tephritidae) on spotted knapweed in northern Idaho. In *Proceedings of the Knapweed Symposium, 4–5 April 1989, Bozeman, Montana,* eds. P. K. Fay and J. R. Lacey, 170–71. Montana State University Extension Bulletin 45.

Lacey, C. 1989. Knapweed management: A decade of change. In *Proceedings of the Knapweed Symposium, 4–5 April 1989, Bozeman, Montana,* eds. P. K. Fay and J. R. Lacey, 1–6. Montana State University Extension Bulletin 45.

McCaffrey, J. P., and R. H. Callihan. 1988. Compatibility of picloram and 2,4-D with *Urophora affinis* and *U. quadrifasciata* (Diptera: Tephritidae) for spotted knapweed control. *Environ. Entomol.* 17:785–88.

McCaffrey, J. P., R. P. Wight, R. L. Stoltz, R. H. Callihan, and D. W. Kidder. 1988. *Collection and redistribution of biological control agents of diffuse and spotted knapweed.* University of Idaho Cooperative Extension Service Bulletin 680.

Maddox, D. M. 1979. The knapweeds: Their economic and biological control in the western states, U.S.A. *Rangelands* 1:139–41.

———. 1982. Biological control of diffuse knapweed (*Centaurea diffusa*) and spotted knapweed (*C. maculosa*). *Weed Sci.* 30:76–82.

Moore, R. J., and C. Frankton. 1974. *The thistles of Canada.* Canada Department of Agriculture Monograph 10. Ottawa: Information Canada.

Müller, H., and D. Schroeder. 1989. The biological control of diffuse and spotted knapweed in North America: What did we learn? In *Proceedings of the Knapweed Symposium, 4–5 April 1989, Bozeman, Montana,* eds. P. K. Fay and J. R. Lacey, 151–69. Montana State University Extension Bulletin 45.

Müller, H., D. Schroeder, and A. Gassmann. 1988. *Agapeta zoegana* (L.) (Lepidoptera: Cochylidae), a suitable prospect for biological control of spotted and diffuse knapweed, *Centaurea maculosa* Monnet de la Marck, and *Centaurea diffusa* Monnet de la Marck (Compositae), in North America. *Can. Entomol.* 120:109–24.

Norambuena, H. L. 1988. Field utilization of food supplements to enhance oviposition of *Urophora affinis* Frauenfeld and *Urophora quadrifasciata* (Meigen) on diffuse knapweed, *Centaurea diffusa* Lam. Master's thesis, Washington State University, Pullman.

Nowierski, R. M. 1985. A new era of biological weed control in the western United States. In *Proceedings of the 6th International Symposium on Biological Control of Weeds, 19–25 August 1984, Vancouver, British Columbia,* ed. E. S. Delfosse, 811–15. Ottawa: Agriculture Canada.

Piper, G. L. 1985. Biological control of weeds in Washington: Status report. In *Proceedings of the 6th International Symposium on Biological Control of Weeds, 19–25 August 1984, Vancouver, British Columbia,* ed. E. S. Delfosse, 817–26. Ottawa: Agriculture Canada.

———. 1989. Release of *Sphenoptera jugoslavica* on diffuse knapweed infestations bordering highway rights-of-way in eastern Washington. In *Proceedings of the Knapweed Symposium, 4–5 April 1989, Bozeman, Montana,* eds. P. K. Fay and J. R. Lacey, 175–79. Montana State University Extension Bulletin 45.

Powell, R. D., and J. H. Myers. 1988. The effect of *Sphenoptera jugoslavica* Obenb. (Col., Buprestidae) on its host plant *Centaurea diffusa* Lam. (Compositae). *J. Appl. Entomol.* 106:25–45.

Roché, B. F., Jr., and C. J. Talbott. 1986. *The collection history of Centaureas found in Washington State.* Washington State University Agricultural Research Center Research Bulletin XB0978.

Roché, B. F., Jr., G. L. Piper, and C. J. Talbott. 1987. *Knapweeds of Washington.* Washington State University Cooperative Service Extension Extension Bulletin 1393.

Roze, L. 1981. The biological control of *Centaurea diffusa* Lam. and *C. maculosa* Lam. by *Urophora affinis* Frauenfeld and *U. quadrifasciata* Meigen. Ph.D. diss., University of British Columbia, Vancouver, Canada.

Schirman, R. 1984. Seedling establishment and seed production of diffuse and spotted knapweed. In *Proceedings of the Knapweed Symposium, 3–4 April 1984, Great Falls, Montana,* eds. J. R. Lacey and P. K. Fay, 7–10. Montana State University Bulletin 1315.

Schroeder, D. 1977. Biotic agents attacking diffuse and spotted knapweed in Europe and their prospective suitability for biological control in North America. In *Proceedings of the Knapweed Symposium, 6–7 October 1977, Kamloops, British Columbia,* 108–31. London: British Ministry of Agriculture.

———. 1985. The search for effective biological control agents in Europe: 1. Diffuse and spotted knapweed. In *Proceedings of the 6th International Symposium on Biological Control of Weeds, 19–25 August 1984, Vancouver, British Columbia,* ed. E. S. Delfosse, 103–19. Ottawa: Agriculture Canada.

Shorthouse, J. D. 1977a. Developmental morphology of *Urophora affinis* galls. In *Proceedings of the Knapweed Symposium, 6–7 October 1977, Kamloops, British Columbia,* 188–95. London: British Ministry of Agriculture.

———. 1977b. Developmental morphology of *Urophora quadrifasciata* galls. In *Proceedings of the Knapweed Symposium, 6–7 October 1977, Kamloops, British Columbia,* 196–201. London: British Ministry of Agriculture.

Stinson, C. S. A. 1987. *Investigations of* Cyphocleonus achates *(Fahr.) (Col.: Curculionidae), a possible biological agent of spotted* knapweed (Centaurea maculosa *Lam.) and diffuse knapweed (C. diffusa Lam.) (Compositae) in North America.* Commonwealth Institute of Biological Control Report.

Story, J. M. 1984. *Collection and redistribution of* Urophora affinis *and U.* quadrifasciata *for biological control of spotted knapweed.* Montana Cooperative Extension Service Circular 308.

———. 1985a. Status of biological weed control in Montana. In *Proceedings of the 6th International Symposium on Biological Control of Weeds, 19–25 August 1984, Vancouver, British Columbia,* ed. E. S. Delfosse, 837–42. Ottawa: Agriculture Canada.

———. 1985b. First report of the dispersal into Montana of *Urophora quadrifasciata* (Diptera: Tephritidae), a fly released in Canada for biological control of spotted and diffuse knapweed. *Can. Entomol.* 117:1061–62.

———. 1989. The status of biological control of spotted and diffuse knapweed. In *Proceedings of the Knapweed Symposium, 4–5 April 1989, Bozeman, Montana,* eds. P. K. Fay and J. R. Lacey, 37–42. Montana State University Extension Bulletin 45.

Story, J. M., and N. L. Anderson. 1978. Release and establishment of *Urophora affinis* (Diptera: Tephritidae) on spotted knapweed in western Montana. *Environ. Entomol.* 7:445–48.

Story, J. M., and R. M. Nowierski. 1984. Increase and dispersal of *Urophora affinis* (Diptera: Tephritidae) on spotted knapweed in western Montana. *Environ. Entomol.* 13:1151–56.

Story, J. M., K. W. Boggs, and W. R. Good. 1988. Optimal timing of 2,4-D applications for compatibility with *Urophora affinis* and *U. quadrifasciata* (Diptera: Tephritidae) for control of spotted knapweed. *Environ. Entomol.* 17:911–14.

Story, J. M., R. M. Nowierski, and K. W. Boggs. 1987. Distribution of *Urophora affinis* and *U. quadrifasciata,* two flies introduced for biological control of spotted knapweed (*Centaurea maculosa*) in Montana. *Weed Sci.* 35:145–48.

Talbott, C. J. 1987. Distribution and ecologic amplitude of selected *Centaurea* species in eastern Washington. Master's thesis, Washington State University, Pullman.

Vogt, E. A. 1986. Influence of adult diet on oogenesis, longevity, and fecundity of *Urophora affinis* Frauenfeld and *U. quadrifasciata* (Meigen) (Diptera: Tephritidae). Master's thesis, University of Idaho, Moscow.

Watson, A. K., and A. J. Renney. 1974. The biology of Canadian weeds. 6. *Centaurea diffusa* and *C. maculosa. Can. J. Plant Sci.* 54:687–701.

Wheeler, G. S. 1985. The bionomics of two spiders, *Dictyna coloradensis* Chamberlin and *D. major* Gertsch (Araneae: Dictynidae): Potential antagonists to the biocontrol of weeds. Master's thesis, University of Idaho, Moscow.

Zwölfer, H. 1965. *Preliminary list of phytophagous insects attacking wild Cynareae (Compositae) in Europe,* 81–154. Commonwealth Institute of Biological Control Technical Bulletin 6.

65 / ITALIAN THISTLE

R. D. GOEDEN

Italian thistle
Carduus pycnocephalus L.
Asteraceae

INTRODUCTION

Biology and Pest Status

In southern and central California, Italian thistle is an erect, herbaceous winter annual; in Oregon it is sometimes a biennial (E. M. Coombs person. commun.). This thistle is readily distinguished by its woolly investiture; its narrow, branched, and spiny-winged stems that reach up to 2 meters in height; and its small, slender, rose-purple capitula, which are generally borne terminally in clusters of three to five. All its leaves are spiny margined, green on top and woolly-white below, lanceolate, and deeply lobed. The thistle reproduces only by windborne or otherwise easily disseminated pappose achenes (Munz and Keck 1959; Robbins, Bellue, and Ball 1970). The slender-flowered thistle, *C. tenuiflorus* L., is a less common, closely related weed that is often found with the Italian thistle and other asteraceous thistles in California and Oregon. Because of their similarity, the Italian and slender-flowered thistles can be treated together; the research on the former described in this chapter applies equally to the latter.

Historical Notes

The Italian thistle is an alien weed of Eurasian origin. It was first reported in California in the early 1930s, and in recent years it has become widely naturalized and common, particularly in coastal counties, on grazing and pasturelands, open woodlands, fallow croplands, and wastelands such as roadsides, railroad rights-of-way, field margins, and ditch banks. The weed can displace more desirable forage or cover plants, but more commonly it colonizes disturbed open habitats where interspecific competition is less intense (Goeden 1974).

In Oregon, most infestations are on hillsides subject to spring livestock grazing. Ranchers often burn hillside pastures in the fall to promote regrowth and to remove thistle stubble, but burning may actually promote thistle growth (Coombs person. commun.).

Between 1968 and 1972 researchers conducted preintroduction surveys of phytophagous insects attacking this weed in the southern and central areas of coastal California (Goeden 1974). In 1971 and 1972 R. D. Goeden (1974) surveyed Italian thistles for natural enemies in central and southern Italy and in Greece, supplementing the limited surveys of this weed by H. Zwölfer (1965) in southern France and by USDA-ARS entomologists and foreign entomologists in federal employ in Italy, Pakistan, and Egypt (Batra et al. 1981). R. D. Goeden (unpub. data) conducted additional surveys in Iran in 1973 and Spain in 1978.

RESULTS

Objective A: The Identification and Introduction of Beneficial Organisms

The weevil *Rhinocyllus conicus*, which was first found by R. D. Goeden (1974) infesting the capitula of Italian thistle in southwestern Italy and Greece, is the only natural enemy of Italian thistle so far to be introduced into western North America. The biology of this univoltine weevil on musk thistle, *Carduus nutans* L., is outlined by L. A. Andres and N. E. Rees in chapter 67 of this volume and was described in detail by H. Zwölfer and P. Harris (1984). *R. conicus* larvae feeding in the small capitula of Italian thistle consume the receptacles and destroy the immature achenes; however, unlike larvae of the weevil biotype on the musk thistle, when their densities are high they do not further mine the peduncles or stems (Goeden 1978; Zwölfer and Harris 1984; Andres and Rees chap. 67).

On March 30, 1973, 1,076 overwintered, sexually immature weevils were collected from rosettes of Italian

My thanks to E. M. Coombs and S. S. Rosenthal for their contributions to this report. I also acknowledge the support and assistance of county collaborators.

thistles near Bari in Italy, carried as hand baggage to California, and processed through quarantine at the University of California at Riverside. On April 2, the surviving 1,029 weevils were released directly on plants in See Canyon (San Luis Obispo County), where Italian thistle was abundant, public access was restricted, and the mild local climate approximated that of the collection area. Researchers recovered eggs later in 1973, and in annual surveys found that by 1977, 91 percent of the capitula sampled at the release site bore eggs (Goeden 1974; Goeden and Ricker 1978).

Redistribution began in 1976, when a total of 1,800 overwintered weevils were released in lots of 400 to 600 at four new sites in southern California. In 1977, another 6,500 overwintered weevils were collected from See Canyon and released in lots of 450 to 560 at 13 new locations in 4 counties throughout the thistle's range in coastal areas of southern and central California (Goeden 1974; Goeden and Ricker 1978). In 1978 and 1979, USDA-ARS entomologists and California Department of Food and Agriculture personnel redistributed *R. conicus* to northern California, where it was augmented by direct releases of additional weevils imported from Italy.

In 1980 *R. conicus* from northern California was first introduced to Douglas County, Oregon; transfers continued each spring until 1983. By 1987 weevil populations had increased enough to allow mass collection and redistribution throughout the thistle's range in southwestern Oregon (Coombs person. commun.).

Objective D: Impact

In field data from evaluation studies at initial colonization sites in southern California, researchers found differences between how *R. conicus* performed on Italian thistle and on milk thistle, *Silybum marianum* (L.) Gaertner. R. D. Goeden (1978) presented evidence that separate biotypes of *R. conicus* were associated with each of these thistles in California. Ten years of data from replicated field trials (Goeden, Ricker, and Hawkins 1985) demonstrated that the Italian thistle biotype of *R. conicus* was clearly better at colonizing Italian thistle than milk thistle, whereas the milk thistle biotype was more versatile. The Italian thistle biotype was much less likely to oviposit on the larger milk thistle capitula than was the milk thistle biotype on Italian thistle capitula. Furthermore, milk thistle capitula did not support larval development by the Italian thistle biotype, even when—after 10 years of colonization had produced "explosive" population densities—these weevils finally began ovipositing on milk thistle. In contrast, the milk thistle biotype—quickly, successfully, and increasingly every year—transferred to and reproduced on Italian thistle capitula.

R. D. Goeden, D. W. Ricker, and B. A. Hawkins (1985) also assayed six polymorphic loci using starch gel electrophoresis to examine possible genetic differences among the Italian, milk, and musk thistle biotypes of *R. conicus*. Only one locus showed a fixed genetic difference, but significant differences in allelic frequencies at four of the other five loci supported the biotype designation for the three weevil populations.

From mid-April to late August 1980—7 years after *R. conicus* became established in See Canyon—30 Italian thistle plants sampled weekly during prebolting to senescence produced 7,735 (a mean of 258) capitula, which potentially contained 25,000 seeds (Goeden and Ricker 1985). *Rhinocyllus conicus* larvae directly and indirectly caused an estimated 55 percent seed loss in these heads, but 11,410 seeds were still produced, mostly after the univoltine weevils stopped ovipositing in late June. The 45 percent larval mortality, mainly in the early instars, was a result of the scramble type of intraspecific competition in the capitula. Unidentified, probably indigenous lepidopterous larvae caused some negligible interspecific competition in the heads. The impact of *Nosema* species infections of the Italian thistle biotype remains undefined (Dunn and Andres 1981). In general, researchers have found very little evidence of parasitism of *R. conicus* larvae or pupae in California (Goeden and Ricker 1978, 1985).

In southern California, 10 years of samplings along line transects at several secondary release sites (Goeden and Ricker unpub. data), as well as visual comparisons, have provided little evidence that high weevil populations reduce the densities of the Italian thistle. This is in stark contrast to the successful biological control of musk thistle by *R. conicus* reported in the eastern United States (Kok and Surles 1975) and Canada (Harris 1984). The difference is due in part to the seasonal asynchrony of the Italian thistle and *R. conicus* (Goeden and Ricker 1985). The lack of interspecific plant competition for Italian thistle in California probably plays a part as well. In more temperate parts of North America higher rainfalls mean lots of grasses, whereas in California infested pastures are more commonly overgrazed, and artificial openings in cleared brushland are usually covered by less dense stands of introduced grass species.

In Oregon the impact of *R. conicus* on the biological control of Italian thistle has been varied. In areas where thistles are burned yearly, little if any change is apparent. In unburned pasture, where vegetation is in good condition, thistle density has been noticeably declining. However, data to support this observation are not yet available (Coombs person. commun.). In coastal thistle infestations, *R. conicus* populations are not as abundant as they are inland.

RECOMMENDATIONS

1. Commonwealth Scientific and Industrial Research Organization research on asteraceous thistles of Mediterranean origin currently being conducted at Montpellier, France, should be followed to see whether natural enemies host-specific to Italian thistle will be found. These new natural enemies, unlike *R. conicus*, should not transfer to native *Cirsium* thistles in California and Oregon (Goeden and Ricker 1986, 1987a,b; Turner, Pemberton, and Rosenthal 1987; Unruh and Goeden 1987).

2. In Oregon, the impact of *R. conicus* on coastal infestations of Italian thistle should be studied and compared with inland populations, and data obtained to support the observed decline of Italian thistle populations on unburned, well-managed pastures.

REFERENCES CITED

Batra, S. W. T., J. R. Coulson, P. H. Dunn, and P. E. Boldt. 1981. *Insects and fungi associated with* Carduus *thistles (Compositae).* USDA Technical Bulletin 1616.

Dunn, P. H. 1976. Distribution of *Carduus nutans, C. acanthoides, C. pycnocephalus,* and *C. crispus* in the United States. *Weed. Sci.* 24:518–24.

Dunn, P. H., and L. A. Andres. 1981. Entomopathogens associated with insects used for biological control of weeds. In *Proceedings of the 5th International Symposium on Biological Control of Weeds, 22–27 July 1980, Brisbane, Australia,* ed. E. S. Delfosse, 241–46. Melbourne: Commonwealth Scientific and Industrial Research Organization.

Goeden, R. D. 1974. Comparative survey of the phytophagous insect faunas of Italian thistle, *Carduus pycnocephalus,* in southern California and southern Europe relative to biological weed control. *Environ. Entomol.* 3:464–74.

———. 1978. Initial analyses of *Rhinocyllus conicus* (Froelich) (Coleoptera: Curculionidae) as an introduced natural enemy of milk thistle (*Silybum marianum* [L.] Gaertner) and Italian thistle (*Carduus pycnocephalus* L.) in southern California, 39–50. USDA Publication 1978-771-106/02. Washington, D.C.: U.S. Department of Agriculture.

Goeden, R. D., and D. W. Ricker. 1974. Imported seed weevils attack Italian and milk thistles in southern California. *Calif. Agric.* 28(1):8–9.

———. 1978. Establishment of *Rhinocyllus conicus* (Coleoptera: Curculionidae) on Italian thistle in southern California. *Environ. Entomol.* 7:787–89.

———. 1985. Seasonal asynchrony of Italian thistle, *Carduus pyc-*

nocephalus, and the weevil, *Rhinocyllus conicus* (Coleoptera: Curculionidae), introduced for biological control in southern California. *Environ. Entomol.* 14:433–36.

———. 1986. Phytophagous insect faunas of the two most common native *Cirsium* thistles, *C. californicum* and *C. proteanum,* in southern California. *Ann. Entomol. Soc. Am.* 79:953–62.

———. 1987a. Phytophagous insect faunas of the native thistles, *Cirsium brevistylum, C. congdonii, C. occidentale,* and *C. tioganum,* in southern California. *Ann. Entomol. Soc. Am.* 80:152–62.

———. 1987b. Phytophagous insect faunas of native *Cirsium* thistles, *C. mohavense, C. neomexicana,* and *C. nidulum,* in the Mojave Desert of southern California. *Ann. Entomol. Soc. Am.* 80:161–75.

Goeden, R. D., D. W. Ricker, and B. A. Hawkins. 1985. Ethological and genetic differences among three biotypes of *Rhinocyllus conicus* (Coleoptera: Curculionidae) introduced into North America for the biological control of asteraceous thistles. In *Proceedings of the 6th International Symposium on Biological Control of Weeds, 19–25 August 1984, Vancouver, British Columbia,* ed. E. S. Delfosse, 181–89. Ottawa: Agriculture Canada.

Harris, P. 1984. *Carduus nutans* L., nodding thistle, and *C. acanthoides* L., plumeless thistle (Compositae). In *Biological control programmes against insects and weeds in Canada 1969–1980,* eds. J. S. Kelleher and M. A. Hulme, 115–26. Farnham Royal, Slough, U.K.: Commonwealth Agricultural Bureaux.

Kok, L. T., and W. W. Surles. 1975. Successful biocontrol of musk thistle by an introduced weevil, *Rhinocyllus conicus. Environ. Entomol.* 4:1025–27.

Munz, P. A., and D. D. Keck. 1959. *A California flora.* Berkeley: University of California Press.

Robbins, W. W., M. K. Bellue, and W. S. Ball. 1970. *Weeds of California.* Sacramento: California State Department of Agriculture.

Turner, C. E., R. W. Pemberton, and S. S. Rosenthal. 1987. Host utilization of native *Cirsium* thistles by the introduced weevil, *Rhinocyllus conicus* (Coleoptera: Curculionidae). *Environ. Entomol.* 16:111–16.

Unruh, T. R., and R. D. Goeden. 1987. Electrophoresis helps to identify which race of the introduced weevil, *Rhinocyllus conicus* (Coleoptera: Curculionidae), has transferred to two native southern California thistles. *Environ. Entomol.* 16:979–83.

Zwölfer, H. 1965. *Preliminary list of phytophagous insects attacking wild Cynareae (Compositae) in Europe,* 81–154. Commonwealth Institute of Biological Control Technical Bulletin 6.

Zwölfer, H., and P. Harris. 1984. Biology and host specificity of *Rhinocyllus conicus* (Froel.) (Coleoptera: Curculionidae), a successful agent for biocontrol of the thistle, *Carduus nutans* L. *Zeit. angew. Entomol.* 97:36–62.

66 / MILK THISTLE

R. D. GOEDEN

| milk thistle |
| Silybum marianum (L.) Gaertner |
| Asteraceae |

INTRODUCTION

Biology and Pest Status

The milk thistle is a robust herbaceous winter annual or biennial with erect, simple, or multibranched stems 1 to 2 meters tall and a deep-seated taproot. Its leaves are spiny margined and spiny tipped, with characteristic white mottling. The basal leaves are 4 to 7 decimeters long and up to 2.5 decimeters wide; the cauline leaves are shorter and narrower. It reproduces only by windborne or otherwise easily disseminated pappose achenes (Munz and Keck 1959; Robbins, Bellue, and Ball 1970).

Historical Notes

This thistle is an alien weed of southwestern Mediterranean origin (Goeden 1976; Goeden and Ricker 1980). It is an occasional pest throughout much of the United States and adjacent parts of Canada, but is widely naturalized only in California and Oregon. In California the milk thistle was first recorded in 1854 (Robbins 1940); it has now spread throughout the state and infests rangeland, open woodland, fallow croplands, and wastelands such as roadsides, railroad rights-of-way, field margins, and ditch banks (Goeden 1971).

This weed shows an affinity for fertile soils near areas where livestock are concentrated—such as vacant corrals, waterholes, and shade sites—and seems to prefer moister areas than do some other thistles (Goeden 1971). For example, in Oregon it has become abundant along the floodplains of several rivers in the Willamette Valley (Coombs person. commun.).

Between 1966 and 1968 researchers conducted preintroduction surveys of phytophagous insects attacking this weed in the southern and central areas of coastal California (Goeden 1971). From 1971 to 1972 R. D. Goeden (1976) surveyed milk thistles for natural enemies in central and southern Italy and in Greece, sup-plementing the limited surveys done by H. Zwölfer (1965) in southern France and by USDA-ARS entomologists in Italy, as well as concurrent surveys conducted in Egypt, Lebanon, and Syria by Egyptian entomologists under contract to the USDA-ARS. Commonwealth Institute of Biological Control entomologists in Pakistan under contract to the University of California, Riverside, conducted subsequent surveys, and R. D. Goeden surveyed Spain in 1978 (Goeden 1976; Goeden and Ricker 1980).

RESULTS

Objective A: The Identification and Introduction of Beneficial Organisms

So far, the only natural enemy of milk thistle that has been introduced into western North America is the flower head–infesting weevil *Rhinocyllus conicus*. The biology of this weevil on musk thistle, *Carduus nutans*, has been outlined by L. A. Andres and N. E. Rees in chapter 67 of this volume and was described in detail by H. Zwölfer and P. Harris (1984). When feeding in the large capitula of milk thistle, *R. conicus* larvae prefer to mine the receptacles and destroy only a portion of the immature achenes, even at high larval densities (Goeden 1978). Unlike larvae of the biotype on musk thistle, even at high densities they do not mine milk thistle's peduncles or stems (Goeden 1978; Zwölfer and Harris 1984; Andres and Rees chap. 67).

In 1971 and 1972, at three and seven southern California sites, respectively, weevils collected from milk thistle near Rome, Italy, were released directly onto milk thistles where the thistles were abundant, public access was restricted, and a range of local climates was represented (Goeden and Ricker 1974, 1977). In 1971 a total of 1,141 weevils were imported and 1,084 released; in

Unpublished data provided by E. M. Coombs and the support and assistance of county collaborators is gratefully acknowledged.

1972, 3,263 weevils were imported and 3,200 released (Goeden and Ricker 1977). R. B. Hawkes, L. A. Andres, and P. H. Dunn (1972) reported that in 1972, after 12 *R. conicus* adults collected in 1969 from *C. nutans* in France failed to become established on *Silybum*, USDA-ARS entomologists released 1,145 weevils collected from southern Italy at a single site in northern California. S. W. T. Batra et al. (1981) reported that between 1972 and 1974, 2,062 weevils were released at five sites, 1,143 at four sites, and 370 at an apparently single site, respectively, in northern California, presumably in cooperation with the California Department of Food and Agriculture.

In 1978 Oregon Department of Agriculture personnel, in cooperation with the USDA-ARS laboratory at Albany, California, and the California Department of Food and Agriculture transferred *R. conicus* from California to Oregon. Records at Oregon's Department of Agriculture show that it became established after about 500 adults were released at each site. This approximated the number of adults previously released at most California sites (Hawkes, Andres, and Dunn 1972; Goeden and Ricker 1977; Batra et al. 1981; E. M. Coombs person. commun.).

Objective D: Impact

In California *R. conicus* that had originated on milk thistle became established at most sites where it was colonized on the same weed (Goeden and Ricker 1977). However, R. D. Goeden (1978) and R. D. Goeden, D. W. Ricker, and B. A. Hawkins (1985) demonstrated that weevils originating on Italian thistle did not readily transfer to milk thistle—partial proof of the existence of biotypes of *R. conicus*, which was subsequently confirmed in Europe by H. Zwölfer and M. Preiss (1983). These published data are apparently at variance with observations by E. M. Coombs (person. commun.) that "seedheads [of milk thistle at three sites in Douglas County, Oregon] that occurred in conjunction with Italian thistle infested with the Italian thistle biotype [of] *R. conicus* showed the highest levels of seed reduction."

R. D. Goeden and D. W. Ricker (1980) reported high mortality of larvae, pupae, and adults in capitula of milk thistle in southern California that was unascribable to adopted natural enemies or to high temperatures but rather was thought to involve imperfect host-plant adaptation. They suggested that milk thistle evolutionarily may be a recently acquired host of *R. conicus*, probably involving a host transfer from *Carduus* species in southern Europe or northeastern Africa.

R. D. Goeden (1978) reported that artificially induced infestations of milk thistle capitula in field cagings demonstrated that *R. conicus* destroyed only part of the achenes, even at mean densities of 82 eggs per capitulum and a maximum density of 134 eggs per

head. These results supported the collective opinion that *R. conicus* is unsuccessful as a biological control agent on milk thistle in California, even though the weevils are well established throughout the range of this weed and often infest more than 90 percent of seedheads (Julien 1987). This again contrasts with unpublished results from Oregon (E. M. Coombs person. commun.), where random samples at three locations in Douglas County indicated that 98 percent of the milk thistle capitula showed evidence of oviposition, and seed production was reduced by 50 to 80 percent.

RECOMMENDATIONS

1. Commonwealth Scientific and Industrial Research Organization research on asteraceous thistles of Mediterranean origins currently being conducted at Montpellier, France, should be followed to see whether natural enemies host-specific to milk thistle will be found in northern Africa, most likely in Morocco or Algeria. These new natural enemies, unlike *R. conicus*, should not transfer to native *Cirsium* thistles in California and Oregon (Goeden and Ricker 1986, 1987a,b; Turner, Pemberton, and Rosenthal 1987; Unruh and Goeden 1987).

2. In Oregon, the impact of *R. conicus* on milk thistle should be studied experimentally and the results described in this chapter should be verified.

REFERENCES CITED

Batra, S. W. T., J. R. Coulson, P. H. Dunn, and P. E. Boldt. 1981. Insects and fungi associated with *Carduus* thistles (Compositae). USDA Technical Bulletin 161.

Goeden, R. D. 1971. The phytophagous insect fauna of milk thistle in southern California. *J. Econ. Entomol.* 64:1101–4.

———. 1976. The Palearctic insect fauna of milk thistle, *Silybum marianum*, as a source of biological control agents for California. *Environ. Entomol.* 5:345–53.

———. 1978. *Initial analyses of* Rhinocyllus conicus (Froelich) (Coleoptera: Curculionidae) *as an introduced natural enemy of milk thistle* (Silybum marianum [L.] Gaertner) *and Italian thistle* (Carduus pycnocephalus L.) *in southern California,* 39–50. USDA Publication 1978-771-106/02.

Goeden, R. D., and D. W. Ricker. 1974. Imported seed weevils attack Italian and milk thistles in southern California. *Calif. Agric.* 28(1):8–9.

———. 1977. Establishment of *Rhinocyllus conicus* on milk thistle in southern California. *Weed Sci.* 25:288–92.

———. 1980. Mortality of *Rhinocyllus conicus* (Coleoptera: Curculionidae) in milk thistle flowerheads in southern California. *Protect. Ecol.* 2:47–56.

———. 1986. Phytophagous insect faunas of the two most common native *Cirsium* thistles, *C. californicum* and *C. proteanum,* in southern California. *Ann. Entomol. Soc. Am.* 79:953–62.

———. 1987a. Phytophagous insect faunas of the native thistles, *Cirsium brevistylum, C. congdonii, C. occidentale,* and *C. tioganum,* in southern California. *Ann. Entomol. Soc. Am.* 80:152–62.

———. 1987b. Phytophagous insect faunas of native *Cirsium* thistles, *C. mohavense, C. neomexicana,* and *C. nidulum,* in the Mojave Desert of southern California. *Ann. Entomol. Soc. Am.* 80:161–75.

Goeden, R. D., D. W. Ricker, and B. A. Hawkins. 1985. Ethological and genetic differences among three biotypes of *Rhinocyllus conicus* (Coleoptera: Curculionidae) introduced into North America for the biological control of asteraceous thistles. In *Proceedings of the 6th International Symposium on Biological Control of Weeds, 19–25 August 1984, Vancouver, British Columbia,* ed. E. S. Delfosse, 181–89. Ottawa: Agriculture Canada.

Hawkes, R. B., L. A. Andres, and P. H. Dunn. 1972. Seed weevil released to control milk thistle. *Calif. Agric.* 26(12):14.

Julien, M. H. 1987. *Biological control of weeds: A world catalogue of agents and their target weeds,* 2nd ed. Wallingford, U.K.: Commonwealth Agricultural Bureaux International.

Munz, P. A., and D. D. Keck. 1959. *A California flora.* Berkeley: University of California Press.

Robbins, W. W. 1940. *Alien plants growing without cultivation in California.* California Agricultural Experiment Station Bulletin 637.

Robbins, W. W., M. K. Bellue, and W. S. Ball. 1970. *Weeds of California.* Sacramento: California State Department of Agriculture.

Turner, C. E., R. W. Pemberton, and S. S. Rosenthal. 1987. Host utilization of native *Cirsium* thistles by the introduced weevil, *Rhinocyllus conicus* (Coleoptera: Curculionidae). *Environ. Entomol.* 16:111–16.

Unruh, T. R., and R. D. Goeden. 1987. Electrophoresis helps to identify which race of the introduced weevil, *Rhinocyllus conicus* (Coleoptera: Curculionidae), has transferred to two native southern California thistles. *Environ. Entomol.* 16:979–83.

Zwölfer, H. 1965. *Preliminary list of phytophagous insects attacking wild Cynareae (Compositae) in Europe,* 81–154. Commonwealth Institute of Biological Control Technical Bulletin 6.

Zwölfer, H., and P. Harris. 1984. Biology and host specificity of *Rhinocyllus conicus* (Froel.) (Coleoptera: Curculionidae), a successful agent for biocontrol of the thistle, *Carduus nutans* L. *Zeit. angew. Entomol.* 97:36–62.

Zwölfer, H., and M. Preiss. 1983. Host selection and oviposition behavior in West European ecotypes of *Rhinocyllus conicus* Froel. (Coleoptera: Curculionidae). *Zeit. angew. Entomol.* 95:113–22.

67 / MUSK THISTLE

L. A. ANDRES AND N. E. REES

> **musk thistle**
> *Carduus nutans* L.
> Asteraceae

INTRODUCTION

Biology and Pest Status

Musk thistle and nodding thistle are both common names for species of the large-headed *Carduus nutans* group, which includes *C. macrocephalus* Desfontaines, *C. nutans* Linneaus, and *C. nutans* sp. *leiophyllus* (Petrovic) (=*theormeri* Weinman). In the United States, this last thistle is the most abundant of the three (McCarty 1982). The musk thistle has become extremely troublesome in pasture, forest, range, and crop areas, as well as along roadsides (Hodgson and Rees 1976). In 1976 it was found in 12 percent of U.S. counties (Dunn 1976); by 1981, it had infested over 730,000 hectares in 40 states (Batra et al. 1981).

The musk thistle reproduces only by seed, with an average plant producing upward of 10,000 seeds (McCarty 1982). It behaves as a biennial or winter annual, with seeds germinating mainly in the fall or between spring and early summer. Seedlings become rosettes in their first year and blossom the following summer (McCarty 1982). The thistle's sharp spines and bracts deter foraging and also protect nearby species (Boldt 1978). J. T. Trumble and L. T. Kok (1982) have determined that a density of one plant per 1.48 square meters reduces forage production by 23 percent.

Historical Notes

A native of Eurasia and North Africa, the musk thistle was accidentally introduced into North America over a hundred years ago (Stuckey and Forsyth 1971). Because it is widespread, and because mowing and chemical control give variable results (McCarty and Hatting 1975), in the 1960s Agriculture Canada initiated a biological control project. In 1968 the receptacle-infesting weevil *Rhinocyllus conicus* Froelich was introduced into Canada (Harris and Zwölfer 1971). In 1969, after additional host-specificity tests with artichoke (*Cynara scolymus* L.) and safflower (*Carthamus tinctorius* L.) at the USDA-ARS Quarantine Laboratory in Albany, California, *R. conicus* was authorized for release in the United States.

Researchers at the Virginia Polytechnic Institute in Blacksburg cleared a second insect, the crown- and root-infesting weevil *Trichosirocalus horridus* (Panzer), from Italy (Ward, Pienkowski, and Kok 1974; Kok 1975). In 1989 a third insect, the stem-infesting syrphid fly *Cheilosia corydon* (Harris) (=*grossa* Fallen), was cleared by the USDA-ARS Laboratory in Rome, Italy.

RESULTS

Objective A: The Identification and Introduction of Beneficial Organisms

Rhinocyllus conicus (Coleoptera: Curculionidae). This European weevil oviposits on the exposed bracts of unopened and partly opened thistle heads and covers each egg with masticated plant material. The larvae enter the bracts directly and tunnel into the developing receptacle. Each larva excavates a cell in the area below the ovules and lines it with fecal and masticated plant materials. After pupating, the new adults remain in these cells for about 2 weeks before they exit through the upper receptacle surface (Zwölfer 1967; Rees 1982b). Adults overwinter in sheltered areas and then congregate on the thistle rosettes the following spring. They feed on the rosettes and on the leaves and stems of bolting plants. Each female produces 100 to 150 eggs. Although adults can cause conspicuous feeding damage, larval feeding is more harmful because it interrupts seed formation, destroys seeds, and reduces the viability of undamaged seeds.

We thank E. M. Coombs (Oregon Department of of Agriculture), J. L. Littlefield (University of Wyoming), G. L. Piper (Washington State University), and J. M. Story (Montana State University) for their contributions.

In 1969 researchers imported *R. conicus* adults from France and released them in Virginia (Surles, Kok, and Pienkowski 1974) and Montana (Hodgson and Rees 1976), where they became successfully established. In 1975 adults collected from Montana field colonies were redistributed to California, Idaho, Kansas, Louisiana, Minnesota, North Dakota, Oklahoma, Utah, Wisconsin, and Wyoming. Redistributions continued in many of those same states, as well as in Oregon and Washington, into the 1980s. The weevil is now established in California, Colorado, Idaho, Iowa, Kansas, Kentucky, Maryland, Minnesota, Missouri, Montana, Nebraska, North Dakota, Oregon, Pennsylvania, South Dakota, Tennessee, Utah, Washington, Wyoming, and perhaps other states (Julien 1987).

Trichosirocalus horridus (Coleoptera: Curculionidae).

The adults of this weevil oviposit into feeding cavities in the midribs of rosette leaves during fall and early spring. Larvae tunnel into the crown, where their feeding and cells weaken the plant and reduce the size of the flower bolts (Kok, Ward, and Grills 1975). Adults, eggs, or larvae can overwinter. Fully grown larvae pupate in the cells they have formed in the plant crown, or in the surrounding soil. Adults emerge in the spring to feed in the shelter of the thistle plants until they again begin ovipositing in the fall (Kok, Ward, and Grills 1975).

In 1977 researchers introduced 600 field-collected adults from Italy to Virginia (Kok and Trumble 1979) and imported another 50 adults each to Kansas, Montana, and Nebraska. However, they halted all direct releases of adults collected in Europe when a *Nosema* species infection was detected in several field-collected weevils. All subsequent releases were restricted to eggs or egg-bearing plants originating from females at the USDA Quarantine Laboratory in Albany, California, and ascertained to be *Nosema*-free. Between 1979 and 1981 researchers made multiple shipments of eggs or larvae to Idaho, Kansas, Maryland, Missouri, Montana, Nebraska, South Dakota, and Wyoming (P. H. Dunn person. commun.). In the 1980s adults collected from established sites in Virginia were redistributed to several states.

So far the establishment of this weevil has been confirmed only in Virginia (Kok and Trumble 1979), Kansas, Missouri (B. Puttler person. commun.), and Wyoming (J. L. Littlefield person. commun.), despite multiple releases in Montana and other states. In Wyoming, 10 out of the 16 releases made between 1980 and 1986 have resulted in establishment. Only one out of four colonizations attempted in Wyoming in 1980 and 1981—with *Nosema*-free larval-infested plants, or with clean eggs and larvae inoculated into plants—was successful (Littlefield person. commun.). Although populations developed slowly during the first 2 to 3 years at all the Wyoming release sites, in some areas up to 78 percent of the rosettes are now infested (Littlefield person. commun.).

Cheilosia corydon (Diptera: Syrphidae).

This phytophagous fly usually deposits its eggs singly on the hirsute leaves and young shoots at the plant centers in March and April (Rizza et al. 1988). Larvae begin feeding in the tender shoots, tunneling up and down the bolting stems. They complete the third and final instar in the crown and root, where they overwinter. Their stem tunneling interrupts water and nutrient transport, so that by the time larvae reach the root the plant has frequently died (Rizza et al. 1988). Field studies at the USDA-ARS Laboratory in Rome, Italy, have demonstrated that larvae limit their feeding to plants of the genus *Carduus* and to one of several *Cirsium* species tested (Rizza et al. 1988).

In Montana attempts to rear and release large numbers of flies from imported pupae failed when refrigerated adults emerged unexpectedly in February, before field plants were available.

Objective D: Impact

Rhinocyllus conicus. The weevil lays its eggs in, and causes most seed damage to, the early-developing terminal heads on the main and secondary stems and on the upper secondary and tertiary heads. Buds that form after the peak oviposition as well as those lower on the stem receive proportionately fewer eggs, so suffer less damage (Wyoming: Littlefield person. commun.; Oregon: E. M. Coombs person. commun.; California: D. B. Joley and L. A. Andres unpub. data). Any environmental factor that intensifies the weevil-plant asynchrony further limits the weevils' impact on seed production.

In Wyoming weevils reduce seed production by an estimated 60 percent in terminal heads, 54 percent in lateral heads, and 24 percent in auxiliary heads, compared with the number of seeds produced at weevil-free sites (Littlefield person. commun.). Studies in California have suggested that the number of viable seeds was reduced by 48 for each larval chamber formed in the receptacle (Joley and Andres unpub. data). In Montana, N. E. Rees (1977) reported that 69 percent of the seed from weevil-free heads was viable the first year, whereas apparently undamaged seeds from heads with four to five larvae were only 45 percent viable, and those from heads with more than nine larvae were less than 2 percent viable. Rees also reported that when egg densities were high, females oviposited on the peduncles and stems of the thistles and began to oviposit on other thistle species. In California gravid females oviposit on the bolting stem meristems, from the time that the weevils cluster on the rosette until the terminal and lateral heads form. This early oviposition results in a scattering of eggs along the thistle stems, especially at the leaf axils. Many of the larvae hatching from these eggs tunnel into the stems but fail to mature (Andres unpub. observation).

Although in Virginia and Montana *R. conicus* provides appreciable control of the musk thistle (Kok and Surles 1975; Hodgson and Rees 1976; Rees 1982b), at some sites in Washington it has only reduced musk thistle density by 40 to 50 percent (G. L. Piper unpub. data). Based on preliminary observations in Idaho, J. P. McCaffrey (person. commun.) notes that the weevil has had no significant impact on the weed, except at one site where ample rainfall allowed grass to compete strongly with the thistle. Although there is a downward trend in the density of musk thistle at the California study site—from 68.7 plants per square meter in 1980 to 5.7 plants in 1985—large annual fluctuations in plant density make it difficult to predict whether the population will decline in any given year (Joley and Andres unpub. data). Studies on plant density are continuing in Oregon and Wyoming.

In Wyoming researchers have found eight species of parasitic Hymenoptera attacking *R. conicus* larvae and pupae. Parasitism rates ranged from incidental to 7 percent for each parasite species. The combined parasitism of weevils infesting thistle peduncles reached 17 percent, compared with 0.6 percent in flower heads. Overall parasitism ranged from 0 to 1.3 percent among release sites, and it did not increase with time (Littlefield person. commun.).

In California 7 out of 13 species of Hymenoptera that emerged from musk thistle heads were believed to be primary parasites of *R. conicus*. Overall parasitism averaged 1.78 percent, but parasitism in stems was 18.9 percent (Wilson and Andres 1986). In Montana N. E. Rees (1982a) reported two species of hymenopterous parasites but noted that although predaceous insects and spiders did attack larvae exposed when plants were accidentally damaged, their effect on the weevil population was insignificant.

The mortality of *R. conicus* did not increase appreciably when weevil-infested plants were sprayed with the herbicide 2,4-D (Miller 1978; Lee and Evans 1980), which prompted studies on integrated thistle control (Trumble and Kok 1982). Even when five insecticides registered for grasshopper control were sprayed on infested seedheads, weevil larvae inside the heads were not significantly affected (Rees and Onsager 1983).

In California *R. conicus* has been reared from 12 nontarget native *Cirsium* species (Turner, Pemberton, and Rosenthal 1987). J. Littlefield (person. commun.) reports that in Wyoming, nine alternative host plants of *R. conicus* have also been identified. Species with larger (greater

than 12-mm-diameter) flowers were infested to a greater extent (between 62 and 88 percent, with a mean number of larval cells of 1.6 to 5.6 per flower) than were species with smaller (less than 11-mm-diameter) flowers (between 3 and 19 percent, with 0.03 to 0.15 larval cells per flower) (Littlefield person. commun.). The weevil was also reported to attack *Carduus acanthoides* L. in Washington (Piper person. commun.) and Idaho (McCaffrey person. commun.), and to attack *Cirsium undulatum* (Nuttall) Sprengel, *C. arvense* L. (Scopoli), and *C. vulgare* (Savi) Tenore in Montana (Rees 1978).

Trichosirocalus horridus. Although in Wyoming this species begins ovipositing in the fall, it lays the majority of its eggs from late March through May. About 91 percent of the eggs are viable. Larvae feed from late September to late June, at which time fully grown larvae abandon the rosettes and move to the soil to pupate. Adult emergence peaks during late June or early July and is followed by several weeks of rosette feeding and then estivation until mid-September. Researchers observed averages of up to 10.5 larvae per plant, though this varied seasonally (Littlefield person. commun.).

In addition to attacking musk thistle, adult weevils feed on *Cirsium arvense, C. vulgare,* and *Onopordum acanthium* L.; in Wyoming the roots of *C. arvense* and *C. vulgare* were also infested with larvae (Littlefield person. commun.).

RECOMMENDATIONS

1. Continue to evaluate and redistribute *R. conicus* and *T. horridus* against musk thistle and to assess their impact on nontarget plants.

2. Further assess the outcome of 1980 to 1981 efforts to establish *T. horridus* through the release of pathogen-free larvae and eggs, in order to ascertain the practicality of this technique.

3. Continue introductions of *Cheilosia corydon.*

4. Assess the host specificities of *Puccinia carduorum* Jacky (Politis, Watson, and Bruckart 1984), a plant pathogen, and of *Psylloides chalcomera* (Illiger), a beetle, and import them if warranted.

REFERENCES CITED

Batra, S. W. T., J. R. Couldon, P. H. Dunn, and P. E. Boldt. 1981. *Insects and fungi associated with* Carduus *thistles (Compositae).* USDA Technical Bulletin 1616.

Boldt, P. E. 1978. Habitat of *Carduus nutans* L. in Italy and two phytophagous insects. In *Proceedings of the 4th International Symposium on Biological Control of Weeds, 30 August–2*

September 1976, Gainesville, Florida, ed. T. E. Freeman, 98–100. Gainesville: University of Florida, Institute of Food and Agricultural Sciences.

Dunn, P. H. 1976. Distribution of *Carduus nutans, C. acanthoides, C. pycnocephalus,* and *C. crispus,* in the United States. *Weed Sci.* 24:518–24.

Harris, P., and H. Zwölfer. 1971. *Carduus acanthoides* L., welted thistle and *C. nutans* L., nodding thistle (Compositae). In *Biological control programmes against insects and weeds in Canada 1959–68,* 76–79. Commonwealth Institute of Biological Control Technical Communication 4.

Hodgson, J. M., and N. E. Rees. 1976. Dispersal of *Rhinocyllus conicus* for biocontrol of musk thistle. *Weed Sci.* 24:59–62.

Julien, M. H. 1987. *Biological control of weeds: A world catalogue of agents and their target weeds,* 2nd ed. Wallingford, U.K.: Commonwealth Agricultural Bureaux International.

Kok, L. T. 1975. Host-specificity studies on *Ceuthorhynchidius horridus* (Coleoptera: Curculionidae) for the biocontrol of musk and plumeless thistles. *Weed Res.* 15:21–25.

Kok, L. T., and W. W. Surles. 1975. Successful biocontrol of musk thistle by an introduced weevil, *Rhinocyllus conicus. Environ. Entomol.* 4:1025–27.

Kok, L. T., and J. T. Trumble. 1979. Establishment of *Ceuthorhynchidius horridus* (Coleoptera: Curculionidae), an important thistle-feeding weevil in Virginia. *Environ. Entomol.* 8:221–23.

Kok, L. T., R. H. Ward, and C. C. Grills. 1975. Biological studies of *Ceuthorhynchidius horridus* (Panzer), an introduced weevil for thistle control. *Ann. Entomol. Soc. Am.* 68:503–5.

Lee, R. D., and J. O. Evans. 1980. The influence of selected herbicides on the development of *Rhinocyllus conicus,* an insect used in biocontrol of musk thistle. *Proc. West. Soc. Weed Sci.* 33:104–10.

McCarty, M. K. 1982. Musk thistle (*Carduus thoermeri*) seed production. *Weed Sci.* 30:441–45.

McCarty, M. K., and J. L. Hatting. 1975. Effects of herbicides or mowing on musk thistle seed production. *Weed Res.* 15:363–67.

Miller, T. J. 1978. The effects of mowing and of spraying musk thistle on the biological control agent *Rhinocyllus conicus.* Master's thesis, Department of Agronomy, Montana State University, Bozeman.

Politis, D. J., A. K. Watson, and W. L. Bruckart. 1984. Susceptibility of musk thistle and related Compositae to *Puccinia carduorm. Phytopathology* 74:687–91.

Rees, N. E. 1977. Impact of *Rhinocyllus conicus* on thistles in southwestern Montana. *Environ. Entomol.* 6:839–42.

———. 1978. Interactions of *Rhinocyllus conicus* and thistles in the Gallatin Valley. In *Biological control of thistles in the genus* Carduus *in the United States: A progress report,* ed. K. E. Frick, 31–38. Washington, D.C.: USDA Science and Education Administration, Agricultural Research Service.

———. 1982a. Enemies of *Rhinocyllus conicus* in southwestern Montana. *Environ. Entomol.* 11:157–58.

———. 1982b. *Collecting, handling, and releasing* Rhinocyllus conicus, *a biological control agent of musk thistle.* USDA-ARS Handbook 579. Washington, D.C.: U.S. Department of Agriculture.

Rees, N. E., and J. A. Onsager. 1983. Effects of selected insecticides on *Rhinocyllus conicus* (Coleoptera: Curculionidae) in musk thistle. *J. Econ. Entomol.* 76:1414–16.

Rizza, A., C. Campobasso, P. H. Dunn, and M. Stazzi. 1988. *Cheilosia corydon* (Diptera: Syrphidae), a candidate for the biological control of musk thistle in North America. *Ann. Entomol. Soc. Am.* 81:225–32.

Stuckey, R. L., and J. L. Forsyth. 1971. Distribution of naturalized *Carduus nutans* (Compositae) mapped in relation to geology in northeastern Ohio. *Ohio J. Sci.* 71:1–15.

Surles, W. W., L. T. Kok, and R. L. Pienkowski. 1974. *Rhinocyllus conicus* establishment for biocontrol of thistles in Virginia. *Weed Sci.* 22:1–3.

Trumble, J. T., and L. T. Kok. 1982. Integrated pest management techniques in thistle suppression in pastures in North America. *Weed Res.* 22:345–59.

Turner, C. E., R. W. Pemberton, and S. S. Rosenthal. 1987. Host utilization of native *Cirsium* thistles (Asteraceae) by the introduced weevil, *Rhinocyllus conicus* (Coleoptera: Curculionidae) in California. *Environ. Entomol.* 16:111–15.

Ward, R. J., R. L. Pienkowski, and L. T. Kok. 1974. Host specificity of first-instar *Ceuthorhynchidius horridus,* a weevil for biological control of thistles. *J. Econ. Entomol.* 67:735–37.

Wilson, R. C., and L. A. Andres. 1986. Larval and pupal parasites of *Rhinocyllus conicus* (Coleoptera: Curculionidae) in *Carduus nutans* in northern California. *Pan-Pacific Entomol.* 62:329–32.

Zwölfer, H. 1967. *The host-range, distribution, and life history of* Rhinocyllus conicus Froel. Delémont, Switzerland: Commonwealth Institute of Biological Control Progress Report 18.

68 / RUSH SKELETONWEED

G. L. PIPER AND L. A. ANDRES

rush skeletonweed
Chondrilla juncea L.
Asteraceae

INTRODUCTION

Biology and Pest Status

Rush skeletonweed is a deeply rooted perennial herb indigenous to the Mediterranean area and to central Asia (McVean 1966). It has been inadvertently introduced into Argentina, Australia, and the United States (Schirman and Robocker 1967; Cullen and Groves 1977; Tortosa and Medan 1977). In the western United States, it has become a major pest of rangeland, semiarid pasture, transportation rights-of-way, and cropland, where the wiry, latex-exuding stems of mature plants hinder the operation of harvest machinery (Cuthbertson 1967).

Various researchers have thoroughly chronicled the biology and ecology of *C. juncea* (McVean 1966; Old 1981; Panetta and Dodd 1987). Its seasonal cycle begins in the fall or early spring, when rains cause seeds to germinate or rosettes to regenerate from established rootstocks. Floral stems develop in late spring, and flower heads begin to form in midsummer and continue until fall frosts. The seeds are produced apomictically (Cuthbertson 1974). A mature plant produces between 15,000 and 20,000 seeds, depending on the plant biotype, its size, and environmental conditions (Piper 1983). The pappus-bearing seeds are readily disseminated by wind, water, animals, and people. The weed can also reproduce vegetatively from injured or fragmented roots (Panetta and Dodd 1987). In the western United States, researchers have identified early- and late-flowering rush skeletonweed biotypes (Schirman and Robocker 1967; Lee 1986).

Historical Notes

Rush skeletonweed was probably first introduced into the eastern United States during the 1870s. From New York to Virginia it is sparsely distributed, and it is not viewed as an important weed in the eastern seaboard states. Western infestations began with contaminated orchard and vineyard rootstocks probably introduced around 1900 from the Mediterranean. In 1938 the plant was first reported in Washington state; then it was found in Idaho in 1960, California in 1965, and Oregon in 1971 (Schirman and Robocker 1967; Coleman-Harrell 1978). In recent surveys researchers have found that the weed infests approximately 1.4 million hectares in southwestern and northwestern Idaho; 809,000 hectares in eastern Washington; 73,000 hectares in western Oregon; and 408,000 hectares in northern California (Lee 1986). A ban on using herbicides to control it on lands owned by the U.S. Bureau of Land Management and U.S. Forest Service in the Pacific Northwest has increased the demand for alternative controls.

During the 1960s Australian scientists in the Commonwealth Scientific and Industrial Research Organization (CSIRO) developed the first effective biological control program against rush skeletonweed. They discovered various natural enemies of the weed in its home range, and successfully introduced several of the most injurious species to Australia. These included the rust fungus *Puccinia chondrillina* Bubak and Sydow; the gall midge *Cystiphora schmidti* Rübsaamen; and the gall mite *Eriophyes chondrillae* (G. Canestrini). The combined effect of these organisms has reduced the incidence of several *C. juncea* biotypes in Australia and has resulted in gains in agricultural production (Cullen 1978). At the request of the California Department of Food and Agriculture, both the USDA-ARS Biological Control of Weeds Laboratory in Albany, California, and the USDA-ARS Plant Disease Research Laboratory in Frederick, Maryland, obtained biotypes of these natural enemies from the CSIRO for testing and release in California. Entomologists, plant pathologists, and weed scientists from Idaho, Oregon, and Washington—with

We thank E. M. Coombs and J. P. McCaffrey for their contributions to this report. We also acknowledge the many state and federal agency personnel and private landowners who have assisted with redistributing the biological control agents in California and the Pacific Northwest.

financial support provided by the Pacific Northwest Regional and Idaho Wheat commissions—joined with California's Department of Food and Agriculture and the USDA to develop an integrated *Chondrilla* management program that relied heavily on the use of the weed's natural enemies. G. A. Lee (1986) has summarized the program's accomplishments.

RESULTS

Objective A: The Identification and Introduction of Beneficial Organisms

After Australian pathotypes of *P. chondrillina* failed to infect North American plants, researchers collected new pathotypes in Europe and screened them for infectivity against biotypes of the plant found in the United States (Emge, Melching, and Kingsolver 1981). Of two pathotypes collected near Eboli, Italy, pathotype PC-1 was highly infective against the southern Idaho biotype, and pathotype PC-16 against the California, Oregon, and southeastern Washington biotypes. The pathogen was first released in California in 1976, in Idaho and Oregon in 1977 (Wallace 1979), and in Washington in 1978 (Adams and Line 1984a). Uredospores were applied by various methods to inoculate rush skeletonweed (Lee 1986). The rust fungus became established in all states, but an early-flowering biotype in northern Idaho and in one county in northeastern Washington remained immune. In California and the Pacific Northwest, the fungus spread rapidly and unassisted from the initial release sites to adjacent infestations of the weed (Adams and Line 1984b; Supkoff, Joley, and Marois 1988). Researchers also redistributed the pathogen extensively in California, Oregon, and Washington. *Puccinia chondrillina* is the first exotic plant pathogen to be successfully employed for the biological control of a weed in North America (Emge, Melching, and Kingsolver 1981).

Researchers have studied the life cycle and epidemiology of *P. chondrillina* in Europe and North America (Hasan 1972; Blanchette and Lee 1981; Adams and Line 1984a). A macrocyclic, autoecious fungus, it remains active throughout the year and may infect all aboveground plant organs. In the fall, eruptive uredia appear on the leaves of seedlings and rosettes that are regenerating from root buds. Fungal growth is slowed during winter but resumes in spring, infecting flower shoots when they bolt. During the summer, several generations of uredospores are produced on stems and flower buds (Blanchette and Lee 1981; Adams and Line 1984a). Telia produced at the base of the flowering shoots yield teliospores that remain dormant until March of the following year, when they germinate to produce basidiospores. Pycnia then produce aecia. Aeciospores germinate and produce uredospores in uredia to contin-

ue the life cycle. In Washington the *P. chondrillina* pathotype PC-16 completes its life cycle by producing all five spore stages (Adams and Line 1984a). Only uredia have been observed in Idaho, where pathotype PC-1 has been used. Thus, rust fungus pathotypes differ not only in the weed biotypes that they infect, but also in their life cycles on the various biotypes.

Cystiphora schmidti was first released in California in 1975 (Supkoff, Joley, and Marois 1988), in Idaho and Washington in 1976 (Littlefield 1980; Piper 1985), and in Oregon in 1977 (R. B. Hawkes person. commun.). The stock originated in Greece via Australia (Sobhian and Andres 1978). The midge readily established itself on all *C. juncea* biotypes and during the last 15 years has been widely redistributed.

L. A. Caresche and A. J. Wapshere (1975), J. L. Littlefield (1980), and R. S. Mendes (1982) have provided detailed accounts of the biology and ecology of *C. schmidti* in Europe and the Pacific Northwest. The insect is active from April until late October, completing four or five generations, and has a life cycle—from egg to adult—lasting 24 to 44 days. Females insert eggs beneath the epidermis of leaves and stems. Anthocyanescent, blisterlike epidermal galls form in 10 to 12 days around larvae feeding on leaf mesophyll or stem parenchyma. The larvae normally pupate within the galls; mature larvae, prepupae, or pupae overwinter within galls on stems or rosettes or as pupae in the soil.

Because the *Eriophyes chondrillae* strain from Greece (via Australia) was unable to form galls on North American plants, researchers imported a new strain from near Vieste, Italy. The first field inoculations were made in California, Idaho, and Oregon in 1977 and in Washington in 1979 (Andres l982). The gall mite readily became established and is now widespread, thanks to its natural dispersal and to extensive redistribution efforts during the last decade. The mite attacks all biotypes, but the damage it does varies with biotype, time of attack, and plant age (Piper unpub. data).

L. A. Caresche and A. J. Wapshere (1974) have investigated the biology of *E. chondrillae*. Adult females overwinter in rosettes without inducing galls or reproducing. Mites invade the axillary and terminal buds of the shoot when the plant bolts in the spring. The mite's feeding causes leafy, hyperplastic galls to form. Under optimal conditions the mite can complete a generation in just 10 days. The insect increases and spreads on a host until the floral shoot growth stops in the fall, when deutonymphs descend to the plant crown to overwinter (Caresche and Wapshere 1974).

Objective D: Impact

When *P. chondrillina* infects rosettes in the fall and spring it often kills the plants, especially seedlings (Lee

1986). It not only greatly inhibits growth rates, floral-stem and bud development, and the weed's regeneration from root buds, but it also reduces seed crops and seed viability (Emge, Melching, and Kingsolver 1981; Adams and Line 1984b). In California the rust fungus is regarded as the most damaging of the triad of introduced natural enemies (Supkoff, Joley, and Marois 1988). In Washington the pathogen has also been effective in reducing the density of the late-flowering biotype, especially on moderately moist sites. The impact of *P. chondrillina* in Idaho and Oregon has not yet been adequately assessed (Idaho: J. P. McCaffrey person. commun.; Oregon: E. M. Coombs person. commun.).

Cystiphora schmidti galls damage or destroy leaf and stem tissues and cause desiccation, premature chlorosis, and seedling death (Lee 1986). Dense aggregations of galls on leaves and stems reduce photosynthesis and weed reproduction. R. S. Mendes (1982) demonstrated that in Washington, flower-head production was reduced by 60 percent. He also found that a decrease in seed weight and viability, as well as an inhibition of branching, was positively correlated with gall formation. Plants that the midge had successively infested for 2 or more years also experienced a 50 percent reduction in stem length (Littlefield 1980). Unfortunately, the impact of *C. schmidti* has been reduced by parasitism from several indigenous species of Hymenoptera (Littlefield 1980; Wehling and Piper 1988; B. Villegas person. commun.), and by earwigs (Coleman-Harrell 1978) and grasshoppers (Littlefield and Barr 1981) preying on immature stages in stem galls. Also, too much moisture in the late winter and early spring can kill pupae overwintering in the soil. Despite these limiting factors, the gall midge is an important natural enemy of rush skeletonweed, especially in Washington (Wehling and Piper 1988).

When *E. chondrillae* infest *C. juncea* buds, they cause a decrease in vegetative shoot production (Spollen 1986), reduce plant biomass and vigor by depleting carbohydrate reserves in the roots (Dimock 1982), diminish rosette regeneration from root buds (Spollen 1986), and decrease or prevent seed production (Caresche and Wapshere 1974; Piper 1985). Heavily galled seedlings and first-year satellite plants show high mortality (Spollen 1986). In California the mite has not had a major influence on *Chondrilla* density, in part because of predation by the indigenous mite, *Typhlodromus pyri* Scheuten (Andres 1982); in Washington, on the other hand, it is regarded as the best natural enemy to have become established to date (Piper 1985).

RECOMMENDATIONS

1. More midges, mites, and rust fungus should be redistributed, particularly in Idaho and Oregon.

2. Joint efforts of the CSIRO and the USDA to acquire and screen more virulent *P. chondrillina* pathotypes in Turkey for use against all rush skeletonweed biotypes found in the western United States should receive continued support.

3. Attention should be given to using root-feeding insects, which could further reduce the weed's vegetative reproductive capacity.

4. The integration of biological, chemical, cultural, and mechanical control practices should receive increased research attention.

REFERENCES CITED

Adams, E. B., and R. F. Line. 1984a. Biology of *Puccinia chondrillina* in Washington. *Phytopathology* 74:742-45.

———. 1984b. Epidemiology and host morphology in the parasitism of rush skeletonweed by *Puccinia chondrillina*. *Phytopathology* 74:745-48.

Andres, L. A. 1982. Considerations in the use of phytophagous mites for the biological control of weeds. In *Biological control of pests by mites*, eds. M. A. Hoy, G. L. Cunningham, and L. Knutson, 53-56. Berkeley: University of California Division of Agriculture and Natural Resources, Special Publication 3304.

Blanchette, B. L., and G. A. Lee. 1981. The influence of environmental factors on infection of rush skeletonweed (*Chondrilla juncea*) by *Puccinia chondrillina*. *Weed Sci.* 29:364-67.

Caresche, L. A., and A. J. Wapshere. 1974. Biology and host specificity of the *Chondrilla* gall mite, *Aceria chondrillae* (G. Can.) (Acarina: Eriophyidae). *Bull. Entomol. Res.* 64:183-92.

———. 1975. The *Chondrilla* gall midge *Cystiphora schmidti* (Rübsaamen) (Diptera: Cecidomyiidae): II. Biology and host specificity. *Bull. Entomol. Res.* 65:55-64.

Coleman-Harrell, M. E. 1978. University of Idaho *Tri-State Skeleton Weed Consortium Newsletter* 1(3):1-5.

Cullen, J. M. 1978. Evaluating the success of the programme for the biological control of *Chondrilla juncea* L. In *Proceedings of the 4th International Symposium on Biological Control of Weeds, 30 August-2 September 1976, Gainesville, Florida*, ed. T. E. Freeman, 117-21. Gainesville: University of Florida, Institute of Food and Agricultural Sciences.

Cullen, J. M., and R. H. Groves. 1977. The population biology of *Chondrilla juncea* L. in south-eastern Australia. *J. Ecol.* 54:345-65.

Cuthbertson, E. G. 1967. *Skeleton weed: Distribution and control.* New South Wales Department of Agriculture Bulletin 68.

———. 1974. Seed development in *Chondrilla juncea* L. *Aust. J. Bot.* 22:13-18.

Dimock, W. J. 1982. Nonstructural carbohydrate analysis of the roots of rush skeletonweed, *Chondrilla juncea* L., affected by the gall mite *Aceria chondrillae* G. Can. (Acarina: Eriophyidae) and picloram. Master's thesis, University of Idaho, Moscow.

Emge, R. G., J. S. Melching, and C. H. Kingsolver. 1981. Epidemiology of *Puccinia chondrillina*, a rust pathogen for the biological control of rush skeleton weed in the United States. *Phytopathology* 71:839-43.

Hasan, S. 1972. Specificity and host specialization of *Puccinia chondrillina*. *Ann. Appl. Biol.* 72:257-63.

Lee, G. A. 1986. Integrated control of rush skeletonweed (*Chondrilla juncea*) in the western U.S. *Weed Sci.* 34 (Suppl. 1):2-6.

Littlefield, J. L. 1980. Bionomics of *Cystiphora schmidti* (Rübsaamen) (Diptera: Cecidomyiidae), an introduced biological control agent of rush skeletonweed, *Chondrilla juncea* L., in Idaho. Master's thesis, University of Idaho, Moscow.

Littlefield, J. L., and W. F. Barr. 1981. Impact of grasshoppers on the rush skeletonweed gall midge in southwestern Idaho. In *Proceedings of the 5th International Symposium on Biological Control of Weeds*, ed. E. S. Delfosse, 595-97. Melbourne: Commonwealth Scientific and Industrial Research Organization.

McVean, D. N. 1966. Ecology of *Chondrilla juncea* L. in south-eastern Australia. *J. Ecol.* 54:345-65.

Mendes, R. S. 1982. Effectiveness of *Cystiphora schmidti* (Rübsaamen) (Diptera: Cecidomyiidae) as a biological control agent of rush skeletonweed, *Chondrilla juncea* L., in eastern Washington. Master's thesis, Washington State University, Pullman.

Old, R. R. 1981. Rush skeletonweed (*Chondrilla juncea* L.): Its biology, ecology, and agronomic history. Master's thesis, Washington State University, Pullman.

Panetta, F. D., and J. Dodd. 1987. The biology of Australian weeds: 16. *Chondrilla juncea* L. *J. Aust. Inst. Agric. Sci.* 53:83-95.

Piper, G. L. 1983. Rush skeletonweed. *Weeds Today* 14:5-7.

———. 1985. Biological control of weeds in Washington: Status report. In *Proceedings of the 6th International Symposium on Biological Control of Weeds, 19-25 August 1984, Vancouver, British Columbia*, ed. E. S. Delfosse, 817-26. Ottawa: Agriculture Canada.

Schirman, R., and W. C. Robocker. 1967. Rush skeletonweed: Threat to dryland agriculture. *Weeds* 15:310-12.

Sobhian, R., and L. A. Andres. 1978. The response of the skeletonweed gall midge, *Cystiphora schmidti* (Diptera: Cecidomyiidae), and gall mite, *Aceria chondrillae* (Eriophyidae), to North American strains of rush skeletonweed (*Chondrilla juncea*). *Environ. Entomol.* 7:506-8.

Spollen, K. M. 1986. Effectiveness of *Eriophyes chondrillae* (G. Canestrini) (Acari: Eriophyidae) as a biological control agent of rush skeletonweed, *Chondrilla juncea* L. (Compositae: Cichoriaceae), seedlings in eastern Washington. Master's thesis, Washington State University, Pullman.

Supkoff, D. M., D. B. Joley, and J. J. Marois. 1988. Effect of introduced biological control organisms on the density of *Chondrilla juncea* in California. *J. Appl. Ecol.* 25:1089-95.

Tortosa, R. D., and D. Medan. 1977. *Chondrilla* L. (Compositae), nuevo género para la Argentina. *Darwiniana* 21:115-19.

Wallace, K. E. 1979. *Chondrilla juncea* in the Pacific Northwest of U.S.A. In *Proceedings of the 7th Asian-Pacific Weed Science Society Conference, Sydney, Australia*, 193-94.

Wehling, W. F., and G. L. Piper. 1988. Efficacy diminution of the rush skeletonweed gall midge, *Cystiphora schmidti* (Diptera: Cecidomyiidae), by an indigenous parasitoid. *Pan-Pacific Entomol.* 64:83-85.

69 / RUSSIAN KNAPWEED

S. S. ROSENTHAL AND G. L. PIPER

Russian knapweed
Centaurea (Acroptilon) repens L.
Asteraceae

INTRODUCTION

Biology and Pest Status

Russian knapweed, *Centaurea repens* L., is a herbaceous perennial weed 0.3 to 0.9 meter tall, with creeping horizontal roots (Robbins, Bellue, and Ball 1951). It reproduces vegetatively from its extensive root system, or by seed. The bitter, unpalatable foliage is somewhat hairy and a grayish blue-green; stem leaves are thinner and smoother than the rosette leaves, and have entirely or slightly toothed margins. The blue, pink, or white florets are borne on heads that are 1.2 centimeters in diameter and subtended by bracts at the stem apices. Outer bracts have thin, papery margins and tips. The weed has an early deciduous white pappus. The seeds can be accidentally spread in contaminated alfalfa or sugarbeet seed, or in infested hay (Maddox, Mayfield, and Poritz 1985).

Historical Notes

Russian knapweed is thought to have originated in Central Asia but now also grows in southeastern Europe, South Africa, North and South America, and Australia (Maddox, Mayfield, and Poritz 1985). First introduced into North America in the early 1900s as a contaminant of Turkestani alfalfa seed (Groh 1940), it is now found in 21 states, particularly in the arid western region (Maddox, Mayfield, and Poritz 1985). The weed is an aggressive pest of grazing land, grainfields, wastelands, and irrigation ditches (Reed and Hughes 1970) and commonly infests such crops as corn, alfalfa, sugarbeets, and forage-seed crops. In what was then the Soviet Union, Russian knapweed reduced wheat and corn yields by up to 75 and 88 percent, respectively (Watson 1980). Livestock typically avoid the plant because of its bitter taste. When eaten fresh or as hay it is toxic to both horses (Young, Brown, and Klinger 1970) and sheep (Everist 1981).

Researchers have done extensive studies on organisms for the biological control of Russian knapweed. A.

K. Watson (1980) has listed 27 arthropods, fungi, and nematodes that are associated with the weed in these countries, of which the nematode *Subanguina* (*Paranguina*) *picridis* (Kirjanova) Brzeski is considered the most promising biological control candidate.

RESULTS

Objective A: The Identification and Introduction of Beneficial Organisms

Researchers in the former Soviet Union (Kirjanova and Ivanova 1969) and in Canada (Watson 1986a) have studied the biology of *S. picridis*. Larvae overwinter in the remains of the galls, on or just below the soil surface. In early spring, as the weed's shoots begin to emerge from the soil, second-instar larvae, activated by soil moisture, leave the disintegrating galls and penetrate the rudimentary leaves and stalks of emerging shoots. Within the hollow galls that form at the infected sites, the nematodes produce two generations during the growing season. By August, infective second-instar larvae predominate in the mature galls, which dry up and fall to the soil as the aboveground parts of the plant become senescent. The larvae are revived by soil moisture and become infective when conditions are cool and moist—especially when nutrients for the plants are at low levels.

In host-specificity tests in Canada (Watson 1986a,b), under optimal conditions of 15° ± 1°C during a 12-hour day and 10° ± 1°C at night, *C. repens* was highly susceptible to the nematode. *Centaurea diffusa* Lam., diffuse knapweed, was moderately susceptible, whereas other *Centaurea* species and related genera proved moderately resistant. In the former Soviet Union, *S. picridis* is sprayed in water suspensions on Russian knapweed infestations (Kovalev, Danilov, and Ivanova 1973).

After the nematode was released and became established at several sites in Canada (Watson and Harris 1984), in 1983 researchers were granted permission to make limited releases of *S. picridis* in north-central

Washington state. The nematode was imported from Canada through the USDA-ARS Biological Control of Weeds Quarantine Laboratory in Albany, California, and released by G. L. Piper about 0.8 kilometers northwest of Evans Lake, Conconully (Okanogan County), in June 1984. However, no *S. picridis* galls were found at that site during visits later that year, nor during visits in 1989.

Before the USDA Animal and Plant Health Inspection Service (APHIS) would permit the nematode to be further distributed, it requested additional testing at temperatures warmer than those used in the Canadian studies, as well as the testing of major crops commonly grown under warmer environmental conditions. This research was completed in 1987 at the USDA laboratory in Albany, California (Rosenthal 1989).

During the first month of this test, average temperatures ranged from 17° to 25°C, which was well above the planned 12.5°C that is considered optimal for gall development and representative of winter months in the warmer areas infested by Russian knapweed. For the next 3 months, daily temperatures ranged from 21° to 28°C—about the 25°C mean during warmer months in such areas. At these temperatures *S. picridis* did not exhibit a wider host range and only attacked *Centaurea* species, including the native North American *C. rothrockii* Greenman and the artichoke (*Cynara scolymus* L.), on which it can form small leaf galls but not stem galls.

The nematode has now been cleared for wider—but still only experimental—release, to determine whether artichokes or the two native North American *Centaurea* species, *C. rothrockii* and *C. americana* Nuttall, will be damaged by it in nature. In 1990 researchers released the nematode on experimental plantings of those species in Washington, Oregon, and Montana.

Objective C: The Conservation and Augmentation of Beneficial Organisms

Because *S. picridis* larvae are known to travel only up to 7 centimeters on their own (though they may be carried greater distances by running water), their most practical use may be in augmentative biological control. O. V. Kovalev, L. G. Danilov, and T. S. Ivanova (1973) have developed such a method to distribute the nematode more effectively in the former Soviet Union.

RECOMMENDATIONS

1. The field studies requested by the Technical Advisory Committee of USDA-APHIS—including experimental releases on Russian knapweed, artichoke, *C. rothrockii*, and *C. americana*—should be completed as soon as possible.

2. Research is needed on other potential biological control agents found in the former Soviet Union, especially the seedhead gall mite *Aceria acroptiloni* V. Shevtchenko & Kovalev; the rust fungus *Puccinia acroptila* Syd, which already occurs in the United States; and the stem gall wasp *Aulacida acroptilonica* Beliz.

3. A method for mass-rearing *S. picridis* should be developed, perhaps involving tissue culture.

REFERENCES CITED

Everist, S. L. 1981. *Poisonous plants of Australia,* rev. ed. London: Angus and Robertson.

Groh, H. 1940. Turkestan alfalfa as a medium of weed introduction. *Sci. Agric.* 21:36–43.

Kirjanova, E. S., and T. S. Ivanova. 1969. New species of *Paranguina* Kirjanova, 1955 (Nematoda: Tylenchidae) in Tadzhikistan (in translation). *Ushchel's Kondara (Akademii Nauk Tadzhikskoi* SSR) 2:200–217.

Kovalev, O. V., L. G. Danilov, and T. S. Ivanova. 1973. *Method of controlling Russian knapweed* (in translation). Opisanie Izobreteniia Kavtorskomu Svidetel'stvu Byulleten 38.

Maddox, D. M., A. Mayfield, and N. H. Poritz. 1985. Distribution of yellow starthistle (*Centaurea solstitialis*) and Russian knapweed (*Centaurea repens*). *Weed Sci.* 33:315–27.

Reed, C. F., and R. Hughes. 1970. *Common weeds of the United States.* USDA-ARS Handbook 366. Washington, D.C.: U.S. Department of Agriculture

Robbins, W. W., M. K. Bellue, and W. S. Ball. 1951. *Weeds of California.* Sacramento: State Department of Agriculture.

Rosenthal, S. S. 1989. Safety of the Russian knapweed nematode for general release in the United States. In *Proceedings of the Knapweed Symposium, 4–5 April 1989, Bozeman, Montana,* eds. P. K. Fay and J. R. Lacey, 190–96. Bozeman: Montana State University Cooperative Extension Service Bulletin 45.

Watson, A. K. 1980. The biology of Canadian weeds: 43. *Acroptilon (Centaurea) repens.* (L.) DC. *Can. J. Plant Sci.* 60:993–1004.

———. 1986a. Biology of *Subanguina picridis,* a potential biological control agent of Russian knapweed. *J. Nematology* 18(2):149–54.

———. 1986b. Host range of, and plant reaction to, *Subanguina picridis. J. Nematology* 18(1):112–20.

Watson, A. K., and P. Harris. 1984. *Acroptilon repens* (L.) DC., Russian knapweed (Compositae). In *Biological control programmes against insects and weeds in Canada 1969–1980,* eds. J. S. Kelleher and M. A. Hulme, 105–10. Farnham Royal, Slough, U.K.: Commonwealth Agricultural Bureaux International.

Young, S., W. W. Brown, and B. Klinger. 1970. Nigropallidal encephalomalacia in horses fed Russian knapweed (*Centaurea repens* L.). *Am. J. Vet. Res.* 31:1393–1404.

70 / SPOTTED KNAPWEED

J. M. STORY

spotted knapweed
Centaurea maculosa Lamarck
Asteraceae

INTRODUCTION

Biology and Pest Status

Spotted knapweed, *Centaurea maculosa* Lamarck, is a purple-flowered, herbaceous weed, 30 to 125 centimeters tall, with 1 to 10 upright stems and a stout taproot. A perennial, it lives an average of 3 to 5 years and frequently up to 9 years (Boggs and Story 1987). The flower heads are enclosed by black-tipped involucral bracts and are borne singly at the terminal ends of branches. The weed flowers between July and October and sheds its seeds as soon as the seedhead matures.

Spotted knapweed is mainly a problem on semiarid rangeland, where it reduces forage by up to 96 percent (French and Lacey 1983). The plant rapidly invades disturbed sites and readily colonizes soils having a wide range of chemical and physical properties (Watson and Renney 1974). Its seeds are spread primarily by vehicles, animals, and water.

Historical Notes

Centaurea maculosa belongs to a poorly studied taxonomic group comprising several species and subspecies native to central and eastern Europe and the southern former Soviet Union (Schroeder 1985). The plant was first reported in North America at Victoria, British Columbia, in 1883 (Groh 1943), probably arriving as a contaminant of alfalfa seed from Turkmenistan (Maddox 1982). Researchers estimate that some 2.8 million hectares in the Pacific Northwest are now infested (Lacey 1989). Although the weed can now be found in all the northwestern states and provinces, the largest infestation is in Montana, where 800,000 hectares are affected, causing annual revenue losses from reduced forage of $4.5 million (French and Lacey 1983).

Although spotted knapweed is easily controlled by herbicides and tillage, these practices are not economically practical on rangeland. Researchers in North America began efforts to use biological control against the weed in 1961, when a study of the phytophagous insects associated with European Cynareae, including *Centaurea* species, was started by the Commonwealth Agricultural Bureaux International Institute of Biological Control of Delémont, Switzerland, on behalf of Agriculture Canada (Müller et al. 1989). D. Schroeder (1985) indicated that in Europe 36 oligophagous insect species attack spotted knapweed, 20 of which were listed as potential biological control agents. The first release of an introduced insect against spotted knapweed in North America occurred in British Columbia, in 1970 (Watson and Renney 1974).

RESULTS

Objective A: The Identification and Introduction of Beneficial Organisms

To date, researchers have screened and subsequently introduced seven insect species into North America (see table 70.1). Six of these were screened by the International Institute of Biological Control (IIBC) under contract with Agriculture Canada, a seventh by the U.S. Department of Agriculture's Agricultural Research Service (USDA-ARS). Montana is currently funding the screening of three new insect species by the International Institute of Biological Control. From 1973 to 1987, all European field-collected material shipped to the United States was processed through the USDA-ARS Biological Control of Weeds Quarantine Laboratory in Albany, California. In 1988 the USDA-ARS knapweed effort was relocated to

I thank E. M. Coombs, R. J. Lavigne, G. L. Piper, J. P. McCaffrey, P. B. McEvoy, and S. S. Rosenthal for their contributions to this report. Appreciation is extended to the USDA-ARS Biological Control of Weeds Laboratory in Albany, California, which provided some of the insects. I also thank the numerous individuals and state and federal agencies that have assisted in various aspects of this regional project, and acknowledge the Province of British Columbia and Agriculture Canada, who funded the screening of most of the insect species mentioned.

Table 70.1. The status of insects introduced against spotted knapweed in the western region of the United States.

Biological control agent	Date introduced and status					
	California	Idaho	Montana	Oregon	Washington	Wyoming
Urophora affinis	1976 (E)[*]	1974 (E)	1973 (E)	1973 (E)	1974 (E)	1984 (E)
U. quadrifasciata		1980 (E)[†]	1980 (E)	1980 (E)	1980 (E)	
Metzneria paucipunctella		1981 (E)	1980 (E)	1981 (E)	1981 (E)	
Agapeta zoegana			1984 (E)	1987 (U)	1987 (E)	
Pelochrista medullana			1984 (N)			
Pterolonche inspersa			1988 (N)			
Cyphocleonus achates			1988 (E)			

[*] E = established; N = not established; U = status unknown.

[†] The dispersal of *U. quadrifasciata* into the United States from Canada is estimated to have occurred in 1980.

the newly constructed quarantine facility at Montana State University in Bozeman. A detailed discussion of the seven introduced species follows.

***Urophora affinis* Frauenfeld.** Biological control work on spotted knapweed in the United States began in 1973, when researchers introduced a European seedhead tephritid fly, *Urophora affinis*, in Montana and Oregon. The fly, collected in France and Austria by USDA-ARS personnel, was released in Idaho and Washington in 1974, in California in 1976, and in Wyoming in 1984. Further releases occurred in some of these states between 1977 and 1980. The fly became established in all six states and in British Columbia (Story and Anderson 1978; Harris 1980a; Maddox 1982; Story 1989). Four years after its release in Montana, *U. affinis* larvae were found in up to 99 percent of the spotted knapweed seedheads sampled within 50 meters of the release site (Story and Nowierski 1984).

Since the early 1980s researchers in Idaho, Montana, Oregon, and Washington have worked hard to redistribute the fly to other spotted knapweed sites, both within and between states. J. M. Story (1984) and J. P. McCaffrey et al. (1988) have described the procedures used in collecting and redistributing *U. affinis* and a closely related fly, *Urophora quadrifasciata*. In 1985 USDA-APHIS (Animal and Plant Health Inspection Service) became involved in redistributing the fly throughout the Pacific Northwest.

H. Zwölfer (1970), J. M. Story and N. L. Anderson (1978), and R. L. Gillespie (1983) have described the biology of the fly. It deposits its eggs into immature flower heads during June and July. Larvae feed in the receptacle and induce the formation of hard, woody galls (one larva per gall). The galls sequester plant nutrients, which results in reduced seed production in both the attacked and unattacked seedheads (Harris 1980b).

Larvae overwinter in the galls and pupate in May. The fly is generally univoltine, though a small percentage emerges in August for a second generation.

***Urophora quadrifasciata* (Meigen).** A second seedhead tephritid, *Urophora quadrifasciata*, was collected in what was then the USSR and introduced into British Columbia in 1972 (Harris 1980a); because of insufficient host-specificity data, it was not introduced into the United States. However, by the early 1980s the fly had dispersed into Idaho, Montana, Oregon, and Washington (Gillespie 1983; Story 1985). By 1985 the fly was established at 95 percent of 88 spotted knapweed sites sampled in Montana, compared with 45 percent for *U. affinis* (Story, Nowierski, and Boggs 1987). *U. quadrifasciata* is also more widely distributed than *U. affinis* in Idaho and Washington (Roché, Piper, and Talbott 1986; McCaffrey et al. 1988). However, though *U. quadrifasciata* is dispersing much more rapidly, *U. affinis* appears to be the more permanent colonizer.

The biology of *U. quadrifasciata* is similar to that of *U. affinis,* except that it forms thin galls in the ovary, attacks larger flower heads than does *U. affinis*, and is generally bivoltine (Harris 1980a; Gillespie 1983).

***Metzneria paucipunctella* Zeller.** In 1973 this seedhead gelechiid moth, collected in Switzerland, was introduced into British Columbia (Harris and Myers 1984). The moth was first released in the United States in Idaho, Montana, Oregon, and Washington in 1980 to 1981 from material collected in British Columbia by USDA-ARS personnel. Since then, state and USDA-APHIS personnel have made numerous collections of the moth in British Columbia for further releases in those four states. In Washington researchers have engaged in extensive in-state collection and redistribution of the moth since 1984. *M. paucipunctella* is now established in all four states (Roché, Piper, and Talbott 1986; McCaffrey et al. 1988; Story 1989).

W. Englert (1973), P. Harris and A. Muir (1986), and J. M. Story et al. (1991) have described the biology of the moth. Females deposit eggs on immature flower heads in June. Early-instar larvae feed on ovules; older larvae feed on mature seeds and mine in the receptacle. The older larvae also bind several seeds together with a silken webbing, which prevents the seeds from dispersing at maturity. Because of strong intraspecific competition, only one larva survives per seedhead. The larvae will also attack and destroy other seedhead insects, including larvae of the two *Urophora* species. The moth overwinters as a larva inside the seedhead and pupates in May; adults emerge in June. The moth has one generation per year.

Agapeta zoegana L. This root-mining cochylid moth, collected in Austria and Hungary by IIBC personnel, was first released in the United States in Montana in 1984. In 1987 the USDA-ARS, in cooperation with state entomologists, released the moth in Oregon and Washington. Establishment of the moth was confirmed in Montana (Story, Boggs, and Good 1991) and Washington in 1987 and 1988, respectively.

H. Müller, D. Schroeder, and A. Gassmann (1988) have described the biology of the moth in Europe. It overwinters as a larva in the root. Adults emerge from mid-June to mid-August; one day later they begin to mate and oviposit on the stems and leaves. Larvae hatch in 7 to 10 days and begin mining the epidermal tissues of the root crown. Later-instar larvae mine the cortex and endodermis. Several larvae may develop in the same plant.

Pelochrista medullana Staudinger. The root-mining tortricid moth *Pelochrista medullana*, collected in Austria and Hungary by IIBC personnel, was released once in Montana in 1984, but it did not become established. Subsequent attempts to collect the moth in Europe have failed to produce enough individuals to make more releases (Schroeder person. commun.). However, a small population is established in British Columbia.

The biology of this moth in Europe, described by A. Gassmann, D. Schroeder, and H. Müller (1982), is nearly identical to that of *A. zoegana*, except that the period of adult emergence is shorter—from June through July. Intraspecific competition usually results in the survival of only one *P. medullana* larva per plant, but the larvae readily coexist with other root insects.

Pterolonche inspersa Staudinger. In 1988 this root-mining pterolonchid moth, collected in Hungary by IIBC personnel, was released in Montana but failed to become established. Previous releases of this moth had been made on diffuse knapweed, *Centaurea diffusa*, by USDA-ARS and state cooperators in Idaho, Oregon, and Washington between 1986 and 1987. USDA-ARS conducted the host-specificity screening of this insect (Dunn et al. 1989).

The moth overwinters as a half-grown larva in the root, feeding again in the spring before pupating in the root. Adults emerge in late July, and begin to mate and oviposit the next day. Eggs are laid on the leaves in early August, and larvae hatch in 8 to 16 days and begin mining in the central vascular tissue. Several larvae may develop in the same plant (Dunn et al. 1989).

Cyphocleonus achates (F.). This root weevil, collected in Austria, was first released in the United States in Montana in 1988. The weevil is now established at several locations in Montana.

C. S. A. Stinson (1987) has described its biology in Europe. The weevil overwinters as a larva in the root. Adults emerge between early August and mid-September and live for 8 to 15 weeks. The female lays eggs singly in a notch she has excavated on the root crown, just below the soil surface. Larvae hatch in 10 to 12 days and mine into the root cortex. Feeding by the older larvae results in a conspicuous root gall. The weevil has one generation per year.

Objective B: Ecological and Physiological Studies

Researchers conducted studies in Idaho and Montana to assess the compatibility of the two *Urophora* species with herbicides. In Idaho they found that a spring application of picloram, or of a picloram–2,4-D combination, did not significantly affect larval-pupal mortality rates of either fly species (McCaffrey and Callihan 1988). Also, low rates of picloram had no long-term effect on the flies' re-establishment in previously treated areas. Spring applications of 2,4-D did not affect *U. affinis* in either state (McCaffrey and Callihan 1988; Story, Boggs, and Good 1988). The herbicide also did not affect *U. quadrifasciata* in Montana, but the results in Idaho were inconclusive. In Montana, applications later in the growing season had no effect on *U. quadrifasciata*, but significantly reduced *U. affinis* emergence.

Researchers in Idaho also studied the diet preference and the effects of diet on longevity of *U. affinis* and *U. quadrifasciata* adults. Both fly species showed a significant preference for honey over a yeast-hydrolysate solution (1:1 yeast to water) and a honey-yeast solution (1:1:1 honey to yeast to water) (Vogt 1986). Flies lived the longest—about 30 days—on a honey-water diet.

In Idaho and Montana researchers found that two spiders, *Dictyna coloradensis* Chamberlin and *D. major* Gertsch, were effective predators of *Urophora* species (Story 1977; Wheeler 1985). Studies on the developmental biology and bionomics of the two spiders were conducted in Idaho (Wheeler 1985; Wheeler, McCaffrey, and Johnson 1990). Of the two spiders, *D.*

coloradensis was the more effective predator, capturing approximately 20 percent of the *Urophora* species striking the webs; nevertheless, only small portions of the fly populations were captured by the spiders (Wheeler 1985).

J. P. McCaffrey, J. B. Johnson, and G. S. Wheeler (1985) conducted laboratory studies on the effects of low temperature (4° to 6°C) on the postdiapause development of overwintering *U. affinis* and *U. quadrifasciata.* The rate of adult fly emergence was inversely related to the duration of the cold treatment. *U. affinis* took more than twice as long to emerge as did *U. quadrifasciata.*

R. M. Nowierski, J. M. Story, and R. E. Lund (1987) developed numerical sampling equations to improve the accuracy and efficiency in future sampling of *U. affinis* and *U. quadrifasciata* in Montana, and to help determine which sites warrant future fly releases.

In Oregon, the gall distribution of *Urophora* species changed from random in 1983 and 1984 to aggregated in 1984 and 1985. The distribution of the two *Urophora* species and *M. paucipunctella* among seedheads was independent in 1984, while the three species were positively associated in 1985. Also, spatial dispersion of *U. affinis* in Montana in 1984 was more clumped than that of *U. quadrifasciata* (Nowierski, Story, and Lund 1987).

Although the *M. paucipunctella* population is increasing, researchers have observed consistently high overwintering mortality in Oregon and Montana. The preliminary data in Montana suggest that *M. paucipunctella* and *U. quadrifasciata* are much less cold-tolerant than *U. affinis,* which can tolerate temperatures down to −34°C (Callan, Story, and Roché 1986). Studies in Oregon indicate that increasing levels of seed destruction have no short-term effect on spotted knapweed density, because of recruitment from seeds already in the soil.

Objective C: The Conservation and Augmentation of Beneficial Organisms

A mass-rearing program on *Agapeta zoegana* has been under way in Montana since 1986. Large numbers of adults were recovered from rearing cages and subsequently released at four sites in the western part of the state in 1988. A similar mass-rearing effort was initiated for *Cyphocleonus achates* in 1988.

Objective D: Impact

J. M. Story, K. W. Boggs, and R. M. Nowierski (1989) have reported that in Montana the two seedhead flies reduced seed production by 36 to 41 percent per seedhead over a 2-year period. However, because the fly galls are nourished at the expense of other plant tissues (Harris 1980b), the total reduction in seed production probably exceeded that amount.

RECOMMENDATIONS

1. Expedite efforts to screen and introduce the remaining five or six promising biological control agents from Eurasia.

2. Develop procedures for redistributing the root-mining insects.

3. Plan and implement studies to measure the impact of both established and newly introduced natural enemies.

4. Expand efforts to integrate biological control with other management practices against spotted knapweed.

5. Continue efforts to locate and collect *P. medullana* in Eurasia.

6. Expand efforts to use plant pathogens against spotted knapweed.

REFERENCES CITED

Boggs, K. W., and J. M. Story. 1987. The population age structure of spotted knapweed (*Centaurea maculosa*) in Montana. *Weed Sci.* 35:194–98.

Callan, N. W., J. M. Story, and R. R. Roché. 1986. Thermoelectric modules and a computerized data acquisition system to determine insect supercooling points. *Ann. Entomol. Soc. Am.* 79:60–61.

Dunn, P., S. S. Rosenthal, G. Campobasso, and S. M. Tait. 1989. Host specificity of *Pterolonche inspersa* (Lepidoptera: Pterolonchidae) and its potential as a biological control agent for *Centaurea diffusa,* diffuse knapweed, and *Centaurea maculosa,* spotted knapweed. *Entomophaga* 34:435–46.

Englert, W. 1973. *Metzneria paucipunctella* Zel. (Lepidoptera: Gelechiidae): A potential insect for the biological control of *Centaurea stoebe* L. in Canada. In *Proceedings of the 2nd International Symposium on Biological Control of Weeds, 4–7 October 1971, Rome, Italy,* ed. P. Dunn, 161–65. Commonwealth Agricultural Bureaux International Miscellaneous Publication 6. Farnham Royal, Slough, U.K.: Commonwealth Institute of Biological Control.

French, R. A., and J. R. Lacey. 1983. *Knapweed: Its cause, effect, and spread in Montana.* Montana Cooperative Extension Service Circular 307.

Gassmann, A., D. Schroeder, and H. Müller. 1982. Investigations on *Pelochrista medullana* (Stgr.) (Lepidoptera: Tortricidae), a possible biocontrol agent of diffuse and spotted knapweed, *Centaurea diffusa* Lam. and *C. maculosa* Lam. (Compositae) in North America. Commonwealth Institute of Biological Control Report. Farnham Royal, Slough, U.K.: CIBC.

Gillespie, R. L. 1983. Bionomics of *Urophora affinis* Frauenfeld, and *U. quadrifasciata* Meigen (Diptera: Tephritidae) in northern Idaho. Master's thesis, University of Idaho, Moscow.

Groh, H. 1943. *Canadian weed survey.* Second Annual Report of the Canadian Department of Agriculture. Ottawa: Agriculture Canada.

Harris, P. 1980a. Establishment of *Urophora affinis* Frfld. and *U. quadrifasciata* (Meig.) (Diptera: Tephritidae) in Canada for the biological control of diffuse and spotted knapweed. *Zeit. angew. Entomol.* 89:504–14.

———. 1980b. Effects of *Urophora affinis* Frfld. and *U. quadrifasciata* (Meig.) (Diptera: Tephritidae) on *Centaurea diffusa* Lam. and *C. maculosa* Lam. (Compositae). *Zeit. angew. Entomol.* 90:190–201.

Harris, P., and A. Muir. 1986. Biological control of spotted knapweed by *Metzneria paucipunctella* (Zeller). *Canadex* 641.613. Ottawa: Agriculture Canada.

Harris, P., and J. H. Myers. 1984. *Centaurea diffusa* Lam. and *C. maculosa* Lam. s. lat., diffuse and spotted knapweed (Compositae). In *Biological control programmes against insect and weeds in Canada 1969–1980*, eds. J. S. Kelleher and M. A. Hulme, 127–37. Farnham Royal, Slough, U.K.: Commonwealth Agricultural Bureaux International.

Lacey, C. 1989. Knapweed management: A decade of change. In *Proceedings of the Knapweed Symposium, 4–5 April 1989, Bozeman, Montana*, 1–6.

Maddox, D. M. 1982. Biological control of diffuse knapweed (*Centaurea diffusa*) and spotted knapweed (*C. maculosa*). *Weed Sci.* 30:76–82.

McCaffrey, J. P., J. B. Johnson, and G. S. Wheeler. 1985. Effects of duration of low-temperature exposure on post-diapause development of overwintering *Urophora affinis* and *U. quadrifasciata*. In *Proceedings of the 6th International Symposium on Biological Control of Weeds, 19–25 August 1984, Vancouver, British Columbia*, ed. E. S. Delfosse, 445. Ottawa: Agriculture Canada.

McCaffrey, J. P., R. P. Wight, R. L. Stoltz, R. H. Callihan, and D. W. Kidder. 1988. *Collection and redistribution of biological control agents of diffuse and spotted knapweed.* Idaho Cooperative Extension Service Bulletin 680.

McCaffrey, J. P., and R. H. Callihan. 1988. Compatibility of picloram and 2,4-D with *Urophora affinis* and *U. quadrifasciata* (Diptera: Tephritidae) for spotted knapweed control. *Environ. Entomol.* 17:785–88.

Müller, H., D. Schroeder, and A. Gassmann. 1988. *Agapeta zoegana* (L.) (Lepidoptera: Cochylidae), a suitable prospect for biological control of spotted and diffuse knapweed, *Centaurea maculosa* Monnet de la Marck and *Centaurea diffusa* Monnet de la Marck (Compositae) in North America. *Can. Entomol.* 120:109–24.

Müller, H., C. S. A. Stinson, K. Marquardt, and D. Schroeder. 1989. The entomofaunas of roots of *Centaurea maculosa* Lam., *C. diffusa* Lam., and *C. vallesiaca* Jordan in Europe. *J. Appl. Entomol.* 107:83–95.

Nowierski, R. M., J. M. Story, and R. E. Lund. 1987. Two-level numerical sampling plans and optimal subsample size computations for *Urophora affinis* and *Urophora quadrifasciata* (Diptera: Tephritidae) on spotted knapweed. *Environ. Entomol.* 16:933–37.

Roché, B. F., G. L. Piper, and C. J. Talbott. 1986. *Knapweeds of Washington.* Washington State University Cooperative Extension Service Bulletin 1393.

Schroeder, D. 1985. The search for effective biological control agents in Europe: 1. Diffuse and spotted knapweed. In *Proceedings of the 6th International Symposium on Biological Control of Weeds, 19–25 August 1984, Vancouver, British Columbia*, ed. E. S. Delfosse, 103–19. Ottawa: Agriculture Canada.

Stinson, C. S. A. 1987. Investigations on *Cyphocleonus achates* (Fahr.) (Coleoptera: Curculionidae), a possible biological agent of spotted knapweed (*Centaurea maculosa* Lam.) and diffuse knapweed (*C. diffusa* Lam.) (Compositae) in North America. Commonwealth Institute of Biological Control Report. Farnham Royal, Slough, U.K.: CIBC.

Story, J. M. 1977. Effect of predation by the spider, *Dictyna major*, on *Urophora affinis* populations in Montana. In *Proceedings of the Knapweed Symposium, 6–7 October 1977, Kamloops, British Columbia*, 206–7. London: British Ministry of Agriculture.

———. 1984. *Collection and redistribution of* Urophora affinis *and* U. quadrifasciata *for biological control of spotted knapweed.* Montana Cooperative Extension Service Circular 308.

———. 1985. First report of the dispersal into Montana of *Urophora quadrifasciata* (Diptera: Tephritidae), a fly released in Canada for biological control of spotted and diffuse knapweed. *Can. Entomol.* 117:1061–62.

———. 1989. The status of biological control of spotted and diffuse knapweed. In *Proceedings of the Knapweed Symposium, 4–5 April 1989, Bozeman, Montana*, 37–42.

Story, J. M., and N. L. Anderson. 1978. Release and establishment of *Urophora affinis* (Diptera: Tephritidae) on spotted knapweed in western Montana. *Environ. Entomol.* 7:445–48.

Story, J. M., and R. M. Nowierski. 1984. Increase and dispersal of *Urophora affinis* (Diptera: Tephritidae) on spotted knapweed in western Montana. *Environ. Entomol.* 13:1151–56.

Story, J. M., K. W. Boggs, and W. R. Good. 1988. Optimal timing of 2,4-D applications for compatibility with *Urophora affinis* and *U. quadrifasciata* (Diptera: Tephritidae) for control of spotted knapweed. *Environ. Entomol.* 17:911–14.

———. 1991. First report of the establishment of *Agapeta zoegana* L. (Lepidoptera: Cochylidae) on spotted knapweed, *Centaurea maculosa* Lamarck, in the United States. *Can. Entomol.* 123:411–12.

Story, J. M., K. W. Boggs, and R. M. Nowierski. 1989. Effect of two introduced seed head flies on spotted knapweed. *Mont. AgResearch* 6(1):14–17.

Story, J. M., R. M. Nowierski, and K. W. Boggs. 1987. Distribution of *Urophora affinis* and *U. quadrifasciata*, two flies introduced for biological control of spotted knapweed (*Centaurea maculosa*) in Montana. *Weed Sci.* 35:145–48.

Story, J. M., K. W. Boggs, W. R. Good, P. Harris, and R. M. Nowierski. 1991. *Metzneria paucipunctella* Zeller (Lepidoptera: Gelechiidae), a moth introduced against spotted knapweed: Its feeding strategy and impact on two introduced *Urophora* spp. (Diptera: Tephritidae). *Can. Entomol.* 123:1001–7.

Vogt, E. A. 1986. Influence of adult diet on oogenesis, longevity, and fecundity of *Urophora affinis* (Frauenfeld) and *U. quadrifasciata* (Meigen) (Diptera: Tephritidae). Master's thesis, University of Idaho, Moscow.

Watson, A. K., and A. J. Renney. 1974. The biology of Canadian weeds: 6. *Centaurea diffusa* and *C. maculosa. Can. J. Plant Sci.* 54:687–701.

Wheeler, G. S. 1985. The bionomics of two spiders, *Dictyna coloradensis* Chamberlin and *D. major* Gertsch (Araneae: Dictynidae): Potential antagonists to the biocontrol of weeds.

Master's thesis, University of Idaho, Moscow.

Wheeler, G. S., J. P. McCaffrey, and J. B. Johnson. 1990. Developmental biology of two *Dictyna* spp. (Araneae: Dictynidae) in the laboratory and at sites of varying season length. *Am. Midl. Nat.* 123:124–34.

Zwölfer, H. 1970. Investigations on the host-specificity of *Urophora affinis* Frfld. (Diptera: Trypetidae). Commonwealth Institute of Biological Control Progress Report 25. Farnham Royal, Slough, U.K.: Commonwealth Institute of Biological Control.

71 / TANSY RAGWORT

C. E. TURNER AND P. B. McEVOY

> **tansy ragwort**
> *Senecio jacobaea* L.
> Asteraceae

INTRODUCTION

Biology and Pest Status

Tansy ragwort, *Senecio jacobaea* L., is a biennial or short-lived perennial herb native to Eurasia and naturalized in North America, South America, Australia, and New Zealand. In North America this weed causes the most severe problems along the western coasts of Canada and the United States and in Canada's eastern maritime provinces. Tansy ragwort contains pyrrolizidine alkaloids, which when ingested can cause severe liver damage in livestock, especially cattle and horses (Fuller and McClintock 1986). The weed also displaces forage species in grasslands, where it can completely dominate other herbs. The weed was first recorded in western North America in 1912 in California (Pemberton and Turner 1990) and in 1922 in Oregon (Isaacson 1973). An aerial survey in 1976 estimated that in western Oregon more than 12,000 square kilometers—more than 3 million acres—were infested (Isaacson 1978a). P. B. McEvoy (1984) and P. B. McEvoy and C. S. Cox (1987) have studied seed dispersal and germination in Oregon, and D. A. Wardle (1987) has provided a review of the weed's ecology.

Historical Notes

E. Cameron (1935) laid some of the groundwork for a biological control program on tansy ragwort by investigating the weed and its natural enemies in Britain. In North America the first work on the weed—the introduction of the cinnabar moth into California in 1959—preceded the inception of the W-84 Project. USDA Agricultural Research Service personnel in Rome, in Paris, and in Albany, California, carried out foreign exploration for natural enemies, conducted host-specificity studies, and introduced biological control agents into the United States.

RESULTS

Objective A: The Identification and Introduction of Beneficial Organisms

Three insects have been introduced into the United States for tansy ragwort control. They are discussed here in the order of their introduction.

***Tyria jacobaeae* (L.) (Lepidoptera: Arctiidae).** J. P. Dempster (1982) has reviewed the biology of *T. jacobaeae*, the cinnabar moth. The moth is univoltine. The brightly colored red-and-black adults emerge, mate, and oviposit in the spring. Females lay eggs in clusters on the undersides of the basal leaves. J. P. Dempster (1982) reported that during 9 years' field measurements in England, mean fecundity ranged from 73 to 295 eggs per female. The developing orange-and-black-banded larvae feed externally on the shoots (especially the flowering shoots) in spring and summer. They eventually leave the host plant as mature fifth-instar larvae, to pupate and overwinter in the soil. Although the larvae are capable of completely stripping the shoots of flowers and leaves, this rarely kills the plant along the Pacific Coast, where fall rains regularly spur regrowth from the base of defoliated and otherwise apparently dead plants. In California the cinnabar moth is subject to attack by predators, parasites, and a microsporidian disease (Hawkes 1973).

E. Cameron (1935) listed tansy ragwort and common groundsel (*Senecio vulgaris* L.) as field hosts of the moth in England. Host tests that focused chiefly on commercially important plants in the Asteraceae were carried out by E. Cameron (1935) and H. L. Parker

We thank E. M. Coombs and G. L. Piper for assistance with the Oregon and Washington information; L. A. Andres, R. D. Goeden, and N. J. Mills for their review of the manuscript; and K. L. Chan for assistance with manuscript preparation.

(1960). G. E. Bucher and P. Harris (1961) conducted host tests designed to further define the host range of the moth and showed that other species of *Senecio* could also support larval feeding and development. The first releases in the United States—using material from France, west of Paris—were of 4,800 larvae in California in 1959, 924 adults in Oregon in 1960, and 237 adults in Washington in 1960 (Frick and Holloway 1964; Hawkes 1968). By 1965 researchers were redistributing populations from northwestern California to other sites in California and to other states. The cinnabar moth is now well established in California, Oregon, and Washington.

Botanophila seneciella (Meade) (Diptera: Anthomyiidae).

E. Cameron (1935) and K. E. Frick (1970a) have described the biology of the ragwort seed fly, *B. seneciella*—formerly referred to as *Hylemyia seneciella* (Meade). The fly is univoltine. Adults emerge in the spring and oviposit in unopened and newly opened capitula. The eggs are laid singly between the involucral bracts and the outermost row of florets. Only one larva develops per capitulum, regardless of the number of eggs oviposited. The larva feeds on the developing seeds and receptacle through the summer and can completely destroy the contents of a capitulum. Mature larvae overwinter in the soil inside puparia.

The only reported field hosts in Europe are tansy ragwort and *Senecio aquaticus* Hill (Frick and Andres 1967). K. E. Frick and L. A. Andres (1967) have carried out host tests in Albany, and K. E. Frick (1969a,b) has discussed the first introductions into the United States. Large numbers of pupae were shipped from France so that adults could emerge in the quarantine laboratory for field release. In 1966, 916 females and 137 males were released near Ft. Bragg in northwestern California, and 1,408 females and 660 males were released in western Oregon. The second release—of 1,721 females and 1,627 males—was made in western Washington in 1968. The fly is now established in California, Oregon, and Washington, but there is relatively little information about its current status.

Longitarsus jacobaeae (Waterhouse) (Coleoptera: Chrysomelidae).

Researchers were spurred to introduce *Longitarsus jacobaeae*, the ragwort flea beetle, because of ragwort's troublesome capability to regrow after being defoliated by the cinnabar moth. The univoltine beetle was successfully introduced from Italy into the United States. In both Italy and California, newly eclosed adults emerge from the soil in the spring and feed externally on foliage before entering summer diapause; the adults resume feeding in the fall when conditions are cool and moist (Frick and Johnson 1973). The adult's summer activity depends on climatic conditions: at moist coastal sites in Oregon, adults can be found feeding throughout the summer (James 1989). In the fall, the beetle lays eggs singly on the base of the plant or in adjacent soil. Throughout the remaining fall, winter, and spring, larvae feed internally on leaves, stem, and—especially—the root crown, and externally on lateral roots. Mature larvae leave the plants in the spring and pupate in the soil. Larval feeding often results in the outright killing of plants, and adult feeding can kill seedlings. Compared with the cinnabar moth, flea beetles disperse and colonize isolated host plants relatively rapidly (Hawkes and Johnson 1978; McEvoy et al. 1993).

H. C. F. Newton (1933) and K. E. Frick (1970b) have described the host specificity of the flea beetle. In Europe the only field host records for the beetle are tansy ragwort and *S. aquaticus* (Frick 1970b). In 1969 the first releases that resulted in establishment were made at the same site near Ft. Bragg, California, with beetles collected from *S. aquaticus* near Rome, Italy, and reared through a generation in the quarantine laboratory (Frick 1970c; Frick and Johnson 1972, 1973; Hawkes and Johnson 1978). These releases consisted of two separate introductions of a total of 54 females and 77 males. Another introduction into northwestern California in 1969—this time with beetles from Switzerland—failed (Frick 1970c; Hawkes and Johnson 1978). In Washington the first release—in 1970—consisted of 200 females and 174 males from Italy (Frick and Johnson 1972). The first release in Oregon, which took place in 1971, consisted of beetles collected from the Ft. Bragg site (Isaacson 1978b). *Longitarsus jacobaeae* is now well established in California, Oregon, and Washington. Based in part on the research cited previously, the beetle was introduced into Australia, Canada, and New Zealand.

Objective B: Ecological and Physiological Studies

In view of the outstanding success of imported natural enemies in suppressing tansy ragwort in California and Oregon (Hawkes and Johnson 1978; McEvoy, Cox, and Coombs 1991), P. B. McEvoy et al. (1993) carried out a perturbation experiment to investigate the interactions between tansy ragwort, disturbance, the natural enemies, and interspecific plant competition at Cascade Head, Oregon—where the weed was reduced to less than 1 percent of its former aboveground biomass. These experiments demonstrated that abundant buried seed—for example, a mean of about 4,000 seeds per square meter in 1986—and local disturbances combined to activate incipient weed outbreaks. When these occurred, interspecific plant competition and the flea beetle combined to inhibit the increase and spread of incipient outbreaks. At the experimental scale examined, all of the tansy ragwort individuals outside of the pool of buried seed were eliminated. These results and other studies have contributed to an understanding of the following key issues and principles in biological control theory and practice.

Local equilibrium versus local extinction. The conventional mathematical theory used to represent and analyze biological control systems has held that natural enemies control pests by reducing their density to a new, low, and stable equilibrium via density-dependent mechanisms (for example, see May and Hassell 1988). However, W. W. Murdoch, J. Chesson, and P. L. Chesson (1985) have claimed that many successful biological control systems are characterized by local extinctions and then reinvasions by both pest and control agents. Despite local instability and extinctions, both pest and agents persist on a regional scale as long as local extinctions are not simultaneous, and as long as the local populations of pest and agents display sufficient migration (Taylor 1990). P. B. McEvoy et al. (1993) supported this local instability and extinction paradigm: their perturbation experiments showed that buried seed and local disturbance combined to initiate local weed outbreaks, which were then eliminated by a combination of ready colonization by insect agents and plant competition.

Single versus multiple agent species. Because different introduced agents may compete among each other and thus have less of an effect than that of the single most efficient species, there is an ongoing controversy about whether it is better to introduce a single "best" agent species or several species. In the tansy ragwort project, the cinnabar moth interferes competitively with and is superior to the fly (Crawley and Pattrasudhi 1988), whereas the cinnabar moth and the flea beetle do not interfere with each other. During the summer cinnabar moth larvae feed on the shoot foliage and flowers, reducing seed production, whereas flea beetle larvae feed on roots during the fall, winter, and spring, causing plant mortality (Hawkes and Johnson 1978; McEvoy et al. 1990). However, despite the apparent complementary nature of the flea beetle and the cinnabar moth, the moth's contribution to the biological control of tansy ragwort remains controversial. Some workers have maintained that the cinnabar moth has no effect on tansy ragwort abundance, and thus has no control value (van der Meijden 1979; Myers 1980; Crawley and Gillman 1989). This would be the case if, despite the fact that moth larvae substantially reduced seed production, the weed was not limited by seed production. It could occur, for example, if weed recruitment were limited by other factors such as "safe sites" for germination and establishment (Crawley and Gillman 1989).

However, the data of P. B. McEvoy et al. (1993) and unpublished data show that the cinnabar moth's reduction of seed production results in a reduced number of rosettes in the next generation. Projections of matrix models based on these results indicate that the cinnabar moth can indeed contribute to the control of experimental ragwort populations.

Herbivory and plant competition. The experiments by P. B. McEvoy et al. (1993) showed that a combination of natural enemies and interspecific plant competition was more effective than either factor alone in suppressing the weed. They found that reduced plant competition—by clipping other vegetation—delayed the elimination of the weed, whereas herbivore pressure by the flea beetle—alone or in combination with the cinnabar moth—accelerated the elimination of the weed and its replacement by other vegetation. Other examples of the combined pressures of plant competition and natural enemies enhancing weed control include prickly-pear cacti in California (Goeden, Fleschner, and Ricker 1967), skeleton weed in Australia (Groves and Williams 1975), and musk thistle in Virginia (Kok, McAvoy, and Mays 1986).

Climate matching. Flea beetles were introduced to California from both Rome, Italy, and Delémont, Switzerland, but only the introductions from Italy became established. One apparently critical difference in the life cycles of the Italian and Swiss flea beetles is that the former estivate as adults and oviposit in the fall, whereas the latter oviposit in the spring and the eggs diapause in summer to develop in the fall (Frick 1971; Frick and Johnson 1972). This egg diapause may be well suited to the humid summers and frequent rainfall of Switzerland (Frick 1970c), but it probably doomed the eggs to desiccation in western California, where summers are more similar to those of Italy (Frick 1971). On the other hand, the life cycle of the Italian biotype is not well suited to the more continental climate of the higher elevations in the Northwest's Cascade Mountains, where the weed tends to have a climatic "refuge" from the flea beetle.

Objective D: Impact

Impact on tansy ragwort. The tansy ragwort biological control project has been a resounding success, at least in California and Oregon. R. B. Hawkes and G. R. Johnson (1978) found a 99 percent or more reduction in ragwort at three coastal study sites near Ft. Bragg, California, by 1976. A follow-up study in 1987 showed sustained control at these same low densities, replaced primarily by more desirable vegetation (Pemberton and Turner 1990).

In western Oregon a 12-year survey of 42 different tansy ragwort populations demonstrated strong and persistent supression of the weed by an average of about 93 percent within 6 years of releasing the beetle (McEvoy, Cox, and Coombs 1991). At the Cascade Head study site, the insect agents reduced tansy ragwort by more than 99 percent over an 8-year period, and the weed was replaced largely by introduced perennial grasses. Perturbation experiments at this site showed that within one tansy ragwort generation—that is, 1 to 3 years—these insects can depress the weed's density, biomass, and reproduction by over 99 percent. The sustained control of the weed in Oregon is due to the persistence of the weed and controlling agents on a regional scale, despite interactions that lead to local extinctions (McEvoy et al. 1993). The successful biological control of the weed in Oregon has an estimated benefit-to-cost ratio of 13:1 to 15:1, with benefits currently valued at $5 million per year (Isaacson and Radtke 1993).

Of all the introduced agents, the flea beetle has clearly had the greatest effect. It is excellent at colonizing and destroying the weed in small or large stands. The cinnabar moth also appears to have contributed to control in Oregon and probably in California, in conjunction with competing vegetation (Hawkes 1968; McEvoy et al. 1993). The impact of the seed fly is largely unknown and, as noted previously, it may be competitively inferior to the cinnabar moth. P. B. McEvoy et al. (1993) reported that at the site of the perturbation experiments in Oregon, the seed fly destroyed only a small fraction—4 percent—of capitula not preempted by the cinnabar moth. G. L. Piper (1985) reported that the seed fly was destroying about 30 percent of the capitula at a site in southwestern Washington in 1981.

Impact on a nontarget plant. For safety, imported agents must be subjected to host-specificity assessment studies, to evaluate their potential impact on nontarget plants (Turner 1985). J. W. Diehl and P. B. McEvoy (1990) have carried out field and laboratory studies on the impact of the cinnabar moth on a nontarget native perennial herb—the arrowleaf groundsel (*Senecio triangularis* Hooker)—that grows in moist habitats, mostly at medium to high elevations in the mountains of western North America. Like tansy ragwort, arrowleaf groundsel contains pyrrolizidine alkaloids (Roitman 1983). Cinnabar moth larvae successfully completed development on arrowleaf groundsel in the laboratory, but their growth was 21 percent slower and pupae were lighter than those reared on tansy ragwort (Diehl and McEvoy 1990). Field studies on Mary's Peak (elevation 1,248 meters) in the Coast Range of Oregon showed that the percentage of arrowleaf groundsel stems attacked varied widely between 1986 and 1987 and among habitats

(roadside, meadow, and forest). In both years the roadside population suffered the highest rate of attack: 77 percent in 1986, 23 percent in 1987. Although in some cases the moth completely defoliated arrowleaf groundsel stems, this neither killed the plant nor affected plant growth or seed production; however, it did reduce seed viability somewhat. On the other hand, native insects on Mary's Peak—particularly *Phyllotrox nubifer* LeConte (Coleoptera: Curculionidae) and *Paroxyna snowi* Hering (Diptera: Tephritidae)—caused extensive damage to arrowleaf groundsel. Adult *P. nubifer* feed on leaves and meristems, the larvae on seeds; *P. snowi* larvae feed on seeds. J. W. Diehl and P. B. McEvoy (1990) concluded that the cinnabar moth had only a very small effect on arrowleaf groundsel's reproductive success. In view of earlier cinnabar moth field records in Europe (Cameron 1935), as well as host tests (Bucher and Harris 1961) that showed the host range of the moth to include other *Senecio* species, the attack of arrowleaf groundsel is not surprising, and is unlikely to be a true evolutionary host shift.

RECOMMENDATIONS

1. Additional agent biotypes and species should be considered for introduction into areas with a more continental climate—that is, higher elevations and sites further inland.

2. The present status of seed fly populations and their impact on tansy ragwort need to be determined.

3. Evaluation studies of the cinnabar moth and the flea beetle should be undertaken in other areas to compare results with the findings in California and Oregon. Particular attention should be paid to the level of control, the amplitude of tansy ragwort fluctuations over time, and the character of the replacement vegetation.

4. An economic evaluation of the costs and benefits of the tansy ragwort biological control program should be made for the Northwest Region.

5. Studies on the impact that biological agents released for tansy ragwort control might have on nontarget plants should be broadened to include other native *Senecio* species.

REFERENCES CITED

Bucher, G. E., and P. Harris. 1961. Food-plant spectrum and elimination of disease of cinnabar moth larvae, *Hypocrita jacobaeae* (L.) (Lepidoptera: Arctiidae). *Can. Entomol.* 93:931–36.

Cameron, E. 1935. A study of the natural control of ragwort (*Senecio jacobaea* L.). *J. Ecol.* 23:265–322.

Crawley, M. J., and M. P. Gillman. 1989. Population dynamics of cinnabar moth and ragwort in grassland. *J. Animal Ecol.* 58:1035–50.

Crawley, M. J., and R. Pattrasudhi. 1988. Interspecific competition between insect herbivores: Asymmetric competition between cinnabar moth and the ragwort seed-head fly. *Ecol. Entomol.* 13:243–49.

Dempster, J. P. 1982. The ecology of the cinnabar moth, *Tyria jacobaeae* L. (Lepidoptera: Arctiidae). *Advances Ecol. Res.* 12:1–36.

Diehl, J. W., and P. B. McEvoy. 1990. Impact of the cinnabar moth (*Tyria jacobaeae*) on *Senecio triangularis*, a non-target native plant in Oregon. In *Proceedings of the 7th International Symposium on Biological Control of Weeds, 6–11 March 1988, Rome, Italy,* ed. E. S. Delfosse, 119–26. Rome: Ministero dell'Agricoltura e delle Foreste / Melbourne: CSIRO.

Frick, K. E. 1969a. Attempt to establish the ragwort seed fly in the United States. *J. Econ. Entomol.* 62:1135–38.

———. 1969b. Tansy ragwort control aided by the establishment of seedfly from Paris. *Calif. Agric.* 23(12):10–11.

———. 1970a. Behavior of adult *Hylemya seneciella*, an anthomyiid (Diptera) used for the biological control of tansy ragwort. *Ann. Entomol. Soc. Am.* 63:184–87.

———. 1970b. *Longitarsus jacobaeae* (Coleoptera: Chrysomelidae), a flea beetle for the biological control of tansy ragwort: 1. Host plant specificity studies. *Ann. Entomol. Soc. Am.* 63:284–96.

———. 1970c. Ragwort flea beetle established for biological control of tansy ragwort in northern California. *Calif. Agric.* 24(4):12–13.

———. 1971. *Longitarsus jacobaeae* (Coleoptera: Chrysomelidae), a flea beetle for the biological control of tansy ragwort: 2. Life history of a Swiss biotype. *Ann. Entomol. Soc. Am.* 64:834–40.

Frick, K. E., and L. A. Andres. 1967. Host specificity of the ragwort seed fly. *J. Econ. Entomol.* 60:457–63.

Frick, K. E., and J. K. Holloway. 1964. Establishment of the cinnabar moth, *Tyria jacobaeae*, on tansy ragwort in the western United States. *J. Econ. Entomol.* 57:152–54.

Frick, K. E., and G. R. Johnson. 1972. *Longitarsus jacobaeae* (Coleoptera: Chrysomelidae), a flea beetle for the biological control of tansy ragwort: 3. Comparison of the biologies of the egg stage of Swiss and Italian biotypes. *Ann. Entomol. Soc. Am.* 65:406–10.

———. 1973. *Longitarsus jacobaeae* (Coleoptera: Chrysomelidae), a flea beetle for the biological control of tansy ragwort: 4. Life history and adult aestivation of an Italian biotype. *Ann. Entomol. Soc. Am.* 66:358–67.

Fuller, T. C., and E. McClintock. 1986. *Poisonous plants of California.* Berkeley: University of California Press.

Goeden, R. D., C. A. Fleschner, and D. W. Ricker. 1967. Biological control of prickly pear cacti on Santa Cruz Island, California. *Hilgardia* 38:579–606.

Groves, R. H., and J. D. Williams. 1975. Growth of skeleton weed (*Chondrilla juncea* L.) as affected by growth of subterranean clover (*Trifolium subterraneum* L.) and infection by *Puccinia chondrilla* Bubak & Syd. *Aust. J. Agric. Res.* 26:975–83.

Hawkes, R. B. 1968. The cinnabar moth, *Tyria jacobaeae*, for control of tansy ragwort. *J. Econ. Entomol.* 61:499–501.

———. 1973. Natural mortality of cinnabar moth in California. *Ann. Entomol. Soc. Am.* 66:137–46.

Hawkes, R. B., and G. R. Johnson. 1978. *Longitarsus jacobaeae* aids moth in the biological control of tansy ragwort. In *Proceedings of the 4th International Symposium on Biological Control of Weeds, 30 August–2 September 1976, Gainesville, Florida,* ed. T. E. Freeman, 193–96. Gainesville: University of Florida, Institute of Food and Agricultural Sciences.

Isaacson, D. L. 1973. Population dynamics of the cinnabar moth, *Tyria jacobaeae* (Lepidoptera: Arctiidae). Master's thesis, Oregon State University, Corvallis.

———. 1978a. *Pacific Northwest Regional Commission land resource inventory demonstration project: Inventory of the distribution and abundance of tansy ragwort in western Oregon.* Salem: Oregon Department of Agriculture.

———. 1978b. The role of biological control agents in integrated control of tansy ragwort. In *Proceedings of the 4th International Symposium on Biological Control of Weeds, 30 August–2 September 1976, Gainesville, Florida,* ed. T. E. Freeman, 189–92. Gainesville: University of Florida, Institute of Food and Agricultural Sciences.

Isaacson, D., and H. Radtke. 1993. An economic evaluation of biological control of tansy ragwort in western Oregon. In *Proceedings of the Oregon Society of Weed Science, 18–19 October 1993, Clackamas, Oregon,* 3–4. Corvallis: Oregon Society of Weed Science.

James, R. R. 1989. The relative impact of single vs. multiple agents on the biological control of tansy ragwort (*Senecio jacobaea*). Master's thesis, Oregon State University, Corvallis.

Kok, L. T., T. J. McAvoy, and W. T. Mays. 1986. Impact of tall fescue grass and *Carduus* thistle weevils on the growth and development of musk thistle (*Carduus nutans*). *Weed Sci.* 34:966–71.

McEvoy, P. B. 1984. Dormancy and dispersal in dimorphic achenes of tansy ragwort, *Senecio jacobaea* L. (Compositae). *Oecologia* 61:160–68.

McEvoy, P. B., and C. S. Cox. 1987. Wind dispersal distances in dimorphic achenes of ragwort, *Senecio jacobaea*. *Ecology* 68:2006–15.

McEvoy, P. B., C. S. Cox, and E. Coombs. 1991. Successful biological control of ragwort. *Ecol. Applications* 1:430–42.

McEvoy, P. B., C. S. Cox, R. R. James, and N. T. Rudd. 1990. Ecological mechanisms underlying successful biological weed control: Field experiments with ragwort, *Senecio jacobaea*. In *Proceedings of the 7th International Symposium on Biological Control of Weeds, 6–11 March 1988, Rome, Italy,* ed. E. S. Delfosse, 55–66. Rome: Ministero dell'Agricoltura e delle Foreste / Melbourne: CSIRO.

McEvoy, P. B., N. T. Rudd, C. S. Cox, and M. Huso. 1993. Disturbance, competition, and herbivory effects on ragwort *Senecio jacobaea* populations. *Ecological Monographs* 63:55–75.

May, R. M., and M. P. Hassell. 1988. Population dynamics and biological control. *Phil. Trans. Royal Soc. London B* 318: 129–69.

Murdoch, W. W., J. Chesson, and P. L. Chesson. 1985. Biological control in theory and practice. *Am. Nat.* 125:344–66.

Myers, J. H. 1980. Is the insect or the plant the driving force in the cinnabar moth–tansy ragwort system? *Oecologia* 47:16–21.

Newton, H. C. F. 1933. On the biology of some species of *Longitarsus* (Col., Chrysom.) living on ragwort. *Bull. Entomol. Res.* 24:511–20.

Parker, H. L. 1960. Starvation tests with the larvae of the cinnabar moth. *J. Econ. Entomol.* 53:472–73.

Pemberton, R. W., and C. E. Turner. 1990. Biological control of *Senecio jacobaea* in northern California, an enduring success. *Entomophaga* 35:71–77.

Piper, G. L. 1985. Biological control of weeds in Washington: Status report. In *Proceedings of the 6th International Symposium on Biological Control of Weeds, 19–25 August 1984, Vancouver, British Columbia,* ed. E. S. Delfosse, 817–26. Ottawa: Agriculture Canada.

Roitman, J. N. 1983. The pyrrolizidine alkaloids of *Senecio triangularis*. *Aust. J. Chem.* 36:1203–13.

Taylor, A. D. 1990. Metapopulations, dispersal, and predator-prey dynamics: An overview. *Ecology* 71:429–33.

Turner, C. E. 1985. Conflicting interests and biological control of weeds. In *Proceedings of the 6th International Symposium on Biological Control of Weeds, 19–25 August 1984, Vancouver, British Columbia,* ed. E. S. Delfosse, 203–25. Ottawa: Agriculture Canada.

van der Meijden, E. 1979. Herbivore exploitation of a fugitive plant species: Local survival and extinction of the cinnabar moth and ragwort in a heterogeneous environment. *Oecologia* 42: 307–23.

Wardle, D. A. 1987. The ecology of ragwort (*Senecio jacobaea* L.): A review. *New Zealand J. Ecol.* 10:67–76.

72 / Yellow Starthistle

C. E. TURNER, J. B. JOHNSON, AND J. P. McCAFFREY

yellow starthistle
Centaurea solstitialis L.
Asteraceae

INTRODUCTION

Biology and Pest Status

Yellow starthistle, *Centaurea solstitialis* L., is a Eurasian annual that is a naturalized weed in North America (Maddox, Mayfield, and Poritz 1985). It is a particular problem in California, where about 3,200,000 hectares are infested (Maddox and Mayfield 1985); Idaho, where 81,000 hectares are infested (Callihan et al. 1989); Oregon, where 400,000 hectares are infested (Coombs person. commun. 1992); and Washington, where 54,000 hectares are infested (Roché and Roché 1988). Its range continues to expand (Maddox and Mayfield 1985). It is primarily a weed of rangelands but also of alfalfa and cereal grains, orchards, vineyards, roadsides, and recreational lands (Robbins, Bellue, and Ball 1970; Maddox and Mayfield 1985). It may have been introduced to North America as a seed contaminant of imported alfalfa seed; multiple introductions are likely (Maddox and Mayfield 1985). The first known specimen in western North America was collected in 1869 in Oakland, California (Maddox 1981). The weed's long, sharp spines deter grazing by livestock, and it can displace more desirable vegetation, including native plants. If ingested by horses, it can cause a chronic and potentially fatal neurological disorder called "chewing disease," or nigropallidal encephalomalacia (Cordy 1978).

Historical Notes

Research on the biological control of yellow starthistle began before the start of the W-84 project. In 1959 a conflict with beekeepers, who value the weed, was resolved at a meeting sponsored by the California State Chamber of Commerce. In it, representatives of the livestock and beekeeping industries unanimously agreed that a weed control project should begin, because the losses caused by the weed outweighed its benefits to beekeepers (Turner 1985). L. A. Andres of the United States Department of Agriculture, Agricultural Research Service (USDA-ARS) began foreign exploration for natural enemies, which was continued in its early phases by H. Zwölfer through a contract with the University of California in the early 1960s. Since then there have been periods of relative inactivity, caused in part by the failure of the first imported insect—*Urophora jaculata* Rondani—to become established. In 1984 the project was assigned a high priority by the USDA-ARS, and in northern Greece D. M. Maddox of the USDA-ARS at Albany carried out field-plot host-specificity tests of *Bangasternus orientalis* (Capiomont) with R. Sobhian of the USDA-ARS at Rome.

RESULTS

Objective A: The Identification and Introduction of Beneficial Organisms

A study by J. B. Johnson, J. P. McCaffrey, and F. W. Merickel (1992) showed that in Idaho the endemic phytophagous insects associated with the yellow starthistle are sparse, are composed of generalist species, and do not have a significant impact on the weed. However, in southern Europe at least 42 insect species use yellow starthistle as a host plant (Clement 1990). As of 1992 six capitulum-feeding insects (three species of tephritid flies and three species of weevils) had been introduced for the control of the weed, and as of 1993 five were established. These insects will be discussed in the order of their introduction. A rust disease, whose host specificity is currently being evaluated, is also briefly discussed.

The California Department of Food and Agriculture and the U.S. Department of the Interior, Bureau of Land Management, provided funding for parts of the research reviewed here. The U. S. Department of Agriculture, Agricultural Research Service, provided funding for research at the University of Idaho. We thank K. L. Chan for assistance with preparing the manuscript, and L. A. Andres and R. D. Goeden for their review of the manuscript.

***Urophora jaculata* Rondani (Diptera: Tephritidae).**
The genus *Urophora* is in the subfamily Myopitinae (White and Korneyev 1989). In a systematic study of the genus (White and Clement 1987; White and Korneyev 1989), researchers resolved the confusion among biological control workers between *U. jaculata* Rondani and *U. sirunaseva* (Hering), which both attack yellow starthistle and are similar in behavior and morphology. *Urophora jaculata* attacks the weed in Italy and in parts of mainland Greece (White and Korneyev 1989), inducing lignified unilocular galls in the receptacle tissue of capitula (White and Clement 1987). H. Zwölfer (1969), who considered the fly to be *U. sirunaseva,* carried out host-specificity and oviposition tests on the fly from Italy. The only field host records for *U. jaculata* are from yellow starthistle (Zwölfer 1969; White and Korneyev 1989).

In 1969, 1970, 1976, and 1977 the fly was imported (as *U. sirunaseva*) from Italy for releases in California, but it did not become established. Although it oviposited on the weed and the eggs hatched, the larvae died as first instars (Ehler and Andres 1983). I. M. White and S. L. Clement (1987) have demonstrated through garden-plot host-specificity tests in Rome, Italy, that *U. jaculata* did not reproduce on yellow starthistle from California, Idaho, or Washington. This indicates that *U. jaculata* has intraspecific host specificity to some, but not all, populations of yellow starthistle, and that the yellow starthistle in the United States originated from outside the geographic range of *U. jaculata*.

***Urophora sirunaseva* (Hering) (Diptera: Tephritidae).**
The biology of *U. sirunaseva* is similar to that of *U. jaculata*. In their discussion of the biology of *U. sirunaseva,* R. Sobhian and H. Zwölfer (1985) included information on both species without distinguishing between them. More recently, R. Sobhian (1993) provided a description of the biology of the true *U. sirunaseva*. *Urophora sirunaseva* appears to be bivoltine in Greece and in the United States. Females oviposit through the top of closed, intermediate-stage capitula that have vertically oriented spines; they deposit their eggs among the immature florets. In a laboratory study, 10 female flies oviposited an average of 136 eggs each (Turner 1994). Lignified, unilocular galls are formed around the developing larvae in the upper layers of the receptacle; mature larvae then overwinter in those galls.

According to I. M. White and V. A. Korneyev (1989), *U. sirunaseva* attacks the weed in northeastern Greece, Turkey, Moldavia, Ukraine, and northern Israel. In Crete the fly attacks the closely related *Centaurea idaea* Boiss. & Heldr., which is also in the subgenus *Solstitiaria* (Dostal 1976). The only field host records for *U. sirunaseva* are yellow starthistle and *C. idaea*

(White and Korneyev 1989). I. M. White and S. L. Clement (1987) showed that *U. sirunaseva*, unlike *U. jaculata*, could develop on yellow starthistle from California and Idaho. C. E. Turner (1994) has carried out laboratory host-specificity tests in Albany, California. The results of field-plot host-specificity tests in northern Greece are reported by K. Groppe, R. Sobhian, and J. Kashefi (1990) and by S. L. Clement and R. Sobhian (1991).

In 1984 the true *U. sirunaseva* was first released into the United States in California using flies from northern Greece, and in Idaho using flies from Turkey. In 1985 more flies from northern Greece were released in California, Idaho, Oregon, and Washington; further releases then ceased until the taxonomic confusion between *U. sirunaseva* and *U. jaculata* was clarified. Researchers thought that the fly had not become established, until populations were discovered in 1989 at each of its only release sites in Loomis, California, and Phoenix, Oregon (Turner et al. 1994). In 1990 field releases commenced again from the California and Oregon populations and from material from northern Greece. As of 1993 *U. sirunaseva* was known to be established in at least California, Oregon, and Washington (Turner et al. 1994).

***Bangasternus orientalis* (Capiomont) (Coleoptera: Curculionidae).** *Bangasternus* is a Palearctic genus in the subfamily Cleoninae and the tribe Lixini; it is native to the area roughly between Greece and the Caucasus region (Ter-Minasyan 1978). R. Sobhian and H. Zwölfer (1985), D. M. Maddox et al. (1986), and R. Sobhian, G. Campobasso, and P. H. Dunn (1992) have described the biology of this univoltine weevil. Overwintered adults emerge in the spring to feed externally on the plant, to mate, and to oviposit. The females prefer to oviposit on the early-stage closed capitula, laying most eggs singly on scale leaves just below the capitula. They then cover the eggs with a mucous cap, which may protect the egg from desiccation. R. Sobhian and H. Zwölfer (1985) have reported that under laboratory conditions the oviposition period can last up to 100 days, and that a single female may oviposit up to 417 eggs. Neonate larvae tunnel through the scale leaves into the peduncle and up into the interior of the capitulum to feed primarily in the receptacle. Mature larvae pupate in capitula and emerge as adults in late summer to overwinter outside the host plant.

D. M. Maddox and R. Sobhian (1987) have carried out field-plot host-specificity tests in northern Greece. In 1985 the weevil, imported from northern Greece, was first released in California, Idaho, Oregon, and Washington (Maddox et al. 1986); it became established in each of those states.

Chaetorellia australis Hering (Diptera: Tephritidae).
Chaetorellia is a Palearctic genus in the tribe Terelliini (White and Marquardt 1989). *Chaetorellia australis* was originally described by E. M. Hering as a subspecies of *Chaetorellia hexachaeta* (Loew) (White and Marquardt 1989); the *C. hexachaeta* referred to by R. Sobhian and H. Zwölfer (1985) and by R. Sobhian and I. S. Pittara (1988) is actually *C. australis* Hering. *Chaetorellia australis* is found in Bulgaria, Greece, Hungary, Moldavia, and Turkey (White and Marquardt 1989).

Females oviposit just beneath the involucral bracts, preferring late-stage closed capitula. Females in cage studies oviposited up to 243 eggs during a maximum ovipositional period of 60 days (Sobhian and Pittara 1988). Each female tends to oviposit one egg per capitulum (Sobhian and Pittara 1988) and may mark that capitulum with an oviposition-deterring pheromone (Pittara and Katsoyannos 1990). Larvae tunnel through the involucre into the interior of the capitulum, where they feed on developing achenes. In a field study in northern Greece, *C. australis* larvae each destroyed a mean of 86.3 percent of the achenes in a capitulum (Sobhian and Pittara 1988). Three generations per year occur in northern Greece (Sobhian and Pittara 1988). The mature larvae of the overwintering generation form cocoons of pappus hairs in the capitula, from which the adults emerge the following spring (Sobhian and Pittara 1988). In northern Greece the overwintered flies that emerge the earliest oviposit on *Centaurea cyanus* L., because this plant—as it does in the United States—generally flowers before yellow starthistle. The summer generations of the fly oviposit on yellow starthistle (Sobhian and Pittara 1988).

The only field-host records for the fly are from *C. solstitialis* and *C. cyanus* in various Eurasian localities, and from *Centaurea depressa* Bieb. in Turkey (White and Marquardt 1989). Laboratory host-specificity tests of *C. australis* from northern Greece were carried out in Albany, California (Maddox, Mayfield, and Turner 1990), and the fly was reared from only *C. solstitialis* and *C. cyanus* in a host-range study of capitulum insects sampled from natural populations of thistles (Cardueae) in mainland Greece (Turner, Sobhian, and Maddox 1989). S. L. Clement and R. Sobhian (1991) have carried out field-plot host-specificity tests in northern Greece.

In 1988 the first North American field release of *C. australis*, imported from northern Greece, was made in California. In 1989 adults were released in California, Idaho, Oregon, and Washington, and there have been additional releases since. As of 1993 the fly was known to be established in at least Oregon (Coombs person. commun.) and Washington (Piper person. commun.), in areas that contain both *C. cyanus*—a naturalized, minor

weed in North America—and yellow starthistle. As in Greece, in western North America there is a partial overlap in the distribution of *C. cyanus* and *C. solstitialis*. The former tends to occur in cooler, moister habitats than does yellow starthistle, and it is more common in Idaho, Oregon, and Washington than in California. If both host-plant species must be present in an area for the fly to become established, this could seriously constrain the use of *C. australis* in the biological control of yellow starthistle.

Eustenopus villosus (Boheman) (Coleoptera: Curculionidae). *Eustenopus* is a Palearctic genus in the subfamily Cleoninae and the tribe Lixini (Ter-Minasyan 1978). The weevil referred to as *Eustenopus hirtus* cf. *abbreviatus* Faust in R. Sobhian and H. Zwölfer (1985) and as *Eustenopus hirtus* (Waltl) by S. L. Clement et al. (1988) is actually *Eustenopus villosus*, which occurs from Greece and Syria to the Caucasus region.

Some aspects of the biology of *E. villosus* have been reported in R. Sobhian and H. Zwölfer (1985), by L. Fornasari, C. E. Turner, and L. A. Andres (1991), and by L. Fornasari and R. Sobhian (1993). The weevil is univoltine. Overwintered adults become active in the spring to feed, mate, and oviposit. Adults feed by chewing through the sides of the involucre into closed capitula, frequently destroying the generally preferred early stages. Adults also feed on the stems, causing them to wilt above the feeding point. Females insert a single egg through a hole chewed through the involucre into the interior of a late-stage closed capitulum. The larva develops inside the capitulum, feeding primarily on the receptacle. The adults emerge in late summer and overwinter outside the host plant. Feeding by both adults and larvae reduces the weed's seed production: adult feeding reduces the number of capitula per plant, and larval feeding reduces the number of achenes per capitulum (Fornasari, Turner, and Andres 1991). In a limited laboratory study with two males and two females per test plant, adult and larval feeding reduced the number of achenes per plant by 98.8 percent (Fornasari, Turner, and Andres 1991).

In Rome S. L. Clement et al. (1988) carried out laboratory host-specificity studies on *E. villosus* from northern Greece, as did L. Fornasari, C. E. Turner, and L. A. Andres (1991) in Rome and in Albany, California. S. L. Clement and R. Sobhian (1991) carried out field-plot host-specificity tests in northern Greece. In 1990 *E. villosus*, imported from northern Greece, was first released in California, Idaho, Oregon, and Washington; the weevil became established in each of those states.

Larinus curtus Hochhut (Coleoptera: Curculionidae). *Larinus*, an Old World genus in the subfamily Cleoninae and tribe Lixini, feeds exclusively on plants in the family Asteraceae (Zwölfer, Frick, and Andres 1971; Ter-Minasyan 1978). *Larinus curtus* occurs in Italy and Egypt eastward to the Caucasus region (Sobhian and Zwölfer 1985). The weevil is univoltine. Adults feed and oviposit on capitula with open flowers, depositing their eggs in and among the flowers. The larvae feed on developing achenes. K. Groppe, R. Sobhian, and J. Kashefi (1990) carried out field-plot host-specificity tests in northern Greece. Laboratory host-specificity tests were conducted by L. Fornasari and C. E. Turner (in press) in Rome and in Albany, California. In 1992 *L. curtus* imported from Greece was first released in California, Idaho, Oregon, and Washington. As of 1993 the weevil was established, at least in California (Turner unpub. data), Washington (Piper person. commun.), and Oregon (Coombs person. commun.).

Puccinia jaceae Otth (Uredinales).

This rust fungus attacks the leaves of the yellow starthistle. Researchers have collected isolates of the rust from the weed in Greece, Italy, Turkey, the former Soviet Union, and Yugoslavia (Bruckart 1989). W. L. Bruckart (1989) carried out host-specificity tests on an isolate from Turkey. With the exception of the minor weed *C. cyanus*, this rust showed a low infectivity of the other test plant species (Bruckart 1989). A biological control agent such as *P. jaceae*, which attacks the vegetative plant body, would be a particularly valuable complement to the capitulum-feeding insects previously described. As of 1993, researchers are still evaluating the rust's host specificity.

Objective D: Impact

The five capitulum insects that have become established appear to have an excellent potential to greatly reduce the production of seeds—the weed's only means of reproduction and dispersal. Together, their oviposition brackets the entire course of the weed's capitulum development: from oviposition on very early, closed capitulum buds by *B. orientalis*; through oviposition on intermediate to late, closed capitula by *U. sirunaseva*, *C. australis*, and *E. villosus*; to oviposition on flowering capitula by *L. curtus*. However, these insects have been established for a relatively short time, and so far do not appear to have reduced the yellow starthistle population.

Urophora sirunaseva and *B. orientalis* have been established the longest—since the mid-1980s. *Urophora sirunaseva* populations are increasing, as illustrated by the data for the two oldest populations—those at Loomis, California, and Phoenix, Oregon. At the Loomis site the infestation rate as measured by capitula with one or more galls increased from 1.1 percent (n = 417 capitula) in 1989 to 22.1 percent (n = 483 capitula) in 1992, and at the Phoenix site from 2.2 percent (n = 311) in

1989 to 22.9 percent (n = 353) in 1992 (Turner et al. 1994). At both sites the mean number of galls per galled capitulum has increased, from 1.0 (n = 5 galled capitula) in 1989 to 2.1 (n = 107 galled capitula) in 1992 at the Loomis site, and from 1.4 (n = 7) in 1989 to 1.8 (n = 81) in 1992 at the Phoenix site (Turner et al. 1994). Through 1992 researchers observed as many as 12 galls in a yellow starthistle capitulum in the field at Loomis (Turner et al. 1994). These data indicate that the fly has the potential to build up to high densities. *Urophora sirunaseva* has also dispersed well. For example, following its 1990 release at Hornbrook, California, it spread 21 kilometers within 3 years (Turner et al. 1994).

Bangasternus orientalis is currently the most widely distributed and most numerous of the established insects. The indigenous wasp *Microdontomerus anthonomi* (Crawford) (Hymenoptera: Torymidae) was found parasitizing *B. orientalis* larvae at four sites in California, but at a rate of less than 1 percent in 1989 at the only site where the parasitization rate was quantified (Turner et al. 1990). A field study at two sites in California showed that *B. orientalis* reduced yellow starthistle seed production in infested capitula by a combined mean (in 1986 and 1987) of 53 percent at one site and 55 percent at the other (Maddox et al. 1991).

RECOMMENDATIONS

1. Obtain biological control agents that attack the vegetative plant body of the yellow starthistle: either the rust *P. jaceae*, if approved for field release, or other agents from further foreign exploration—for example, in Turkey or the Caucasus region of the former Soviet Union.

2. Elucidate the nature of the presumably vast seed bank of the weed in the soil in terms of seed location, viability, and germination.

3. Investigate management practices to see whether they can increase the pressure of interspecific plant competition on the weed.

4. Determine whether the capitulum insect species compete with or complement each other.

5. Determine whether the lignified galls of *U. sirunaseva* act as a resource sink on the weed, reducing the whole plant's overall production of capitula.

6. Elucidate the nature of the first generation of *C. australis* relative to host plants in Greece: Does the entire first generation reproduce on *C. cyanus*? What happens with the first generation in localities with yellow starthistle but without *C. cyanus*? Do they migrate from areas occupied by *C. cyanus*?

REFERENCES CITED

Bruckart, W. L. 1989. Host range determination of *Puccinia jaceae* from yellow starthistle. *Plant Disease* 73:155–60.

Callihan, R. H., F. E. Northam, J. B. Johnson, E. L. Michalson, and T. S. Prather. 1989. *Yellow starthistle: Biology and management in pasture and rangeland.* University of Idaho Cooperative Extension Service, Current Information Series 634.

Clement, S. L. 1990. Insect natural enemies of yellow starthistle in southern Europe and the selection of candidate biological control agents. *Environ. Entomol.* 19:1882–88.

Clement, S. L., and R. Sobhian. 1991. Host-use patterns of capitulum-feeding insects of yellow starthistle: Results from a garden plot in Greece. *Environ. Entomol.* 20:724–30.

Clement, S. L., T. Mimmocchi, R. Sobhian, and P. H. Dunn. 1988. Host specificity of adult *Eustenopus hirtus* (Waltl) (Coleoptera: Curculionidae), a potential biological control agent of yellow starthistle, *Centaurea solstitialis* L. (Asteraceae, Cardueae). *Proc. Entomol. Soc. Wash.* 90:501–7.

Cordy, D. R. 1978. *Centaurea* species and equine nigropallidal encephalomalacia. In *Effects of poisonous plants on livestock,* eds. R. F. Keeler, K. R. Van Kampen, and L. F. James, 327–36. New York: Academic Press.

Dostal, J. 1976. 138: *Centaurea* L. In *Flora europaea,* vol. 4 , eds. T. G. Tutin, V. H. Heywood, N. A. Burges, D. M. Moore, D. H. Valentine, S. M. Walters, and D. A. Webb, 254–301. Cambridge: Cambridge University Press.

Ehler, L. E., and L. A. Andres. 1983. Biological control: Exotic natural enemies to control exotic pests. In *Exotic plant pests and North American agriculture,* eds. C. L. Wilson and C. L. Graham, 395–418. New York: Academic Press.

Fornasari, L., and R. Sobhian. 1993. Life history of *Eustenopus villosus* (Coleoptera: Curculionidae), a promising biological control agent for yellow starthistle. *Environ. Entomol.* 22:684–92.

Fornasari, L., and C. E. Turner. In press. Host specificity of the Palaearctic weevil *Larinus curtus* Hochhut (Coleoptera: Curculionidae), a natural enemy of *Centaurea solstitialis* L. (Asteraceae: Cardueae). In *Proceedings of the 8th International Symposium on Biological Control of Weeds, 2–7 February 1992, Canterbury, New Zealand,* eds. E. S. Delfosse and R. R. Scott. Melbourne: DSIR/CSIRO.

Fornasari, L., C. E. Turner, and L. A. Andres. 1991. *Eustenopus villosus* (Coleoptera: Curculionidae) for biological control of yellow starthistle (Asteraceae: Cardueae) in North America. *Environ. Entomol.* 20:1187–94.

Groppe, K., R. Sobhian, and J. Kashefi. 1990. A field experiment to determine host specificity of *Larinus curtus* Hochhut (Col., Curculionidae) and *Urophora sirunaseva* Hg. (Dipt., Tephritidae), candidates for biological control of *Centaurea solstitialis* L. (Asteraceae), and *Larinus minutus* Gyllenhal, a candidate for biological control of *C. maculosa* Lam. and *C. diffusa* Lam. *J. Appl. Entomol.* 110:300–306.

Johnson, J. B., J. P. McCaffrey, and F. W. Merickel. 1992. Endemic phytophagous insects associated with yellow starthistle in northern Idaho. *Pan-Pacif. Entomol.* 68:169–73.

Maddox, D. M. 1981. *Introduction, phenology, and density of yellow starthistle in coastal, intercoastal, and Central Valley situations in California.* USDA-ARS, ARR-W-20.

Maddox, D. M., and A. Mayfield. 1985. Yellow starthistle infestations are on the increase. *Calif. Agric.* 39(11–12):10–12.

Maddox, D. M., and R. Sobhian. 1987. Field experiment to determine host specificity and oviposition behavior of *Bangasternus orientalis* and *Bangasternus fausti* (Coleoptera: Curculionidae), biological control candidates for yellow starthistle and diffuse knapweed. *Environ. Entomol.* 16:645–48.

Maddox, D. M., A. Mayfield, and N. H. Poritz. 1985. Distribution of yellow starthistle (*Centaurea solstitialis*) and Russian knapweed (*Centaurea repens*). *Weed Sci.* 33:315–27.

Maddox, D. M., A. Mayfield, and C. E. Turner. 1990. Host specificity of *Chaetorellia australis* (Diptera: Tephritidae) for biological control of yellow starthistle (*Centaurea solstitialis,* Asteraceae). *Proc. Entomol. Soc. Wash.* 92:426–30.

Maddox, D. M., D. B. Joley, A. Mayfield, and B. E. Mackey. 1991. Impact of *Bangasternus orientalis* (Coleoptera: Curculionidae) on achene production of *Centaurea solstitialis* (Asterales: Asteraceae) at a low and high elevation site in California. *Environ. Entomol.* 20:335–37.

Maddox, D. M., R. Sobhian, D. B. Joley, A. Mayfield, and D. Supkoff. 1986. New biological control for yellow starthistle. *Calif. Agric.* 40(11–12):4–5.

Pittara, I. S., and B. I. Katsoyannos. 1990. Evidence for a host-marking pheromone in *Chaetorellia australis. Entomol. Exp. Appl.* 54:287–95.

Robbins, W. W., M. K. Bellue, and W. S. Ball. 1970. *Weeds of California.* Sacramento: State Department of Agriculture.

Roché, C. T., and B. F. Roché, Jr. 1988. Distribution and amount of four knapweed (*Centaurea* L.) species in eastern Washington. *Northwest Sci.* 62:242–53.

Sobhian, R. 1993. Life history and host specificity of *Urophora sirunaseva* (Hering) (Dipt., Tephritidae), a candidate for biological control of yellow starthistle, with remarks on the host plant. *J. Appl. Entomol.* 116:381–90.

Sobhian, R., and I. S. Pittara. 1988. A contribution to the biology, phenology and host specificity of *Chaetorellia hexachaeta* Loew (Dipt., Tephritidae), a possible candidate for the biological control of yellow starthistle (*Centaurea solstitialis* L.). *J. Appl. Entomol.* 106:444–50.

Sobhian, R., and H. Zwölfer. 1985. Phytophagous insect species associated with flower heads of yellow starthistle (*Centaurea solstitialis* L.). *Zeit. angew. Entomol.* 99:301–21.

Sobhian, R., G. Campobasso, and P. H. Dunn. 1992. Contribution to the biology of *Bangasternus orientalis* (Capiomont) (Col., Curculionidae). *J. Appl. Entomol.* 113:93–102.

Ter-Minasyan, M. E. 1978. *Weevils of the subfamily Cleoninae in the fauna of the USSR* (in translation). New Delhi: Amerind.

Turner, C. E. 1985. Conflicting interests and biological control of weeds. In *Proceedings of the 6th International Symposium on*

Biological Control of Weeds, 19–25 August 1984, Vancouver, British Columbia, ed. E. S. Delfosse, 203–25. Ottawa: Agriculture Canada.

———. 1994. Host specificity and oviposition of *Urophora sirunaseva* (Hering) (Diptera: Tephritidae), a natural enemy of yellow starthistle. *Proc. Entomol. Soc. Wash.* 96:31–36.

Turner, C. E., R. Sobhian, and D. M. Maddox. 1989. Host-specificity studies of *Chaetorellia australis* (Diptera: Tephritidae), a prospective biological control agent for yellow starthistle, *Centaurea solstitialis* (Asteraceae). In *Proceedings of the 7th International Symposium on Biological Control of Weeds, 6–11 March 1988, Rome, Italy,* ed. E. S. Delfosse, 231–36. Rome: Ministero dell'Agricoltura e delle Foreste / Melbourne: CSIRO.

Turner, C. E., E. E. Grissell, J. P. Cuda, and K. Casanave. 1990. *Microdontomerus anthonomi* (Crawford) (Hymenoptera: Torymidae), an indigenous parasitoid of the introduced biological control insects *Bangasternus orientalis* (Capiomont) (Coleoptera: Curculionidae) and *Urophora affinis* Frauenfeld (Diptera: Tephritidae). *Pan-Pacific Entomol.* 66:162–66.

Turner, C. E., R. Sobhian, D. B. Joley, E. M. Coombs, and G. L. Piper. 1994. Establishment of *Urophora sirunaseva* (Hering) (Diptera: Tephritidae) for biological control of yellow starthistle in the western United States. *Pan-Pacif.. Entomol.* 70:206–11.

White, I. M., and S. L. Clement. 1987. Systematic notes on *Urophora* (Diptera, Tephritidae) species associated with *Centaurea solstitialis* (Asteraceae, Cardueae) and other Palaearctic weeds adventive in North America. *Proc. Entomol. Soc. Wash.* 89:571–80.

White, I. M., and V. A. Korneyev. 1989. A revision of the western Palaearctic species of *Urophora* Robineau-Desvoidy (Diptera: Tephritidae). *Syst. Entomol.* 14:327–74.

White, I. M., and K. Marquardt. 1989. A revision of the genus *Chaetorellia* Hendel (Diptera: Tephritidae) including a new species associated with spotted knapweed, *Centaurea maculosa* Lam. (Asteraceae). *Bull. Entomol. Res.* 79:453–87.

Zwölfer, H. 1969. Urophora siruna-seva (Hg.) (*Dipt.: Trypetidae*), *a potential insect for the biological control of* Centaurea solstitialis *L. in California.* Commonwealth Institute of Biological Control Technical Bulletin 11.

Zwölfer, H., K. E. Frick, and L. A. Andres. 1971. *A study of the host plant relationships of European members of the genus* Larinus (*Coleoptera: Curculionidae*). Commonwealth Institute of Biological Control Technical Bulletin 14.

73 / RUSSIAN THISTLE

R. D. GOEDEN AND R. W. PEMBERTON

Russian thistle
Salsola australis R. Brown
Chenopodiaceae

INTRODUCTION

Biology and Pest Status

Russian thistle is a profusely branched annual herb, bushy and rounded in habit, that is 1.5 to 12 decimeters tall and 3 to 20 decimeters in diameter. Its efficient spreading taproot, abundant seed production, reduced leaf surface, and photosynthesizing branches adapt this weed well to disturbed semiarid agricultural environments such as overgrazed rangeland, grain fields, fallow land, and field margins, as well as to nonagricultural areas such as vacant residential lots and railroad or highway rights-of-way. Russian thistle is also damaging as a seed contaminant, as a forage weed that may contain toxic levels of soluble oxalates and nitrates, and as an alternative host of several economically significant insect species (Goeden 1968; Goeden and Ricker 1968; Reed and Hughes 1970). The mature plants break off at ground level, creating spiny, brittle, windblown tumbleweeds that fill drainage canals, catchments, and swimming pools and that pile up against fences and dwellings, becoming objectionable sights, fire hazards, and foci for the accumulation of other windblown debris (Goeden and Ricker 1968).

Historical Notes

As its name implies, Russian thistle is native to southern Russia and western Siberia. It was originally introduced to the United States as a contaminant of flax seed in South Dakota in 1873 (Robbins, Bellue, and Ball 1970) but was probably also present in many shipments of Turkestani alfalfa imported into North America (Harris 1984). Within 20 years it had spread to 16 western states and several Canadian provinces (Robbins, Bellue, and Ball 1970). It is now widespread throughout western North America (Reed and Hughes

1970) and occurs on a limited scale in Hawaii (Markin person. commun.).

RESULTS

Objective A: The Identification and Introduction of Beneficial Organisms

A biological control program was initiated in 1965, when L. A. Andres (person. commun.) found several apparently stenophagous species of insects attacking annual species of *Salsola* in the south-central Soviet Union during a survey of the natural enemies of *Halogeton glomeratus* (M. von Bieberstein) C. A. Meyer. In a 1965 and 1966 preintroduction survey of insects attacking Russian thistle in southern and central California, researchers demonstrated that the weed lacked effective natural enemies (Goeden and Ricker 1968). In the late 1960s, Public Law 480 research projects on the biological control of Russian thistle were begun in Egypt and Pakistan. In 1970 R. D. Goeden surveyed the insects on annual *Salsola* species in Turkey and detected several promising natural enemies (Goeden 1973).

One difficulty in evaluating the insect enemies of Russian thistle was that many of the species also fed on cultivated Chenopodiaceae such as beets, spinach, and Swiss chard. Eventually two promising candidate agents were selected for intensive study, both moths in the genus *Coleophora* (Lepidoptera): the stem and branch borer, *C. parthenica* Meyrick, and the case bearer, *C. klimeschiella* Toll. In field and laboratory studies in Egypt, Pakistan, and California, researchers demonstrated that both moth species were confined to hosts in the genera *Salsola* and *Halogeton* (Baloch and Mushtaque 1973; Hawkes and Mayfield 1976, 1978a,b; Khan and Baloch 1976; Baloch and Khan 1978). The latter genus

We thank R. J. Lavigne, J. P. McCaffrey, G. P. Markin, G. L. Piper, and J. M. Story for their contributions to this report. We also acknowledge the support and assistance of state and federal collaborators.

contains *H. glomeratus*, which is also a rangeland weed in the western United States (Reed and Hughes 1970).

The life history of *C. parthenica* has been described by G. M. Baloch and M. Mushtaque (1973), by R. B. Hawkes and A. Mayfield (1976, 1978b), and by G. M. Baloch and A. G. Khan (1978). The last-instar larvae overwinter inside open mines in the dried branches and stems of dead Russian thistles. They pupate in early spring, and moths emerge in the late spring through thin epidermal windows cut by the fully grown larvae. The moths mate shortly after emerging, and females soon begin to oviposit, gluing eggs singly to leaves on terminal growth of young plants. The newly hatched larvae chew directly through the chorions to enter the leaves and then mine downward into the branches, where they largely confine their feeding to the central pith. In Egypt, Pakistan, and California the moth completes three generations annually; the last instars of the third generation enter diapause in the late fall and early winter within senescent plants, where they overwinter. R. B. Hawkes and A. Mayfield (1976) have reported on the host range of the moth.

A. G. Khan and G. M. Baloch (1976) as well as R. B. Hawkes and A. Mayfield (1978a) have described the life history of *C. klimeschiella*. The last-instar larvae overwinter inside elongated cylindrical cases that are fashioned from hollowed-out leaves and attached to dead Russian thistles. They pupate within the cases, and adults emerge in the spring. Newly emerged females mate and then oviposit in leaf axils; eggs hatch after 1 week. The first two of five instars are leaf miners; the last three are mobile case bearers, which move between and mine a succession of up to 15 leaves. In Pakistan one to three overlapping generations are produced annually.

In the spring of 1973 researchers released the first insectary-reared adults of *C. parthenica* at several locations in southern California (Hawkes et al. 1975; Goeden, Ricker, and Hawkes 1978; Hawkes and Mayfield 1978b), with one release each made in Idaho, Nevada, and Utah (Hawkes and Mayfield 1978b). These initial releases were made on actively growing Russian thistles in field cages. The cages were removed after 1 or 2 weeks (Goeden, Ricker, and Hawkes 1978) or when the insects had completed one or two generations (Hawkes and Mayfield 1978b). Most of the released moths were reared from stocks imported from Pakistan, but in southern California Egyptian and Turkish moths were also released (Goeden, Ricker, and Hawkes 1978). The initial release in Idaho took place in Owyhee County and involved 173 adults, with 99 more adults released in 1979; neither release resulted in establishment (McCaffrey person. commun.).

From 1975 on, new releases in California were made primarily in the southern half of the state, in coopera-

tion with the state's Department of Food and Agriculture and Department of Transportation. The releases involved numerous transfers of dead plants infested with overwintering larvae or of green plants bearing F_1 pupae. These came from a field colony of *C. parthenica* near Kettleman City, in central California, that had been established from Pakistani stocks (Goeden, Ricker, and Hawkes 1978; Hawkes and Mayfield 1978b).

Between 1976 and 1981, as part of the W-84 project, *C. parthenica* was transported from the USDA Biological Control of Weeds Laboratory at Albany, California, to several western states. Overwintering larvae were shipped to Arizona for release in the Phoenix and Tucson areas in 1976 (Hawkes and Mayfield 1978b). In 1979, 88 *C. parthenica* were released at Winifred, Montana, but did not become established (Story person. commun.). In 1979 and 1980, 134 and 50 adults moths were released in Laramie and Platte counties, Wyoming, respectively, but also failed to become established (Lavigne person. commun.). Similarly, in 1979, 1980, and 1981, 133, 48, and 64 adults, respectively, were released on caged plants near Almota, Washington—again, the release was unsuccessful (Piper person. commun.). In 1980, 84 *C. parthenica* were released on Hawaii (Lai, Funasaki, and Higa 1982), and again establishment failed (Markin person. commun.). From 1977 to 1980, *C. parthenica* was also released in Colorado, Minnesota, Nebraska, Nevada, South Dakota, Texas, Utah, and Saskatchewan, Canada (Pemberton unpub. data). The moth has become established in California, Texas, and Arizona, where it produces three generations annually (Julien 1987).

After extensive study and testing at Albany, California, during 1975 and 1976 (Hawkes and Mayfield 1978a), researchers released laboratory-reared *C. klimeschiella* larvae at three locations in southern California. These and additional releases made in 1978 each involved about 400 larvae placed on uncaged, succulent, actively growing plants (Hawkes 1979; Goeden, Ricker, and Müller 1987).

In 1979 researchers began distributing *C. klimeschiella* adults from Albany to other Project W-84–affiliated states. During 1979, 1980, and 1981, 353, 46, and 1,000 individuals, respectively, were released in field-caged plants near Almota, Washington, but did not become established because the plants were accidentally sprayed with herbicide and killed (Piper person. commun.). During 1980 and 1981, 84 and 160 individuals, respectively, were released on Hawaii (Lai, Funasaki, and Higa 1982; Lai and Funasaki 1984), but establishment failed (Markin person. commun.). In 1980, 1981, and 1982, 50 adults, 500 larvae, and 214 adults were shipped to Wyoming and released in Platte, Sweetwater, and Sublette counties, respectively, but there was no evidence that they became established (Lavigne person. commun.). Similarly, a total of 621 and 150 moths were released at Moccasin, Montana, in 1981 and 1982, respectively, but they failed to become established (Story person. commun.). However, between 1978 and 1982, a total of over 1,354 moths were released at three locations in Idaho, where they did become established (McCaffrey person. commun.). From 1980 to 1982, *C. klimeschiella* was also shipped to Colorado, Kansas, Minnesota, Nebraska, Nevada, South Dakota, and Texas (Pemberton unpub. data). According to M. H. Julien (1987), *C. klimeschiella* became established in California, Idaho, and Washington. However, the Washington listing, due to the reasons previously described, was erroneous.

Objective D: Impact

So far researchers have conducted evaluation studies only in California. R. B. Hawkes and A. Mayfield (1976, 1978b) and L. A. Andres (1977) considered the impact of *C. parthenica* on Russian thistle to be quite damaging; however, R. D. Goeden, D. W. Ricker, and R. B. Hawkes (1978), R. D. Goeden and D. W. Ricker (1979), and R. W. Pemberton (1986) have presented field and laboratory data demonstrating that *C. parthenica* had little impact on the Russian thistle and was ineffective as a biological control agent. For example, at mean densities three times those that R. B. Hawkes and A. Mayfield (1978b) considered damaging, R. D. Goeden and D. W. Ricker (1979) determined that neither the number of larvae nor the amount of pith tunneling influenced plant size, mortality, or maturation (that is, incidence of fruiting). Likewise, R. W. Pemberton (1986) found no significant correlations between the measured levels of water stress, translocation, or various reproductive indices and the levels of larval infestation and damage.

In southern California's Coachella Valley, researchers studied the relative importance of host-plant synchronization and native predators and parasitoids in limiting the population growth of *C. parthenica*, from 1979 to 1980 (Nuessly and Goeden 1983, 1984a,b) and from 1985 to 1986 (Müller and Goeden 1990; Müller, Nuessly, and Goeden 1990).

The oviposition period of the overwintering F_3 generation coincided with a high mortality of young plants caused by moisture stress. This density- and herbivore-independent plant mortality, which also caused the death of the young larvae living in the plants, was mainly responsible for the drastic population decrease of *C. parthenica* in the early spring. During the two summer generations in 1985, larvae suffered minimal losses from parasitism and rodent predation; however, spider predation was estimated to account for 28.7 percent of the losses of F_1 moths and 30.4 percent of F_2 moths (Müller, Nuessly, and Goeden 1990). Of the seven species of spi-

ders observed feeding on *C. parthenica* adults, the two most common were *Dictyna reticulata* Gertsch & Ivie (Dictynidae) and *Diguetia mojavea* Gertsch (Diguetidae) (Nuessly and Goeden 1983, 1984a). The house mouse, *Mus musculus* L., preyed on the overwintering larvae (Nuessly and Goeden 1984b).

The *C. parthenica* population increased slightly during the summer of 1979 and was highest in mature plants during the winter of 1979/80, when the highest larval mortality was also recorded. During that winter intrinsic larval mortality was 8.5 percent, with rodents removing 25 percent of the overwintering larvae, and larval parasitism reaching 42.1 percent, compared with only 13.5 percent in 1984/85, and with 11.0 percent in 1985/86 (Müller, Nuessly, and Goeden 1990). H. Müller and R. D. Goeden (1990) described the parasitoid complex that *C. parthenica* acquired 10 years after being introduced into southern California. The principal parasitoids were *Norbanus perplexus* Ashmead (Pteromalidae) and *Eurytoma strigosa* Bugbee (Eurytomidae), both solitary ectoparasitoids.

Rodent predation—which is limited by their need for the physical support offered by branches—was most significant in the secondary branches arising from the primary branches on the central stem. Parasitoids predominantly attacked larvae in the thinner, tertiary branches, which the rodents could not reach. Thus, poor host-plant synchronization, generalist predators, and parasitoids together considerably limited the population growth of *C. parthenica*. The moth's failure to control the Russian thistle can be partly explained by the extent of these mortality factors, together with the limited impact of larval tunneling (Goeden, Ricker, and Hawkes 1978; Goeden and Ricker 1979; Pemberton 1986).

Little is known about the effectiveness of *C. klimeschiella*, other than reports that it is well established in Idaho and central California (Julien 1987; McCaffrey person. commun.). R. D. Goeden, D. W. Ricker, and H. Müller (1987) reported on the difficulties the moth has had in becoming established in southern California. Those difficulties are due in part to suspected poor host-plant acceptance and in part to indigenous parasitoids and predators—for example, *Phylabaenus atriplexus* (Foster) (Coleoptera: Cleridae) and *Macroneura* sp. (Hymenoptera: Eupelmidae), which transferred from the gall midges *Neolasioptera* and *Ophiomyia* species (Diptera: Cecidomyiidae) on native *Atriplex* species (Chenopodiaceae) (Hawkins and Goeden 1984).

RECOMMENDATIONS

1. Introductions of additional species of Russian thistle insect enemies from the former Soviet Union—for example, *Careopalpis davletshinae* Marjkovski and

Asiodiplosis propria Marjkovski (Diptera: Cecidomyiidae) (Andres person. commun.)—should be undertaken.

2. Phytopathogens recorded from *Salsola* species—for example, *Diplodia salsolae*, *Phoma salsolae*, and particularly *Uromyces salsolae*—should also be investigated as possible biological control agents.

3. Root-feeding insects should be sought in Eurasia to fill the one critical empty niche on this xerophytic weed in North America.

4. Evaluation studies of *C. klimeschiella* should be conducted in Idaho and central California, where this agent has been established.

REFERENCES CITED

Andres, L. A. 1977. The biological control of weeds. In *Integrated control of weeds,* eds. J. D. Fryer and S. Matsunaka, 153–76. Tokyo: University of Tokyo Press.

Baloch, G. M., and A. G. Khan. 1978. Possibilities for the biological control of Russian thistles, *Salsola* spp. (Chenopodiaceae). In *Proceedings of the 4th International Symposium on Biological Control of Weeds, 30 August–2 September 1976, Gainesville, Florida,* ed. T. E. Freeman, 108–12. Gainesville: University of Florida, Institute of Food and Agricultural Sciences.

Baloch, G. M., and M. Mushtaque. 1973. Insects associated with *Halogeton* and *Salsola* in Pakistan with notes on the biology, ecology, and host specificity of the important enemies. In *Proceedings of the 2nd International Symposium on Biological Control of Weeds, 4–7 October 1971, Rome, Italy,* ed. P. H. Dunn, 103–13. Commonwealth Agricultural Bureaux International Miscellaneous Publication 6. Farnham Royal, Slough, U.K.: Commonwealth Institute of Biological Control.

Goeden, R. D. 1968. Russian thistle as an alternate host to economically important insects. *Weed Sci.* 16:102–3.

———. 1973. Phytophagous insects found on *Salsola* in Turkey during exploration for biological weed control agents for California. *Entomophaga* 18:439–48.

Goeden, R. D., and D. W. Ricker. 1968. The phytophagous insect fauna of Russian thistle (*Salsola kali* var. *tenuifolia*) in southern California. *Ann. Entomol. Soc. Am.* 61:67–72.

———. 1979. Field analyses of *Coleophora parthenica* (Lepidoptera: Coleophoridae) as an imported natural enemy of Russian thistle, *Salsola iberica,* in the Coachella Valley of southern California. *Environ. Entomol.* 8:1099–1101.

Goeden, R. D., D. W. Ricker, and R. B. Hawkes. 1978. Establishment of *Coleophora parthenica* (Lepidoptera: Coleophoridae) in southern California for the biological control of Russian thistle. *Environ. Entomol.* 7:294–96.

Goeden, R. D., D. W. Ricker, and H. Müller. 1987. Introduction, recovery, and limited establishment of *Coleophora klimeschiella* (Lepidoptera: Coleophoridae) on Russian thistles, *Salsola australis,* in southern California. *Environ. Entomol.* 16:1027–29.

Harris, P. 1984. *Salsola pestifer* A. Nels., Russian thistle (Chenopodiaceae). In *Biological control programmes against insects and weeds in Canada 1969–1980,* eds. J. S. Kelleher and M. A. Hulme, 191–93. Farnham Royal, Slough, U.K.: Commonwealth Agricultural Bureaux International.

Hawkes, R. B. 1979. *Biological control of Russian thistle along California highways.* California Department of Transportation Publication T 900 B55.

Hawkes, R. B., and A. Mayfield. 1976. Host specificity and biological studies of *Coleophora parthenica* Meyrick, an insect for the biological control of Russian thistle. In *Commemorative volume in entomology,* 37–43. Moscow: University of Idaho, Department of Entomology.

———. 1978a. *Coleophora klimeschiella,* biological control agent for Russian thistle: Host-specificity testing. *Environ. Entomol.* 7:257–61.

———. 1978b. *Coleophora* spp. as biological control agents against Russian thistle. In *Proceedings of the 4th International Symposium on Biological Control of Weeds, 30 August–2 September 1976, Gainesville, Florida,* ed. T. E. Freeman, 113–16. Gainesville: University of Florida, Institute of Food and Agricultural Sciences.

Hawkes, R. B., R. D. Goeden, A. Mayfield, and D. W. Ricker. 1975. Biological control of Russian thistle. *Calif. Agric.* 29(4):3–4.

Hawkins, B. A., and R. D. Goeden. 1984. Organization of a parasitoid community associated with a complex of galls on *Atriplex* spp. in southern California. *Ecol. Entomol.* 9:271–92.

Julien, M. H. 1987. *Biological control of weeds: A world catalogue of agents and their target weeds,* 2nd ed. Wallingford, U.K.: Commonwealth Agricultural Bureaux International.

Khan, A. G., and G. M. Baloch. 1976. *Coleophora klimeschiella* (Lepidoptera: Coleophoridae) a promising biocontrol agent for Russian thistles, *Salsola* spp. *Entomophaga* 21:425–28.

Lai, P. Y., and G. Y. Funasaki. 1984. Introductions for biological control in Hawaii: 1981 and 1982. *Proc. Hawaii. Entomol. Soc.* 25:83–88.

Lai, P. Y., G. Y. Funasaki, and S. Y. Higa. 1982. Introductions for biological control in Hawaii: 1979 and 1980. *Proc. Hawaii. Entomol. Soc.* 24:108–13.

Müller, H., and R. D. Goeden. 1990. Parasitoids acquired by *Coleophora parthenica* (Lepidoptera: Coleophoridae), ten years after its introduction into southern California for the biological control of Russian thistle. *Entomophaga* 35:257–68.

Müller, H., G. S. Nuessly, and R. D. Goeden. 1990. Natural enemies and host-plant asynchrony contributing to the failure of the introduced moth, *Coleophora parthenica* Meyrick (Lepidoptera: Coleophoridae), to control Russian thistle. *Agric. Ecosyst. and Environ.* 32:133–42.

Nuessly, G. S., and R. D. Goeden. 1983. Spider predation on *Coleophora parthenica* (Lepidoptera: Coleophoridae), a moth imported for the biological control of Russian thistle. *Environ. Entomol.* 12:1433–38.

———. 1984a. Aspects of the biology and ecology of *Diguetia mojavea* Gertsch (Araneae: Diguetidae). *J. Arachnol.* 12: 75–85.

———. 1984b. Rodent predation on larvae of *Coleophora parthenica* (Lepidoptera: Coleophoridae), a moth imported for the biological control of Russian thistle. *Environ. Entomol.* 13:1433–38.

Pemberton, R. W. 1986. The impact of a stem-boring insect on the tissues, physiology, and reproduction of Russian thistle. *Entomol. Exp. Appl.* 42:169–77.

Reed, C. F., and R. O. Hughes. 1970. *Selected weeds of the United States,* 138–39. USDA-ARS Handbook 366. Washington, D.C.: U.S. Department of Agriculture.

Robbins, W. W., M. K. Bellue, and W. S. Ball. 1970. *Weeds of California.* Sacramento: State Printing Office.

74 / St. Johnswort

J. P. McCaffrey, C. L. Campbell, and L. A. Andres

> ## St. Johnswort
> ### *Hypericum perforatum* L.
> ### Hypericaceae

INTRODUCTION

Biology and Pest Status

St. Johnswort, also known as St. John's-wort, Klamath weed, and goatweed, is an erect, multistemmed perennial herb 3 to 10 decimeters tall that reproduces by rhizomes or by seeds—of which each plant produces 15,000 to 33,000 (Sampson and Parker 1930; Tisdale, Hironaka, and Pringle 1959; Crompton et al. 1988). The simple, opposite, ovate leaves and the perfect yellow flowers borne in cymes are characterized by the presence of small, black glands that contain the photodynamic pigment hypericin (Crompton et al. 1988). Hypericin is toxic to livestock; light-colored animals are particularly affected by this toxin's photodynamic properties. St. Johnswort is primarily a weed of rangelands, but it also occurs in poorly managed pastures, roadsides, waste places, forest clearings, and other nonarable sites (Sampson and Parker 1930; Crompton et al. 1988).

Historical Notes

St. Johnswort is native to Europe, Asia, and northern Africa. Although first reported in the United States in Pennsylvania in 1793, it was not until the early 1900s that the weed spread to the western states. By the mid-1940s, about 2 million hectares in the western United States and Canada were infested; almost half of this acreage was in California (Smith 1951; Holloway 1958, 1964; Goeden 1978).

Researchers in Australia were the first to develop biological control programs aimed at St. Johnswort; the initial survey of insects associated with the plant was done in England in 1926. Several natural enemies were eventually introduced to Australia, with only limited success. However, the Australian efforts laid the groundwork for the successful efforts in the United States that followed, from the mid-1940s through the 1950s.

California researchers initiated the St. Johnswort biological control program in the western United States in 1944 (Holloway 1948); it culminated in the release and establishment within the western region, including Hawaii, of these four insects: *Chrysolina quadrigemina* (Suffrian), *Chrysolina hyperici* (Forster) (Coleoptera: Chrysomelidae), *Agrilus hyperici* (Creutzer) (Coleoptera: Buprestidae), and *Zeuxidiplosis giardi* Kieffer (Diptera: Cecidomyiidae).

Since natural enemies were first released in the western states, during the mid- to late 1940s and early 1950s, St. Johnswort infestations have been reduced to 1 to 3 percent of 1940s levels (Huffaker and Kennett 1959; Tisdale 1976). The weed continues to exist along roadsides and occasionally in stands of up to several hectares.

This report will not address those initial biological control efforts, which predate the W-84 project. Those interested in earlier activities should see the excellent reviews by R. D. Goeden (1978) and M. H. Julien (1987). The information reported here is from studies conducted since the 1960s.

RESULTS

Objective A: The Identification and Introduction of Beneficial Organisms

***Chrysolina quadrigemina* and *C. hyperici*.** Both of these chrysomelids have become established in the western United States, but *C. quadrigemina* is by far the dominant species. C. L. Campbell and J. P. McCaffrey (1991a) reported that *C. quadrigemina* dominates northern Idaho grassland sites; however, they found *C. hyperici* to be the

We thank W. F. Barr (Professor Emeritus, University of Idaho), G. L. Piper (Associate Professor of Entomology, Washington State University), E. M. Coombs (Oregon State Department of Agriculture), N. Rees (USDA-ARS Biological Weed Control Laboratory, Bozeman, Montana), G. Markin (USDA-Forest Service, Hilo, Hawaii), and E. Yoshioka (Hawaii State Department of Agriculture) for their contributions to this report.

only species present at a more mesic, forest meadow site they studied. Both beetles have contributed to the successful control of St. Johnswort in the western United States, but both large- and small-scale redistribution programs are still used in many areas to supplement low populations or to establish the beetles in newly infested areas. For example, over 350,000 adult *C. quadrigemina* and *C. hyperici* were collected and redistributed to 11 counties in Washington state during the late 1970s through the early 1980s (Piper 1985). In 1965 over 3,000 beetles—listed as *C. quadrigemina*—were collected in California and shipped by the USDA-ARS Biological Weed Control Laboratory to the island of Hawaii for release on Mt. Hualalai (Anonymous 1966). As both *C. quadrigemina* and *C. hyperici* became established in Hawaii, both species were probably present in this initial shipment (Anonymous 1976). C. L. Campbell and J. P. McCaffrey (1988) outlined specific collection and redistribution protocols for these insects, to facilitate the involvement of various weed management personnel.

Agrilus hyperici. The United States has had mixed results with the root-boring buprestid *A. hyperici*, imported from France in 1950. The beetle displayed an ability to destroy the weed in California (Holloway and Huffaker 1953), but it was displaced by *C. quadrigemina* and consequently persisted only in several areas (Holloway 1964). By 1956 the insect had become established in its original southeastern Washington release area (Johansen 1957), but by 1985 it had spread no more than 100 kilometers (Neighbors 1986). In 1953 the beetle was released throughout northern Idaho, where it was observed periodically thereafter. Because it was later thought to have died out, it was reintroduced into northern Idaho in 1981 (Barr unpub. data), where recent surveys indicate that it is now widely distributed (Campbell and McCaffrey 1991a). In 1988 over 1,500 adult beetles were redistributed from northern Idaho collection sites to Grant, Okanogan, and Spokane counties in Washington (Piper unpub. data). A recent release in Montana in 1976 led to a limited recovery of the beetle (Story 1979). *Agrilus* is apparently not established in Oregon: in 1986 adults collected from Idaho were released near Salem, Oregon, but the site was lost to cultivation (Coombs person. commun.).

Zeuxidiplosis giardi. Researchers collected this gall-forming midge in California and released it in Idaho, Oregon, and Washington in 1981, but it failed to become established. Subsequent releases in Oregon and Idaho in 1982, and in Washington in 1982 and 1983, were also unsuccessful. In 1965 the midge was introduced to Mt. Hualalai, Hawaii, from New Zealand; it has become well established (Anonymous 1966).

Aplocera (=Anaitis) plagiata. In 1988 this geometrid moth was cleared for release in the United States. In 1989 researchers made three releases of 400 larvae each in Montana, and one release of 400 larvae in Oregon. All of those larvae came from a lab culture initiated from an established field population in British Columbia. The moth is now established in Idaho, Montana, Oregon, and Washington.

Objective B: Ecological and Physiological Studies

Chrysolina spp. C. L. Campbell and J. P. McCaffrey (1991a) described the population dynamics of *C. quadrigemina* and *C. hyperici* in northern Idaho. They found that the egg was the primary overwintering stage, with some larvae and adults successfully overwintering. Given that few larvae were present in the fall and early winter, substantial larval feeding on procumbent growth did not occur until early spring, when the overwintering eggs hatched. This situation differs from that in California, where larvae feed throughout the winter and are nearly mature by spring (Holloway 1948; Holloway and Huffaker 1951). In Idaho larvae fed on procumbent and bolting stems until they entered the soil to pupate, between mid-May and early June. During June and July, adults fed on upright stem and bloom-phase foliage and then left the plants by the end of July to estivate in litter and soil. In California adult feeding and estivation occurred about a month earlier (Holloway and Huffaker 1951) than in northern Idaho, where the adults began mating and ovipositing about mid-September, following rains that terminated their estivation.

C. L. Campbell and J. P. McCaffrey (1991a) found that *C. quadrigemina* was the dominant species at most grassland sites they studied; however, the ratio of *C. quadrigemina* to *C. hyperici* varied depending on the specific habitat. Generally, their results supported those of P. Harris and D. P. Peschken (1971), P. Harris and M. Maw (1984), and K. S. Williams (1985), who concluded that *C. quadrigemina* is better adapted to drier areas, *C. hyperici* to moister areas.

F. Wilson (1943) and D. P. Peschken (1972) documented the presence of bronze, blue, green, and purple morphs of *C. quadrigemina*. D. P. Peschken (1972) correlated the ratio of bronze morphs to other color morphs in British Columbia with winter temperatures; the higher the winter temperatures, the higher the proportion of bronze morphs. C. L. Campbell and J. P. McCaffrey (1991a) reported similar results in northern Idaho.

Agrilus hyperici. The population dynamics of *A. hyperici* in the United States have only recently been described. C. L. Campbell and J. P. McCaffrey (1991a) found that in Idaho, overwintering larvae pupated from late May to early July. Adult beetles were most common from late June to mid-July. The females deposited the majority of eggs during July; first-instar larvae were found by mid-July, mature larvae by September. At the sites the two researchers studied, the proportion of stems infested with eggs ranged from 0 to 92 percent. More than 87 percent of the plants containing larvae held only one individual larva, pupa, or teneral adult beetle. Mean larval numbers were always less than one per plant. V. M. Neighbors (1986) found a similar situation in eastern Washington, where over 90 percent of the infested plants contained only one larva. The rates of infestation by all *A. hyperici* life stages (except eggs) ranged from 8 to 87 percent in eastern Washington (Neighbors 1986) and from 1 to 23 percent in northern Idaho (Campbell and McCaffrey 1991a). C. L. Campbell and J. P. McCaffrey (1991a) noted that it is also important to check apparently dead plants for *A. hyperici* infestations—for example, at one site over 44 percent of the plants assumed dead were infested with live *A. hyperici* larvae, pupae, or teneral adult beetles.

F. Wilson (1943) questioned whether the weakening of the plants by *A. hyperici* would reduce the availability of foliage for *C. quadrigemina*, particularly during the important period of adult feeding and oviposition in the fall, when the only available foliage would be from seedlings. C. L. Campbell and J. P. McCaffrey (1991a) reported that they found no evidence to support this idea; they suggested that abiotic factors such as a fall drought were probably more important in limiting the availability of procumbent growth.

Because the adult stage of *A. hyperici* depends on St. Johnswort terminal foliage as a food source during its summer ovipositional period, it was not known whether defoliation by the adult chrysomelids negatively influenced oviposition by *A. hyperici*. V. M. Neighbors (1986) found that *A. hyperici* oviposited on identical proportions of flowering plants that were 0 to 50 percent defoliated by *Chrysolina* species or by hand, but that they oviposited on a significantly smaller proportion of totally defoliated plants. At one out of four study sites having a relatively high *Chrysolina* species population, C. L. Campbell and J. P. McCaffrey (1991a) reported a significant negative correlation (r = −0.28) between the presence of *A. hyperici* eggs and adult *Chrysolina* species during bloom; this suggests that the chrysomelids could negatively affect *A. hyperici* oviposition. However, the two researchers never observed complete defoliation of individual plants at that particular site. Thus, as adult *A. hyperici* require a relatively small quantity of foliage (Wilson 1943), oviposition should rarely be affected—particularly since *A. hyperici* will oviposit on completely defoliated stems (Neighbors 1986; Campbell and McCaffrey 1991a).

Objective C: The Conservation and Augmentation of Beneficial Organisms

C. L. Campbell and J. P. McCaffrey (1991b) conducted a survey of the potential arthropod parasites and predators of *Chrysolina* species associated with St. Johnswort. After collecting *Chrysolina* from four representative St. Johnswort infestations in northern Idaho, they recovered no parasitoids from a total of 6,247 eggs, 1,340 third- and fourth-instar larvae, 162 pupae, and 610 adults. Sweepnet sampling of the foliage yielded few predaceous insect species, and their numbers were generally very low. The most common ground-dwelling species collected by pitfall traps were the carabids, *Harpalus fraternus* LeConte, *Anisodactylus similis* LeConte, *Amara littoralis* Mannerheim, and *Zacotus matthewsi* LeConte; the gryllid, *Gryllus assimilis* F.; and several species of lycosid wolf spiders. None of these ground-dwelling predators was present in the St. Johnswort stands at the times when larvae and adult *Chrysolina* species were active—that is, spring and early summer—or when adults entered estivation. Although these ground-dwelling predators occur together in time and space with the chrysomelids, it is not known whether they actually prey on the *Chrysolina* species. This issue needs further study. In fact, *Chrysolina* species feeding on *Hypericum* produce polyoxygenated steroids that may chemically deter predators (Daloze, Braekman, and Pasteels 1985). This chemical defense also warrants further study.

Between 1985 and 1987 populations of *C. quadrigemina* in Oregon declined to the point at which it was impractical to collect and redistribute them. In 1988, 5 percent of the adult beetles collected near Salem were infected with the microsporidian parasite *Nosema* (Coombs person. commun.), which may partly explain the earlier population crash.

Objective D: Impact

The biological control results obtained in California and portions of Oregon have been called spectacular (Ritcher 1966; Goeden 1978). In California, St. Johnswort was reduced by over 99 percent after the beetles were introduced (Huffaker and Kennett 1959), and it was eventually removed from the state's noxious weed list (Huffaker 1967). In Washington *Chrysolina* species reportedly produced a substantial reduction of the weed (Johansen 1957; Piper 1985). In most areas of Montana, *C. quadrigemina* populations apparently stabilized at too low a level to reduce the plant population adequately (Story 1979). In northern Idaho, the overall abundance of St. Johnswort fluctuates around an estimated 3 percent of the 1948 level (Tisdale 1976).

In California the most effective species of beetle was

C. quadrigemina (Huffaker 1967), although *C. hyperici* apparently continued to persist at a low level. *Agrilus hyperici* was displaced by *C. quadrigemina* (Holloway 1964), and the buprestid's current status is unclear, even though all of the insect's life stages have been observed on *H. perforatum* in the Sierra foothills and the Mt. Shasta area. *Agrilus hyperici* has also been reported interfering with attempts to determine *C. quadrigemina*'s impact on a native *Hypericum* species, *H. concinnum* Benth. (Andres 1985). In Washington *A. hyperici* appeared to be maintaining an important presence—with infestation levels reaching up to 87 percent—but it has spread no more than about 100 kilometers from its original release point (Neighbors 1986). After several introductions, the status of the buprestid in Montana was still uncertain (Story 1979). Despite renewed attempts in British Columbia, it has not become successfully established there (Harris and Maw 1984), although the fact that large amounts of the weed persist in the cooler, moister areas of the province dominated by *C. hyperici* (Williams 1985) might be partly explained by the absence of *A. hyperici*. In northern Idaho the cooler, moister areas dominated by *C. hyperici* (Campbell and McCaffrey 1989a) have the buprestid but no persistent St. Johnswort stands.

Northern Idaho has experienced an excellent reduction of St. Johnswort. This has been largely due to *C. quadrigemina*, but biological control in northern Idaho may have the distinction of including the most widespread—and perhaps the most successful—involvement of *A. hyperici*, as indicated by the beetle's distribution and infestation levels. It has been found in all areas of northern Idaho that have been surveyed (Campbell and McCaffrey 1991a).

In Hawaii St. Johnswort control is also good, with one researcher noting that ". . . it is completely absent from open areas and found in the shade of taller shrubs and trees" (Yoshioka person. commun.).

In California *C. quadrigemina*, *Z. giardi*, and *A. hyperici* have been reported reproducing on *Hypericum concinnum* (Andres 1985), and *C. quadrigemina* is also reported on *H. calycinum* (Andres 1985). In Hawaii *Chrysolina hyperici* has been reported on *H. degeneri* Fosberg (Anonymous 1971), but other than some occasional severe damage to *H. calycinum*, its impact on the other *Hypericum* species is unclear.

The stability and value of the plant communities that have replaced St. Johnswort have differed from area to area. In California the void created by the reduction in St. Johnswort was largely filled by a desirable perennial grass, *Danthonia californica* Bol., plus annual grasses, legumes, and forbs of varying forage value (Huffaker 1951; Huffaker and Kennett 1959). In northern Idaho, however, the forage species that replaced the weed have not solved

the problem of unstable plant communities. For example, E. W. Tisdale (1976) reported that St. Johnswort populations were largely replaced by the same forbs and seral grasses—that is, *Bromus* species—that had been dominant before St. Johnswort invaded the area in the early 1900s. Unfortunately, many of these replacement species were introduced annual grasses. Compared with St. Johnswort they were an improvement in forage, but they also provided an unstable plant community that was subject to invasion by other weedy forbs and grasses (Tisdale 1976). Thus, in Idaho nonforage species such as yellow starthistle (*Centaurea solstitialis* L.) and spotted knapweed (*C. maculosa* Lam.) have replaced St. Johnswort in many areas (Campbell and McCaffrey 1991a). This situation illustrates how important it is to have a strategy for successfully ensuring the re-establishment of stable plant communities after a weed has been controlled.

RECOMMENDATIONS

1. Introduce *Aplocera plagiata* into all of the western states.

2. Initiate studies on *Aplocera plagiata* once it has become established.

3. Redistribute *Agrilus hyperici* in Oregon and Montana.

4. Clarify the role that *Agrilus hyperici* plays in the control of St. Johnswort.

5. Further evaluate the role of abiotic factors in regulating bioagent populations and influence.

6. Evaluate St. Johnswort biological control within the framework of a vegetation management system for intermountain grassland habitats.

7. Follow studies by Australian scientists on the testing of other biological control agents, and take action as necessary.

REFERENCES CITED

Andres, L. A. 1985. Interaction of *Chrysolina quadrigemina* and *Hypericum* spp. in California. In *Proceedings of the 6th International Symposium on Biological Control of Weeds, 19–25 August 1984, Vancouver, British Columbia*, ed. E. S. Delfosse, 235–39. Ottawa: Agriculture Canada.

Anonymous. 1966. Notes and exhibitions. *Proc. Hawaii. Entomol. Soc.* 19:134–35.

———. 1971. Notes and exhibitions. *Proc. Hawaii. Entomol. Soc.* 21:60–62.

———. 1976. Notes and exhibitions. *Proc. Hawaii. Entomol. Soc.* 22:177.

Campbell, C. L., and J. P. McCaffrey. 1988. *Collection and redistri-*

bution of biological control agents of St. Johnswort. University of Idaho Cooperative Extension Service Current Information Series 798.

———. 1991a. Population trends, seasonal phenology, and impact of *Chrysolina quadrigemina, C. hyperici* (Coleoptera: Chrysomelidae), and *Agrilus hyperici* (Coleoptera: Buprestidae) associated with *Hypericum perforatum* in northern Idaho. *Environ. Entomol.* 20:303–15.

———. 1991b. Survey of potential arthropod parasites and predators of *Chrysolina* spp. associated with St. Johnswort in northern Idaho. *Pan-Pacific Entomol.* 66:217–26.

Crompton, C. W., I. V. Hall, K. I. N. Jensen, and P. D. Hilderbrand. 1988. The biology of Canadian weeds: 83. *Hypericum perforatum* L. *Can. J. Plant Sci.* 68:149–62.

Daloze, D., J. C. Braekman, and J. M. Pasteels. 1985. New polyoxygenated steroidal glucosides from *Chrysolina hyperici* (Coleoptera: Chrysomelidae). *Tetrahaedron Lett.* 26:2311–14.

Goeden, R. D. 1978. Biological control of weeds. In *Introduced parasites and predators of arthropod pests and weeds: A world review,* ed. C. P. Clausen, 357–414. USDA-ARS Handbook 480. Washington, D.C.: U.S. Department of Agriculture.

Harris, P., and M. Maw. 1984. *Hypericum perforatum* L., St. John's-wort (Hypericaceae). In *Biological control programmes against insects and weeds in Canada 1969–1980,* eds. J. S. Kelleher and M. A. Hulme, 171–77. Farnham Royal, Slough, U.K.: Commonwealth Agricultural Bureaux International.

Harris, P., and D. P. Peschken. 1971. *Hypericum perforatum* L., St. Johns-wort (Hypericaceae). In *Biological control programmes against insects and weeds in Canada 1959–1968,* 89–94. Commonwealth Institute of Biological Control Technical Communication 4.

Holloway, J. K. 1948. Biological control of Klamath weed: Progress report. *J. Econ. Entomol.* 41:56–57.

———. 1958. The biological control of Klamath weed in California. *Proc. 10th Internat. Congr. Entomol.* 4:557–60.

———. 1964. Projects in biological control of weeds. In *Biological control of insect pests and weeds,* ed. P. DeBach, 650–70. New York: Reinhold.

Holloway, J. K., and C. B. Huffaker. 1951. The role of *Chrysolina gemellata* in the biological control of Klamath weed. *J. Econ. Entomol.* 44:244–47.

———. 1953. Establishment of a root borer and a gall fly for control of Klamath weed. *J. Econ. Entomol.* 46:65–67.

Huffaker, C. B. 1951. The return of native perennial bunchgrass following the removal of Klamath weed (*Hypericum perforatum* L.) by imported beetles. *Ecology* 32:443–58.

———. 1967. A comparison of the status of biological control of St. Johnswort in California and Australia. *Mushi* (Tokyo) 39:51–73.

Huffaker, C. B., and C. E. Kennett. 1959. A ten-year study of vegetational changes associated with biological control of Klamath weed. *J. Range Manage.* 12:69–82.

Johansen, C. A. 1957. History of biological control of insects in Washington. *Northwest Sci.* 31:57–79.

Julien, M. H. 1987. *Biological control of weeds: A world catalogue of agents and their target weeds,* 2nd ed. Wallingford, U.K.: Commonwealth Agricultural Bureaux International.

Neighbors, V. M. 1986. Distribution and abundance of *Agrilus hyperici* (Creutzer) (Coleoptera: Buprestidae), a biocontrol agent of *Hypericum perforatum* L., in Washington. Master's thesis, Washington State University, Pullman.

Peschken, D. P. 1972. *Chrysolina quadrigemina* (Coleoptera: Chrysomelidae) introduced from California to British Columbia against the weed *Hypericum perforatum*: Comparison of behaviour, physiology, and colour in association with post-colonization adaptation. *Can. Entomol.* 104:1689–98.

Piper, G. L. 1985. Biological control of weeds in Washington: Status report. In *Proceedings of the 6th International Symposium on Biological Control of Weeds, 19–25 August 1984, Vancouver, British Columbia,* ed. E. S. Delfosse, 817–26. Ottawa: Agriculture Canada.

Rees, C. J. C. 1969. Chemoreceptor specificity associated with choice of feeding site by the beetle *Chrysolina brunsvicensis* on its foodplant, *Hypericum hirsutum. Entomol. Exp. Appl.* 12:565–83.

Ritcher, P. O. 1966. *Biological control of insects and weeds in Oregon.* Oregon Agricultural Experiment Station Technical Bulletin 90.

Sampson, A. W., and K. W. Parker. 1930. *St. Johns-wort on rangelands of California.* Berkeley: California Agricultural Experiment Station Bulletin 503.

Smith, J. M. 1951. Biological control of weeds in Canada. In *Proceedings of the 5th Meeting, Eastern Section, National Weed Committee, 7–9 November 1951, MacDonald College, Quebec,* 95–97.

Story, J. 1979. *Biological weed control in Montana.* Montana Agricultural Experiment Station Bulletin 717.

Tisdale, E. W. 1976. Vegetational responses following biological control of *Hypericum perforatum* in Idaho. *Northwest Sci.* 50:61–75.

Tisdale, E. W., M. Hironaka, and W. L. Pringle. 1959. Observations on the autecology of *Hypericum perforatum. Ecology* 40:54–62.

Williams, K. S. 1985. Climatic influences on weeds and their herbivores: Biological control of St. John's wort in British Columbia. In *Proceedings of the 6th International Symposium on Biological Control of Weeds, 19–25 August 1984, Vancouver, British Columbia,* ed. E. S. Delfosse, 127–32. Ottawa: Agriculture Canada.

Wilson, F. 1943. *The entomological control of St. John's wort* (Hypericum perforatum *L.), with particular reference to the insect enemies of the weed in southern France.* Australian Council of Science Industrial Research Bulletin 169.

75 / FIELD BINDWEED

S. S. ROSENTHAL

field bindweed
Convolvulus arvensis L.
Convolvulaceae

INTRODUCTION

Biology and Pest Status

Convolvulus arvensis L., field bindweed, is a prostrate or climbing perennial herb that occurs in all but the southernmost areas of the United States (Reed and Hughes 1970). The stems reach up to 3 meters in length and bear an abundance of white or pink funnel-shaped flowers up to 2.3 centimeters wide. The roots and rhizomes can extend over an area up to 6 meters in diameter and can penetrate 9 meters below the surface. The weed reproduces by long-lived seeds or by sprouts that rise from the lateral roots. Pieces of root that are broken off the main plant—as happens when the soil is cultivated—can form new plants. *Convolvulus arvensis* shows great variation in both form and physiological characteristics (Whitworth 1964; Garcia-Baudin and Darmency 1979).

Field bindweed is mainly a pest of cultivated land, but it is also found along roadsides, in fallow fields, and in other noncultivated areas. Its extensive root system makes it highly competitive with other plants for nutrients and water, and its long stems twine about the stems of cultivated plants, interfering with their growth and harvest. Field bindweed also serves as an alternative host for arthropod pests and several crop viruses (Holm et al. 1977).

Historical Notes

Although it is now found throughout the temperate regions of the world, *C. arvensis,* along with other members of the tribe Convolvuleae, is thought to have originated in the Mediterranean region and the adjoining part of western Asia (Sa'ad 1967). Present in North America by 1739 (Wiese and Phillips 1976), by 1965 it was considered the 25th most important weed in the United States (Danielson et al. 1968). It is a more serious problem in the arid western states than in the east. In California between 1965 and 1980, infested land increased by almost 500,000 acres (202,347 hectares), to an estimated 1.9 million acres (768,920 hectares)—despite the new herbicides developed for its control (Rosenthal 1983a). During the 1980 survey, California counties reported spending $4.9 million for field bindweed control and calculated that the losses it caused to various crops amounted to $25.2 million.

In North America field bindweed is such a serious weed that there has long been considerable interest in developing some kind of biological control. In Kansas researchers conducted an early field study to see which insects were attacking it (Smith 1938). A recent literature review of the research conducted in Canada, the United States, and Eurasia mentions hundreds of arthropods associated with *Convolulus* and *Calystegia* species in different parts of the world (Wang and Kok 1985). However, much of this work on *Convolvulus arvensis* predates the W-84 project or is not within its scope.

RESULTS

Objective A: The Identification and Introduction of Beneficial Organisms

Beginning in 1970, researchers conducted surveys in the Mediterranean area to find natural enemies of *C. arvensis* and closely related plants. This effort was part of a joint USDA–University of California project funded by the California Department of Transportation and Department of Food and Agriculture. Of the 155 species of arthropods and diseases found to be associated with field bindweed (Rosenthal 1981; Rosenthal and

I am very thankful to C. J. DeLoach for information on recent releases of insects against field bindweed in Texas, Oklahoma, Missouri, and New Jersey. I am also very grateful to the members of the European Biological Control Laboratory who collected and sent herbivores to the United States for release against field bindweed.

Buckingham 1982), the field surveys and subsequent literature review indicated that about 20 were host specific. However, preliminary laboratory testing of the most promising species indicated that all but three would develop on the sweet potato, *Ipomoea batatas* Monnet Lamarck, which is an important American crop plant. The exceptions were the defoliating chrysomelid beetle *Galeruca rufa* Germar (Rosenthal and Carter 1977); the defoliating noctuid moth *Tyta luctuosa* (Denis & Schiffermueller) (Rosenthal 1978); and the leaf-galling eriophyid mite, *Aceria malherbe* Nuzzaci (Rosenthal 1983b).

Galeruca rufa. In Italy *G. rufa* completes two to three generations between March and September every year (Rosenthal and Carter 1977; Rosenthal and Hostettler 1980) and overwintered as an adult. Both larvae and adults feed on the leaves of *C. arvensis*, *C. althaeoides* L., and *Calystegia sepium* (L.). When not feeding, the adults and larvae, along with pupae and clumps of yellow eggs, are generally found beneath the plants or debris on surface litter. In Italy the beetle's natural enemies include a eulophid larval parasite, probably *Asecodes* sp.; two fungal pathogens, *Beauveria bassiana* (Balsamo) Vuillemin and *Entomophthora* sp.; and the protozoan *Nosema* species.

In subsequent studies in the USDA quarantine laboratory at Albany, California, researchers found that *G. rufa* could utilize sweet potato as a host. This change in feeding specificity was believed to be caused by eliminating *Nosema* from the laboratory colony (Rosenthal; Krueger and Geyer unpub. data). As a result *G. rufa* has never been released in the United States, despite its ability to severely defoliate field bindweed in field cages in Italy (Rosenthal and Hostettler 1980).

Tyta luctuosa. This moth, which also defoliates field bindweed during most of the growing season in Europe (Rosenthal et al. 1988), is widely distributed in Eurasia and also occurs in North Africa. It undergoes two generations and perhaps a partial third between May and September each year; it overwinters as a pupa in the soil. This species is oligophagous on *Convolvulus* and *Calystegia* species (Rosenthal 1978) and, in laboratory tests, was able to complete its development on native North American *Calystegia* species (Clement et al. 1984). In 1986 the USDA Animal and Plant Health Inspection Service (APHIS) cleared the moth for release in North America.

The first releases were of eggs and larvae originating in Rome, Italy. In 1987 S. S. Rosenthal, in collaboration with R. Eikenbary, A. Wiese, and J. Cuda, placed the moth on field-caged bindweed plants in Enid, Oklahoma, and Amarillo, Texas. In 1988 and 1989 C. J.

Deloach imported more *T. luctuosa* from Italy to the USDA quarantine laboratory at the Grassland Forage Research Center at Temple, Texas. These moths were released in collaboration with B. Puttler at Columbia, Missouri, as well as at the original release sites in Texas and Oklahoma.

Aceria malherbe. In Italy, Spain, and Greece researchers found these gall mites distorting and galling field bindweed leaves and buds (Rosenthal 1981; Rosenthal and Buckingham 1982). In Greece the mite is multivoltine, active between May and early November, and overwinters belowground in field bindweed buds (Rosenthal 1983b). In the laboratory it produced a new generation every 10 days.

Although in California an indigenous *Aceria* species forms leaf galls on *Calystegia fulcrata* (Gray) Brummitt, researchers found neither this nor other gall mites attacking field bindweed in North America (Rosenthal, Andres, and Huffaker 1983). In preliminary host-specificity tests conducted in Thessaloniki, Greece (Rosenthal 1983b), and in Rome, Italy (Clement et al. 1984), *Aceria malherbe* Nuzzaci (=*A. convolvuli* in earlier reports) was unable to live on sweet potato but could reproduce on some North American *Calystegia* species. During 1986 researchers at the USDA quarantine laboratory in Albany, California, conducted host-specificity tests on 48 plant species representing 21 plant families (Rosenthal and Platts 1990). They concentrated on the Convolvuleae—including *Convolvulus*, many native North American species of *Calystegia*, *Evolvulus*, and *Jacquemontia*—and representatives of three other Convolvulacea tribes. Four ecotypes of field bindweed from different parts of the United States, three varieties of sweet potato, and native representatives of *Dichondra*, *Ipomoea*, and *Turbinia* were also tested.

This research confirmed that *A. malherbe* is oligophagous on *Convolvulus* and *Calystegia* species. Bindweed susceptibility to *A. malherbe* varied from high in New Jersey to intermediate in California and Texas and low in Nebraska. C. J. Deloach of the USDA in Temple, Texas, and state collaborators in Texas, Oklahoma, New Jersey, and Missouri are making releases of mites collected in Greece.

So far researchers have not found a natural enemy that is strictly host specific to field bindweed: even the most specialized natural enemies are able to use related species in the genera *Convolvulus* and *Calystegia*. Because there are few commercially valuable plants in these genera, however, such oligophagous organisms are unlikely to become economic pests. There is concern, though, that once they become established they might damage native North American *Calystegia* species. Because the 11 native North American *Calystegia* species are restricted to California and adjacent states, it has been possible to release *A. malherbe* east of the Rocky Mountains with reduced risk. If the risk to *Calystegia* species does not materialize, *A. malherbe* would be a good candidate for integrating into the pest management system now used in vineyards along the California coast (Rosenthal 1985).

RECOMMENDATIONS

1. Research is needed on integrating the biological control of field bindweed with the pest management systems used in orchards, vineyards, and field crops.

2. Pathogens, particularly the smut *Thecophora seminisconvolvuli*, need to be studied to develop bioherbicides for this weed.

REFERENCES CITED

Clement, S. L., S. S. Rosenthal, T. Mimmocchi, M. Cristofaro, and G. Nuzzaci. 1984. Concern for U.S. native plants affects biological control of field bindweed. In *Proceedings of the 10th International Congress on Plant Protection, 20–25 November 1983,* 775. Publication 5A–R3.

Danielson, W., B. Ennis, Jr., J. T. Holstun, Jr., L. L. Jansen, D. L. Klingman, F. L. Timmons, J. R. Paulling, and A. S. Fox. 1968. *Extent and cost of weed control with herbicides and an evaluation of important weeds: 1965.* USDA-ARS H–1. Washington, D.C.: U.S. Department of Agriculture.

Garcia-Baudin, J. M., and H. Darmency. 1979. Différences intraspécifiques chez *Convolvulus arvensis* L. *Weed Res.* 19:212–24.

Holm, L. G., D. L. Pluncket, J. V. Pancho, and J. P. Herberger. 1977. *The world's worst weeds.* Honolulu: University Press of Hawaii.

Reed, C. F., and R. O. Hughes. 1970. *Common weeds of the United States.* USDA-ARS Handbook 366. Washington, D.C.: U.S. Department of Agriculture.

Rosenthal, S. S. 1978. Host specificity of *Tyta luctuosa* (Lepidoptera: Noctuidae), an insect associated with *Convolvulus arvensis* (Convolvulaceae). *Entomophaga* 23:367–70.

———. 1981. European organisms of interest for the biological control of *Convolvulus arvensis* in the United States. In *Proceedings of the 5th International Symposium on Biological*

Control of Weeds, 22–27 July 1980, Brisbane, Australia, ed. E. S. Delfosse, 537–44. Melbourne: Commonwealth Scientific and Industrial Research Organization.

———. 1983a. Field bindweed in California: Extent and cost of infestation. *Calif. Agric.* 37(9–10):16–17.

———. 1983b. Current status and potential for biological control of field bindweed with *Aceria convolvuli.* In *Biological control of pests by mites,* eds. M. A. Hoy, G. L. Cunningham, and L. Knutson, 57–60. Berkeley: University of California Division of Agriculture and Natural Resources, Special Publication 3304.

———. 1985. The place of biological control of field bindweed in California's coastal vineyards. *Agric. Ecosyst. and Environ.* 13:43–58.

Rosenthal, S. S., and G. R. Buckingham. 1982. Natural enemies of *Convolvulus arvensis* in western Mediterranean Europe. *Hilgardia* 50(2):1–19.

Rosenthal, S. S., and J. Carter. 1977. Host specificity and biology of *Galeruca rufa*, a potential biological control agent for field bindweed. *Environ. Entomol.* 6:155–58.

Rosenthal, S. S., and N. Hostettler. 1980. *Galeruca rufa* (Coleoptera: Chrysomelidae) seasonal life history and the effect of its defoliation on its host plant, *Convolvulus arvensis* (Convolvulaceae). *Entomophaga* 25:381–88.

Rosenthal, S. S., and B. E. Platts. 1990. Host specificity of *Aceria (Eriophyes) malherbe*, a biological control agent for the weed *Convolvulus arvensis* (Convolvulaceae). *Entomophaga* 35:459–63.

Rosenthal, S. S., L. A. Andres, and C. B. Huffaker. 1983. Field bindweed in California: The outlook for biological control. *Calif. Agric.* 37(9–10):18–22.

Rosenthal, S. S., S. L. Clement, N. Hostettler, and T. Mimmocchi. 1988. Biology of *Tyta luctuosa* (Lepidoptera: Noctuidae) and its potential value as a biological control agent for the weed *Convolvulus arvensis. Entomophaga* 33:185–92.

Sa'ad, F. 1967. *The Convolvulus species of the Canary Isles, the Mediterranean region, and the Near and Middle East.* Rotterdam: Bronder-Offset.

Smith, R. C. 1938. A preliminary report on the insects attacking bindweed with special reference to Kansas. *Trans. Kansas Acad. Science* 41:183–91.

Wang, R., and L. T. Kok. 1985. Bindweeds and their biological control. *Biocontrol News and Information* 6(4):303–10.

Whitworth, J. W. 1964. The reaction of strains of field bindweed to 2,4-D. *Weeds* 15:275–80.

Wiese, A. F., and W. M. Phillips. 1976. Field bindweed. *Weeds Today* 7:22–23.

76 / LEAFY SPURGE

R. W. PEMBERTON

leafy spurge
Euphorbia esula L.
Euphorbiaceae

INTRODUCTION

Biology and Pest Status

Leafy spurge is a deep-rooted perennial herb native to Eurasia from Spain to Japan (Ohwi 1965; Radcliffe-Smith and Tutin 1968). The plant spreads through vigorous lateral root growth and forms large, coalescing patches that can dominate rangeland vegetation in the Great Plains region of North America. The 0.1- to 1.0-meter-high stems bear blue-green leaves and showy yellow-green inflorescences that produce an average of 140 seeds per stem (Lacey et al. 1985). The seeds, ejected by explosive capsules up to 5 meters from the plants (Bakke 1936), are dispersed by water, birds, grazing animals, ants, and humans (Bowes and Thomas 1978; Pemberton 1988).

All parts of the plant produce a latex that can cause dermatitis in humans and cattle (Lacey et al. 1985). More important, leafy spurge can also reduce rangeland's cattle-carrying capacity—through the loss of grasses by competition, and the tendency of cattle to avoid infested grass (Lacey et al. 1985)—by 50 to 70 percent (Alley et al. 1984) and, in some cases, by 100 percent (Watson 1985). At medium and high densities, leafy spurge can eliminate both grass and forb species (Nowierski and Harvey unpub. data).

Historical Notes

Leafy spurge was first recorded in North America at Newbury, Massachusetts, in 1827 (Britton 1921), but it has probably been introduced to the continent many separate times (Dunn 1979). G. W. Selleck, R. T. Coupland, and C. Frankton (1962) reported that leafy spurge occurred in 20 states and all Canadian provinces except Newfoundland. P. H. Dunn (1979) surveyed American weed scientists in the late 1970s, who reported leafy spurge in 26 states. By 1979 researchers estimated that the weed had infested more than 1 million hectares in North America (Nobel, Dunn, and Andres 1979). By 1983 the plant had infested more than 320,000 hectares in North Dakota (Messersmith and Lym 1983), and by 1985 more than 220,000 hectares in Montana (Lacey et al. 1985).

P. H. Dunn (1985) has theorized that the leafy spurge that infests the Great Plains region was introduced from the Ukraine and Penza Province in Russia's Volga Valley by Mennonite immigrants along with imported smooth broom-grass (*Bromus inermis* Leysser) and seed grains. In 1987 I collected leafy spurges from northern China and Inner Mongolia that are very similar to the Great Plains leafy spurge (Pemberton unpub. data). Introductions from those areas may have been made around the turn of the century through the cold-hardy seed grain collections of F. N. Meyer (Cunningham 1984).

The morphological variation in North American leafy spurge has created considerable confusion and disagreement concerning the plant's taxonomy. A. Radcliffe-Smith (1985) has created a key to 21 leafy spurge taxa—11 species and 10 hybrids—in North America that had been previously identified by L. Croizat (1945) and by P. H. Dunn and A. Radcliffe-Smith (1980). S. J. Harvey et al. (1988) examined the leaf morphology and latex triterpenoid composition of the Montana and four related European leafy spurges and concluded that all the Montana leafy spurge and three of the five European species were *Euphorbia esula*.

In the early 1960s, Agriculture Canada—spurred by the spread of the plant and the cost and ineffectiveness of chemical controls—initiated a biological control program against the weed (Harris 1984). In the mid-1970s

I thank R. B. Carlson, E. M. Coombs, P. Harris, R. J. Lavigne, R. M. Nowierski, G. L. Piper, N. E. Rees, and J. M. Story for their contributions to this report. I also wish to acknowledge G. R. Johnson for technical assistance with host-specificity research at the USDA-ARS Albany laboratory; P. Pecora and other USDA-ARS Rome, Italy, laboratory personnel for prequarantine evaluations of candidate insects, and for collecting insects for research and release; and the many state and federal collaborators who helped release leafy spurge insects.

the USDA-ARS Biological Control of Weeds Laboratory at Albany, California, and the USDA-ARS Laboratory in Rome, Italy, joined the effort—first in order to use insects cleared for introduction by Canada, and second to discover and test new insects (Pemberton 1985).

The biological control of leafy spurge has been, and still is, the subject of several conflicts of interest. *Euphorbia lathyris* L., a related spurge that belongs to the same subgenus (*Esula*) as leafy spurge, has been studied as a potential energy crop (Calvin 1978); however, even when irrigated and fertilized, its oil yields were not economically competitive with those of other oil sources (Sachs et al. 1981). The poinsettia (*Euphorbia pulcherrima* Wildenow) and the candelilla plant (*E. antisyphilitica* Zuccarini) of northern Mexico are both commercially important *Euphorbia* species (Harris et al. 1985), but since they are actually fairly distant relatives of leafy spurge—belonging to different subgenera—most of the insects that feed on leafy spurge find them unacceptable as hosts.

Another conflict concerns the large number (113) of native *Euphorbia* species in North America north of Mexico that could become hosts for leafy spurge biological control agents (Pemberton 1985). I employed a program strategy to select biological control insects that have host ranges no broader than the subgenus *Esula*. This excludes most native *Euphorbia* as potential hosts, including the legally protected species (Pemberton 1985). Some scientists think that introduced biological control agents pose little danger to native plants but agree that conflicts of interest can be minimized by using narrowly specialized agents (Harris et al. 1985).

RESULTS

Objective A: The Identification and Introduction of Beneficial Organisms

Hyles euphorbiae L. (Lepidoptera: Sphingidae). This Eurasian hawkmoth feeds on the leaves and flowers of *Euphorbia* species in the subgenus *Esula* (Harris 1984). The Canadian Department of Agriculture did the original research on its biology and use as a biological control agent (Harris and Alex 1971; Harris 1984). Each female lays 70 to 110 eggs singly or in clusters on the plant surface, and a generation can be completed in six weeks at 32°C. The hawkmoth pupates under rocks or below the soil surface, and overwinters in the pupal stage. In Europe, depending on the latitude, the moth has one to three generations per year; in Canada it has one.

In 1964 releases began in the United States from stocks originating from the cypress spurge (*Euphorbia cyparissias* L.) and *E. seguieriana* Necker in Switzerland,

but collected from an established population on cypress spurge in Braeside, Ontario (Harris 1984; Julien 1987). Between 1973 and 1980 the Braeside hawkmoth was imported to the USDA-ARS laboratory in Albany, California, where it was reared and released in six states: California, Idaho, Nebraska, Nevada, Oregon, and Montana. Of the 18 releases during this period, 14 were made with larvae, 3 with eggs, and 1 with adults. Thirteen of the releases involved fewer than 200 insects, and the other five, all of which took place in 1974, involved more than 1,000 insects—1,030 and 1,600 larvae in Oregon, 1,000 eggs in Nebraska, and 1,391 and 1,286 eggs in Montana. The hawkmoth only became established in Montana, perhaps because of the large number of eggs—2,677—released there. Between 1980 and 1985, Montana received additional releases of eggs and larvae (1,500 and 350 per site), pupae (143), and adults (9), from stocks originating on *E. virgata* Waldstein-Wartemberg & Kitaibel (*E. esula* complex) in Debrecen, Hungary. Again, the releases were successful. In Oregon, however, a release of 31 pupae from the same stocks failed.

In a major redistribution effort in 1985, larvae collected from the established hawkmoth population near Bozeman, Montana, were released in batches of 100 at 32 locations in Montana, 5 in Wyoming, 6 in North Dakota, 1 in Colorado, and 1 in Alberta, Canada (Nowierski unpub. data). As of 1991, none of these releases had resulted in establishment. Releases of Montana material were also made in Washington in 1985 (457 larvae) and 1986 (204 larvae), but again the moth failed to become established (Piper unpub. data). Researchers have attributed the poor establishment rate to predation by ants and carabids (Harris et al. 1985). P. Harris (1984) has also suggested that it could be related to the possible existence of different host strains.

Hawkmoths from Braeside, Ontario, were also released in several places in the eastern United States. In 1977, 180 larvae were released in Chestertown, New York. By 1982 the population had increased to about 1 million, and in some areas had totally defoliated the spurge (*E. cyparissias* and *E. pseudoesula* Schur.) (Batra 1984).

Chamaesphecia empiformis (Esp.) (Lepidoptera: Sesiidae). This clearwing moth from Europe bores in the roots of *Euphorbia* species. Researchers at the Commonwealth Institute of Biological Control Laboratory in Delémont, Switzerland, have studied its biology and conducted host-specificity tests (Harris 1984). The moth was thought to have two strains, one feeding in the stem and roots of *E. cyparissias* and the other in the roots of *E. esula*. Researchers later separated these into sibling species with the leafy spurge species *C. tenthrediniformis* Denis & Schiffermuller (Naumann and Schroeder 1980). This univoltine moth lays its eggs on the lower part of the stem, the root crown, and the leaves. Most stems receive only one or two eggs. The hatching larvae enter and mine the crown or roots. Mature larvae spin loose, silky webs, generally near the root collar, where they pupate (Schroeder 1968, 1969).

In Austria the main insect controlling *E. esula* was *C. tenthrediniformis*, which had attack rates ranging from 10 to 100 percent of the plants in a stand (Harris 1984). Releases of this moth began in Canada in 1971 (Harris 1984), and in the United States in 1975. The USDA-ARS Biological Control of Weeds Laboratory in Albany, California, released 47 and 36 larvae in 1975 and 1976 in Idaho, and 80 adults in 1977 in Montana, using Austrian stocks collected from *E. esula* (Story unpub. data). In 1979, 150 eggs from moths that were also collected in Austria were released in Oregon. None of these releases, nor any of the Canadian releases (Harris 1984) involving larger numbers of eggs, was successful.

Laboratory tests later showed that Canadian leafy spurge was an unacceptable host for the moth (Harris 1984). This caused releases to be suspended in Canada and the United States, and stimulated taxonomic research on leafy spurge. Scientists began searching for expected sibling *Chamaesphecia* species (Naumann and Schroeder 1980) that might accept North American leafy spurge. In Romania they found *C. crassicornis* (Barte) on *E. virgata*, which was evaluated by the USDA-ARS laboratory in Rome. It feeds on North American leafy spurge and is thought to be a narrow specialist (Pecora unpub. data).

Oberea erythrocephala (Schrank) (Coleoptera: Cerambycidae).
This longhorn beetle is native to Eurasia, where it feeds on the stems and roots of *Euphorbia* species. F. Bin (1973) and D. Schroeder (1980) have studied the life history and host range of the beetle in Italy, Switzerland, and Austria. Adults emerge when the host plant flowers, and the females feed on the young leaves and flowers for about 2 weeks before beginning to oviposit. They do so by girdling the upper stems and cutting a hole in the center of the stem, into which they lay an egg; the puncture is then covered with latex. Usually there is one egg or larva per shoot. In the laboratory females laid an average of 20 fertile eggs. The larvae bore down through the stem and into the crown and roots in less than 1 month, and feed there until March or April of the following year. In early May they pupate in a cell in the root crown. Researchers found the insect quite destructive locally, and thought it was able to reduce flowering and growth. Host-specificity tests found the beetle to be restricted to the genus *Euphorbia* excluding *E. lathyris*, poinsettia, and some other European species.

Researchers at the USDA-ARS laboratory in Albany, California, released the beetle in Oregon and Wyoming in 1980, 1981, and 1982 with material from *E. esula* and *E. cyparissias* in northern Italy. Adults were released (except in Oregon, where plants with larvae inside were transplanted—in numbers ranging from 40 to 302 individuals, but the beetle did not become established. Releases were made in Montana in 1982, 1983, and 1986 using the same Italian source material, and in 1983, 1984, and 1985 with beetles collected from Hungary on *E. virgata*. The beetle became established at two Montana sites, where 820 Italian adults and 50 Hungarian adults had been released (Rees et al. 1986). Releases in North Dakota in 1985 and 1986—of 68 and 193 adults collected in Italy—also resulted in establishment.

Although it is now established at some sites, the beetle is present in very small numbers. Canadian researchers have also experienced difficulties: the beetle became established at one of six release sites, but that colony was subsequently lost (Harris 1984; Julien 1987).

Aphthona flava Guill. (Coleoptera: Chrysomelidae).
One of a complex of 40 *Aphthona* species recorded to feed on *Euphorbia* species in Europe (Harris et al. 1985) and Asia (Pemberton and Wang 1989), this orange 3- to 4-millimeter-long flea beetle was the first of four *Aphthona* species released in the United States. Like most other *Aphthona* species that feed on *Euphorbia*, it has one generation per year. The adults feed on leaves and flower bracts, the larvae on the root hairs and roots. The adults usually emerge in June, feed, and lay about 225 eggs between the stem and soil during the several months of life. The mature larvae overwinter and pupate in late spring or early summer. This flea beetle is native to Europe from northern Italy east and north through Yugoslavia, Hungary, Czechoslovakia, Bulgaria, Romania, and Russia (Sommer and Maw 1982).

G. Sommer and E. Maw (1982) and I (Pemberton and Rees 1990) have evaluated *A. flava* as a candidate for the biological control of leafy spurge. In Europe it has been recorded on *E. esula* and a few allied *Euphorbia*. Host-specificity testing indicated that its potential host range is restricted to *Euphorbia* species belonging to the subgenus *Esula*. Three of the six North American subgenus *Esula* species tested, including two rare species, did not appear to be potential hosts.

We released *A. flava* in Montana between 1985 and 1987, and in Idaho and North Dakota in 1986 (Pemberton and Rees 1990). All of the releases involved field-collected adults from *E. esula* and *E. cyparissias* in northern Italy, because a high mortality rate among overwintering larvae in laboratory rearings greatly reduced the number of available beetles. Before release, each collection was checked for pathogens and normal oviposition. Neither of the two Idaho releases, which involved an estimated 200 and 210 beetles, appeared to result in establishment. However, in 1987 the 260 *A. flava* released in North Dakota were found to have become established, and by 1988 workers counted 14 beetles per square meter at the release site. In Montana *A. flava* has been recovered at four of eight release sites; establishment occurred from releases of 50, 57, 106, and 2,077 beetles, but not from releases of 46, 59, 150, and 240 beetles. Two years later, an average of 31 beetles per square meter was found at the site where 106 had been released. The beetle is also established in Alberta, Canada, where it appears to be causing stress to the leafy spurge (Julien 1987).

Aphthona cyparissias **(Koch).** This was the second flea beetle introduced for leafy spurge control. The adults are medium brown and 2 to 3 millimeters long. The beetle's life history (Sommer and Maw 1982) is similar to that of *A. flava* and other *Aphthona* that feed on *Euphorbia* species, but it has a higher fecundity than *A. flava*, with females laying an average of 285 eggs. The beetle occurs at high altitudes and in areas with cool, rainy summers, whereas *A. flava* is distributed in areas with drier and warmer summers. Its native range is from southern Spain and France through central and eastern Europe to western Russia, where it feeds on *E. esula* and a few other *Euphorbia* species. Host-specificity tests have confirmed its narrow host range (Sommer and Maw 1982; Pemberton 1986), which is restricted to the subgenus *Esula* species of the genus *Euphorbia*. However, the beetle did not accept *Euphorbia purpurea* (Rafinesque) Fernald or *E. telephiodes* Chapman, two rare North American species in the subgenus *Esula*.

We began introducing material collected in Austria in 1986, releasing 173 and 175 adults at two Wyoming sites and 130 adults in North Dakota. In 1987, 300, 740, and 1,470 adults were released at three sites in Montana, 248 adults were released in Wyoming, and 180 adults were released in North Dakota. In 1988 a few adult beetles were found at two of the Montana sites and also in North Dakota (Carlson unpub. data). The flea beetle is also established in Manitoba (Julien 1987).

Aphthona czwalinae **Weise.** This 2- to 3-millimeter-long blue-black flea beetle is native to central and eastern Europe—Germany, Poland, Austria, the lower Danube, and Russia—and to central Asia and eastern Siberia (Gassmann 1984). It is most frequently found at mesic sites where *Euphorbia* is intermixed with other vegetation. The beetle's biology and host range in nature is similar to that of *A. flava* and *A. cyparissias*. Host-specificity tests have confirmed the narrowness of its range, finding it to be limited to fewer species in the subgenus *Esula* than either *A. flava* or *A. cyparissias* (Gassmann 1984; Pemberton 1987). The beetle is thought to have the potential to colonize mesic sites such as stream margins, where leafy spurge is often most abundant.

In 1987, 360 adults collected in Austria were released at a Montana site, but researchers have found no evidence that the beetle has become established (Rees unpub. data). The beetle has also been released in Canada, but again establishment has not been confirmed (Julien 1987).

Aphthona nigriscutis **Foudras.** The fourth flea beetle to be introduced against leafy spurge in the United States, this is a 2- to 3-millimeter-long brown species that often has a black scutellum. The beetle is native to Europe, where it feeds on leafy spurge and related *Euphorbia* species (Gassmann 1985). It has the same life history pattern as the other *Euphorbia*-feeding *Aphthona* species, but it shows a preference for drier sites. Host-specificity tests have demonstrated that the beetle is restricted to a few *Euphorbia* species in the subgenus *Esula* (Gassmann 1985; Pemberton 1989a). Releases in the United States began in 1989, with 6,000 adults collected from an established population in Manitoba, Canada, that originated in Hungary (Julien 1987). These were released at eight Montana sites, two Nebraska sites, and one site each in North Dakota and Idaho (Rees person. commun.).

Aphthona nigriscutis and *A. cyparissias* are difficult to distinguish—in fact, the *Aphthona* species that has significantly reduced the leafy spurge in Manitoba is *A. nigriscutis*, not *A. cyparissias* as reported in M. H. Julien (1987) (Harris person. commun.).

Several other *Aphthona* species show promise as potential biological control agents, including two Chinese species, *A. chinchihi* Chen and *A. seriata* Chen, which are abundant in Inner Mongolia (Pemberton and Wang 1989), where adult feeding defoliates the plants.

Spurgia (Bayeria) esulae **Gagne (Diptera: Cecidomyiidae).** In Europe this midge causes stem-tip galls on *Euphorbia* species, which reduces seed production since galled shoots do not flower. M. Solinas and P. Pecora (1984) have described its biology. Overwintering larvae pupate in the spring. Females lay their eggs in groups on the leaves of the shoot tips, and newly hatched larvae stimulate gall formation. In northern Italy, where four to five generations occur every year, the first galls appear in April. Host-specificity tests at the USDA-ARS laboratories in Rome, Italy, and Albany, California, found the midge to be limited to some *Euphorbia* species in the subgenus *Esula* (Pecora 1983; Pemberton 1984; Pecora et al. 1991).

In 1985, 1986, and 1987, adult midges from stocks originating on *E. esula* in northern Italy were released in

Montana. Adults from the same source material were released in North Dakota in 1986 and 1987, in Wyoming in 1987, and in Oregon with galls in 1988. In 1987 the midge was found to be well established in Montana; in North Dakota it was well established by 1988 (Carlson unpub. data). At one 1985 release site in Montana, the density of galls in the center of the colony reached 135 per square meter by 1988 (Rees and Pemberton unpub. data).

Dasineura species near capsulae. This univoltine midge also forms galls on *Euphorbia* species in Europe. It emerges in the spring to lay eggs in flower buds; in the laboratory, it lays a mean of 36 eggs per bud (Pecora, Cristofaro, and Stazi 1989). The galls are formed in the individual ovaries (capsules), thus preventing seed formation. Mature larvae leave the galls during periods of high humidity and enter the soil. The larvae overwinter in the soil and pupate in the spring a few days before the adults emerge.

Host-specificity tests with material from *E. esula* at San Rossore in northern Italy have confirmed that this gall midge only uses the genus *Euphorbia* as hosts (Pecora, Cristofaro, and Stazi 1989). In additional testing with 12 native North American *Euphorbia*, the gall midge accepted only leafy spurge and one species in the subgenus *Esula* (Pemberton 1989b). Its release has been delayed because it lacks a name, and because of confusion concerning the identities of the midges that form galls on *Euphorbia* species. The midge will probably be approved for release after a description of it has been written.

The USDA-ARS Rome and Albany laboratories have evaluated two other insects—*Oncochila simplex* H. S. (Hemiptera: Tingidae) and *Lobesia euphorbiana* Frr. (Lepidoptera: Tortricidae)—as candidates for release against leafy spurge. Researchers have not petitioned for their release because their potential host ranges appeared to be too broad for safe use (Pemberton 1985, unpub. data).

RECOMMENDATIONS

1. Collect and release USDA-approved cold-hardy strains of insects from countries in the former Soviet Union and from Inner Mongolia. Cold-hardy strains could have more vigor and better synchrony with leafy spurge in North America than the European species previously released.

2. Clarify the taxonomy of *Dasineura* species near *capsulae*, and write a description for the gall midge so that it can be released.

3. Evaluate the stem-mining flies *Pegomya transveraloides* Schnabel and *P. virgatae* species nov. (Diptera: Antho-

myiidae) that were recently released in Canada for potential release in the United States.

4. Complete the host-specificity testing of *Chamaesphecia crassicornis*, a root-boring sesiid moth.

5. Evaluate the potential of *Aphthona chinchihi* and *A. seriata* from Inner Mongolia, which have the ability to develop very large populations that defoliate plants and damage the roots.

6. Obtain *Eurytoma euphorbiae* Zerova (Hymenoptera: Eurytomidae), a seed chalcid from the countries in the former Soviet Union, for host-specificity testing. This species could complement *Dasineura* species near *capsulae* in destroying seed.

7. Undertake research on diapause in *Aphthona* species to facilitate the development of mass-rearing systems for these flea beetles. This would allow the establishment and spread of large numbers of beetles without direct releases of large amounts of material collected overseas.

8. Employ narrow specialists with potential host ranges at or below the level of the subgenus *Esula* of the genus *Euphorbia*, to avoid damage to most nontarget *Euphorbia* species native to North America. Insects or diseases that can use members of the subgenus *Chamaesyce* as hosts could reduce populations and possibly even cause the loss of legally protected species.

REFERENCES CITED

Alley, H. P., N. Humburg, J. K. Fornstrom, and M. Ferell. 1984. Leafy spurge repetitive herbicide treatments. University of Wyoming *Agricultural Experiment Station Research Journal* 192:90–93.

Bakke, A. L. 1936. Leafy spurge, *Euphorbia esula* L. Iowa State College *Agricultural Experiment Station Research Bulletin* 198:208–46.

Batra, S. W. T. 1984. Establishment of *Hyles euphorbiae* (L.) (Lepidoptera: Sphingidae) in the United States for control of the weedy spurges *Euphorbia esula* L. and *E. cyparissias* L. *J. N. Y. Entomol. Soc.* 91:304–11.

Bin, F. 1973. Osservazioni bio-etologiche su *Oberea erythrocephala* Schrank (Coleoptera: Cerambycidae, Lamiinae). *Boll. Zool. Agrària Bachicoltura, Ser. II*, 11:141–49.

Bowes, G. G., and A. G. Thomas. 1978. Longevity of seeds in soils after various controls. *J. Range Management* 31:137–40.

Britton, N. L. 1921. The leafy spurge becoming a pest. *J. New York Bot. Gard.* 22:73–75.

Calvin, M. 1978. Green factories. *Chem. and Eng. News* 56:30–36.

Croizat, L. 1945. *Euphorbia esula* in North America. *Am. Midl. Nat.* 33:231–43.

Cunningham, I. S. 1984. *Frank N. Meyer, plant hunter in Asia.* Ames: Iowa State University Press.

Dunn, P. H. 1979. The distribution of leafy spurge (*Euphorbia esula*) and other weedy *Euphorbia* spp. in the United States. *Weed Sci.* 27:509–16.

———. 1985. Origins of leafy spurge in North America. In *Leafy spurge*, ed. A. K. Watson, 7–13. Weed Science Society of America Monograph Ser. 3.

Dunn, P. H., and A. Radcliffe-Smith. 1980. The variability of leafy spurge (*Euphorbia* spp.) in the United States. *Res. Rep. N. Centr. Weed Contr. Conf.* 37:48–53.

Gassmann, A. 1984. *Aphthona czwalinae* Weise (Coleoptera: Chrysomelidae): A candidate for the biological control of leafy spurge in North America. Commonwealth Institute of Biological Control Unpublished Report. Delémont, Switzerland: CIBC.

———. 1985. *Aphthona nigriscutis* Foudras (Coleoptera: Chrysomelidae): A candidate for the biological control of cypress spurge and leafy spurge in North America. Commonwealth Institute of Biological Control Unpublished Report. Delémont, Switzerland: CIBC.

Harris, P. 1984. *Euphorbia esula-virgata* complex, leafy spurge, and *E. cyparissias* L., cypress spurge (Euphorbiaceae). In *Biological control programmes against insects and weeds in Canada 1969–1980*, 159–69. Farnham Royal, U.K.: Commonwealth Agricultural Bureaux.

Harris, P., and J. Alex. 1971. *Euphorbia esula* L., leafy spurge, and *E. cyparissias*, cypress spurge (Euphorbiaceae). In *Biological control programmes against insects and weeds in Canada 1959–1968*, 83–88. Farnham Royal, U.K.: Commonwealth Agricultural Bureaux.

Harris, P., P. H. Dunn, D. Schroeder, and R. Vonmoos. 1985. Biological control of leafy spurge in North America. In *Leafy spurge*, ed. A. K. Watson, 79–92. Weed Science Society of America Monograph Ser. 3.

Harvey, S. J., R. M. Nowierski, P. Mahlberg, and J. M. Story. 1988. Taxonomic evaluation of leaf and latex variability of leafy spurge (*Euphorbia* spp.) for Montana and European accessions. *Weed Sci.* 36:726–33.

Julien, M. H. 1987. *Biological control of weeds: A world catalogue of agents and their target weeds*, 2nd ed. Wallingford, U.K.: Commonwealth Agricultural Bureaux International.

Lacey, C. A., P. K. Fay, R. G. Lym, C. G. Messersmith, B. Maxwell, and H. P. Alley. 1985. *Leafy spurge distribution, biology, and control*. Montana State University Cooperative Extension Service Circular 309.

Messersmith, C. G., and R. G. Lym. 1983. Distribution and economic impacts of leafy spurge in North Dakota. *North Dakota Farm Res.* 40:8–13.

Naumann, C. M., and D. Schroeder. 1980. Ein weiteres zwillingsarten-paar Mitteleuropaischer Sesiiden: *Chamaesphecia tenthrediniformis* ([Denis & Schiffermuller], 1775) und *Chamaesphecia empiformis* (Esper, 1783) (Lepidoptera, Sesiidae). *Zeit. Arbeitsgem. Oesterr. Entomol.* 32:29–46.

Nobel, D. L., P. H. Dunn, and L. A. Andres. 1979. The leafy spurge problem. In *Proceedings of the Leafy Spurge Symposium, Fargo, North Dakota*, 8–15.

Ohwi, J. 1965. *Flora of Japan.* Washington, D.C.: Smithsonian Institution.

Pecora, P. 1983. A petition for the introduction into quarantine for further testing of *Bayeria capitigena* (Bremi) (Diptera: Cecidomyiidae), potential biocontrol agent of leafy spurge (*Euphorbia esula-virgata* "complex"). On file at the USDA-ARS Biological Control Documentation Center, Beltsville, Maryland.

Pecora, P., M. Cristofaro, and M. Stazi. 1989. *Dasineura* sp. near *capsulae*: A candidate for biological control of leafy spurge. *Ann. Entomol. Soc. Am.* 82:693–700.

Pecora, P., R. Pemberton, M. Stazi, and G. Johnson. 1991. Host specificity of *Spurgia esulae* Gagné (Diptera: Cecidomyiidae), a gall midge introduced into the United States for control of leafy spurge (*Euphorbia esula* L. complex). *Eviron. Entomol.* 20:283–87.

Pemberton, R. W. 1984. Petition for the release of *Bayeria capitigena* against leafy spurge (*Euphorbia esula* complex) in the United States. On file at the USDA-ARS Biological Control Documentation Center, Beltsville, Md.

———. 1985. Native plant considerations in the biological control of leafy spurge. In *Proceedings of the 6th International Symposium on Biological Control of Weeds, 19–25 August 1984, Vancouver, British Columbia,* ed. E. S. Delfosse, 365–90. Ottawa: Agriculture Canada.

———. 1986. Petition for the release of *Aphthona cyparissiae* against leafy spurge (*Euphorbia esula* complex) in the United States. On file at the USDA-ARS Biological Control Documentation Center, Beltsville, Maryland.

———. 1987. Petition for the release of *Aphthona czwalinae* Weise against leafy spurge (*Euphorbia esula*) in the United States. On file at the USDA-ARS Biological Control Documentation Center, Beltsville, Md.

———. 1988. Myrmecochory in the introduced range-weed leafy spurge. *Am. Midl. Nat.* 119:431–35.

———. 1989a. Petition to release *Dasineura* sp. near *capsulae* Kieffer against leafy spurge (*Euphorbia esula* L.) in the United States. On file at the USDA-ARS Biological Control Documentation Center, Beltsville, Maryland.

———. 1989b. Petition to release *Aphthona nigriscutis* Foudras against leafy spurge (*Euphorbia esula* L.) in the United States. On file at the USDA-ARS Biological Control Documentation Center, Beltsville, Md.

Pemberton, R. W., and N. E. Rees. 1990. Host specificity and establishment of *Aphthona flava* Guill. (Chrysomelidae), a biological control agent for leafy spurge (*Euphorbia esula* L.) in the United States. *Proc. Entomol. Soc. Wash.* 92:351–57.

Pemberton, R. W., and R. Wang. 1989. Survey for natural enemies of *Euphorbia esula* L. in northern China and in Inner Mongolia (in Chinese with English summary). *Chinese J. Biol. Control* 5:64–67.

Radcliffe-Smith, A. 1985. Taxonomy of North American leafy spurge. In *Leafy spurge*, ed. A. K. Watson, 14–25. Weed Science Sciety of America Monograph Ser. 3.

Radcliffe-Smith, A., and T. G. Tutin. 1968. *Euphorbia.* In *Flora europea,* vol. 2, ed. T. G. Tutin, 213–16. Cambridge: Cambridge University Press.

Rees, N. E., R. W. Pemberton, A. Rizza, and P. Pecora. 1986. First recovery of *Oberea erythrocephala* on the leafy spurge complex in the United States. *Weed Sci.* 34:395–97.

Sachs, R. M., C. B. Low, J. D. MacDonald, A. R. Awad, and M. J. Sully. 1981. *Euphorbia lathyris*: A potential source of petroleum-like products. *Calif. Agric.* 35(7–8):29–32.

Schroeder, D. 1968. Studies on phytophagous insects of *Euphorbia* spp.: *Chamaesphecia empiformis* (Esp.). Commonwealth Institute of Biological Control Unpublished Progress Report 21. Delémont, Switzerland: CIBC.

———. 1969. Studies on phytophagous insects of *Euphorbia* spp.: *Chamaesphecia empiformis* (Esp.). Commonwealth Institute of Biological Control Unpublished Progress Report 22. Delémont, Switzerland: CIBC.

———. 1980. Investigations on *Oberea erythrocephala* (Schrank) (Coleoptera: Cerambycidae), a possible biological control agent of leafy spurge, *Euphorbia* spp. (Euphorbiaceae) in Canada. *Zeit. angew. Entomol.* 90:237–54.

Selleck, G. W., R. T. Coupland, and C. Frankton. 1962. Leafy spurge in Saskatchewan. *Ecol. Monographs* 32:1–29.

Solinas, M., and P. Pecora. 1984. The midge complex (Diptera: Cecidomyiidae) on *Euphorbia* spp. I. *Entomologica* 19: 167–213.

Sommer, G., and E. Maw. 1982. *Aphthona cyparissiae* (Koch) and *A. flava* Guill. (Coleoptera: Chrysomelidae): Two candidates for the biological control of cypress and leafy spurge in North America. Commonwealth Institute of Biological Control Unpublished Report. Delémont, Switzerland: CIBC.

Watson, A. K. 1985. Introduction: The leafy spurge problem. In *Leafy spurge,* ed. A. K. Watson, 1–6. Weed Science Society America Monograph Ser. 3.

77 / MEDITERRANEAN SAGE

L. A. ANDRES, E. M. COOMBS, AND J. P. McCAFFREY

Mediterranean sage
Salvia aethiopis L.
Lamiaceae

INTRODUCTION

Biology and Pest Status

Mediterranean sage is a grayish green, woolly-leafed herbaceous biennial that grows well on soil types ranging from dry, gravelly upland to heavy meadowland. The weed—which has first- and second-year rosettes and flowering stalks formed in the spring of the second year—commonly infests rangeland and pastures; it can also invade alfalfa and grain fields and occasionally lawns. The plant reproduces only by seeds, which it produces in quantity. Seeds may be spread by contaminated soils, infested hay, and agricultural equipment; more commonly, however, fall winds fracture the 20- to 90-centimeter-tall flower stalks, causing them to tumble and the seeds to scatter (Bellue 1950; Robbins, Bellue, and Ball 1970). The seeds germinate in the fall or spring; as they imbibe water, a mucilaginous layer forms around the seed coat and limits water loss, just as a covering of soil might do (Young, Evans, and Martinelli 1970).

Historical Notes

Mediterranean sage infests approximately 600,000 hectares of dryland range in seven eastern and southern counties of Oregon; however, the bulk of the infestation occurs in Lake County (Hawkes person. commun.). The weed infests over 3,000 hectares in northeastern California and 1,600 hectares in Idaho County, Idaho (Crabtree person. commun.). It is also found in Colorado (Weber 1976), Arizona, Texas (Kearney and Peebles 1951), and Washington, and it may spread through much of Idaho's Salmon River Canyon, the Great Basin, and northern California. Although it is not toxic to livestock, Mediterranean sage is unpalatable, hence worthless as forage; it causes an estimated 60 per-

cent forage loss or, in better pastures, a loss of $109 per hectare per year. Aerial application of 2,4-D costs approximately $15 per hectare; although effective, it must be repeated (Oregon State University Cooperative Extension Service, Lake County, person. commun. 1994). The weed is not known to harbor any insect pests or phytopathogens. In some areas its abundance appears to be correlated with overgrazing or other poor management practices; infestations on poorer land serve as seed sources for other areas. Isolated new infestations are becoming more common along roadsides throughout eastern Oregon.

The plant occurs sparsely in Spain, France, and Italy, and is somewhat more common in Yugoslavia, Bulgaria, Greece, Turkey, and Iran. Mediterranean sage also grows in Uzebekistan and in the Crimea, in the former USSR (Andres unpub. data). However, in none of those areas does its density approach that of the western United States. *Salvia* is believed to have been introduced into western North America before 1882, probably in contaminated alfalfa seed (Hawkes, Whitson, and Dennis 1985). It was first recorded in California, but the exact point of origin is unknown.

RESULTS

Objective A: The Identification and Introduction of Beneficial Organisms

The biological control program for *Salvia aethiopis* began in 1962 with a survey of its natural enemies in Italy, Turkey, and Iran (Andres and Drea 1963). Researchers studied two new species of weevils, *Phrydiuchus spilmani* Warner and *P. tau* Warner (Warner 1969), and then introduced them into North America. A subsequent survey in Yugoslavia detected additional natural enemies

We wish to acknowledge the information provided by E. Johansen, a graduate student at the Department of Entomology, University of Idaho, Moscow, and by R. Hawkes of the Oregon Department of Agriculture at Salem. We also acknowledge the role that the former USDA Biological Control of Weeds Laboratory in Italy played in clearing and introducing the *Phrydiuchus* species weevils to the United States.

(Bogavac and Mitic-Muzina 1972), but no further introductions were made.

After *P. tau* was discovered in Turkey in 1962, *P. spilmani* was observed feeding under the rosettes of *Salvia verbenacea* L. in central Italy. Because *P. spilmani* also fed and oviposited on *S. aethiopis* and was easily accessible from laboratory facilities in Italy, researchers conducted biological and host-specificity studies on this natural enemy first. The life history and specificity of *P. tau* were later found to closely parallel that of *P. spilmani*.

Both species are univoltine. The adults have an estival diapause that commonly ends when the fall rains start. Adults feed on rosette foliage and oviposit on the undersides of leaves and petioles. When the eggs hatch, the larvae mine the petioles and midribs of the leaves and then tunnel into the crown, where they form cavities and complete their development. Adults, eggs, and larvae may overwinter. The immature stages complete their development in the spring, when the mature larvae exit the plant and pupate in the soil. The adults emerge in late spring or early summer to feed on the foliage and flower stalks before entering estivation sites (Andres and Rizza 1965; Bogavac and Mitic-Muzina 1971).

Researchers tested the adult weevils on 42 species of Labiatae. Although the weevils fed or nibbled on 25 of these plants, only three of the species supported ovigenesis. Oviposition and larval entry occurred only on Mediterranean sage and closely related herbaceous species that were morphologically suited to support larval development (Andres 1966).

Phrydiuchus spilmani adults were collected from around rosettes of *Salvia verbenacea* in the Alban Hills near Tuscolo, about 33 kilometers east of Rome, Italy. *Phrydiuchus tau* was collected from several sites in Serbia (Bogavac and Mitic-Muzina 1971). In the fall of 1969 and 1970, researchers released a total of 631 *P. spilmani* adults at four sites in Lake County, Oregon. In the fall of 1971, 1972, and 1973 and in the spring of 1972, a total of 1,610 *P. tau* adults were released at five sites in the same county.

Phrydiuchus spilmani has not been recovered from any release site; however, *P. tau* became established at four sites. Between 1979 and 1981, 2,800 adults were re-collected from those Oregon sites and released at 28 new sites in the state; in 1976 and 1977, 1,850 adults were re-collected and released in California; and in 1979, 418 adults were re-collected and released at two sites in Idaho. There have been subsequent redistributions in all three states.

Objective D: Impact

As the larvae eat their way down a leaf petiole of a rosette, they frequently destroy the axial bud at the juncture of the petiole and the plant crown. When larval populations are high and the crowns are heavily mined, bolting is reduced or even prevented. Adult foliar feeding in the spring and fall has little impact.

In one Oregon pasture, larval feeding reduced *Salvia* cover from 27.7 percent in 1976 to 1.2 percent in 1980. This reduction was preceded by an increase in *P. tau* adults and then followed by a drop, as plant abundance decreased. Despite this documented reduction and the impression that there has been a slight decrease in the size and density of *Salvia*, the plant continues to spread in other areas of its range. In Idaho the weevils also appear to have reduced the percentage of bolting plants, but specific data are lacking. Most researchers believe that additional agents will be needed to control this weed.

RECOMMENDATIONS

1. A detailed evaluation is needed of the impact that *P. tau* has on the weed under different management practices in California, Oregon, and Idaho.

2. Additional agents should be sought, studied, and imported to improve biological control.

REFERENCES CITED

Andres, L. A. 1966. Host-specificity studies of *Phrydiuchus topiarius* and *Phrydiuchus* sp. *J. Econ. Entomol.* 59:69–76.

Andres, L. A., and J. Drea, Jr. 1963. *Exploration for natural enemies of* Salvia aethiopis L. *and* Linaria dalmatica *Miller in the Mediterranean and Near East, May–June, 1962.* Special Report of the USDA-ARS Entomology Research Division, Insect Identification and Foreign Parasite Introduction Research Branch. Beltsville, Md.: U.S. Department of Agriculture.

Andres, L. A., and A. Rizza. 1965. Life history of *Phrydiuchus topiarius* (Coleoptera: Curculionidae) on *Salvia verbenacea* (Labiatae). *Ann. Entomol. Soc. Am.* 58:314–19.

Bellue, M. K. 1950. Mediterranean sage moves. *Calif. Dept. Agric. Bull.* 39:43–46.

Bogavac, M., and N. Mitic-Muzina. 1971. *Phrydiuchus tau* R. E. Warner (Coleoptera: Curculionidae) as reducing agent of Mediterranean sage (in Serbo-Croatian). *Zastita Bilja* 22(114):233–46.

———. 1972. Phytophagous insects of weed flora from the genera *Linaria* and *Salvia* (in Serbo-Croatian). *Zastita Bilja* 23(119–120):217–30.

Hawkes, R. B., T. D. Whitson, and L. J. Dennis. 1985. *A guide to selected weeds of Oregon.* Salem: Oregon Department of Agriculture.

Kearney, T. H., and R. H. Peebles. 1951. *Arizona flora.* Berkeley and Los Angeles: University of California Press.

Robbins, W. W., M. K. Bellue, and W. S. Ball. 1970. *Weeds of*

California. Sacramento: California State Department of Agriculture.

Warner, R. E. 1969. The genus *Phrydiuchus* with the description of two new species (Coleoptera: Curculionidae). *Ann. Entomol. Soc. Am.* 62:1293–1302.

Weber, W. W. 1976. *Rocky Mountain flora.* Boulder: Colorado Associated University Press.

Young, J. A., R. A. Evans, and P. C. Martinelli. 1970. Mucilaginous coatings of weed seeds. *Proc. West. Soc. Weed Sci.* 23:35.

78 / GORSE

G. P. MARKIN, E. R. YOSHIOKA, AND R. E. BROWN

gorse
Ulex europaeus L.
Fabaceae

INTRODUCTION

Biology and Pest Status

Gorse, *Ulex europaeus* L., is a spiny, dense-growing shrub native to Western Europe. It has become a serious pest at higher elevations on two Hawaiian islands, in coastal northern California and Washington (where it infests about 1,000 hectares), and on a few islands off the coast of British Columbia. It is most abundant on the southern coast of Oregon in Coos, Curry, and Lane counties, where it infests about 15,378 hectares (Hermann and Newton 1968). Although also a weed in Oregon's pastures, gorse primarily plagues forests, where it interferes with management efforts during the reforestation of logging areas. Its extreme flammability makes dense stands a serious fire hazard to nearby towns and recreation areas: gorse is charged with carrying a wildfire that destroyed an entire town in 1932 (Holbrook 1943).

In Hawaii gorse is generally restricted to elevations of between 630 and 2,250 meters on the islands of Maui, where 5,985 hectares are infested, and of Hawaii, where 8,262 hectares are affected (Markin et al. 1988). The infested areas are prime pastureland, where uncontrolled gorse quickly forms impenetrable patches that prevent grazing or farming. On the island of Hawaii it has invaded the native forests, where it interferes with natural regeneration. Gorse is also found in subdevelopments on Maui and in a major watershed on Hawaii, where its combustibility makes it a fire hazard.

Historical Notes

Before barbed wire was invented, European farmers planted gorse in hedges to contain livestock. It was also used as feed for sheep and goats, which readily graze on the plant's soft new foliage (Jobson and Thomas 1964). Early European emigrants introduced gorse as a useful plant to more than 15 countries or island groups. Now, however, it is seen as an undesirable weed and is causing problems in the United States, Chile, Australia, and New Zealand (Holm et al. 1979).

In the United States chemical control has been the main method of gorse management for 30 years, and it has effectively contained the two major infestations in Hawaii (Markin et al. 1988). Researchers have estimated that if chemical control in Hawaii were relaxed, however, gorse's range would expand 10- to 20-fold, primarily on valuable grazing land. Since seeds can remain viable in the soil for up to 30 years, herbicide control usually requires first cutting or burning the existing stand and then applying two or more follow-up treatments to eliminate regeneration.

Another control method that has been tried successfully in Oregon and in New Zealand, and to a limited extent in Hawaii, is planting fast-growing trees that form a closed canopy and shade out the gorse (Hermann and Newton 1968). This method requires extensive site preparation to eliminate existing gorse plants, as well as follow-up treatments to kill the regenerating gorse until the trees are tall enough to form a canopy. Once a closed canopy is formed, however, the gorse quickly dies out.

RESULTS

Objective A: The Identification and Introduction of Beneficial Organisms

***Apion ulicis* (Forster) (Coleoptera: Apionidae).** In England this weevil is univoltine; overwintering females mate in spring and lay eggs in developing gorse pods. The larvae consume the seeds and pupate within the

We thank E. M. Coombs of the Oregon Department of Agriculture and G. L. Piper of Washington State University for their contributions to this report, as well as G. Funasaki for giving us access to the historical record in the Hawaii Department of Agriculture's files, and for reviewing this manuscript.

pods. The new adults emerge in summer when the mature pods open and remain active during the summer, feeding on the plant. They overwinter on the gorse and emerge the following spring during flowering.

In 1927 *A. ulicis* collected in England was first introduced to Hawaii on the island of Maui. This release was unsuccessful, as were subsequent releases in 1949, 1951, and 1952; the latter releases used weevils from a New Zealand population that had been successfully introduced from England in 1931 (Miller 1970). Suspecting that the English strain might not be climatically compatible with Hawaii, researchers in 1955 and 1956 released insects collected from southern France. Those releases were successful, and by 1960 the weevil was established on Maui. By 1972 it had spread over the entire range of gorse, albeit in very low numbers and attacking only 1.5 percent of the pods. A 1984 survey, however, indicated that weevils were attacking 52 percent of the pods.

On the island of Hawaii, the weevil became established in 1960, but it was restricted to an isolated 5-hectare stand several kilometers from the main infestation. During a control program in 1976 and 1977, this gorse stand was treated with herbicides and then burned, apparently eradicating the population. In 1984 the weevil was reintroduced, using 10,000 adults collected from Maui and released at nine sites. By June 1988 the weevils were well established at seven of the nine sites and were attacking an average of 43 percent of the pods. Weevils were also abundant—attacking 15 percent of the pods—100 meters from the 7 release sites, and researchers recovered adult weevils at 7 of 27 randomly sampled sites over the 2,500-hectare core of the gorse infestation.

On the mainland biological control efforts began in 1948, when the seed weevil was shipped from France to the USDA Quarantine Laboratory at Albany, California; however, it could not be propagated successfully. Researchers conducted subsequent host testing in France. In 1952 researchers received clearance to release the weevil, and 95 adults collected from southern England were released in July 1953, in California's Mendocino County. A second release of 168 adults was made in March 1954, in Mendocino and San Mateo counties. By 1959 the weevils could be readily found in Mendocino County, and by 1956 they were well established in San Mateo County (Holloway and Huffaker 1957). In 1956 weevils collected in California were released at Bandon, Coos County, Oregon, where they became established by 1959. They were then redistributed to other coastal Oregon counties (Anonymous 1974). By 1961 the weevils were well established on the mainland (Holloway 1961). An Oregon Department of Agriculture survey in 1988 estimated that the weevil

had become established in over 95 percent of the coastal infestation.

The weevil was first introduced in Washington state in 1962, but it apparently did not become established. In 1986, 1987, and 1988 the Pacific County Noxious Weed Control Board released adults from Oregon; however, a 1988 survey did not recover any weevils.

Apion species (Coleoptera: Apionidae). During Hawaiian efforts to find a climatically suitable strain of *A. ulicis*, in 1957 researchers collected another weevil in Spain and Portugal. This weevil was subsequently identified as a closely related species, possibly *A. uliciperda* Pandelle. In 1958, 136 adults were released on Maui, but they had not become established by 1962, when the release area was invaded by *A. ulicis* (Davis 1959). Subsequent surveys have found no evidence of this weevil.

Apion scutellare Kirby (Coleoptera: Apionidae). *A. scutellare* is a gall-forming weevil that is widely—but never very abundantly—distributed in Europe from England to Portugal. Although it will attack the new shoots of mature plants, it prefers the young growth of crowns that are regenerating after burning or cutting. It also appears to prefer the low-growing gorse species, *Ulex minor* Roth, which in Europe is found intermixed with stands of *U. europaeus;* in these mixed stands galls are usually 5 to 10 times more abundant on *U. minor.* In southern England *A. scutellare* adults emerge from the galls in early June and lay eggs in the succulent new shoots. Larvae produce a visible gall by fall. Prepupae are present in the hard, woody galls by March, and adults emerge in early June.

In 1961, 160 *A. scutellare* adults were reared from galls in Portugal, shipped to Hawaii, and directly released in the field; a second shipment of 10 adults followed in 1962 (Davis and Krauss 1962, 1963). Researchers never recovered this weevil and do not know whether it failed to become established or was exterminated when the release site was treated and burned as part of a control program in 1963.

Agonopterix ulicetella (Stainton) (Lepidoptera: Oecophoridae). After the major effort to establish biological control of gorse in Hawaii in the mid-1950s and early 1960s, work was discontinued around 1965, except for occasional surveys. In 1984 a new cooperative program was undertaken between the Hawaii Department of Agriculture, the USDA Forest Service, the Oregon Department of Agriculture, and the Division of Science and Industrial Research of New Zealand. At the time New Zealand had already begun its own program, and in 1982 R. L. Hill, working at the Council on International Biological Control in Silwood Park in

England, had identified *A. ulicetella* as one of the most promising agents, along with the mite *Tetranychus lintearius* Dufour and the tingid *Dictyonota strichnocera* Fieber (Hill 1983). After researchers in England had studied the moth's biology and done preliminary field-host testing, a colony of *A. ulicetella* was established in quarantine in New Zealand in 1984 for additional host-specificity tests. However, the moth was not released in New Zealand because it proved capable of developing on tree lucerne or taganaste—*Chamaecytisus palmensis* (Christ) Hutch.—an introduced plant closely related to gorse that was being promoted for grazing. In November 1986 a colony of *A. ulicetella* from New Zealand was brought into quarantine in Hawaii (Markin and Yoshioka 1990). Fortunately *C. palmensis* has not been introduced into Hawaii, and *A. ulicetella* passed host-testing on Hawaiian plants. In September 1988 the Hawaii Board of Agriculture granted permission to release *A. ulicetella*, and introductions began in November. Researchers established a mass-rearing facility in Hilo on the island of Hawaii, and made several more releases in the winter of 1988 to 1989 to study the insect's response to the new habitat and climate. Between June and August of 1989, when the newly mated adults were synchronized with the new flush of foliage, researchers released 10,000 adults.

In England *A. ulicetella* overwinters as unmated adults that emerge from diapause in April or May, mate, and lay eggs on the older spines and stems. New larvae migrate to the ends of the shoots to feed on developing spines. They pupate in July, and adults emerge in August or September. In the fall adults move to the center of the gorse plant to overwinter. In the milder Hawaiian winter, the phenology of gorse is considerably different: the major flush of new flowers occurs in winter, rather than spring, and new foliage appears over a much longer period in summer. This difference apparently is not a problem, because *A. ulicetella* is now well established on gorse at all higher-elevation sites. It has, however, failed to establish at the low-elevation Maui site of between 630 and 1,0000 meters. *A. ulicetella* is also being considered for release on the mainland, but this may be complicated by the presence of an accidentally introduced related species, *A. nervosa* (Haworth), which is already established throughout the range of gorse (Hodges 1974).

Sericothrips staphylinus Haliday (Thysanoptera: Thripoidae).

The newest biological control agent, *S. staphylinus*, was released in 1991 and is well established at higher elevations on the islands of Hawaii and Maui. This small thrips is multivoltine and feeds on new growth after it has hardened. Because this thrips is flightless, a major effort at redistributing it through the gorse infestations on both islands is under way.

RECOMMENDATIONS

1. In early 1989 New Zealand released the tetranychid gorse mite *T. lintearius*. The mite should be field-host tested in New Zealand on selected plant species from Oregon and Hawaii to determine whether it is suitable for release in the United States.

2. The cooperative program between Hawaii, Oregon, and New Zealand that supports the work of the International Institute of Biological Control in England—identifying and testing other potential biological control agents—needs to be continued.

3. The gorse infestations on Maui are located at elevations of 630 to 2,250 meters, in an area that is climatically very different from southeastern England. Surveys in Spain and Portugal to obtain a more southern genotype of *A. ulicetella* are needed.

4. To date, efforts to find biological control agents have concentrated on insects that defoliate gorse. The program should be expanded to include insects that attack the reproductive parts of the plant—flowers, flower buds, pods, and seeds—and that will supplement the seed weevil in reducing the threat of further spread.

5. The Hawaii Department of Agriculture has constructed an insect plant pathogen quarantine facility, and present plans include studying possible pathogens of gorse as one of its first priorities. This program, which is operating in cooperation with plant pathologists in Europe, should be encouraged.

REFERENCES CITED

Anonymous. 1974. *Gorse*. Oregon State University Cooperative Extension Service PNW Bulletin 107.

Davis, C. J. 1959. Recent introductions for biological control in Hawaii IV. *Proc. Hawaii. Entomol. Soc.* 17:620–66.

Davis, C. J., and N. L. H. Krauss. 1962. Recent introductions for biological control in Hawaii VII. *Proc. Hawaii. Entomol. Soc.* 18:125–29.

———. 1963. Recent introductions for biological control in Hawaii VIII. *Proc. Hawaii. Entomol. Soc.* 18:245–49.

Hermann, R. K., and M. Newton. 1968. *Tree planting for control of gorse in the Oregon coast*. Oregon State University Forest Resource Laboratory, Resource Paper 9.

Hill, R. L. 1983. Prospects for the biological control of gorse. In *Proceedings of the 36th New Zealand Weed and Pest Control Conference*, 56–58.

Hodges, R. W. 1974. *The moths of America north of Mexico*. Fasc. 6.2., Gelechioidea: Oecophoridae. London: E. W. Classey.

Holbrook, S. H. 1943. The gorse of Bandon. In *Burning an empire*, 147–55. New York: Macmillan.

Holloway, J. K. 1961. Biological control of weeds. In *Proceedings of the 13th Annual California Weed Conference,* 116–17.

Holloway, J. K., and C. B. Huffaker. 1957. Establishment of the seed weevil *Apion ulicis* Forst. for suppression of gorse in California. *J. Econ. Entomol.* 50:498–99.

Holm, L. G., J. V. Pancho, J. P. Herberger, and D. L. Plucknett. 1979. *Ulex europaeus.* In *A geographical atlas of world weeds,* 373. New York: Wiley-Interscience.

Jobson, H. T., and B. Thomas. 1964. The composition of gorse (*Ulex europaeus*). *J. Sci. Fed. Agric.* 15:652–56.

Markin, G. P., and E. R. Yoshioka. 1990. Present status of biological control of the weed gorse (*Ulex europaeus* L.) in Hawaii. In *Proceedings of the 7th International Symposium on Biological Control of Weeds, 6–11 March 1988, Rome, Italy,* ed. E. S. Delfosse, 357–62. Rome: Ministero dell'Agricoltura e delle Foreste / Melbourne: CSIRO.

Markin, G. P., L. A. Dekker, J. A. Lapp, and R. F. Nagata. 1988. Distribution of the weed gorse (*Ulex europaeus* L.) in Hawaii. *Bull. Hawaii. Bot. Soc.* 27:110–17.

Miller, D. 1970. Biological control of weeds in New Zealand 1927–48. *New Zealand Department of Scientific and Industrial Research Information Series* 74:37–58.

79 / SCOTCH BROOM

L. A. ANDRES AND E. M. COOMBS

Scotch broom
Cytisus scoparius (L.) Link
Fabaceae

INTRODUCTION

Biology and Pest Status

Scotch broom, *Cytisus scoparius* (L.) Link, is a naturalized woody perennial 1 to 2 meters tall that is often admired for its bright yellow blossoms that flower from April through June. The plant favors full sunlight and dry sandy sites, and does well in both acid and alkaline soils (pH 4.5 to 7.5) (Gill and Pogge 1974). The fruit is a flattened pod, somewhat hairy along the margins, and contains 2 to 10 or more seeds that can remain viable for 81 years (Gill and Pogge 1974).

Scotch broom invades pastures, cultivated fields, native grasslands, and waste areas along roadsides, displacing desirable forage and vegetation (Gilkey 1957). It does not do well in forested areas, but after logging, land clearing, and burning, it may invade rapidly and compete with young conifers in areas that are being reforested (Mobley 1954). Found in inland valleys from British Columbia to central California (Hitchcock and Cronquist 1973), the weed has not yet reached the full extent of its range in the western United States; it will become an increasing problem. Three related broom species also occur as pests in areas of California and western Oregon.

Historical Notes

Scotch broom is native to Europe. Its Old World range includes the British Isles, central and southern Europe, and the Canary Islands (Clapham, Tutin, and Warburg 1957). It was first introduced to the East Coast of North America in Nova Scotia, and from New York to Georgia. In 1861 it was being sold as an ornamental in California (Butterfield 1964), and by 1900 it was naturalized on British Columbia Island (Bailey 1906).

RESULTS

Objective A: The Identification and Introduction of Beneficial Organisms

The biological control of Scotch broom in the United States began in 1960 and 1961 with the introduction of approximately 6,000 twig-mining moths, *Leucoptera spartifoliella* Hübner (Lepidoptera: Lyonetidae), from Paris, France. Later, researchers found that this moth was already present in Alameda and Del Norte counties in California, and in parts of Washington and Oregon. While describing the 1960 to 1961 releases, K. E. Frick (1964) noted that *Leucoptera* was already present and surmised that unnoticed larvae mining the twigs of imported broom plants were the source of the moth population at Tacoma, Washington.

Adults of the univoltine *L. spartifoliella* emerge and oviposit on the tips of the current year's foliar twigs. The larvae tunnel the twigs during the summer and then feed and remain in the mines until spring. They then emerge and spin cocoons on the undersides of the mined twigs. Their life history is similar to that described by H. L. Parker (1964), who also tested the host specificity of this moth using 12 annual and perennial legumes. In these tests the female moths oviposited on nine species of plants other than *Cytisus scoparius*, but no larvae tunneled more than 3 millimeters.

Researchers released 642 to 2,860 adults at each of five California coastal and Sierra foothill locations; the moth became established at all of the sites (Frick 1964). At the Sierra foothill sites researchers at first recovered only limited numbers of cocoons, and only on plants growing in shaded areas. These populations remained at several dozen cocoons per plant for several years, after which regular monitoring was discontinued. However, on a return visit in 1973 to the Georgetown site in the

We express our appreciation to the USDA Biological Control Laboratories in Italy and France for their role in clearing *Leucoptera* and *Apion* for introduction to the United States, and for collecting and shipping material for release.

Sierra foothills, the cocoon densities had increased into the hundreds per plant. Cocoons were even abundant on plants in full sun. Over the next 5 years, progressively higher numbers of cocoons per stem appeared on plants to the north and south of Georgetown (Andres unpub. data).

Leucoptera spartifoliella was first released in Oregon in 1970; however, the moth was later determined to have pre-existed there—probably as a result of accidental introductions on ornamental plants (Ritcher 1966). The moth now occurs throughout most of the areas where broom grows at elevations below 800 meters. *Leucoptera* is also present in Washington, presumably having been accidentally introduced before 1941 (Frick 1964).

In 1964 about 10,000 adult seed weevils, *Apion fuscirostre* F. (Coleoptera: Curculionidae), were introduced into California from broom near Ascoli, Italy, east of Rome (Andres, Hawkes, and Rizza 1967). The adult weevils hibernate in crevices along the branches and trunk and in the partially opened pods that remain on the plant. Weevils have also been found on the bark of nearby chestnut, oak, and other woody perennials that are associated with broom in Europe (Andres unpub. data). As the spring weather becomes warmer, adults fly to and congregate on branches bearing young buds; they begin feeding when the first blossoms appear. This feeding precedes mating and ovarian maturation, and synchronizes oviposition with the appearance of the young pods that form shortly after blossom-fall (Andres 1963). Each larva feeds on and destroys one or two of the developing seeds before it pupates in the pod. The adults emerge in summer as the pods split open. In pods that fail to split, adults can live until the following spring, if they are not killed by summer heat. Emerging adults feed on the current season's twigs before hibernating.

Researchers conducted laboratory starvation tests with overwintered adult weevils collected directly from plants as well as with newly formed adults removed from unopened pods. They found typical feeding damage on three legumes—*Ulex europaeus* L., *Spartium junceum* L., and *Cytisus villosus* Pourret (*C. triflorus* L'Her)—though the number of scars on each was less than half the amount on Scotch broom. More importantly, only females that fed on the floral buds and blossoms of Scotch broom developed ovaries (Andres 1963). The weevils did no significant damage to any of the economically important legumes tested. Researchers later confirmed this high degree of specificity by sampling the pods of 21 species of related woody legumes; though they found other *Apion* species in many of those pods, only *A. fuscirostre* was reared from the pods of Scotch broom (Andres 1963).

In the spring of 1964, 100 to 1,842 weevils were released at each of 13 northern California sites and became established at all locations. In 1983, after they had been approved for release, weevils were transferred from California to Polk County, Oregon, 6 kilometers west of Salem. In 1986 researchers began limited redistribution of weevil adults from this site. By 1988 the population level was high enough to allow the transfer of 250 adults to each of 14 new sites in eight Oregon counties. A survey in 1988 indicated that the weevils were established at four of those sites. Researchers are planning to release the seed weevil in Washington, even though it may already be present there.

In addition to these planned natural enemy releases, at least eight other broom-feeding arthropods have been accidentally introduced to North America (Waloff 1966).

Objective D: Impact

In California the numbers of *L. spartifoliella* pupae per plant are frequently in the hundreds. In the more heavily mined twigs, only the proximal or basal buds remain viable, which causes clusters of blossoms and new foliage to form at the bases of the previous season's mined twigs. Although *Leucoptera* distorts and stunts the plant, producing an unthrifty appearance, it is difficult to measure its effect on plant survival.

A survey near Salem, Oregon, indicated that 80 percent of 30 Scotch broom plants were infested with an average of 6.7 pupae per plant (ranging from 1 to 15 per plant). Of pupae held in the laboratory, 7 out of 25 were parasitized by hymenopterous parasitoids.

Although at some California sites it has damaged 60 percent of seeds, *A. fuscirostre* has had no observable effect on existing populations of the weed. In 1988 a random sample of 100 seed pods near Salem, Oregon, showed that 31 percent were infested by weevils. Damage to the infested pods ranged from 15 to 100 percent, with an average of 85 percent of the seed destroyed.

Of the natural enemies that were apparently accidentally introduced to North America (Waloff 1966), one insect has drawn recent attention: the Eurasian oecophorid moth, *Agonopterix nervosa* (Haworth) (Lepidoptera: Oecophoridae). Believed to have been introduced into southern Vancouver Island between 1915 and 1920, the moth has since spread to Washington, Oregon, California, and Nevada (Hodges 1974; Piper person. commun.). It was first reared from Scotch broom near Salem, Oregon, in 1960 or 1961. Larvae were observed on Scotch broom in Oregon's Polk County in 1988 and were also found at all of the major gorse infestations surveyed. In Europe both plants are known hosts of *A. nervosa* (Hodges 1974).

RECOMMENDATIONS

l. New agents are needed to control Scotch broom, but first, decisions must be made about the value of closely related woody legumes that may also be attacked.

2. The impact of *Leucoptera, Apion,* and the naturalized natural enemies of Scotch broom should be studied.

3. A compilation is needed of the known natural enemies of Scotch broom in its native range.

REFERENCES CITED

Andres, L. A. 1963. *Summary of observations on the host range and development of* Apion fuscirostre *(Curculionidae), seed weevil on Scotch broom* (Cytisus scoparius [L.] *Link: Leguminosae).* USDA-ARS, Entomology Research Division, Insect Identification and Foreign Parasite Introduction Research Branch Special Report.

Andres, L. A., R. B. Hawkes, and A. Rizza. 1967. *Apion* seed weevil introduced for biological control of Scotch broom. *Calif. Agric.* 21(8):13.

Bailey, L. H. 1906. *The American cyclopedia of horticulture.* New York: Macmillan.

Butterfield, H. M. 1964. Dates of introductions of trees and shrubs to California. Landscape Horticulture Department mimeo. Davis: University of California.

Clapham, A. R., T. G. Tutin, and W. F. Warburg. 1957. *Flora of the British Isles.* Cambridge: Cambridge University Press.

Frick, K. E. 1964. *Leucoptera spartifoliella*, an introduced enemy of Scotch broom in the western United States. *J. Econ. Entomol.* 57:589–91.

Gilkey, H. M. 1957. *Weeds of the Pacific Northwest.* Corvallis: Oregon State University Press.

Gill, J. D., and F. L. Pogge. 1974. *Cytisus scoparius*, Scotch broom. In *Seeds of woody plants in the United States,* ed. C. S. Shopmeyer, 370–71. USDA-ARS Handbook 450. Washington, D.C.: U.S. Department of Agriculture.

Hitchcock, C. L., and A. Cronquist. 1973. *Flora of the Pacific Northwest.* Seattle: University of Washington Press.

Hodges, R. W. 1974. *The moths of America north of Mexico.* Fasc. 6.2., Gelechioidea. London: E. W. Classey.

Mobley, L. 1954. Scotch broom, a menace to forest, range, and agricultural land. In *Proceedings of the 6th Annual California Weed Conference,* 39–42.

Parker, H. L. 1964. Life history of *Leucoptera spartifoliella* with results of host-transfer tests conducted in France. *J. Econ. Entomol.* 57:566–69.

Ritcher, P. O. 1966. *Biological control of insects and weeds in Oregon.* Corvallis: Oregon State University Technical Bulletin 90.

Waloff, N. 1966. Scotch broom, *Sarothamnus scoparius* (L.) Wimmer, and its insect fauna introduced into the Pacific Northwest of America. *J. Appl. Ecol.* 3:293–311.

80 / CLIDEMIA (KOSTER'S CURSE)

G. P. MARKIN AND R. M. BURKHART

> **clidemia (Koster's curse)**
> *Clidemia hirta* (L.) D. Don
> **Melastomataceae**

INTRODUCTION

Biology and Pest Status

Clidemia hirta is an aggressive, dense-growing perennial shrub 1 to 2 meters tall that is native to central and tropical South America. The hairy, papery leaves are dark green and contain veins that are slightly recessed, giving the leaf a distinct netted or cell-like appearance. Flowers are white, and the purple- or black-fleshed fruit is small (less than 1 cm in diameter), has many seeds, and grows in clusters in the upper leaf axils (Haselwood, Moter, and Hirano 1983). The fruit is relished by birds and mongeese (Hosaka and Thistle 1954), which, along with hikers' boots, are primarily responsible for the plant's dispersal in Hawaii (Wester and Wood 1977).

Clidemia foliage is not palatable to livestock. After being introduced to the Fiji Islands, where it was left unchecked, the weed readily invaded pastures and rendered them unfit for grazing (Simmonds 1933). In Hawaii—where it was probably introduced before 1940—*C. hirta* is mostly a forest pest, and is one of the few introduced weeds that readily invades dense, undisturbed stands of native forests below 1,000 meters, where it crowds out and replaces the native understory. Its rapid spread represents a serious threat to the long-term survival of these unique ecosystems.

Historical Notes

First observed on the island of Oahu in 1941, in 1952 *Clidemia hirta* was still limited to a single area of infestation of less than 100 hectares (Anonymous 1954). However, it soon began spreading, and by 1975 it covered 20,000 hectares on both of the island's major mountain ranges (Wester and Wood 1977).

The weed had been accidentally introduced to the island of Hawaii by 1972, to Molokai by 1973, to Maui by 1977, and to Kauai by 1982 (Nakahara, Burkhart, and Funasaki 1992). The most recent figures estimate that on Oahu alone it infests over 40,000 hectares, and it is found on all of Hawaii's major islands except Lanai (Trujillo, Latterell, and Rossi 1988).

On the Fiji Islands clidemia was a major weed until a very successful biological control program based on the introduction of the thrips *Liothrips urichi* Karny was begun in 1930 (Simmonds 1933). *L. urichi* successfully controlled *C. hirta* in pastures and open areas, where the plant was exposed to direct sunlight. However, it was less successful in the forests (Parham 1958; Wester and Wood 1977).

RESULTS

Objective A: The Identification and Introduction of Beneficial Organisms

Early introductions. In 1953, after realizing that clidemia was well established and rapidly expanding its range, researchers began a biological control program that used *L. urichi* from Fiji (Fullaway 1954; Weber 1954). This thrips feeds on and destroys the plant's young, tender leaves and growing shoot tips. Although the thrips became established and spread quickly, the weed continued its rapid spread, but only in the forest understory. In Hawaii *C. hirta* has never been a significant problem in open areas, and *L. urichi* is generally credited with preventing its spread into pastures and cultivated lands (Wester and Wood 1977). N. J. Reimer's (1985) recent studies, which showed that the fecundity of the thrips is significantly reduced in shaded compared with sunny areas, generally supports this conclusion.

We thank C. J. Davis (retired), E. Yoshioka, and P. Conant of the Hawaii Department of Agriculture for providing information for this report. In particular, we want to thank G. Funasaki of the Head Plant Pest Control Branch, Hawaii Department of Agriculture, for help in gaining access to historical records and for reviewing this manuscript.

Because *C. hirta* continued its spread unabated in forested areas, researchers began to search for additional control agents. In 1966, 41 adults of the pyralid moth, *Ategumiam atulinalis* Guenée (originally referred to as *Blepharomastix ebulealis* Guenée), from Trinidad, were introduced on the island of Oahu (Davis 1971). Although researchers propagated the moth artificially and released thousands more adults over the next 4 years, the moth was not recovered in the field until 1974 (Nakao and Funasaki 1976). Populations of this moth have never built up to any significant level, and researchers generally believe that it is suppressed by several hymenopterous parasites (Fujii 1977). In fact, N. J. Reimer (1985) found that the moth suffered a 43 percent parasitism rate by four species of hymenopterous parasites, two of which—the chalcid *Brachymeria obscurata* (Walker) and the braconid *Meteorus laphygmae* Vierech—had been purposely introduced in 1895 and 1942, respectively, as biological control agents for other lepidopterous pests (Funasaki et al. 1988). L. M. Nakahara, R. M. Burkhart, and G. Y. Funasaki (1992) have compiled a review of the earlier efforts at the biological control of *C. hirta*.

Recent introductions. In 1980 a new effort to find biological control agents was undertaken jointly by the Hawaii Department of Agriculture and the Hawaii Department of Land and Natural Resources. R. M. Burkhart, an exploratory entomologist, was assigned to study *C. hirta* in part of its native range in Trinidad, West Indies. By 1983, R. M. Burkhart had identified 12 promising insect species; preliminary host testing indicated were all restricted to *C. hirta* or a few closely related Melastomas (Burkhart 1983). In 1988, after additional field work in Trinidad, the Hawaii Board of Agriculture issued a permit to release two of those insects—the leaf-feeding buprestid beetle, *Lius poseidon* Napp, and the noctuid moth, *Antiblemma acclinalis* Hübner.

Lius poseidon (Coleoptera: Buprestidae). This beetle is multivoltine and requires 40 days to complete a generation. The early-instar larvae are serpentine leafminers, but later instars form blotches, two or three of which can destroy a normal *C. hirta* leaf. Adults also feed on clidemia leaves but prefer the older, more mature ones, and do not compete with the larvae, which restrict their feeding to younger leaves. The beetle prefers plants in light shade, and does not survive on plants exposed to long periods of direct sunlight. In Trinidad the beetle undergoes extreme population fluctuations, depending on the weather. In late 1988 three releases of *L. poseidon* were made on Oahu. In visits to the release site 9 months later, small numbers of all stages were found

(Conant person. commun.), confirming that the buprestid had become established. It is now well established on Oahu and has also been introduced and established on the islands of Hawaii, Kauai, and Maui.

Antiblemma acclinalis (Lepidoptera: Noctuidae). The larvae of this noctuid are defoliators that feed on the leaves of younger plants growing in moderate to heavy shade. The moth is multivoltine, and requires less than 40 days to complete a generation. Although permission to release this insect was given in October 1988, funding restrictions and a redirection of efforts interrupted the program before any releases were made.

In October 1986 the Hawaii Board of Agriculture issued a permit for a third agent, a plant pathogen, the leaf-spot fungus *Colletotrichum gloeosporioides* F. S. clidemiae (Cda.) (Melanconiaceae: Fungi Imperfecti) (Trujillo, Latterell, and Rossi 1988). Discovered in Panama in 1985, the fungus was shipped to the USDA Plant Pathogen Quarantine Facility at Frederick, Maryland, for pathogenicity testing, and then transferred to the University of Hawaii at Manoa for host-specificity testing (Trujillo 1986). In late October 1986, researchers made the first field application, using a suspension of spores hand-sprayed on plants on Oahu; in March 1987 they also introduced the pathogen on Kauai. Although the pathogen quickly defoliated the treated plants (killing some), most subsequently regenerated. To date, none of the treated areas has shown a reduction in the amount of clidemia, though diseased plants are common, and the fungus periodically causes repeated defoliation. The pathogen has spread up to 0.8 kilometers from its release points and has also been dispersed artificially to several new locations.

Objective D: Impact

The thrips *L. urichi* has apparently been successful in preventing clidemia from becoming established in pastures and open areas, but the combined efforts of the thrips and *A. atulinalis* have not halted or slowed the rapid spread of this pest in native forests. Both *L. poseidon* and the plant pathogen have become established and are causing noticeable damage locally. However, it is still too early to determine whether either will be effective.

RECOMMENDATIONS

1. A single Hawaiian state agency should assume responsibility for overseeing and promoting the *C. hirta* program. During the Hawaii Legislature's 1989 budget session, the Department of Agriculture and the Department of Land and Natural Resources each felt that the other was taking the lead in the program.

Accordingly, no budget for *C. hirta* was passed, and the program is on hold.

2. The necessary funds should be obtained to collect and introduce the noctuid moth *Antiblemma acclinalis,* which has already been cleared for release.

3. Work in Trinidad should be reinitiated on the 10 additional natural enemies that have been identified (Nakahara, Burkhart, and Funasaki 1992). Because three defoliating agents have already become established, the new efforts should concentrate on the fruit- and seed-feeders, which might restrict the spread of this weed.

4. In the 1970s, a very successful biological control program was conducted in Hawaii on the weed *Ageratin riparia* (Regel) Kashad R. The weed came under nearly complete control after researchers introduced two insects and a plant pathogen. Unfortunately, there were no detailed follow-up studies, and there is local controversy concerning whether the pathogen or the insects were the effective agents. To prevent this problem from reccurring, detailed monitoring studies on the spread, interaction, and importation of control agents for clidemia should be started as soon as possible.

REFERENCES CITED

Anonymous. 1954. Notes and exhibitions. *Proc. Hawaii. Entomol. Soc.* 15(2):263–65; 279–81; 282–83.

Burkhart, R. M. 1983. Progress report on exploratory studies on *Clidemia hirta* in Trinidad, West Indies, June 1982–June 1983. Unpublished report. Honolulu: Hawaii Department of Agriculture.

Davis, C. J. 1971. Recent introductions for biological control in Hawaii XVI. *Proc. Hawaii. Entomol. Soc.* 21(1):59–62.

Fujii, J. K. 1977. Notes and exhibitions. *Proc. Hawaii. Entomol. Soc.* 22(3):394–95.

Fullaway, D. T. 1954. Notes and exhibitions. *Proc. Hawaii. Entomol. Soc.* 15(2):280.

Funasaki, G. Y., P. Y. Lai, L. M. Nakahara, J. W. Beardsley, and A. K. Ota. 1988. A review of biological control introductions in Hawaii: 1890 to 1985. *Proc. Hawaii. Entomol. Soc.* 28:105–60.

Haselwood, E. L., G. G. Motter, and R. T. Hirano, eds. 1983. *Clidemia hirta.* In *Handbook of Hawaiian weeds,* 2nd ed., 284–85. Honolulu: University of Hawaii Press.

Hosaka, E. Y., and A. Thistle. 1954. *Noxious plants of the Hawaiian ranges.* University of Hawaii Cooperative Extension Service Bulletin 62.

Nakahara, L. M., R. M. Burkhart, and G. Y. Funasaki. 1992. Review and status of biological control of *Clidemia hirta* in Hawaii. In *Alien plant invasion in native ecosystems of Hawaii: Management and research,* eds. C. P. Stone, C. W. Smith, and T. T. Tunison, 452–65. Honolulu: University of Hawaii Press.

Nakao, H. K., and G. Y. Funasaki. 1976. Introductions for biological control in Hawaii, 1974. *Proc. Hawaii. Entomol. Soc.* 22(2):329–31.

Parham, J. W. 1958. *The weeds of Fiji.* Fiji Department of Agriculture Bulletin 35. Suva: Government Press.

Reimer, N. J. 1985. An evaluation of the status and effectiveness of *Liothrips urichi* Karny (Thysanoptera: Phlaeothripidae) and *Blepharomastix ebulealis* (Guenée) (Lepidoptera: Pyralidae) on *Clidemia hirta* (L.) D. Don in O'ahu forests. Ph.D. diss., University of Hawaii, Honolulu.

Simmonds, F. W. 1933. The biological control of the weed *Clidemia hirta* D. Don, in Fiji. *Bull. Entomol. Res.* 24:345–48.

Trujillo, E. E. 1986. Host-specificity test with *Colletotrichum gloeosporioides* F. S. *clidemiae* (Cda.) (Melanconiaceae: Fungi Imperfecti): A biological control candidate for *Clidemia hirta* (L.) D. Don. Unpublished report (May 16, 1986). Honolulu: Hawaii Board of Agriculture.

Trujillo, E. E., F. M. Latterell, and A. E. Rossi. 1988. *Colletotrichum gloeosporioides,* a possible biological control agent for *Clidemia hirta* in Hawaiian forests. *Plant Dis.* 70:974–76.

Weber, P. W. 1954. Recent liberations of beneficial insects III. *Proc. Hawaii. Entomol. Soc.* 15:371–72.

Wester, L. L., and H. B. Wood. 1977. Koster's curse (*Clidemia hirta*), a weed pest in Hawaiian forests. *Environmental Conservation* 4(1):35–41.

81 / Banana Poka

G. P. MARKIN AND R. W. PEMBERTON

banana poka *Passiflora mollissima* (HBK) Bailey Passifloraceae

INTRODUCTION

Biology and Pest Status

Passiflora mollissima, a climbing liana vine native to the northern Andes of South America, is currently established in native koa-ohia forests on three Hawaiian islands. When abundant, this vine spreads through the forest canopy, killing trees by shading them, breaking them with its added weight, or making them more susceptible to being blown over in heavy winds. In open areas in the forest, thick mats of the vines quickly smother seedlings and interfere with forest regeneration (Markin and Nagata 1989). Recent studies also indicate that large masses of vines can interfere with nutrient cycling in the native rain forests (Scowcroft, U. S. Forest Service, Honolulu, person. commun.). Researchers estimate that the vine is now established on over 40,000 hectares of native Hawaiian forest (Warshauer et al. 1983). Consequently, it has been declared the major threat to the existence of this unique forest ecosystem.

Historical Notes

First introduced to Kauai around the turn of the century, before 1920 the plant had reached the island of Hawaii. Most recently, it has shown up on Maui (La Rosa 1984; Markin, Nagata, and Taniguchi 1989). The plant was probably imported for its bright, showy flowers or its large, elongate, yellow fruit, which led to its common names—banana poka or banana passion fruit. In the Andes—in Colombia, Ecuador, and Peru—the banana poka is cultivated at elevations above 3,000 meters (Martin and Nakasone 1970), both for local human consumption and in small commercial plantings as a cash crop. The fruit has never been exploited commercially in Hawaii, and visiting South American scientists claim that the Hawaiian form is less flavorful than the commercial South American variety. The plant has also been introduced into California, New Zealand, and the Portugese Madeira Islands in the Atlantic. In all three islands it has escaped from cultivation but is not considered a problem.

The plant growing in Hawaii has been identified as a variety of *P. mollissima* (Neal 1965), one of several closely related commercial varieties that have been developed as agricultural crops (Escobar 1980). In South America, however, some taxonomists recognize each variety as an individual species, and refer to the form found in Hawaii as *P. tripartita* (La Rosa 1987).

In the early 1980s, after researchers recognized the rapid expansion of the weed's range in native forests and the failure of conventional control methods, several state and federal agencies began a cooperative biological control program. In 1982 a USDA-ARS exploratory entomologist visited South America to survey the plant's native enemies (Pemberton 1982). In 1983 and 1984, a new quarantine facility was built at Hawaii Volcanoes National Park on the island of Hawaii, primarily for work on this weed; it received its first shipment of potential biological control agents from South America in December 1985.

RESULTS

Objective A: The Identification and Introduction of Beneficial Organisms

R. W. Pemberton (1982) has identified approximately 100 species of insects feeding on or associated with this plant in South America. Based on his observations and the recommendations of a local Colombian entomologist, M. Rojas de Hernández of the University of Cali, who is familiar with the insect pests of commercial plantings, three insects were selected as potential biological control agents: *Cyanotricha necyria* Felder, *Pyrausta perelegans* Hampson, and *Zapriothica salebrosa* Wheeler.

***Cyanotricha necyria* (Lepidoptera: Dioptidae).** In Colombia this moth is a minor defoliator of *P. mollissima*.

However, because during outbreaks it can totally denude commercial vineyards of this plant, it was the first insect to be considered. Between 1985 and 1987 researchers in Hawaii conducted host tests of it in quarantine (Markin, Nagata, and Taniguchi 1989). In 1987 the state of Hawaii issued a permit for its release (Markin and Nagata 1989).

After mass-rearing the moth, the Hawaii Department of Lands and Natural Resources first released it in the Laupahoehoe Forest Reserve on the island of Hawaii in February 1988. By September 1, 1988, 10,000 larvae, eggs, or pupae had been released. The release areas are being monitored, but it appears that the moth has failed to become established.

Pyrausta perelegans (Lepidoptera: Pyralidae). This moth was chosen because it is an important enemy of *P. mollissima* throughout its range in South America (Rojas de Hernández and Chación 1982). The first-instar larvae are shoot-tip miners, but later instars move down the stem and attack the growing flower buds. If it were to become established, this insect would not affect established banana poka plants, but it would reduce fruit and seed production, thus slowing the spread of the weed. Researchers completed host testing in quarantine in 1989, and approval for its release was granted in 1990. The first release was made in 1991. Establishment has been confirmed at one location each on the islands of Hawaii and Maui, but the populations remain very low and do not appear to be increasing.

Zapriothica salebrosa (Diptera: Drosophilidae). The larvae of this drosophilid fly attack the anthers in developing flower buds, causing abortion of the flowers. Several Hawaiian attempts to establish colonies of this insect in quarantine foundered when adults failed to mate. Consequently, M. Rojas de Hernández was contracted to conduct biological studies and host-range testing of this insect in the field in Colombia. As of September 1989, the work was still in progress.

Besides insects, researchers have identified several pathogens that affect *P. mollissima* in its home range in South America and in Hawaii (Gardner and Davis 1982; Pemberton 1982). These pathogens are being studied by E. Trujillo of the University of Hawaii at Manoa and D. Gardner of the Cooperative Parks Studies Unit of the National Park Service, University of Hawaii at Manoa. They have not yet identified any highly specific or virulent pathogens, though a number of general fungi do attack *P. mollissima* seedlings in Hawaii and New Zealand.

Forest land managers in Hawaii consider banana poka to be the most important of forest weeds, and feel that biological control offers the only long-term solution (Waage, Smiley, and Gilbert 1981; Gardner and Davis 1982). The present program against *P. mollissima* is expected to continue for at least 5 more years. At the moment, work in Colombia is difficult: insurgents, drug growers, and political instability make many of the major *P. mollissima*-growing areas in Colombia inaccessible even to local entomologists. Work has therefore shifted to Venezuela and Ecuador.

With the release of *C. necyria* under way, and the release of *P. perelegans* pending, efforts will continue on *Z. salebrosa*. If conditions allow, researchers will begin studying crown- and stem-boring insects, as well as other defoliators.

RECOMMENDATIONS

1. Accelerate work in South America before the areas where banana poka is cultivated become even more inaccessible due to deteriorating local conditions.

2. Look for alternative methods of control, because if access to the necessary areas in South America is lost, it might terminate the program's progress. These methods could include preventing further spread by fencing out the wild pigs that are a major carrier of the seeds; managing existing stands by grazing them with cattle; and using fungal mycoherbicides from pathogens that are already known to attack the plant locally.

REFERENCES CITED

Escobar, L. K. 1980. Interrelationships of the edible species of *Passiflora* centering around *Passiflora mollissima* (HBK) Bailey, subgenus *Tacsonia*. Ph.D. diss., University of Texas, Austin.

Gardner, D. E., and C. J. Davis. 1982. *The prospects for biological control of non-native plants in Hawaiian National Parks.* Cooperative National Park Studies Unit, University of Hawaii at Manoa, Department of Botany Technical Report 45.

La Rosa, A. M. 1984. *The biology and ecology of* Passiflora mollissima *in Hawaii*. Cooperative National Park Studies Unit, University of Hawaii at Manoa, Department of Botany Technical Report 50.

———. 1987. Note on the identity of the introduced passionflower vine "Banana poka" in Hawaii. *Pac. Sci.* 39(4):369–71.

Markin, G. P., and R. F. Nagata. 1989. *Host preference of* Cyanotricha necyria *Felder (Lepidoptera: Dioptidae), a potential biological control agent of the weed* Passiflora mollissima *(HBK) Bailey in Hawai'i forests*. Cooperative National Park Resources Studies Unit, University of Hawaii at Manoa, Department of Botany Technical Report 67.

Markin, G. P., R. F. Nagata, and G. Taniguchi. 1989. Biological and behavior of the South American moth, *Cyanotricha necyria* (Felder and Rogenhofer) (Lepidoptera: Dioptidae), a potential biocontrol agent in Hawaii of the forest weed *Passiflora mollis-*

sima (HBK) Bailey. *Proc. Hawaii. Entomol. Soc.* 29:115–23.

Martin, F. W., and H. Y. Nakasone. 1970. The edible species of *Passiflora. Econ. Bot.* 24:333–43.

Neal, M. C. 1965. *Gardens of Hawaii.* Bernice P. Bishop Museum Special Publication 50. Honolulu: Bishop Museum Press.

Pemberton, R. W. 1982. Exploration for natural enemies of *Passiflora mollissima* in the Andes. Unpublished report. Albany, Calif.: USDA Biological Control of Weeds Laboratory.

———. 1989. Insects attacking *Passiflora mollissima* and other *Passiflora* species: Field survey in the Andes. *Proc. Hawaii. Entomol. Soc.* 29:71–84.

Rojas de Hernández, M., and P. Chacón de Ulloa. 1982. Contri-bución a la biología de *Pyrausta perelegans* Hampson (Lepidoptera: Pyralidae). *Brenesia* 19/20:325–31.

Waage, J. K., J. T. Smiley, and L. E. Gilbert. 1981. The *Passiflora* problem in Hawaii: Prospects and problems of controlling the forest weed *P. mollissima* (Passifloraceae) with Helicioniiae butterflies. *Entomophaga* 26:275–84.

Warshauer, F. R., J. D. Jacobi, A. M. La Rosa, J. M. Scott, and C. W. Smith. 1983. *The distribution, impact, and potential management of the introduced vine,* Passiflora mollissima *(Passifloraceae), in Hawaii.* Cooperative National Park Resources Studies Unit, University of Hawaii at Manoa, Department of Botany Technical Report 48.

82 / DALMATIAN TOADFLAX

R. M. NOWIERSKI

Dalmatian toadflax
Linaria genistifolia ssp. *dalmatica* (L.) Maire & Petitmengin
Scrophulariaceae

INTRODUCTION

Biology and Pest Status

Dalmatian toadflax is a broad-leaved perennial herb native to the Mediterranean region from the former Yugoslavia to northern Iran (Tutin et al. 1972; Robocker 1974). The weed has glaucous green foliage, bright yellow snapdragonlike flowers, and ovoid to nearly spherical fruit (Alex 1962). Large plants are capable of producing half a million seeds, which can remain viable in the field for up to 10 years (Robocker 1970, 1974). In addition to reproducing sexually, new plants can arise from the secondary crown points along the lateral root system, which also can develop floral stems. The survival of new seedlings, combined with the yearly extension of lateral roots and the production of floral stems from those roots, allows toadflax to persist in areas where interspecific competition is low (Robocker 1974).

In North America this robust herb is adapted to cool, semiarid climates and coarse-textured soils; it occurs most often on sparsely vegetated soils and depleted rangelands (Alex 1962; Robocker 1974). Compared with established perennials and fast-maturing winter annuals, Dalmatian toadflax seedlings are poor competitors for soil moisture; however, established plants can be extremely competitive and can have a significant effect on the yearly composition of annual vegetation, the production of other perennial herbs, and its own seedlings' survival (Robocker 1974).

O. Polunin (1969) noted that Dalmatian toadflax is toxic to livestock; however, cattle, possibly horses, and other animals have reportedly used the plant as forage (Robocker 1974). C. F. Reed (1970) reported that cattle browse the tips of the floral stalks and may help spread the seed via their feces. Deer and pocket gophers reportedly also eat the plant (Robocker 1974).

Historical Notes

Dalmatian toadflax was cultivated as an ornamental in Europe as early as 1594, and was probably intentionally introduced into North America around 1894 (Alex 1962). It escaped from cultivation and, because of its adaptiveness and high reproductive capacity, spread to become an important weed of pastures, cultivated fields, and disturbed rangelands throughout the northern and western United States and Canada (Coupland 1954; Lange 1958; Montgomery 1964; Reed 1970).

Herbarium records have shown that by 1980 Dalmatian toadflax could be found in 58 northwestern U.S. counties (Forcella and Harvey 1980). A plot of the number of infested counties against time (fig. 82.1) shows the weed's dramatic and accelerated increase

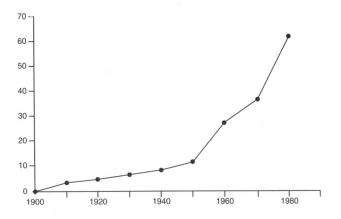

Figure 82.1. Chronological plot of the number of counties infested with Dalmatian toadflax in Washington, Oregon, Idaho, Montana, and Wyoming (Forcella and Harvey 1980).

Appreciation is extended to the following for contributing information to this chapter: L. Andres, T. Breitenfeldt, E. Coombs, P. Harris, R. Lavigne, J. Littlefield, J. McCaffrey, G. McDermott, G. Piper, J. Story, A. Sturko, and C. Turner. This chapter was supported by the Noxious Weed Trust Fund of the Montana State Department of Agriculture; the Bureau of Indian Affairs in Billings, Montana; Big Horn and Missoula counties, Montana; Fremont County Weed and Pest Control District, Wyoming; and the Montana Agricultural Experiment Station.

between 1920 and 1980. Since 1980 the weed has continued to increase rapidly in distribution (fig. 82.2), according to a survey by the Cooperative Agricultural Pest Survey (Lajeunesse et al. 1993).

RESULTS

Objective A: The Identification and Introduction of Beneficial Organisms

In North America three accidentally introduced natural enemies attack yellow toadflax, *Linaria vulgaris* Miller, and to a lesser degree, Dalmatian toadflax (Smith 1959). These include an ovary-feeding nitidulid, *Brachypterolus pulicarius* (L.), and two seed capsule–feeding weevils, *Gymnetron antirrhini* (Paykull) and *G. netum* (Germar). In some regions of Canada *B. pulicarius* has reportedly reduced yellow toadflax seed production by 80 to 90 percent (Harris 1961), and its presence has coincided with the decline of the weed in western Canada (Harris 1984). J. M. Smith (1959) reported that this beetle will also attack the broad-leaved form of Dalmatian toadflax, as has been seen in the East Kootenays of British Columbia (Harris 1988). In Canada the redistribution of *B. pulicarius* collected from Dalmatian toadflax is continuing (Harris 1988). In the United States the beetle has been recorded on *L. vulgaris* in Oregon, Idaho, and Montana (Smith 1959; Barr unpub. data; Coombs person. commun.; McCaffrey person. commun.).

In June 1992 I traveled to Kamloops, British Columbia, to collect *B. pulicarius* from Dalmatian toadflax. Approximately 2,000 beetles were collected (Nowierski and Sturko unpub. data), and 500 beetles each were subsequently released at three Dalmatian toadflax sites in Montana. Beetles were recovered at all three sites in 1993, indicating that they had successfully overwintered. Preliminary greenhouse studies examining the impact of the beetle on the growth and reproduction of Dalmatian toadflax suggest that larval and adult feeding may reduce seed production on Dalmatian toadflax by as much as 39 percent (Grubb and Nowierski unpub. data). In addition, adult feeding on the young growing tips was found to cause increased branching of flowering stems, primary branches, and secondary branches. This sort of feeding damage in the field could conceivably result in more prostrate plants that cannot compete as well with other plant species.

The two weevils, *G. antirrhini* and *G. netum*, though readily attacking *L. vulgaris*, apparently only accept the narrow-leaved forms of Dalmatian toadflax, *L. genistifolia genistifolia* (L.) Miller (Smith 1959). J. M. Smith (1959) reported that in Canada's eastern provinces, in British Columbia, and in the northwestern United States, *G. antirrhini* was the major factor in the natural control of

L. vulgaris, and that where the weevil was found, the weed was usually not damaging at economic levels. As a result, researchers in Canada are continuing to redistribute *G. antirrhini* against yellow toadflax (Harris 1988).

In the United States *G. antirrhini* has been recorded in Washington, Oregon, Idaho, Montana, and Wyoming (Smith 1959; Piper, Coombs, McCaffrey, Lavigne person. commun.; McDermott and Nowierski unpub. data). In Oregon researchers have reported that *G. antirrhini* substantially defoliates and damages the flowers of *L. vulgaris*; however, its effect on Dalmatian toadflax was not apparent (Coombs person. commun.). In Washington field observations suggest that the weevil can destroy 85 to 90 percent of the potential seed crop of *L. vulgaris* (Piper person. commun.). Host strains of this weevil may exist, since it is common on Dalmatian toadflax in the former Yugoslavia (Harris 1988). Researchers are currently conducting electrophoretic studies on both the common and Dalmatian toadflax strains of the weevil to identify any strain-specific enzymes (McDermott, Nowierski, and FitzGerald unpub. data).

The second weevil species, *G. netum*, has been recorded in a number of eastern states, but in western North America it has been found only in Washington and British Columbia (Smith 1959). This species appears to be displaced by *G. antirrhini* in Canada and seems to be of little biological control value (Harris 1988). The fact that the two *Gymnetron* species will readily attack only the narrow-leaved variety of Dalmatian toadflax is disappointing, as the broad-leaved form is more common and widely distributed (Coupland and Alex 1954; Smith 1959).

To date only one insect species has been screened and released against Dalmatian toadflax in the United States—the defoliating moth, *Calophasia lunula* Hufnagel. This moth was first released in Canada in 1965 (Darwent et al. 1975) and in the United States in 1968 (Barr unpub. data; McCaffrey and Piper person. commun.) to control both Dalmatian toadflax and yellow toadflax. Since then multiple releases have been made in Washington, Oregon, Idaho, Montana, Wyoming, and Colorado. The material for the releases came mainly from colonies at the USDA-ARS Biological Control of Weeds Laboratory in Albany, California; however, three of the seven releases in Washington state were made from colonies obtained in Ottawa, Ontario (Piper person. commun.).

In Ontario *C. lunula* is well established on yellow toadflax and defoliates up to 20 percent of the stems (Harris 1988). This is not surprising, because the initial strains of the moth were collected on yellow toadflax in Europe (Harris 1988). Additional *C. lunula* populations have been found on yellow toadflax at two sites in northern Idaho (McCaffrey and Piper person. commun.). Although both sites contained a mixed stand of *L. vulgaris* and *L. genistifolia* ssp. *dalmatica*, no larvae were observed on Dalmatian toadflax. In Saskatchewan and British Columbia, another strain of *C. lunula*, collected from Dalmatian toadflax in the former Yugoslavia, has been released on both toadflax species; the moth developed equally well on both (Harris 1988).

In 1989 researchers found the first evidence that *C. lunula* had become established on Dalmatian toadflax in North America (McDermott, Nowierski, and Story 1990). That July, 26 larvae were collected from Dalmatian toadflax approximately 11 kilometers southeast of Missoula, Montana. This population apparently came from releases made near Missoula between 1982 and 1985 (Story person. commun.). No yellow toadflax plants were observed in the area. Researchers have attributed the insect's failure to become established on Dalmatian toadflax at other release sites to predation by ants and birds (Harris 1984; Otten person. commun.), poor tolerance of extremely low temperatures (FitzGerald and Nowierski unpub. data), and the lack of a *Calophasia* strain adapted to Dalmatian toadflax. The apparent host strain differences in *C. lunula* are being examined with electrophoresis (McDermott and Nowierski unpub. data).

Federal and international agencies have participated in the screening of additional insects. Personnel at the former USDA-ARS Biological Control of Weeds Laboratory in Albany, California, contributed to screening a leaf-feeding chrysomelid, *Chrysolina gypsophilae* Kuster, that had been partially screened a number of years earlier (Rizza and Pecora 1984; Turner unpub. data). Unfortunately, preliminary screening results suggested that the host-plant range for this insect was too broad to warrant continuation of the screening tests; hence, work on this insect was discontinued.

The Commonwealth Agricultural Bureaux of the International Institute of Biological Control in Delèmont, Switzerland, conducted screening tests of two new insects that attack Dalmatian toadflax, as well as of a Dalmatian toadflax–adapted strain of an insect already established in North America on yellow toadflax. These were a stem-boring weevil, *Mecinus janthinus* Germar; a root-boring moth, *Eteobalea intermediella* (Riedle) (Lepidoptera: Cosmópterygidae); and a strain of *G. antirrhini* that is adapted to Dalmatian toadflax. A fourth insect, *Eteobalea serratella* (Treitschke), was also screened by IIBC for release against yellow toadflax in Canada. Screening tests by IIBC are under way on two root-galling weevils, *Gymnetron linariae* (Panzer) and *G. collinin* (Gyllenhal), and on two stem-galling weevils, *G. hispidum* (Brullé) and *G. thapsicola* (Germar). The addition of stem- and root-feeding insects should greatly improve the chances for successful biological control of Dalmatian toadflax in North America.

Objective C: The Conservation and Augmentation of Beneficial Organisms

Although *C. lunula* has been successfully reared on Dalmatian toadflax in the greenhouse and laboratory (Hoel 1984; McDermott and Nowierski unpub. data; Coombs, Lavigne, Piper, and Story person. commun.), certain behavioral aspects of the larvae have frustrated rearing efforts. For example, the larvae's tendency to drop off plants when disturbed, or after encountering other individuals, caused considerable mortality (Story unpub. data). To resolve this, researchers placed potted Dalmatian toadflax plants in circular holes cut in plywood, so that the tops of the pots were nearly flush with the plywood surface. This helped the dislodged larvae to relocate a host plant and resulted in the production of thousands of individuals (Story person. commun.).

Calophasia larvae reared at the Albany laboratory were released in Washington in 1968 and 1970 (Piper person. commun.), in Oregon in 1983 (Coombs person. commun.), in Idaho in 1968, 1969, 1982, and 1983 (Barr unpub. data; McCaffrey person. commun.), in Montana in 1972, 1975, 1982, and 1983 (Story 1979; Story person. commun.), and in Wyoming in 1982 and 1984 (Lavigne person. commun.). In 1982 and 1983 larvae reared in Washington from the colony originally obtained from Ottawa (Piper person. commun.) were redistributed within that state; in 1984 and 1985 larvae reared in Montana were redistributed (1,502 larvae were also sent to Oregon during those two years; Story person. commun.); and in 1984 there was an intrastate redistribution of larvae reared in Wyoming (Lavigne person. commun.). The population of the moth near Missoula, Montana, still provides the only definitive evidence that *C. lunula* can become established on Dalmatian toadflax in North America.

Approximately 500 *C. lunula* larvae, collected from Dalmatian toadflax near Missoula in 1989, were successfully reared to later instars. Nearly half were released at two field sites in Montana during late summer, 1989 (Nowierski and McDermott unpub. data). The remaining individuals were used in a mass-rearing effort for new Dalmatian toadflax sites in Montana in 1990. Larvae of *C. lunula* were collected at the Missoula field site from 1990 to 1993 and used in the mass-rearing effort. Release efforts for the moth in those years expanded to include 14 Dalmatian toadflax release sites in Montana. In addition, larvae reared at Bozeman were sent to cooperators in Colorado, Wyoming, and Idaho for release in those states. Preliminary observations suggest that the moth has become established in Montana at three release sites besides the Missoula site (Nowierski and FitzGerald unpub. data). Researchers are currently

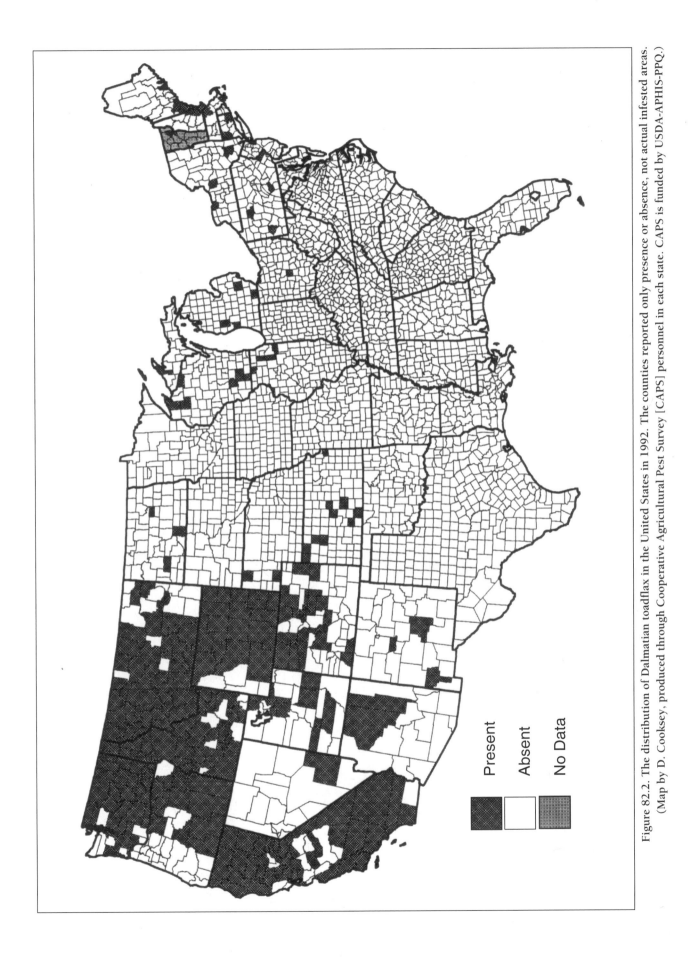

Figure 82.2. The distribution of Dalmatian toadflax in the United States in 1992. The counties reported only presence or absence, not actual infested areas. (Map by D. Cooksey, produced through Cooperative Agricultural Pest Survey [CAPS] personnel in each state. CAPS is funded by USDA-APHIS-PPQ.)

Present

Absent

No Data

developing an artificial diet for *C. lunula,* to augment the rearing effort (FitzGerald and Nowierski unpub. data).

In July 1989 my laboratory in Bozeman, Montana, received 20 *C. lunula* pupae from a rearing colony at the Agriculture Canada Research Station in Regina, Saskatchewan; the population was originally from Dalmatian toadflax in Yugoslavia (Harris unpub. data). From the nine moths that emerged from this shipment, approximately 150 larvae were successfully reared to the pupal stage. Unfortunately, they became infected with a cytoplasmic polyhedrosis virus infection (Breitenfeldt person. commun.), which made it impossible to redistribute the strain. The source of the virus has never been determined.

RECOMMENDATIONS

1. Survey Dalmatian toadflax sites to determine whether *B. pulicarius* is present. Redistribute *B. pulicarius* collected from Dalmatian toadflax in Canada to sites in the United States where it does not already occur.

2. Acquire *M. janthinus, E. intermediella,* and the Dalmatian toadflax–adapted strain of *G. antirrhini.* Following approval by the Technical Advisory Group and APHIS-PPQ, release those control agents at Dalmatian toadflax sites in the United States.

3. Continue to rear the Dalmatian toadflax–adapted strain of *C. lunula* and target new areas for its release.

4. Establish permanent vegetative plots at all release sites to monitor plant biomass and cover—of all grasses and forbs, including Dalmatian toadflax—as well as the density of Dalmatian toadflax before, during, and after the biological control agents become established.

5. Finish screening *C. linariae, G. collinin, G. hispidum,* and *G. thapsicola.*

6. Begin screening the other biological control candidates that USDA and IIBC records indicate occur on toadflax in Eurasia.

7. Continue the search in Eurasia for additional natural enemies.

REFERENCES CITED

Alex, J. F. 1962. The taxonomy, history, and distribution of *Linaria dalmatica. Can. J. Bot.* 40:295–307.

Coupland, R. T. 1954. *The Saskatchewan weed survey.* Department of Plant Ecology, College of Agriculture, University of Saskatchewan Circular 34.

Coupland, R. T., and J. F. Alex. 1954. Growth development at Saskatoon of *Linaria dalmatica* from various sources. *Res. Rep. Nat. Weed Comm. (West. Sect.)* 1954:104. Ottawa: Canada Department of Agriculture, National Weed Committee.

Darwent, A. L., W. Lobay, W. Yarish, and P. Harris. 1975. Distribution and importance in northwestern Alberta of toadflax and its insect enemies. *Plant Sci.* 55:157–62.

Forcella, F., and S. J. Harvey. 1980. *New and exotic weeds of Montana. I: Recent introductions.* Bozeman: Montana Weed Survey, Montana State University.

Harris, P. 1961. Control of toadflax by *Brachypterolus pulicarius* (L.) (Coleoptera: Nitidulidae) and *Gymnaetron antirrhini* (Payk.) (Coleoptera: Curculionidae) in Canada. *Can. Entomol.* 93:977–81.

———. 1984. *Linaria vulgaris* Miller, yellow toadflax, and *L. dalmatica* (L.) Mill., broad-leaved toadflax (Scrophulariaceae). In *Biological control programmes against insects and weeds in Canada 1969–1980,* eds. J. S. Kelleher and M. A. Hulme, 179–82. Farnham Royal, Slough, U.K.: Commonwealth Agricultural Bureaux International.

———. 1988. *The biocontrol of Dalmatian toadflax: A cooperative project report for 1988.* Agriculture Canada, British Columbia (Ministry of Agriculture, British Columbia Forest Service, and the Cattlemen's Association), and the state of Montana.

Hoel, C. 1984. A preliminary evaluation of mass-rearing of *Calophasia lunula* (Lepidoptera: Noctuidae) for biological control of toadflax, with emphasis on diapause. Master's thesis, University of Wyoming, Laramie.

Lajeunesse, S. E., P. K. Fay, J. R. Lacey, R. M. Nowierski, D. Zamora, and D. Cooksey. 1993. *Dalmatian toadflax—A weed of pasture and rangeland: Identification, biology, and management.* Montana Agricultural Extension Bulletin 115.

Lange, A. W. 1958. Dalmatian toadflax, a possible rival of goatweed as a serious range weed. *Weeds* 6:68–70.

McDermott, G. J., R. M. Nowierski, and J. M. Story. 1990. First report of establishment of *Calophasia lunula* Hufn. (Lepidoptera: Noctuidae) on Dalmatian toadflax, *Linaria genistifolia* ssp. *dalmatica* (L.) Maire & Petitmengin, in North America. *Can. Entomol.* 122:767–68.

Montgomery, F. H. 1964. *Weeds of the northern United States and Canada.* New York: Frederick Warne.

Polunin, O. 1969. *Flowers of Europe: A field guide.* Oxford: Oxford University Press.

Reed, C. F. 1970. *Selected weeds of the United States.* USDA-ARS Handbook 366. Washington, D.C.: U.S. Department of Agriculture.

Rizza, A., and P. Pecora. 1984. *Chrysolina gypsophilae* (Coleoptera: Chrysomelidae), a potential biocontrol agent of Dalmatian toadflax, *Linaria dalmatica. Ann. Entomol. Soc. Am.* 77:182–87.

Robocker, W. C. 1970. Seed characteristics and seedling emergence of Dalmatian toadflax. *Weed Sci.* 18:720–25.

———. 1974. *Life history, ecology, and control of Dalmatian toadflax.* Washington Agricultural Experiment Station Technical Bulletin.

Smith, J. M. 1959. Notes on insects, especially *Gymnaetron* spp.

(Coleoptera: Curculionidae) associated with toadflax, *Linaria vulgaris* Mill. (Scrophulariaceae) in North America. *Can. Entomol.* 91:116–21.

Story, J. M. 1979. *Biological weed control in Montana.* Montana State University Agricultural Experiment Station Bulletin 717.

Tutin, T. G., V. H. Heywood, N. A. Burges, D. M. Moore, D. H. Valentin, S. M. Walters, and D. A. Webb. 1972. *Flora Europaea,* vol. 3. Cambridge: Cambridge University Press.

83 / PUNCTUREVINE

L. A. ANDRES AND R. D. GOEDEN

puncturevine
Tribulus terrestris L.
Zygophyllaceae

INTRODUCTION

Biology and Pest Status

This cosmopolitan weed of Old World origin is a prostrate annual herb that bears an abundance of small, yellow flowers and troublesome spiny fruit. At maturity the fruit separate into five bony, one- to four-seeded segments, each of which is studded with two or four sharp, rigid, and divergent spines. The seeds are believed to survive in the soil for as long as 20 years. The spines lodge in tires, shoes, and the feet and fur of animals, all of which help disseminate the seeds (Johnson 1932; Goeden and Ricker 1973).

As an agricultural weed, puncturevine's spiny fruit interfere with hand-harvesting; injure livestock; and contaminate seed, feed, and wool (Johnson 1932). Puncturevine is also found in disturbed residential and industrial land and, like crabgrass, is a weed that many city dwellers recognize.

Historical Notes

The natural range of puncturevine apparently includes the Mediterranean regions of Europe and Africa and the drier parts of Asia (Andres and Angelet 1963). Some authorities believe that the weed originated in what is now the Sahara Desert (Kluge 1975). It was accidentally introduced into the midwestern United States with livestock, particularly sheep, imported from the Mediterranean area. Puncturevine now occurs from coast to coast, but it is most common in the Southwest. It arrived in California about 1900, apparently as a contaminant of railroad ballast, and spread rapidly along railroad and highway rights-of-way. *Tribulus terrestris* appeared in Hawaii about 1900; the congeneric *T. cistoides* L. is listed as native or endemic to Hawaii, though

it also occurs in the "coastal areas of all tropical regions of the world" (Markin person. commun.). As an annual, nonwoody, nonrangeland weed, puncturevine represents a departure from the traditional type of weed targeted for biological control. Before the efforts discussed here, there had been no biological control studies on this weed in North America or elsewhere.

RESULTS

Objective A: The Identification and Introduction of Beneficial Organisms

Work on this project predated the W-84 project, beginning in 1957 and 1958 with a survey of the natural enemies of puncturevine in India, southern France, and Italy. Researchers selected the seed-feeding weevil *Microlarinus lareynii* (Jacquelin duVal) (Coleoptera: Curculionidae) and the stem- and crown-mining weevil *M. lypriformis* (Wollaston) as the most promising candidates for biological control. Field and laboratory studies in France, Italy, and California from 1959 to 1961 demonstrated that, though the adults of both weevil species fed on a wide range of plant species, they were only able to reproduce successfully on puncturevine, other species of *Tribulus*, and a few herbaceous Zygophyllaceae native to the southwestern United States—namely, *Kallstroemia* species (Andres and Angelet 1963).

L. A. Andres and G. W. Angelet (1963) and R. L. Kirkland and R. D. Goeden (1978a) have described the biology of *M. lareynii*, and L. A. Andres and G. W. Angelet (1963) and R. L. Kirkland and R. D. Goeden (1978b) that of *M. lypriformis*. The immature stages of both weevils have been described in detail by R. L. Kirkland and R. D. Goeden (1977).

Special thanks to G. P. Markin for information supplied from Hawaii. We also would like to acknowledge the role of the former USDA Biological Control of Weeds Laboratory in Italy in carrying out much of the research that led to the clearance of *Microlarinus* species weevil shipments to the United States, as well as its role in collecting and shipping the weevils.

Briefly, an *M. lareynii* female deposits an egg in a pit chewed in the pericarp of an immature fruit or, occasionally, in a floral bud or flower, and then caps it with an anal secretion often stained dark with feces. The larva feeds on the seeds and surrounding tissues, destroying seeds directly, as well as indirectly by inducing abortion. The larva pupates in an open cell in the fruit, and the adult emerges through an exit hole it chews between two carpels. The eggs hatch in 2 to 3 days; larval development lasts 13 to 16 days; and, in southern California, the pupal stage lasts 4 to 5 days. The biology of *M. lypriformis* is similar, but it oviposits mostly on the undersides of the central, older parts of the prostrate, matlike plants—the root crowns, primary branches, and stem bases. The young larvae tunnel into the pith, to which they largely confine their feeding, and eventually pupate in open cells. The adults emerge from circular holes chewed in the upper surfaces of stems, branches, and crowns. Both weevil species are multivoltine and produce a generation each month in the summer by reinfesting plants and attacking new plants as dispersed adults. Both species also overwinter as adults in reproductive diapause among surface debris and plant litter, and in the shelter of adjacent perennial plants.

Although there was minor concern that the weevils might have a detrimental effect on native species of *Kallstroemia*, this conflict of interest was resolved by weighing the potential benefits of the biological control of puncturevine against the potential damage to those elements in the native flora. Due to pressure for immediate action, both weevils were approved for release in compliance with federal regulatory procedures.

Stem weevil adults were collected at Bari and Catania, Italy, in July and August of 1961, and seed weevil adults were collected at Fiumicino, near Rome, as well as at Catania, Italy. Although researchers recorded a number of parasitoids attacking the immature stages of the weevils in France, Italy, and India (Angelet and Andres 1965), these field-collected adult weevils were parasitoid-free. The weevils were then processed through the University of California quarantine facility at Berkeley and released directly in the field.

In July and August 1961, the first releases of stem weevils were made in Arizona, California, Colorado, Nevada, and Washington; they became established in Arizona, California, and Nevada. Also in July and August 1961, seed weevils were released in Arizona, California, Colorado, Nevada, Utah, and Washington, but they only became established in Arizona, California, and Nevada (Huffaker, Ricker, and Kennett 1961; Andres and Angelet 1963; Maddox 1976; Andres unpub. data).

In California the weevils spread rapidly and widely, aided by extensive transfers of field-collected adults (Goeden and Ricker 1967). In 1962 and 1963, adults collected from field colonies near Manteca in San Joaquin County and Brentwood in Contra Costa County were shipped to other areas of California, as well as to Colorado, Idaho, Kansas, Montana, Nebraska, Nevada, New Mexico, Oklahoma, Oregon, Texas, Utah, Washington, and Wyoming. The seed weevils became established in New Mexico, Texas, and—temporarily—Washington, as well as in Arizona, California, and Nevada, as noted previously. D. M. Maddox (1976) later reported the establishment of both weevils in Colorado, Kansas, New Mexico, Oklahoma, Texas, and Utah, and of the stem weevil in Florida, as well as the spread of both weevils into Mexico.

In 1962, 375 adult seed weevils collected in southern California were shipped to Hawaii and released on the island of Kauai; in 1963, 460 stem weevils from southern California and Arizona were released on the same island. Both species became established within 2 months of their release (Markin person. commun.).

Under the auspices of the W-84 project, researchers have continued making periodic redistributions and evaluating the weevils in the western United States.

Objective D: Impact

Soon after both weevils became established in southern California, R. D. Goeden and D. W. Ricker (1967, 1970) reported substantial egg predation by native Heteroptera, and larval and pupal parasitism by indigenous chalcidoid Hymenoptera. R. D. Goeden and R. L. Kirkland (1981) assessed this predation in irrigated and nonirrigated field plots, and determined that egg predators killed about half of the seed weevil eggs infesting puncturevine fruit, which reduced fruit infestation rates from 50 to 25 percent. Using the insecticidal-check method, R. L. Kirkland and R. D. Goeden (1978c) assessed the effects of both weevils acting in concert on irrigated and nonirrigated plants in field plots. Their results showed that water stress was the main cause of early-season plant mortality; however, weevil attacks on the surviving nonirrigated plants reduced flower production by 60 percent. Moreover, on the nonirrigated plants only half of the flowers produced fruit late in the growing season. Using insecticidal-check plots, D. M. Maddox (1981) demonstrated that the stem weevils had a greater impact than did the seed weevils, as measured by stem-growth rates, metered water stress, and whole-plant biomass comparisons. The seed weevils also drastically reduced seed germination in infested fruit. Furthermore, D. M. Maddox (1981) reported—and C. B. Huffaker, J. Hamai, and R. M. Nowierski (1983) agreed—that the seed weevils, acting largely alone in the atypical experimental field plots, increased the weed's flower production when their feeding on the underside of the stem tips caused a shortening of the internodes and increased the number of nodes—with flowers and seeds—per length of stem.

C. B. Huffaker, J. Hamai, and R. M. Nowierski (1983) reported that 15 years after the weevils' introduction, puncturevine coverage and seed production declined in more than 80 percent of 1,200 field plots in northern and southern California. They attributed the decline to the actions of both weevil species. The various participants in this project have agreed that the biological control of puncturevine in California is, in general, at least a partial success, and is a substantial success under field conditions where weevils attack nonirrigated plants (Maddox and Andres 1979; Kirkland and Goeden 1978c; Julien 1987). However, in years of ample or ill-timed rainfall, puncturevine may temporarily resurge.

In Hawaii researchers believed that by 1966 the weevils had controlled both species of puncturevine on Kauai. The weevils were then redistributed to the other islands, where their effect was equally dramatic. In fact, the control of puncturevine is considered to be Hawaii's most rapid and successful weed biological control project (Julien 1987; Markin person. commun.).

In 1966 stem weevils were shipped from Hawaii to the island of St. Kitts in the West Indies, and in 1969 seed weevils from southern California were also sent to St. Kitts. The species from California failed to become established, but those from Hawaii alone provided complete biological control of *T. cistoides* (Julien 1987).

Since their introduction into North America, *Microlarinus* weevil adults have been collected from a wide array of cultivated plants—for example, alfalfa, chrysanthemum, and grapes (Andres 1978)—as well as from weeds and uncultivated native plants, such as *Chenopodium album* L., *Bromus rigidus* Roth, and *Ambrosia* sp. (Goeden and Ricker 1970, 1976, and unpub. data). But as predicted, these weevils have only been reported to breed on *Tribulus* and *Kallstroemia* species (Andres and Angelet 1963; Turner 1985; Hawkes unpub. data), and the weevils' effect on nontargeted plants has apparently been negligible (Andres 1978 and unpub. data; Goeden unpub. data). The weevils' effect on *Kallstroemia* remains undefined, however, since their attacks may be masked by the attacks of a stem-boring cerambycid beetle (Hawkes unpub. data) and other native arthropod species.

Although at first the puncturevine weevils failed to become established in several states with cold winters—for example, Colorado, Kansas, and Oklahoma—D. M. Maddox (1976) reported that the weevils are now present in several of those states. This suggests that cold-tolerant weevil strains may have evolved from the original Italian material. That would not be surprising, because researchers have collected the stem weevil in

Karaj, Iran, and Uzbekistan (Andres unpub. data), both cold-winter areas.

In various parts of the world other species of natural enemies have been observed on, or reported from, *Tribulus* species. For example, in southern India an eriophyid mite attacks puncturevine (Andres unpub. data), and in Greece an unidentified fungal pathogen has been observed on the weed (Clement unpub. data). In South Africa puncturevine is also galled or otherwise heavily attacked by at least four species of insect enemies (Kluge 1975; Goeden unpub. data).

RECOMMENDATIONS

1. Collect, verify for cold hardiness, and redistribute *M. lareynii* and *M. lypriformis* from cold-winter states to similar states where they do not occur.

2. Introduce additional natural enemies from around the world.

3. Evaluate studies of the impact of puncturevine weevils in other states, to compare them with the published findings reported from California.

4. Compare the current guilds of adopted predators and parasitoids in California and other states with earlier published findings; this may provide useful information on the environmental resistance to, and natural enemy acquisition by, colonizing insect species.

5. Study the impact of *M. lypriformis* on *Kallstroemia* species under field as well as laboratory conditions.

REFERENCES CITED

Andres, L. A. 1978. Biological control of puncturevine, *Tribulus terrestris* (Zygophyllaceae): Post-introduction collection records of *Microlarinus* spp. (Coleoptera: Curculionidae). In *Proceedings of the 4th International Symposium on Biological Control of Weeds, 30 August–2 September 1976, Gainesville, Florida*, ed. T. E. Freeman, 132–36. Gainesville: University of Florida, Institute of Food and Agricultural Sciences.

Andres, L. A., and G. W. Angelet. 1963. Notes on the ecology and host specificity of *Microlarinus lareynii* and *M. lypriformis* (Coleoptera: Curculionidae) and the biological control of puncturevine, *Tribulus terrestris*. *J. Econ. Entomol.* 56:333–40.

Angelet, G. W., and L. A. Andres. 1965. Parasites of two weevils, *Microlarinus lareynii* and *M. lypriformis*, that feed on the puncturevine, *Tribulus terrestris*. *J. Econ. Entomol.* 58:1167–68.

Goeden, R. D., and R. L. Kirkland. 1981. Interactions of field populations of indigenous egg predators, imported *Microlarinus* weevils, and puncturevine in southern California. In *Proceedings of the 5th International Symposium on Biological Control of Weeds, 22–27 July 1980, Brisbane, Australia*, ed. E. S.

Delfosse, 515–27. Melbourne: CSIRO.

Goeden, R. D., and D. W. Ricker. 1967. *Geocoris pallens* found to be predaceous on *Microlarinus* spp. introduced to California for the biological control of puncturevine, *Tribulus terrestris*. *J. Econ. Entomol.* 60:725–29.

———. 1970. Parasitization of introduced puncturevine weevils by indigenous Chalcidoidea in southern California. *J. Econ. Entomol.* 63:827–31.

———. 1973. A soil-profile analysis for puncturevine fruit and seed. *Weed Sci.* 21:504–7.

———. 1976. The phytophagous insect fauna of the ragweed, *Ambrosia psilostachya*, in southern California. *Environ. Entomol.* 5:1169–77.

Huffaker, C. B., J. Hamai, and R. M. Nowierski. 1983. Biological control of puncturevine, *Tribulus terrestris*, in California after twenty years of activity of introduced weevils. *Entomophaga* 28:387–400.

Huffaker, C. B., D. W. Ricker, and C. Kennett. 1961. Biological control of puncture-vine with imported weevils. *Calif. Agric.* 15(12):11–12.

Johnson, E. 1932. *The puncturevine in California.* University of California College of Agriculture Experiment Station Bulletin 528.

Julien, M. H. 1987. *Biological control of weeds: A world catalogue of agents and their target weeds,* 2nd ed. Wallingford, U.K.: Commonwealth Agricultural Bureaux International.

Kirkland, R. L., and R. D. Goeden. 1977. Descriptions of the immature stages of imported puncturevine weevils, *Microlarinus lareynii* and *M. lypriformis*. *Ann. Entomol. Soc. Am.* 70:583–87.

———. 1978a. Biology of *Microlarinus lareynii* (Coleoptera: Curculionidae) on puncturevine in southern California. *Ann. Entomol. Soc. Am.* 71:13–18.

———. 1978b. Biology of *Microlarinus lypriformis* (Coleoptera: Curculionidae) on puncturevine in southern California. *Ann. Entomol. Soc. Am.* 71:65–69.

———. 1978c. An insecticidal-check study of the biological control of puncturevine (*Tribulus terrestris*) by imported weevils, *Microlarinus lareynii* and *M. lypriformis* (Coleoptera: Curculionidae). *Environ. Entomol.* 7:349–54.

Kluge, R. L. 1975. Observations on insects associated with *Tribulus terrestris* L. in southern Africa. Master's thesis, University of Pretoria, Pretoria, Republic of South Africa.

Maddox, D. M. 1976. History of weevils on puncturevine in and near the United States. *Weed Sci.* 24:414–19.

———. 1981. Seed and stem weevils of puncturevine: A comparative study of impact, interaction, and insect strategy. In *Proceedings of the 5th International Symposium on Biological Control of Weeds, 22–27 July 1980, Brisbane, Australia,* ed. E. S. Delfosse, 447–67. Melbourne: CSIRO.

Maddox, D. M., and L. A. Andres. 1979. Status of puncturevine weevils and their host plant in California. *Calif. Agric.* 33(6):7–8.

Turner, C. A. 1985. Conflicting interests and biological control of weeds. In *Proceedings of the 6th International Symposium on Biological Control of Weeds, 19–25 August 1984, Vancouver, British Columbia,* ed. E. S. Delfosse, 203–25. Ottawa: Agriculture Canada.

Target pests	Natural enemies	States where released	States where established
Dialeurodes citri	*Encarsia lahorensis*	CA	CA
(citrus whitefly)	*Encarsia strenua* (Japan)	CA	
	Encarsia strenua (Puerto Rico)	CA	
	Eretmocerus sp.	CA	
Dialeurodes citrifolii	*Encarsia strenua* (Puerto Rico)	CA	
(cloudy-winged whitefly)			
Siphoninus phillyreae	*Encarsia inaron*	AZ, CA, NV	AZ, CA, NV
(ash whitefly)	*Clitostethus arcuatus*	CA	CA
PSYLLIDS			
Acizzia uncatoides	*Diomus pumilio*	CA, HI	CA
(acacia psyllid)	*Harmonia conformis*	CA, HI	HI
Cacopsylla pyricola	*Prionomitus mitratus*	CA, OR, WA	
(pear psylla)	*Trechnites psyllae*	CA, OR, UT, WA	
	Trechnites sp.	CA, OR, UT, WA	
	Anthocoris nemoralis	WA	(CA), WA
	Anthocoris nemorum	WA	
	Calvia quaturodecim-guttata	WA	
	Chrysoperla (=*Chrysopa*) *carnea*	WA	
	Coccinella septem-punctata	WA	
	Diomus pumilio	CA, OR	
	Harmonia axyridis	WA	
	Harmonia conformis	CA, CO, OR, UT, WA	
	Harmonia dimidiata	CA, UT, WA	
	Menochilus quadripla-giatus	WA	
	Oenopia conglobata	CA, OR, WA	
	Propylaea quaturodecim-punctata	WA	
Calophya rubra	*Tamarixia* sp. nr. *triozae*	CA	CA
(pepper tree psyllid)			
Heteropsylla cubana	*Psyllaephagus yaseeni*	HI	HI
(leucaena psyllid)	*Curinus coeruleus*	GU	GU
TRUE BUGS			
Lygus spp.	*Peristenus digoneutis*	AZ, CA	
(lygus bugs)	*Peristenus rubricollis*	AZ, CA	
	Peristenus stygicus	AZ, CA	

Target pests	Natural enemies	States where released	States where established
Nezara viridula (southern green stink bug)	*Trichopoda pennipes*	(HI)	(HI)
	Trichopoda pilipes	(HI)	(HI)
	Trissolcus basalis	CA, (HI)	CA, (HI)
THRIPS			
Heliothrips haemorrhoidalis (greenhouse thrips)	*Goetheana parvipennis*	CA	
	Thripobius semiluteus	CA	CA
BEETLES			
Brontispa spp. (coconut flat beetles)	*Tetrastichus brontispae*	GU, HI, SA	GU, HI, SA
Hypera brunneipennis (Egyptian alfalfa weevil) and	*Bathyplectes anurus*	CA, ID, NM, UT, WY	CA, NM
	Bathyplectes curculionis	AZ, CA, ID, NM, UT	AZ, CA, (CO), ID, NM, (UT)
Hypera postica (alfalfa weevil)	*Bathyplectes stenostigma*	CA, CO, ID, NM	CO, NM
	Dibrachoides dynastes	CA, ID, UT, WY	CA
	Habrocytus sp.	CA	
	Microctonus aethiopoides	AZ, CA, ID, NM, UT, WY	CA, NM
	Microctonus colesi	CA, UT, WY	
	Microctonus sp.	CA	
	Patasson sp.	CA	
	Tetrastichus incertus	AZ, CA, CO, ID, NM, UT	CA, NM, UT
Leptinotarsa decemlineata (Colorado potato beetle)	*Edovum puttleri*	ID, NY, WA	
Oulema melanopus (cereal leaf beetle)	*Anaphes flavipes*	UT	
	Diasparsus sp.	UT	
	Tetrastichus julis	UT	UT
Rhabdoscelis obscurus (New Guinea sugarcane weevil)	*Lixophaga beardsleyi*	HI	
Xanthogaleruca luteola (elm leaf beetle)	*Aprostocetus brevistigma*	CA	CA
	Erynniopsis antennata	CA	CA
	Oomyzus gallerucae	CA, NM, OR	CA, NM
TRUE FLIES			
Liriomyza spp. (serpentine leafminers)	*Ganaspidium utilis*	GU, HI	GU, HI
	Diglyphus begini	GU	
Rhagoletis completa (walnut husk fly)	*Biosteres juglandis*	CA	
	Biosteres sublaevis	CA	CA
	Coptera evansi	CA	CA

Target pests	Natural enemies	States where released	States where established
MOTHS AND CATERPILLARS			
Amyelois transitella	*Bracon* sp.	CA	
(navel orangeworm)	*Chelonus mccombi*	CA	
	Copidosomopsis plethorica	CA	CA
	Diadegma sp.	CA	CA
	Goniozus legneri	CA	CA
	Goniozus sp. nr. *emigratus*	CA	CA
	Goniozus spp.	CA	
	Phanerotoma flavitestacea	CA	
	Phanerotoma sp.	CA	
	Trathala sp.	CA	
	Trichogrammatoidea annulata	CA	
	Haplothrips sp.	CA	
	Phyllobaenus discoideus	CA	
Artogeia (=*Pieris*) *rapae*	*Cotesia* (=*Apanteles*) *rubecula*	CA, KS, OR, WA	
(imported cabbageworm)	*Trichogramma evanescens*	CA	
Keiferia lycopersicella	*Apanteles gelechidiivoris*	CA	
(tomato pinworm)	*Apanteles subandimus*	CA	
	Elachertus sp.	CA	
	Hyssopus sp.	CA	
	Microchelonus sp.	CA	
	Orgilus sp.	CA	
	Parasierola sp.	CA	
	Trichogramma sp.	CA	
Pectinophora gossypiella	*Apanteles angaleti*	CA	
(pink bollworm)	*Apanteles oenone*	CA	
	Bracon gelechiae	CA	
	Bracon kirkpatricki	AZ, CA	
	Bracon mellitor	CA	
	Chelonus blackburni	AZ, CA	
	Chelonus sp. nr. *curvimaculatus*	AZ, CA, NM	
	Exeristes roborator	CA	
	Goniozus aethiops	CA	
	Goniozus emigratus	AZ, CA	
	Goniozus pakmanus	AZ, CA, NM	
	Pristomerus hawaiiensis	AZ, CA	
	Trichogrammatoidea bactrae	AZ, CA, NM	
Penicillaria jocosatrix	*Aleiodes* sp. nr. *circumscriptus*	GU	
(mango shoot caterpillar)	*Blepharella lateralis*	GU	GU
	Euplectrus sp. nr. *parvulus*	GU	GU
	Trichogramma platneri	GU	

Target pests	Natural enemies	States where released	States where established
Pericyma cruegeri (poinciana looper)	Brachymeria lasus	GU	GU
Phthorimaea operculella (potato tuberworm)	Agathis unicolor	CA	
	Apanteles subandinus	CA	
	Chelonus curvimaculatus	CA	
	Copidosoma koehleri	CA	CA
	Orgilus lepidus	CA	CA
	Orgilus parcus	CA	
	Campoplex haywardi	CA	
	Diadegma stellenboschense	CA	
	Nythobia sp.	CA	
	Temelucha sp.	CA	
Plutella xylostella (diamondback moth)	Cotesia plutella	WA	
Rhyacionia frustrana (Nantucket pine tip moth)	Campoplex frustranae	CA	CA
Spodoptera exigua (beet armyworm)	Telenomus sp.	CA	
Yponomeuta malinellus (apple ermine moth)	Ageniaspis fuscicollis	WA	WA
	Diadegma armillata	WA	
	Eurysthaea scutellaris	WA	
	Herpestomus brunnicornis	WA	
MITES			
Oligonychus punicae (avocado brown mite)	Amblyseius chiapensis	CA	
	Amblyseius herbicolus	CA	
	Euseius quetzali	CA	
	Typhlodromus annectens	CA	
	Typhlodromus helveolus	CA	
	Typhlodromus porresi	CA	
Panonychus citri (citrus red mite)	Amblyseius colimensis	CA	
	Amblyseius deloni	CA	
	Amblyseius largoenis	CA	
	Amblyseius potentillae	CA	
	Amblyseius swirskii	CA	
	Euseius addoensis	CA	CA
	Euseius citri	CA	
	Euseius citrifolius	CA	
	Euseius concordis	CA	
	Euseius elinae	CA	

Target pests	Natural enemies	States where released	States where established
	Euseius scutalis	CA	
	Euseius stipulatis	CA	
	Euseius victoriensis	CA	
	Iphiseius degenerans	CA	
	Neoseiulus californicus	CA	
	Phytoseiulus persimilis	CA, HI	CA
	Typhlodromus athiasae	CA	
	Typhlodromus persianus	CA	
	Typhlodromus rickeri	CA	CA
Panonychus ulmi (European red mite)	*Stethorus histrio*	CA	CA
Tetranychus urticae (twospotted spider mite)	*Phytoseiulus persimilis* *Stethorus histrio*	CA CA	CA CA

Notes: This table was prepared by J. W. Beardsley from edited project summaries. The abbreviations used in this table for states and territories include American Samoa (SA), Arizona (AZ), California (CA), Colorado (CO), Guam (GU), Hawaii (HI), Idaho (ID), Kansas (KS), Montana (MT), Nevada (NV), New Mexico (NM), New York (NY), Oregon (OR), Utah (UT), Washington (WA), and Wyoming (WY).

[*]Parentheses enclosing a state indicate introduction or establishment before the W-84 project was initiated in 1964, or that the data are from a source not part of the W-84 project.

[†]Predators are listed alphabetically after parasitoids.

[‡]Question marks indicate probable establishment. However, insufficient time has passed, or further recovery and identification research is needed, in order to confirm establishment.

APPENDIX 2

Summary of the introduction and establishment of natural enemies for the biological control of weeds investigated within the W-84 research project region.

Target weeds	Natural enemies	States where released[*]	States where established
Carduus nutans (musk thistle)	Rhinocyllus conicus	CA, CO, ID, KS, MT, OR, UT, WA, WY	CA, CO, ID, KS, MT, OR, UT, WA, WY
	Trichosirocalus horridus	ID, KS, MT, WY	KS, WY
Carduus pycnocephalus (Italian thistle)	Rhinocyllus conicus	CA, OR	CA, OR
Centaurea diffusa (diffuse knapweed)	Agapeta zoegana	ID, MT, OR, WA	MT
	Cyphocleonus achates	MT	MT
	Metzneria paucipunctella	ID, MT, OR, WA	ID, MT, OR, WA
	Pelochrista medullana	MT	
	Pterolonche inspersa	ID, MT, OR, WA	
	Sphenoptera jugoslavica	CA, ID, MT, OR, WA	ID, MT, OR, WA
	Urophora affinis	CA, ID, MT, OR, WA, WY	CA, ID, MT, OR, WA, WY
	Urophora quadrifasciata[†]	WY	ID[†], MT[†], OR[†], WA[†]
Centaurea maculosa (spotted knapweed)	Agapeta zoegana	MT, OR, WA	MT, WA
	Cyphocleonus achates	MT	
	Metzneria paucipunctella	ID, MT, OR, WA	ID, MT, OR, WA
	Pelochrista medullana	MT	
	Pterolonche inspersa	MT	
	Urophora affinis	CA, ID, MT, OR, WA, WY	CA, ID, MT, OR, WA, WY
	Urophora quadrifasciata[†]	ID[†], MT[†], WA[†]	ID[†], MT[†], OR[†], WA[†]
Centaurea repens (Russian knapweed)	Subanguina picridis	MT, WA[‡]	MT
Centaurea solstitialis (yellow starthistle)	Bangasternus orientalis	CA, ID, OR, WA	CA, ID, OR, WA
	Chaetorellia australis	CA, ID, OR, WA	OR, WA
	Eustenopus villosus	CA, ID, OR, WA	CA, ID, OR, WA
	Larinus curtus	CA, ID, OR, WA	CA, OR, WA
	Urophora jaculata	CA	
	Urophora sirunaseva	CA, ID, OR, WA	CA, OR, WA
Chondrilla juncea (rush skeletonweed)	Cystiphora schmidti	CA, ID, OR, WA	CA, ID, OR, WA
	Eriophyes chondrillae	CA, ID, OR, WA	CA, ID, OR, WA
	Puccinia chondrillina	CA, ID, OR, WA	CA, ID, OR, WA
Cirsium arvense (Canada thistle)	Altica carduorum	CA, CO, ID, MT, NV, OR, WA	
	Ceutorhynchus litura	CA, CO, ID, MT, OR, WA, WY	ID, MT, OR, WY

Target weeds	Natural enemies	States where released[*]	States where established
	Urophora cardui	CA, CO, ID, MT, NV, OR, WA, WY	CA, MT, NV, OR, WA, WY
Cirsium vulgare (bull thistle)	*Urophora stylata*	CO, MT, OR, WA	WA
Clidemia hirta (clidemia, Koster's curse)	*Ategumiam atulinalis*	HI	HI
	Colletotrichum gloeosporioides	HI	HI
	Liothrips urichi	HI	HI
	Lius poseidon	HI	HI
Convolvulus arvensis (field bindweed)	*Aceria malherbe*[§]	OK, TX	
	Tyta luctuosa[§]	AR, IA, MD, OK, TX	
Cytisus scoparius (Scotch broom)	*Apion fuscirostre*	CA, OR	CA, OR
	Leucoptera spartifoliella	CA, OR	CA, OR[‖], WA[‖]
Euphorbia esula (leafy spurge)	*Aphthona cyparissias*	MT, WY	MT
	Aphthona czwalinae	MT	
	Aphthona flava	ID, MT	MT
	Aphthona nigriscutis	ID, MT, ND	MT
	Chamaesphecia hungarica	MT	
	Chamaesphecia tenthrediniformis (=empiformis)	ID, MT, OR	
	Hyles euphorbiae	CA, CO, ID, MT, ND, NE, NV, NY, OR, WA, WY	MT, NY
	Oberea erythrocephala	MT	MT
	Spurgia capitigena	MT, OR, WY	MT
Hypericum perforatum (St. Johnswort)	*Agrilus hyperici*	CA, ID, MT, OR, WA	CA, ID, MT, OR, WA
	Aplocera plagiata	MT, OR	
	Chrysolina hyperici	CA, HI, ID, MT, NV, OR, WA	HI, ID, MT, NV, OR, WA
	Chrysolina quadrigemina	CA, HI, ID, MT, OR, WA	CA, HI, ID, MT, OR, WA
	Chrysolina varians	CA	
	Zeuxidiplosis giardi	CA, HI, ID, MT, OR, WA	CA, HI
Linaria genistifolia (Dalmatian toadflax)	*Calophasia lunula*	CO, ID, MT, OR, WA, WY	ID, MT, WA
Passiflora mollissima (banana poka)	*Cyanotricha necyria*	HI	
	Pyrausta perelegans	HI	HI
Salsola australis (Russian thistle)	*Coleophora klimeschiella*	CA, CO, HI, ID, KS, MT, NV, WA, WY	CA, ID, KS
	Coleophora parthenica	AZ, CA, CO, HI, ID, MT, NV, UT, WA, WY	AZ, CA

Target weeds	Natural enemies	States where released	States where established[*]
Salvia aethiopis (Mediterranean sage)	*Phrydiuchus spilmani*	OR	
	Phrydiuchus tau	CA, ID, OR	CA, ID, OR
Senecio jacobaea (tansy ragwort)	*Botanophila seneciella*	CA, OR, WA	CA, OR, WA
	Longitarsus jacobaeae	CA, OR, WA	CA, OR, WA
	Tyria jacobaeae	CA, OR, WA	CA, OR, WA
Silybum marianum (milk thistle)	*Rhinocyllus conicus*	CA, OR	CA, OR
Tribulus terrestris (puncturevine)	*Microlarinus lareynii*	AZ, CA, CO, HI, ID, KS, MT, NM, NV, OR, UT, WA, WY	AZ, CA, CO, HI, KS, NM, NV, UT
	Microlarinus lypriformis	AZ, CA, CO, HI, ID, KS, MT, NM, NV, OR, UT, WA, WY	AZ, CA, CO, HI, KS, NM, NV, UT
Ulex europaeus (gorse)	*Agonopterix ulicetella*	HI	
	Apion scutellare	HI	
	Apion sp.	HI	
	Apion ulicis	CA, HI, OR, WA	CA, HI, OR, WA
	Sericothrips staphylinus	HI	HI
	Tetranychus lintearius	CA, OR	OR

Notes: Prepared by R. D. Goeden from edited project summaries. The abbreviations used in this table for states and territories include American Samoa (SA), Arizona (AZ), Arkasas (AR), California (CA), Colorado (CO), Guam (GU), Hawaii (HI), Idaho (ID), Iowa (IA), Kansas (KS), Maryland (MD), Montana (MT), Nebraska (NE), Nevada (NV), New Mexico (NM), New York (NY), North Dakota (ND), Oklahoma (OK), Oregon (OR), Texas (TX), Utah (UT), Washington (WA), and Wyoming (WY). Weed-feeding insects were shipped to and released in states outside of the project.

[*]Does not include releases too recent to assess as established.

[†]The natural enemy was released in Canada and spread unaided to the United States as indicated. It was subsequently redistributed in these states.

[‡]Trial releases only.

[§]Not released in states west of the Rocky Mountains.

[||]Accidentally introduced to Oregon and Washington from Europe before its release in California.

GENERAL INDEX

INDEX OF SCIENTIFIC NAMES

INDEX OF COMMON NAMES

NOTE: Names of edible and ornamental plants appear in the general index.